Ultrasonic Measurements for Process Control

Theory, Techniques, Applications

Ultrasonic Measurements for Process Control

Theory, Techniques, Applications

LAWRENCE C. LYNNWORTH

Panametrics, Inc.
Waltham, Massachusetts

ACADEMIC PRESS, INC.

Harcourt Brace Jovanovich, Publishers

Boston San Diego New York
Berkeley London Sydney
Tokyo Toronto

ACADEMIC PRESS, INC.
1250 Sixth Avenue, San Diego, CA 92101

United Kingdom Edition published by
ACADEMIC PRESS INC. (LONDON) LTD.
24–28 Oval Road, London NW1 7DX

Library of Congress Cataloging-in-Publication Data

Lynnworth, Lawrence C.
  Ultrasonic measurements for process control : theory, techniques,
applications / Lawrence C. Lynnworth.
    p. cm.
  Bibliography: p.
  Includes index.
  ISBN 0-12-460585-0
  1. Process control. 2. Ultrasonic waves—Industrial applications.
I. Title.
TS156.8.L95 1989   88-14595
660.2′81—dc19   CIP

Printed in the United States of America
89 90 91 92   9 8 7 6 5 4 3 2 1

To
Family
and
Friends

Contents

Preface

The ability of ultrasound to interrogate noninvasively, nondestructively and rapidly the surface and internal regions of human subjects or material objects is clearly desirable. In the field of process control, one usually wants to maintain constant process conditions and consistent product properties in a batch or continuous process. It is necessary to remain informed about the process without perturbing the process or its boundaries. Information needs to reach the controllers fast enough to prevent errors. Sometimes the necessary information can indeed be obtained from outside the process boundary, e.g., from outside a pipe. But there are many cases where higher accuracy or data on certain properties or at remote or internal locations can be obtained only by penetrating a boundary to accommodate the sensor.

This book examines the theory, techniques and applications of ultrasound with respect to control of industrial processes. The book's perspective favors intuitive explanations over rigorous mathematical derivations, although references to such material are generally included. Measurement techniques, such as coupling and acoustic isolation, are included. Guidelines are given for selecting the best mode of vibration from among longitudinal, shear, Lamb, Rayleigh and other waves on the menu. This kind of information is useful to a new user of ultrasound, whether that user plans to conduct an experiment, wants to understand an equipment problem or needs to predict the outcome of a measurement proposed to control a process. Recent advances in transducer materials, microprocessors, measurement techniques, and sometimes in the basic understanding of wave propagation, wave/material and wave/process interactions, now permit one to obtain data rapidly, reliably and economically in applications that would have appeared "off limits" to ultrasound only a few years ago.

Most of the applications are garnered from industry. Discussion of these applications will help engineers, scientists, technologists and educators in what are broadly termed the process industries. In other words, the book is directed to process and instrument engineers and their colleagues whose background is typically chemical, petroleum, aerospace, industrial, mechanical or electrical engineering, and also to R&D personnel whose applications are not conventional. Persons responsible for recommending or approving purchase of equipment to measure flow, temperature, density, liquid level and other interfaces, or any other important process measurand derivable from an ultrasonic measurement may also find applications discussed that are relevant to their own situation.

There also seems to be a growing academic community of graduate students, scientists, engineers and other researchers whose laboratory pursuits in physics, engineering or materials science, for example, could benefit from the latest ultrasonic technology available to industry. It is hoped that this book will acquaint these individuals too with the ultrasonic choices now available. Accordingly the book may be helpful to educators and students involved in "individual study" programs or "special topics" courses in engineering and science curricula, and occasionally in liberal arts programs.

A number of illustrations, tables, brief passages or extracts from earlier publications are reproduced as in the original works, except for minor editing. This essentially preserves their technical content and manner of expression.

Acknowledgments

The field of ultrasonic measurements for process control owes its breadth and depth to the contributions of many workers around the world. Likewise, the present attempt to describe this field owes a sincere debt of gratitude to the many contributors who have generously permitted extended passages from their own publications to be reproduced herein, or who, in a few cases, have edited or authored new sections expressly for this volume. Via "private communications" a number of individuals supplied historical or technical information beyond that which one might expect to find in searches of the literature. In the foregoing categories, the author thanks: T.T. Anderson, A.E. Arave, H.H. Bau, G. Bryant, D.W. Cannon, V. Chatigny, N. Dam, E. Dieulesaint, G.S. Doble, G.P. Erickson, W. Evans, T.G. Fain, R.S. Flemons, J. Galkowski, Jr., M. Greenspan (deceased), R.J. Hansman, Jr., T.R. Henderson, L.J. Hollander, P. Jackman, G.V. Jeskey, R.D. Joy, M. Katsoulakos, D. Kehrhahn, J.O. Kim, J. Kritz, S. Krupski, D. Kupperman, M.D. Kyser, H. Kytomaa, A. Laenen, C.B. Leffert, E. Little, V. Mágori, W.P. Mason (deceased), D.P. Massa, R.W. Miller, M.E. Motamedi, F. Nadeau, S.P. Nuspl, J. O'Brien, J. Paros, C.R. Peterson, E. Powers, G.R. Robinson, R. Roy, J.S. Schoenwald, W. Smith (for J. Waller, deceased), R.C. Swengel (deceased), R.H. Tancrell, S.E. Tehon, H.E. Van Valkenburg, H.N.G. Wadley, W. Welkowitz, R.M. White, D.H. Willson, E. Winston and E.M. Zacharias, Jr., and their firms or institutions.

My co-authors and colleagues now or formerly at Panametrics who contributed to theory, techniques or applications reported in this book

xvii

include: J.E. Bradshaw, J.H. Bradshaw, C.A. Carey, E.H. Carnevale, D. Chleck, G.M. Elfbaum, S.S. Fam, K.A. Fowler, F.A. Hanser, S.A. Jacobson, S.L. Klaidman, J.M. Korba, G.S. Larson, R.N. Lawson, J.E. Matson, M.S. McDonough, W.F. McGrath, W.C. Mellish, P.R. Morel, T.B. Moss, T.C. Nicholson, T.H. Nguyen, T.J. Nelligan, E.P. Papadakis, D.R. Patch, N.E. Pedersen, R.C. Pierce, S.E. Rehn, M.J. Scelzo, B.J. Spencer, J.A. True and D.R. Wallace.

Non-Panametrics co-authors whose work is reported herein include: L.D. Braswell, W.D. Munk, J.W. Smalling and A.W. Warburton.

In addition, S.R. Desrosiers, helped by B.A. Chiacchio, typed the manuscript, L. Spaiser assisted in electronic manuscript preparation and C.D. Smart drafted most of the new artwork.

Beyond specific contributions, I want to acknowledge the mentoring of the late C.H. Hastings of Avco under whose inspiring supervision, between 1959 and 1962, I first became acquainted with the art, science and NDT applications of ultrasonics. At Avco, S.A. LoPilato was my principal tutor in the art of coupling and other practical aspects of his craft.

Starting in late 1962 and continuing for over twenty-five years, Panametrics' co-founder E.H. Carnevale has encouraged me to pursue my own interests in ultrasonics, including the publication of articles and, especially, of this book. His permission and that of his fellow co-founder D. Chleck, to publish information beyond minimum technical details, is appreciated. Without such information this book would have been more like an outline drawing than a picture, with respect to work at Panametrics.

The support of Government and private organizations that sponsored much of the R&D herein reported is gratefully acknowledged.

The following publishers granted permission for portions of their copyrighted works to be reproduced: Academic Press, American Gas Association (AGA), American Foundrymen's Society, American Institute of Aeronautics and Astronautics (AIAA), American Institute of Physics (AIP), American Meteorological Society, American Society of Mechanical Engineers (ASME), American Institute of Mining, Metallurgical and Petroleum Engineers (AIME), American Society for Nondestructive Testing (ASNT), American Society for Testing and Materials (ASTM), Butterworths, Chemical Processing, Helmers Publishing, Inc., Institute of Electrical and Electronics Engineers (IEEE), Instrument Society of America (ISA), McGraw-Hill, Measurements & Control, National Association of Corrosion Engineers (NACE), Plenum Press, Society of Mining Engineers (SME), and Texas A&M University (Chemical Engineering Department).

Numerous companies kindly supplied brochures or photographs of their equipment, as acknowledged in the respective captions.

The author also wishes to acknowledge the cooperation and patient support of the reviewers, editors and staff at Academic Press, particularly Carolyn Artin, Amy Strong, and Robert Kaplan. The author especially appreciates the encouragement of his parents and his wife, Marianne.

1. Introduction

When I began the writing part of my career as a freshman cub reporter for the New York University *Heights Daily News* in 1954, the editor-in-chief told me that a good newspaper story tells the reader right in the first paragraph the answers to: Who? What? Where? How? When? Why? As an introduction to this book, and to the subject of ultrasonic measurements for process control, the journalist's six-word outline may be interpreted to mean:

- For whom is this book written? Whose work is reported?

- What theories, principles, methods, sensors and equipment are used in ultrasonic measurements? What results have been obtained? What are the limitations? What about microprocessors and computers?

- Where is the proper place for ultrasonic measurements, within the general field of industrial process control? Where are the examples taken from—the research laboratory or from applications proven in industry?

- How does one apply theory and techniques to new applications?

- When was the work done—so long ago that it is well established, or so recently that it can not yet be said to have withstood the test of time?

- Why was the reported measurement made by ultrasonic means? Why should I, the reader, choose to measure Parameter X by ultrasonics instead of using another non-acoustic discipline like optics, mechanics or electronics?

The answers to these questions are given below in short form. More complete answers form the body of subsequent chapters. The reader may gain an overview of the subject from Chapter 2, "Scope of Applications." That chapter summarizes some twenty-five measurement or application areas where ultrasound has been applied successfully.

The rest of the book *could* have expanded upon each of these twenty-five areas, giving equal weight to each area. Following Chapter 3, "Theory and Measurement Techniques," the book indeed deals with a number of application areas but concentrates most heavily on flow in Chapter 4, the major "applications" chapter. Subsequent chapters on applications cover temperature, liquid level and interface detection, density, and elastic moduli. Nondestructive testing (NDT) applications, including thickness gaging, flaw detection and acoustic emission, are included to the extent necessary to illustrate certain principles or measuring techniques. Despite their technological importance, however, NDT topics are not treated in depth because it is felt that they are covered adequately in other books.

Two reasons for devoting so many of the application pages to flow are: (1) Many principles and techniques of ultrasonic measurement, regardless of application, can be demonstrated in depth and in combination within the flow category; and (2) the number of flow applications, the variety of requirements and conditions under which flow must be measured, and the importance of flow measurements both technically and economically justifies special treatment.

Returning to the journalist's six-word outline:

For Whom Is This Book Written?

This book is designed for use by an individual whose training may be in science, engineering, technology or a liberal arts field, who is *responsible for designing, selecting, operating or troubleshooting* a system to measure one or more process control parameters (like flow, liquid level, viscosity). Usually this individual has zero or only limited experience with ultrasound. Sometimes this individual may be quite expert with the use of ultrasound in a technically related field like NDT, or may be experienced with industrial measurements using other wave phenomena like optics or microwaves. Such experience, while helpful, is not a prerequisite for using this book.

Whose Work Is Reported?

Where theories, principles or certain methods or apparatus are concerned, the original author or inventor is usually identified. Where applications are concerned, it is not always possible to identify the "author," because, as principles mature to the applied stage, the individuals involved may be unmotivated to report their work or, in fact, may be prevented from reporting in the open literature because of proprietary, security or testimonial restrictions. Stated another way, there are numerous applications that are not reported here due to lack of supporting documentation in the archival literature.

In the past twenty-five years, the author and his colleagues at Panametrics have been fortunate in having worked on projects sponsored by the government or private industry, where the sponsors have encouraged publication of the results in many cases. Friends and acquaintances of the author have also been kind enough to supply new material for this book. Within the space and scope constraints of this book, the author assumes responsibility for including or excluding particular works and references.

What Theories, Principles, Methods and Equipment Are Used?

With closed-loop process control as the main objective, the relevant theories or correlations generally relate sound speed and attenuation to controllable parameters or to specified properties. For example, in an ideal gas, the speed of sound is proportional to the square root of the absolute temperature, and the absorption of sound is proportional to viscosity. The sound speed or attenuation is often measured by using the processed material as its own sensor; i.e. *its* acoustic properties become part of the circuit that generates correction/error signals. In other cases a *different* material (invasive waveguide, probe or transducer) acts as the sensor to indicate some aspect of a process. Methods include pulse and resonance, continuous or modulated waves, pressure coupling and non-contact coupling, etc. Equipment varies from general-purpose instruments used in the laboratory to specialized equipment optimized for one particular measurement in the field. Transducers and sensors likewise may be general-purpose or specific. Emphasis is on *ultrasonic frequencies*, but in a few cases devices or measurements operating below 20 kHz are reported because they illustrate important principles or emerging applications. Audible tone bursts, for example, have been used to measure the temperature of hot gas over long paths, tens of meters. Some sensors utilize mechanical resonators or edge tone ("fluidic") resonators that are structured so that the resonance frequency is audible.

What Results Have Been Obtained?

Fundamental results include measurement of sound speed and attenuation coefficients in materials as functions of temperature, pressure, heat treatment and the like. Examples of such results form the database at the end of Chapter 3. Results also include the practical application of ultrasound to sense, by passive or active means, measurands of industrial significance—flow, temperature, pressure, etc.

What Are the Limitations?

Caveat emptor—let the buyer beware! Limitations need to be discussed not only to protect the buyer but to protect the seller too. Those who make unrealistically optimistic assumptions about the applicability of ultrasound to a particular problem usually end up embarrassed. The opposite is true too—if one pessimistically disregards ultrasonics because it is erroneously believed that ultrasound could not propagate across a particular medium, then an alternative technology might be selected that is unnecessarily complicated or expensive.

What About Microprocessors (μPs) and Computers?

Ultrasonic measurements typically involve a transducer or sensor and an electronic instrument that processes and interprets waveform data and generates or computes values of a measurand and generates alarms. In years past, the complexity of these electronic tasks was substantially greater for ultrasonic systems than for many nonultrasonic systems. Now, however, the electronic differences are buried or masked in one or a few chips, or compensated for by a computer. As a specific example, consider three temperature sensors: thermocouple (TC), platinum resistance temperature detector (RTD), and a resonant ultrasonic thermometer. The TC would generally be judged as simplest. But when millidegree sensitivity is required, the TC yields to the RTD and the resonant thermometer. Which of these two is "simpler" or "better"? Chapter 4 will help answer that question.

The subject of thickness gaging provides another example of the influence of μPs. By the mid-1980s ultrasonic thickness gages of about the same size as a machinist's micrometer were available. Which is simpler to use? The answer depends in part on whether access to a part is restricted to one side or two and on how the readings are to be used, stored (memorized), recorded or interfaced to other equipment.

In the area of flow measurement, μPs enable one to design a "smart" flowmeter, capable of automatically selecting the best mode from a menu that may include transmission, reflection, tag correlation and noise detection. A μP-based smart flowmeter can now be designed that uses a transmission mode to obtain the line-averaged flow velocity V along one particular path (e.g., tilted diameter) and then uses a reflection mode such as range-gated Doppler to obtain profile data that is used by the flowmeter to correct its first estimate and obtain a better measure of the area-averaged flow $\mathbb{V}$ in a conduit.

Computers also make practical the on-line quality control of rapidly-manufactured parts, only some of which are inspected, through statistical process control (Papadakis, 1988; Pfeifer, 1988; Yehling et al., 1988).

Where Is the Proper Place for Ultrasonic Measurements Within the General Field of Industrial Process Control?

This question involves comparing competing technologies for specific measurement capabilities or specific applications. For example, which of the following is the "best" way to measure flow: ultrasonics, venturi, pitot tube, magnetic flowmeter, rotameter, vortex shedder, etc.? The answer, in general, is not easily determined. Does "best" mean most accurate, least expensive to purchase, least expensive to maintain, most reliable, least invasive (mininvasive), etc.?

The answer may also depend on how parameters other than flow are being measured. In other words, it may be easier to maintain or multiplex a number of like-technology sensors.

In summary, ultrasonics ought to be selected when it is better, faster or cheaper than the alternatives. More precisely, ultrasonics may be selected for process control in order to obtain one or more of the following benefits:

- Noninvasive or mininvasive measurement

- High accuracy (measurands usually transformed to time or frequency measurements)

- Reliability (no moving parts, in the usual sense)

- Fast response (can be <1 ms)

- Remote sensing, sometimes with no physical contact

- Average reading over an extended region

- Profile information (point by point, or small-path average)

- Computer compatibility of time, frequency or, say, 8-bit amplitude data
- Low cost, especially for multiplexed and/or mass-produced sensors
- Small size, small mass
- Avoidance of problem(s) associated with competing (nonultrasonic) technologies
- Data or results unobtainable any other way

Where Are the Book's Examples Taken from—The Research Laboratory or from a Process Control Application Already Proven in Industry?

From both. One advantage of the research laboratory is that it allows the investigator to isolate one parameter from noise due to unwanted variables and, in many cases, to publish the basic results without proprietary restrictions. In this way physical principles may be understood clearly and communicated to the researcher's audience. But there is also a disadvantage—the ivory tower work often neglects the noise and practical constraints that exist in the real world of applications. So examples from the smokestack industries are needed too. A few examples are also taken from the fields of NDT and biomedical engineering to illustrate theory, techniques or applications. Biomedical uses have also focused attention on safety-related aspects of ultrasound (exposimetry, exposure field, and dosimetry). A special issue on ultrasonic exposimetry appeared in the March 1988 issue of the *IEEE Transactions on Ultrasonics, Ferroelectrics, and Frequency Control.*

How Does One Apply Theory and Techniques to New Applications?

Start by defining the problem, listing objectives and listing known interfering variables. Search the literature. Communicate with manufacturers or other sources of knowledge in the disciplines involved. Use expert system software. Conduct feasibility tests. Divide the project into steps.

When Was the Work Done?

Was the work done so long ago that it is well established or so recently that it can not yet be said to have withstood the test of time? The beginnings of ultrasonic echo ranging in air and water are found in nature [Fig. 1-1(a), (b)]. Primitive pottery workers probably recognized quality by the sound of their vessels [Fig. 1-1(c)]. Even today we speak of a "sound"

(a)   (b)

(c)

FIG. 1-1. (a) Active and passive methods used in air by bat and moth, respectively. (b) Echolocation and communication underwater by dolphins. (c) Early test for soundness of pottery. (Illustrations by Margaret C. Lynnworth.)

casting to mean a good one. Passive techniques as well as active techniques yield information to the alert listener.

Among the many important events recounted by Lindsay, 1966, in his history of acoustics, we want to draw attention to two of these, the measurement of the speed of sound in air in 1635, and that in water in 1826. Even today in these media, ultrasonic experiments and applications meet with success more often than in less familiar, less homogeneous media.

Turning now to events of the last hundred years or so, piezoelectricity was discovered in 1880 by the Curie brothers. Their discovery, which underlies the majority of ultrasonic sensors in use today in process control, occurred before Roentgen's discovery of X-rays in 1895—before the ether drift controversy was resolved by the Michelson-Morley experiment in 1887. Lord Rayleigh's first edition of his classic work, *The Theory of Sound*, was published in 1877, before atomic theory, before modern physics, before quantum mechanics, before relativity theory. *Ultrasonics evidently has a long history.* Electronic aspects of ultrasonic measuring systems got a big boost from radar and sonar developments in the 1939–1945 period. It is therefore not surprising to find historic milestones in ultrasonic NDT and flowmetering in the late 1940s. Coinciding with and perhaps because of developments in space and military projects in the 1960s, ultrasonic instrumentation advanced again in that decade in terms of practical demonstrations, e.g., clamp-on flowmetering of large steel pipes. Thus, some manufacturers of ultrasonic flowmeter and NDT products can justifiably boast of their twenty and forty years' experience in these fields, respectively. Such time frames imply mature product lines. Younger companies, as proponents of new products, may emphasize different advantages, e.g., "superior" performance that uses the "latest" technology.

As one way to protect themselves from what they consider "unproven" technology, some organizations specify, when about to purchase ultrasonic equipment, that the manufacturer shall supply evidence of five years of continuous operation of a specific type of equipment. However, if all organizations adopted this rule, the prospects for emerging technology would be discouraging.

New materials for transducers, for couplants, new semiconductor chips, new integrated circuits and μPs all combine to offer ultrasonic manufacturers and their customers new opportunities for potentially improved measurement. In critical cases, however, one must weigh the advantages of the potential improvement against the risks associated with a system whose performance in the field, or whose mean time to failure, has not yet been determined.

Why Was the Reported Measurement Made by Ultrasonic Means?

Reasons include noninvasiveness, high accuracy, fast response, etc.

Why Should I, the Reader, Choose to Measure Parameter X by Ultrasonics Instead of Using Another Non-Acoustic Discipline Like Optics, Mechanics or Electronics?

If life were simple, this important question could be answered by one short phrase, like "because it's less expensive," or "because it's noninvasive." Sometimes the best solution to a problem requires the combination of two (or more) techniques (for example, laser generation of ultrasound; fiberoptic transmission of ultrasonic data). Cielo, 1988, covers optical techniques for a number of the measurands that in the present book are addressed by ultrasonic techniques.

In general, a *complete* specification for equipment to measure Parameter X may involve one or more pages of requirements. The length of the spec may be in proportion to the cost if the equipment were to fail. (For example, compare the specs for a gage that measures fuel remaining in the gas tank of an automobile, in an airplane, and in a space-flight vehicle or space laboratory.)

Given a lengthy spec, the evaluator of proposals who is trying to find the best technology may need to assign weighting factors to different objectives to arrive at a figure of merit for each candidate. Another approach is to only consider bids that take no exception to any part of the spec, and choose the lowest bidder. It does not take too much imagination, however, to foresee cases where this latter approach does not necessarily serve the best interests of the purchaser.*

Apart from all the perfectly logical reasons that one could list for selecting an ultrasonic system over some particular nonultrasonic system, emotional factors must be considered too. The purchaser wants to feel comfortable with the selection. One of the purposes of this book is to remove the mystery shrouding some types of ultrasonic equipment, so that, through understanding, fear of the unknown can be conquered. Other purposes include clarifying limitations, so that expectations will not be unrealistic. The same information that leads one to select ultrasound to measure Parameter X in one case may, in a different set of circumstances, dictate that a nonultrasonic approach be selected.

---

* It's unwise to pay too much, but it's unwise to pay too little. When you pay too much you lose a little money, that is all. When you pay too little, you sometimes lose everything, because the thing you bought was incapable of doing the thing you bought it to do. The common law of business balance prohibits paying a little and getting a lot. It can't be done. If you deal with the lowest bidder, it's well to add something for the risk you run. And if you do that, you will have enough to pay for something better.

—John Ruskin (1819–1900)

2. Scope of Applications

The main purpose of this chapter is to identify the breadth, depth, practicality and limitations of industrial applications of small-signal ultrasound. Additionally, we will attempt to identify patterns of emerging ultrasonic technology. In many respects, this chapter is a short version of the book.

In general, the scope of this book and this chapter is limited to industrial applications wherein the emission, transduction or propagation of low-intensity ultrasound responds to the properties, state or quality of the medium or part in question. By generally restricting the scope to "industrial" applications, we choose to omit numerous interesting and important applications in research and those in medical, dental and biological areas. "Low-intensity" avoids macrosonic and nonlinear acoustic areas such as ultrasonic cleaning, machining, wire welding, atomizing, cavitating, emulsifying, influencing of chemical reactions, shock-wave measurements and therapy. By limiting the scope to cases in which one measures ultrasound emission, transduction or propagation to determine the value of the measurand, we intend to detour around devices such as quartz clocks, ultrasonic garage door openers, TV-channel selectors, delay lines, filters, and signal processors, despite the obvious industrial significance of such devices. In view of all these omissions, the reader may rightfully ask, "What's left?" For the answer see Tables 2-1 and 2-2.

Standard commercial equipment, particularized for a specific application, is available for almost every measurand in Table 2-1. Additionally, since virtually any ultrasonic measurement can be analyzed in terms of

Table 2-1.   Industrial Measurement, Test, and Process-Control
Applications, Partial List, Adapted from
Lynnworth, 1975

Item No.	Parameter
1	Flowmetry
2	Thermometry
3	Density, Porosity
4	Pressure
5	Dynamic force, vibration, acceleration
6	Viscosity in fluids
7	Other transport properties
8	Level
9	Location of low-reflectivity interfaces
10	Phase, microstructure, nodularity
11	Thickness
12	Position
13	Composition
14	Anisotropy, texture
15	Nondestructive testing
16	Grain size in metals
17	Stress and strain
18	Acoustic emission
19	Imaging, holography, microscopy
20	Elastic properties
21	Bubbles and particles
22	Gas leaks
23	Interrupted sound beam
24	Surface acoustic wave (SAW) sensors and membrane transducer appplications
25	Other applications (see also Table 2-2)

observations related to transit time or wave amplitude, general-purpose electronic measuring equipment, such as digital oscilloscopes, computing counters, time intervalometers, peak detectors, etc., may also be used to perform the industrial measurements or tests to be discussed below.

The items in Table 2-1 could have been categorized into two major groups in terms of instrument response being associated primarily with sound speed c or attenuation coefficient α.* Such a categorization, without a further breakdown into specific applications, might be useful aca-

---

* In some industrial applications it is required to measure c or α per se. Special oceanographic probes and readout equipment have been developed for measuring c versus depth. Backings or isolation members for flaw-detection transducer search units are characterized by α. Delay-line materials are partly characterized by both c and α.

demically to individuals who are not necessarily responsible for solving a specific industrial measurement problem. However, if one were to shrink Table 2-1 down to these two main groups, without exposing the contents of each group, the present scope of ultrasonic test and measuring equipment would remain obscure. Guidance on the selection of standard equipment for particular applications (flow, temperature, density, etc.) would still be lacking. For these reasons, this chapter is organized mainly in terms of measurands, i.e., in terms of the industrial user's language. Space allows only a small sampling of available ultrasonic equipment to be illustrated.

Since this chapter is intended to provide an overview of ultrasonic measurements for process control, it was decided to group its topics more or less as in the applications chapters 4 through 9. Accordingly, the major topics are flowmetry, thermometry, density, interface sensing, nondestructive testing, and lastly, other measurements, other applications and special topics. In addition, most sections include a brief passage on theory, so that the subsequently-presented technique, equipment or application will be understandable.

Readers desiring more detail on particular items than is contained in this chapter are referred to the corresponding chapters 4 through 9 or subsections therein, or to the literature. For example, for a more complete picture of flow measurement than contained in Section 2.1, Flowmetry, refer to Chapter 4. For more information on Thermometry than in Section 2.2, refer to Chapter 5.

For information on *non*ultrasonic methods used in process control, one can refer to Considine, 1985, Cheremisinoff and Cheremisinoff, 1987, or Noltingk, 1988. *Optical* alternatives for addressing some of the measurands in this chapter are described in Cielo, 1988. With regard to *flow*, see Miller, 1989 for a comprehensive general treatment.

In the following sections, we introduce the parameters listed in Table 2-1 and/or a typical measurement or test situation and then present a brief explanation of the measurement or test in terms of the influence of the parameter upon the sound wave's emission, propagation or transduction. Then, rather than elaborate on physical principles and electronic processing of received ultrasonic waves, we choose instead to identify probes or equipment specifically designed to measure the parameters listed. In this way it is intended to clarify the extent to which theory and experiment have been reduced to commercial practice. (A somewhat different perspective is afforded by categorizing applications as in Table 2-2.) To provide a measure of historical perspective, some equipment is illustrated that was available ten or more years ago. New equipment representing the state of the art at press time inevitably becomes obsolete. Accordingly,

Table 2-2. Summary of Papers in a Special Issue on Acoustic Sensors, According to a Classification by Motamedi and White, 1987. © 1987 IEEE

Area[a] Authors[b]	Measurand	Mode[c]	Principles Employed	Key Characteristics
General: White	Sensor classification			
Biological and Chemical:				
Thompson et al.	Biological components	B S	Resonator (bulk); ultrasonic SAW delay-line oscillator	Sensitivity; selectivity; interfacial processes
Ito	Concentration (water vapor)	B	Resonator	Linearity; low hysteresis
Martin et al.	Components and concentrations (organic vapors)	S	Ultrasonic SAW delay-line oscillator	Measuring sorption isotherm; thermal desorption spectroscopy
Venema et al.	Concentration (NO_2)	S	Ultrasonic SAW delay-line oscillator	Sensitivity; modularity
Vetelino et al.	Concentration (H_2S)	S	Ultrasonic SAW delay-line oscillator	Sensitivity; stability; regenerability
White et al.	Components and concentrations	S P	Ultrasonic SAW or plate delay-line oscillator	Sensitivity
Wohltjen et al.	Components and concentrations	S	Ultrasonic SAW delay-line oscillator	Sensitivity; high-frequency operation
Electric: Stearns and Kino	Surface recombination velocity; surface structure	B	Photoacoustics	Resolution; noncontact

	Magnetic:ᵃ	Modeᶜ		
Hanna	Magnetic field	S	Ultrasonic SAW delay-line oscillator	Accuracy; simplicity
Mechanical:				
Adler and Desmares	Position	S	Attenuation, reflection, timing (SAW)	Resolution; economics
Mágori and Walker	Position	B	Wave reflection, timing	Reliability; economics
Jacobson et al.	Position; flow rate	B R	Wave reflection, timing (torsional and other modes)	Accuracy; speed
Schoenwald et al.	Position	B	Wave reflection, timing	Accuracy
Yano et al.	Position; surface characterization	B	Wave reflection, transmission, timing	Resolution; speed
Motamedi	Acceleration	B S	Resonator (bulk, SAW, cantilever beam)	Sensitivity; dynamic range; stability
Clayton et al.	Force; pressure	B	Resonator (tuning-fork)	Increased Q design
Hauden	Force; pressure; flow rate; temperature	B S	Resonator; ultrasonic SAW delay-line oscillator	Sensitivity; selectivity; economics
Braeuel et al.	Surface characterization	B	Emitted sound	Sensitivity; reliability
Cohen-Tenoudji et al.	Viscosity	B	Wave reflection, timing (shear mode)	Ability to follow resin curing
Thermal:				
Nakazawa et al.	Temperature	B	Resonator	Sensitivity; linearity; stress compensation

ᵃ *Area* indicates the field of application of the sensors described in each paper. Because of their similarity, biological and chemical sensors are listed together.

ᵇ Authors may be found in Motamedi and White, 1987.

ᶜ *Mode* indicates the mode of vibration or propagation involved—bulk wave (B), surface wave (S), plate wave (P), or rod wave (R).

the reader is referred to periodically-revised sources such as those cited at the end of this chapter in Table 2-7.

At the outset, the reader is cautioned that, despite the widespread use of a particular product, a full understanding of the wave/parameter interaction and electronic instrumentation details is probably the best insurance against wasting time and money attempting to use that product beyond its inherent limitations which may be unknown even to the vendor. That is to say, one must strive to understand the relationship between the measurand and c or α. Furthermore, the influence of undesired variables must also be assessed properly.

2.1   Flowmetry

Industrial applications for ultrasonic flowmeters include:

- gas, liquid and solid (including particulate) matter

- flow magnitudes up to a few meters per second for most liquid cases, and beyond Mach 1 for gases in wind tunnels or outside supersonic aircraft

- conduit sizes from <1 mm diameter to >10 m diameter

- >1 km paths across rivers

- liquid level in weirs, flumes or other open-channel flow situations

- temperature extremes from cryogenic levels (liquid oxygen, liquid natural gas) to potentially at least 1500°C (liquid steel)

- pressure from near-zero (suction-pumped polymers) to >700 atmospheres (70 MPa), e.g., oil, gas and geothermal downhole fluids, deep-sea water

- response times from the order of 1 ms or less (flow transients, engine control) to hours (exhaust stack monitoring)

- single- and two-phase media

- batch and continuous processes

- dedicated and survey meters

- clamp-on, wetted and hybrid transducers for single path and multipath interrogations, depending on accuracy required and access constraints

- installations ranging from clamp-on, insert probe, spoolpiece, to hot-tapped ports

An excellent review of early ultrasonic flowmeter developments is due to Herrick, 1977. The earliest reference she found on the use of sound to measure fluid flow in a pipe is Rütten's German patent (1928; issued in 1931), for the transit-time method. Out-of-pipe experiences, such as Chilowski and Langevin's Doppler methods for measuring boat speed, date back to 1916. These two methods are represented by the first two entries in Fig. 2.1.

Another historical review of interest is that contained in McCullough and Graeper, 1979.

2.1.1   Physical Principles

Flow can be measured by both *passive* and *active* methods. Noise monitoring, for example, is a *passive* way to sense the presence or absence of flow. Examples include flow switches and leak detectors. One *could* think of turbulence-induced noise as a mechanism that "modulates" silence (Table 2-3). (In Table 2-3 we extend the concept of modulation slightly beyond its usual connotation to facilitate the comparison of a number of passive and active ultrasonic flow-measurement techniques.) From the measurement of noise amplitude one usually cannot determine flow velocity with precision. But, timing at two separate transducers, the arrival of noise pulses due to a sudden leak in a pressurized pipe enables one to locate the source of that leak fairly precisely.

Active methods may be grouped according to the way that flow modulates one or more acoustic transmissions over one or more paths in or near the flow. The three principal active methods used to measure flow in closed conduits in use in the mid-1980s were transit time (contrapropagation), Doppler, and vortex-shedding. In open channels and partly full pipes, active measurements of liquid level yield data that can be interpreted in terms of flow velocity.

Algebraically, one can summarize certain interactions of ultrasound and flow quite easily. Consider the contrapropagation method, for example. For simplicity, assume that flow velocity is low (small Mach number M_s < 0.1) and consider sound waves propagating a distance L axially upstream and downstream in a pipe. The time difference for the two opposite directions is

$$\Delta t = t_2 - t_1 = 2LV/c^2, \qquad (2\text{-}1)$$

where V = flow velocity and c = sound speed. In most cases, the fluid

Table 2-3.   Passive, Active and Combination Methods of Sonic or Ultrasonic Flow
Measurement, Partial List

Passive Methods	What Happens?/ What is Modulated?	Remarks/Type of Modulation
Noise monitoring		
Amplitude only	Silence disturbed by turbulent flow	am nonlinearly related to flow velocity
Spectrum analysis	Screech frequencies excited	Detects flow transitions, cavity resonances; am and fm of flow noise
Vibration analysis	Strut vibrates at frequency of shed vortices	fm
Timing of pulses generated by sudden leak; cross-correlation of noise	Expansion waves are generated by leak impulse; noise signatures are generated by continuous noisy leak	Locates source of leak from difference in arrival times at two detectors
Active Methods		
Transit time upstream and downstream	Transit time increased or decreased by flow	Phase or "transit time" modulation; pulse rates are generated in sing-around method at frequencies proportional to flow velocity
Transit time across turbulent flow	Transit time modulated by eddies	pm; rms value of phase noise approximately proportional to flow velocity
Transit time across turbulent wake of a strut	Amplitude modulated at the vortex-shedding frequency	am at frequency proportional to flow velocity
Amplitude of sound beam transmitted across flow	Beam drifts downstream	am
Transit time to surface of liquid in open channel and partly filled pipe	Liquid level is related to flow velocity and/or volumetric flowrate	Phase or "transit-time" modulation
Correlation over parallel paths	Tags modulate transmission	am and/or pm
Hot wire thermoacoustic (temperature profiling) waveguide	Transit time in electrically heated zone(s) affected by local flow	Phase or "transit-time" modulation
Doppler	Sound is reflected from eddies or particulates at Doppler-shifted frequencies	fm

temperature T, and therefore c, varies so much that V cannot be calculated to 1% accuracy if one assumes $c = $ const. in Eq. (2-1). To eliminate errors due to uncertainty or variations in c, the usual remedy is to eliminate c. One way to eliminate c utilizes the fact that the sum of the transit times, $t_1 + t_2$, is independent of V. Accordingly, V is found to be proportional to $\Delta t/(t_1 + t_2 - 2t_w)^2$, where $t_w = $ delays other than in the fluid. At higher Mach numbers, but with M_s still <1, a more rigorous derivation yields V proportional to $(1/t_1 - 1/t_2)$ or, equivalently, to $\Delta t/t_1 t_2$.

Another way to eliminate c requires that one either measure or generate quantities proportional to the reciprocals of t_1 and t_2:

$$\Delta f = f_2 - f_1 = |\, 1/t_2 - 1/t_1 \,| = 2V/L. \tag{2-2}$$

Many transit-time flowmeters take advantage of this latter equation, using sing-around circuits or servoed-frequency approaches to measure f_1, f_2 and/or Δf. (The origin of the sing-around method is recounted by Greenspan in Chapter 10.)

It is common industrial practice to place a pair of Doppler transducers either near each other in one assembly or else separated by 90° or 180° around the pipe. The Doppler shift Δf of the waves scattered off eddies or particulates is fundamentally proportional to the scatterer's local Mach number $M_s = V/c$. But in the usual clamp-on case, Δf is proportional to V, not V/c, because changes in c are compensated by changes in the refracted angle in the fluid. The velocity of scatterers can also be determined in the time domain from the change in their distance from a transducer between two successive interrogations.

The aforementioned equations usually need to be modified to take into account the angle between the ultrasonic path and the flow axis, to correct for conduit curvature in nonrectilinear conduits, and, often most importantly, to convert from a velocity measurement V_d averaged over a single path to the desired area average $\mathbb{V}$. To compensate for flow profile, the path-averaged reading V_d is multiplied by a meter factor K defined by $K = \mathbb{V}/V_d$. K is usually either fixed empirically or calculated dynamically as a function of the Reynolds number Re, or better yet, calculated from profile data.

2.1.2 Flowcell and Transducer Configurations

Figures 2-2 through 2-26 illustrate flowcell and transducer designs used with transmission, reflection, vortex-shedding, liquid-level and noise-detecting ultrasonic flowmeter instruments. Designs for other methods symbolized in Fig. 2-1 appear in Chapter 4. Figure 2-2 shows typical 45° recessed transducer ports, single path; Fig. 2-3 shows a rectilinear flow

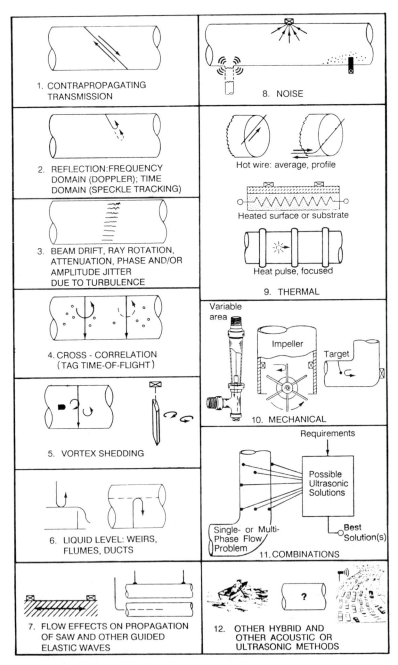

FIG. 2-1. Twelve categories of acoustic or ultrasonic flow measurement principles and methods, with examples. Flow generally from left to right.

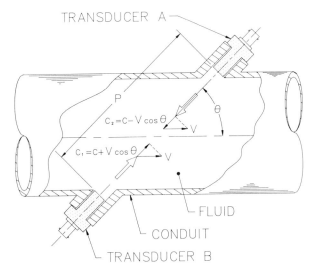

FIG. 2-2.   Sound bursts are propagated alternately in opposite directions between a pair of transducers situated diagonally along the pipeline. The upstream signal is delayed and the downstream signal is speeded up by the moving fluid. Illustration adapted from Crouzet.

FIG. 2-3.   One area-averaging flowcell that evolved in the early 1970s included a rectilinear flow channel, rectilinear transducer channel, screens over transducer ports to minimize eddy generation therein, and a gradual inlet transition. Transducers interrogate 100% of flow channel cross section, sometimes utilizing a zigzag path. The area-averaging approach provides approximately linear response over laminar, transitional, and turbulent flow regimes.

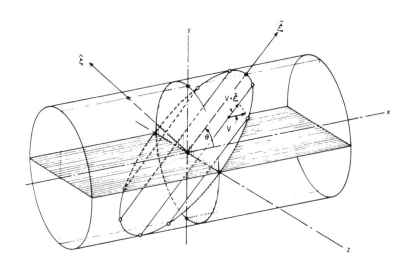

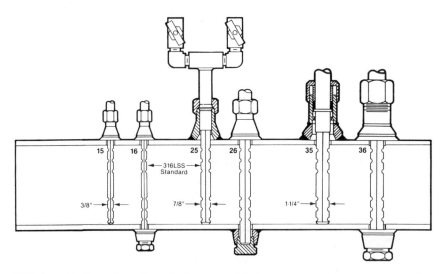

FIG. 2-4. To determine volumetric flow rate when flow velocity is not uniform, it is common practice to sample the flow at several points or along several chords and integrate. Top: Gaussian quadrature *ultrasonic* flow velocity measurement chords. Each ellipse is defined by the intersection of the measurement plane with the pipe wall, and volumetric flow rate is given by the integral of the normal component of the fluid velocity vector over the area of the ellipse. The transducers are normally recessed. For axial flow (no crossflow, no swirl) measurements in one plane suffice. After Fisher and Spink, 1972. © 1972 Peregrinus. Bottom: *Multiport pitot* tube available from Dieterich Standard. Facing page, top: *Multiport thermal anemometers* available from Kurz.

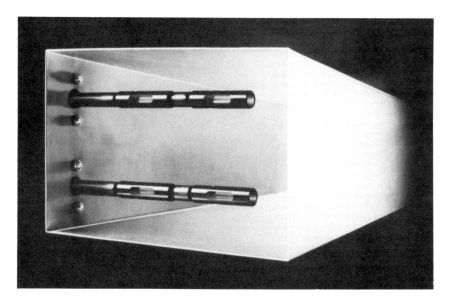

FIG. 2-4.   (Continued)

channel and beam for linear response (area-averaging), substantially independent of profile. Figure 2-4 shows a four-chord quadrature method
for area-averaging. This method has provided the highest proven accuracy
achievable with ultrasonics in pipes large enough to accommodate the
four paths. (An additional set of four paths is required if cross flow or
swirl is significant.)

Figure 2-5 is an externally mounted refraction method first reported
in 1964. Separate clamped-on wedges launch *longitudinal*, *shear* or *plate*
waves in the pipe wall depending, in general, on the geometry, the type

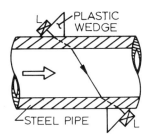

FIG. 2-5.   Use of externally coupled wedge for introducing longitudinal or shear wave at
oblique incidence.

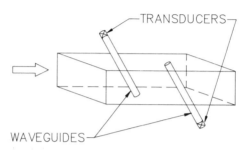

FIG. 2-6. Top: Incident wave launches plate wave (e.g., flexural wave) in sheet between two fluids. As plate wave propagates along sheet it radiates (leaks) into both adjacent fluids. (Sanders, 1939; Deighton et al., 1981). Bottom: Principle of extended-source leaky waveguide used in ultrasonic flowmeter (Swengel, 1956).

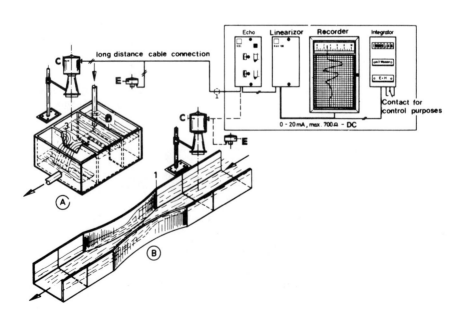

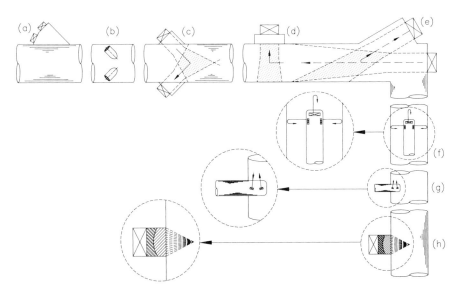

FIG. 2-8.   Schematic of Doppler arrangements. (a) Two externally-mounted transducers on same side of pipe (Fowlis, 1973). (b) Two transducers 90° apart interrogate axial region. (c) As in (b) except transducers are 180° apart. (d) Orthogonal arrangement near corner interrogates entire cross section. (e) Oblique arrangement near corner. (f) Catheter concept measures flow and duct area (Martin and Watkins, 1980). (g) Two-transducer Doppler probe penetrates pipe orthogonally. (h) Focused orthogonal transducer interrogates flow oblique to flow axis (Newhouse et al., 1985, 1986).

of incident wave, the acoustical properties of the materials, and boundary conditions. A plate wave travelling axially provides an extended source of radiation (Sanders, 1939). One form of extended radiation source was first used in a contrapropagation flowmeter by Swengel, 1950. He utilized leaky compressional waves in a parallel pair of rods that he installed near the side walls of a large rectangular cross section duct conveying water to a turbine (see Fig. 2-6).

Figure 2-7 is applicable to weirs, flumes and venturi channels in which liquid level is related to flow velocity. Figure 2-8 represents Doppler ar-

---

FIG. 2-7.   Open channel measurements of flow can utilize the reproducible relation between flow and liquid level attending the proper use of a weir, flume, or venturi channel. Noncontact sonic measurement of water level is shown above, where the sensor or transducer C aims a pulse at the surface. The round trip time must be temperature-compensated. Illustration courtesy Endress & Hauser, Aquatot DMU 160 Manual 6.74.12-0. A-Measurement with measuring weirs (3); B-Measurement with Venturi channels (1); C-Sensor; D-Baffle plate; E-Temperature probe.

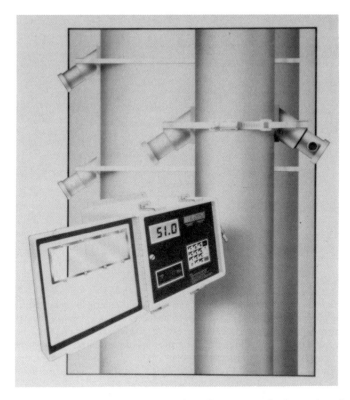

FIG. 2-9.   Doppler clamp-ons from Texas Nuclear. Strap-on toggle clamps (not shown) are available from Texas Nuclear for single-side or dual-side mounting of transducers on pipes from 30 to 180 mm diameter. Texas Nuclear's Doppler transducers are dry-coupled through contoured solid cured Silastic, thereby avoiding the need for grease or epoxy. (L. Missman, priv. comm., 1988).

rangements schematically. Figures 2-9 through 2-14 show examples of Doppler equipment and details on Doppler catheter and range-gating methods. Both flush-mounted and intrusive Doppler wetted probes (Fig. 2-11) were introduced by EDO Corporation in the early 1970s, following their earlier (1966) interests in Doppler for shipboard navigation (Winston, 1974) and in Doppler R&D (research and development) pipe flow measurements that began around 1969 (Robinson, priv. comm., 1988; EDO,

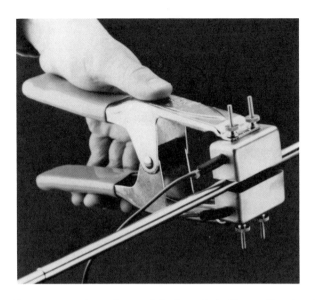

FIG. 2-10. Clamp-on transducer for conduits from 6- to 19-mm diameter. Courtesy of Omega Engineering, an Omega Technologies Company.

1970). The 1970 flush probe shown in Fig. 2-11a was used in what appears to be the *first* commercially-available Doppler flowmeter for measuring fluid flow in pipes. The intrusive design, Fig. 2-11b, transmitting typically near 3 MHz, used two transducers aimed at a point 30 cm upstream, in a selectably representative region of flow. That is to say, the probe penetration was adjustable, so one could profile the flow and then position the probe optimally. For example, in some cases the preferred location was near $r = 0.2a$ for turbulent flow in a pipe of radius a. The dual-transducer Doppler probes were dimensioned about 2 cm in diameter for pipes up to 60 cm in diameter and 5 cm in diameter for pipes up to about 2.5 m diameter. The probes were cleaned by a wiper assembly on a timed automatic cycle, a useful feature in sewage applications. Intrusive Doppler probes were introduced by other manufacturers about ten years later (see example in Fig. 2-12). A Doppler catheter for intravascular measurement of blood flow, due to Martin and Watkins, 1980 and shown in Fig. 2-13, consists of a cylindrical ultrasonic transducer array of six ele-

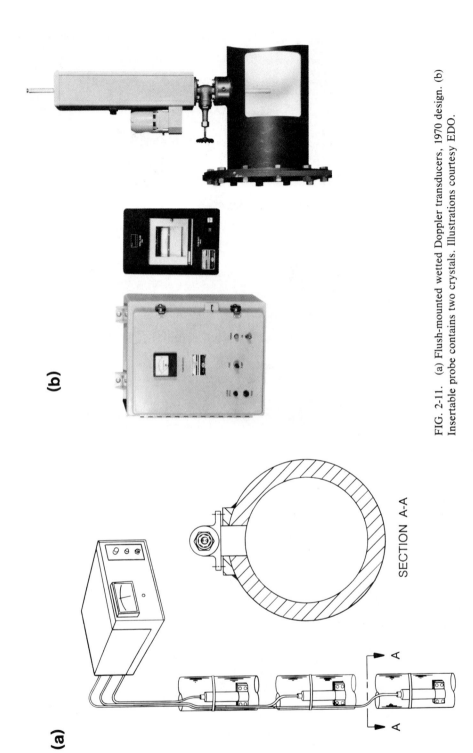

FIG. 2-11. (a) Flush-mounted wetted Doppler transducers, 1970 design. (b) Insertable probe contains two crystals. Illustrations courtesy EDO.

SECTION A-A

28

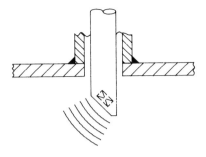

FIG. 2-12. K-Flow Doppler Model 710 is an insertion angled flow sensor designed to measure muds, drilling fluids, cements, abrasives, suspended solids and other slurries in pipes from 2- to 18-inch (approx. 50- to 450-mm) diameter. Sensor is rated to 20 bar and ± 100°C.

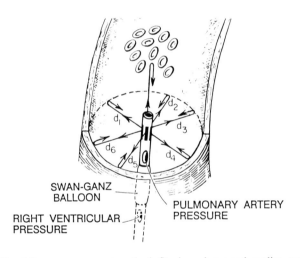

FIG. 2-13. Blood flow measurement method. Stroke volume and cardiac output are measured by integrating pulmonary artery blood flow. Instantaneous pulmonary artery flow (Q) is measured by 1) determining vessel sectional area A (area of the wall outlined by a plane that passes through the vessel at any arbitrary angle) and (ideally) 2) multiplying it by the average blood flow velocity V that flows perpendicular to the plane. Hence $Q = AV$. In order to accomplish this measurement, an ultrasonic catheter is positioned so the tip is in the pulmonary artery. Vessel wall to catheter distances d_i (where $i = $ 1-6) are detected by scanning the radially transceiving transducer array. Sectional vessel area is computed from this information by 1) solving for the constants of an equation that describes an ellipse that best fits the wall contour detected, and 2) computing the area of the ellipse using these constants. Blood velocity is detected via the tip transducer from the ultrasound backscattered energy from the blood cells using pulsed Doppler techniques. The balloon is inflated only during catheter insertion to allow the blood flow to assist in directing the catheter tip to the pulmonary artery. Pressure lumens are used in part for guidance to position the tip just distal to the pulmonary artery valve. When pressures are obtained at the two ports as indicated, the desired position is confirmed. After Martin and Watkins, 1980. © 1980 IEEE.

29

ments for detecting lumen area, along with a single element for detecting blood velocity, both located at the tip of a 2-mm diameter catheter. Flow is derived from the area and velocity information. A 128-gate range-gated Doppler system due to Brandestini, 1978 is shown in Fig. 2-14. Novatec of Switzerland manufactures an Ultrasonic Velocity Profile meter (UVP)

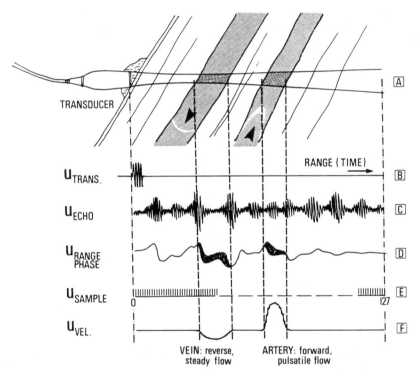

FIG. 2-14. Transducer, blood vessel and signal relationships for 128 range gates. A piezoelectric transducer is coupled to the skin with a jelly compound and aimed at the vessel at an oblique angle (A). The diagram illustrates the division of the entire observation range along the sound beam path into 128 velocity channels (E). The various electrical signals are shown according to their time and range relationships. The top signal (B) is a burst of sine waves that excites the crystal at a repetition rate of approximately 6.4 kHz to yield a 10-cm depth of view. The echo (C) shows the location of the tissue boundaries yet does not reveal any Doppler shift. Mixing the echo with the master oscillator (coherent with the transmitted bursts) produces the range phase information (D). The bulge at the vessel depth illustrates how phase varies with time (Doppler frequency) when moving particles are present. The sampling train (E) points out where the phase signal is sampled into the various range bins for separate processing. The output of the instrument (F) will produce a real-time velocity profile indicating the depth and velocity of the blood flow (after Brandestini, 1978). © 1978 IEEE.

having the following main characteristics: f, 0.5 to 10 MHz; prf, adjustable for ranging from 1 to >100 cm; axial resolution, 0.5 to 25 mm; velocity resolution $\approx$ 1/(time to make a measurement); number of gates, 3 to 128 (Bonifay and Frens-Cart, priv. comm., 1988). Examples of its use appear in Takeda, 1986 and 1987.

A vortex shedder is shown in Fig. 2-15. Vortex-shedding applications are discussed in Section 4.4. It is interesting to observe that while vortex shedding appears to be overlooked in some reviews of ultrasonic flow-meters, Joy's remarks in Chapter 10 indicate that this is probably the most commonly-used ultrasonic method in industry today, even if one counts only its application in measuring air flow in automobile engines.

Returning now to the contrapropagation transit-time method, standard transit-time clamp-on flowmeters are available commercially for small pipes (diameter $\geq$2 cm) to large pipes (diameter $\leq$10 m). Clamping pressure is usually developed by tightening conventional clamps or setscrews. Examples of some transit-time and clamp-on flowmeters and accessories are shown in Figs. 2-16 through 2-20.

In contrast to these noninvasive transducer mounts, it is interesting to note that the long-abandoned in-the-flow transducer concept of Rütten may be revitalized for specific applications. One motivation is to trade noninvasiveness for axial propagation. Frankenberger et al., 1974 con-

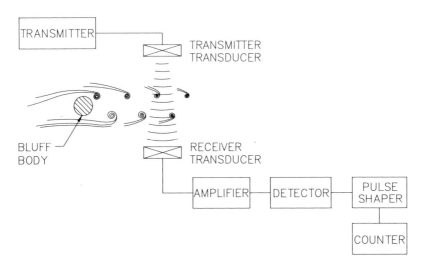

FIG. 2-15. Vortex shedding flowmeter concept, according to Joy and Colton, 1972. Shed vortices modulate the ultrasonic beam at a frequency proportional to the flow velocity, for $Re > 10^4$.

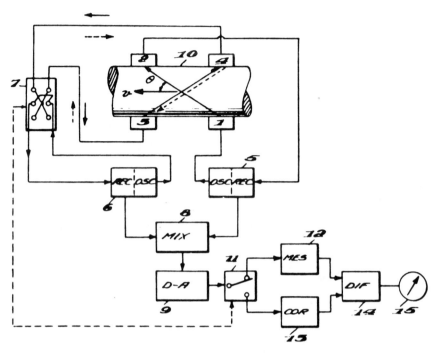

FIG. 2-16. Ultrasonic transit time flowmeter due to Yamamoto and Ito, 1966, and having a Japanese patent application claims priority dated Aug. 24, 1962, appears to be the *first transit-time clamp-on* patent. Flow is computed from the difference in upstream and downstream sing-around frequencies. In practice the four transducers are replaced by two, often on the same side of the pipe.

centrically mounted two pulsed PZT (lead zirconate titanate) annular transducers within a tubular flowcell, to measure laminar air flow (Fig. 2-21). A relatively small flowcell, inserted as a probe to scan or to measure flow at a fixed distance from the wall of an air duct, was manufactured by Novex starting in 1987. Downhole flowtools of outside diameter ap-

$\longrightarrow$

FIG. 2-18. Multiplexed and survey clamp-on transit time flowmeters introduced since the mid-1980s by Controlotron. Left: Four channel System 960 has a flow range of ± 12 m/s (including zero) with a resolution of about 3 mm/s. Claimed calibration accuracy is 2% of actual flow (or better). Manufacturer states that the meter can measure flow in pipe sizes from 12.7-mm to about 4 m OD (outside diameter) with wall thicknesses from about 0.75 to 38 mm. Each of the four channels in the multiplexed model illustrated independently deal with different pipe sizes, liquids and flow ranges. Right: Portable survey flowmeter is user-programmable for pipes from about 12.7-mm to 2-m OD. Liquid crystal graphics display the flow rate, total flow, setup and status information. Strip chart, analog and digital outputs are also provided. Illustrations courtesy Controlotron.

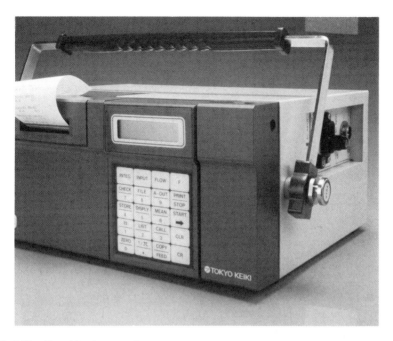

FIG. 2-17. Portable clamp-on flowmeter UFP-1000 made by Tokyo Keiki allows user to enter site parameters for a wide variety of pipes. Instrument "sets itself up" and instructs user on how far to separate transducers on the pipe. Printer provides permanent record.

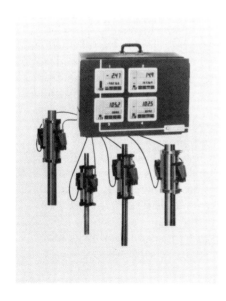
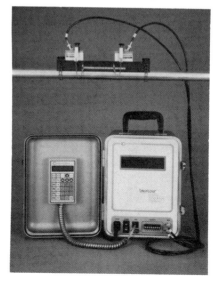

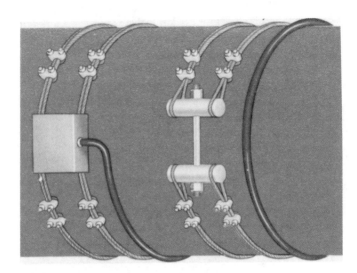

FIG. 2-19. Clamp-on flowmeter available commercially in the US in the early 1970s for pipe diameters larger than 30 cm. For pipe diameters as small as 15 cm, spoolpieces were available from the manufacturer with transducers clamped on the outside. Illustration courtesy Badger Meter.

proximately 40 to 80 mm are now being developed for oil, gas and geothermal applications in wellbores (Lynnworth, 1988d; Chapter 9).

Other intrusive flowmeter probes may be mentioned. In the measurement of flare gas flow rate, either high attenuation or access constraints often force the flowmeter designer to locate the transducers intrusively in the flow where they interrogate the flowing gas along a chord segment (Fig. 2-22). In principle, to measure the flow rate of a fluid whose temperature, corrosiveness, or other attributes provides an intolerable environment for transducers, one could use transducers buffered by solid rods or flowing gas. One example of wetted buffer rods is given in Fig. 2-23. In this diagram, shear waves generated at the external transducer, and polarized with a particle displacement component parallel to the flow direction (SV), propagate to the chamfer, where they are mode converted and transmitted into the fluid as longitudinal (L) waves. Part of the L-wave energy reflects internally off the chamfer, emerging from the side of the probe adjacent to the chamfered end, either substantially or in preferred cases precisely perpendicular to the major axis of the buffer rod. These emerging L waves have been used to measure the flow velocity of convectively-driven, high-temperature, high-pressure water.

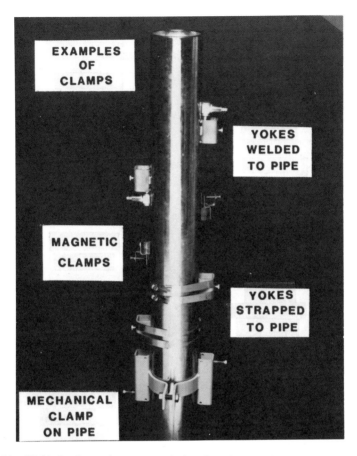

FIG. 2-20.   Welded yokes and strap-on and pipe riser clamps allow transducers to be re-moved without moving clamps, in these designs from Panametrics. Later, transducers can be installed again in essentially the same locations, as is required for repeatable flow mea-surements.

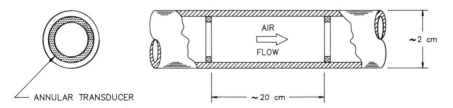

FIG. 2-21.   Use of intrusive transducers to beam the interrogating waves essentially along axial path. After Frankenberger et al., 1974.

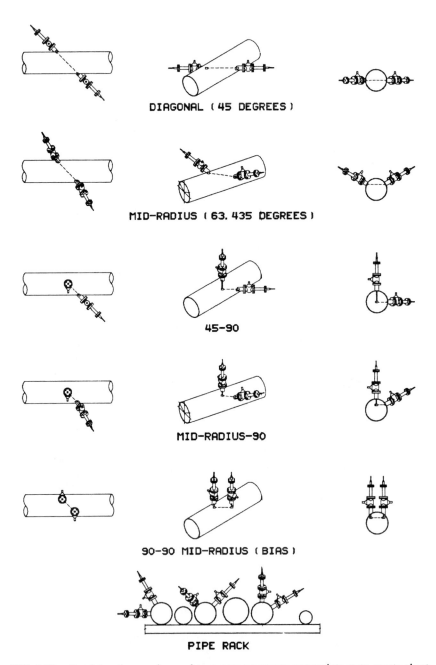

DIAGONAL (45 DEGREES)

MID-RADIUS (63. 435 DEGREES)

45-90

MID-RADIUS-90

90-90 MID-RADIUS (BIAS)

PIPE RACK

FIG. 2-22. Special paths may be used to overcome access constraints or to create short path in large pipe in which the fluid has a high attenuation. The "local" flow velocity in short paths is related to the area-average flow velocity by the meter factor K. K depends on the flow profile. Preferred paths minimize this dependence. After Smalling et al., 1984. © 1984 Texas A&M, Dept. of Chemical Engineering.

36

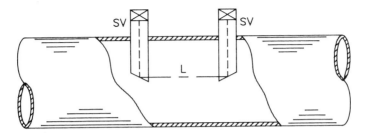

FIG. 2-23. Intrusive flowmeter mode-converting probes with external shear wave transducers (Lynnworth, 1967a,b, 1969b, 1974b).

2.1.3 Leak Detection

As pressurized gas escapes through an aperture it creates considerable ultrasonic noise, in particular, within the band 36 to 44 kHz. For over ten years Hewlett-Packard's Delcon Division manufactured instruments which detect this characteristic sound with a directional barium titanate microphone and shift the signal to audio by mixing it with a 40-kHz local oscillator signal. The audio signal is then amplified and monitored on a speaker and sound level meter. Similar equipment became available around 1986 from Communication Technology Corp., e.g., their model C4918A ultrasonic leak detector. See Fig. 2-24.

To detect leaks in aerial cables, the equipment operator merely scans the cable from the ground with the flashlight-size microphone, listening for the characteristic hissing sounds of a leak. By simultaneously observing the level meter, one can "peak in" on the leak and determine its exact location. Pole-mounted accessories are also available for closer scanning of the cable.

Leaks in ducted underground systems are located with a "Duct Probe" accessory. Consisting of a miniature microphone connected to a system of aluminum rods, the Duct Probe can be used to explore up to ~150 m into a cable conduit. The leak is thereby pinpointed, permitting repair of the damage with a minimum of excavation.

About 1973, Dukane introduced their model 42A15 ultrasonic leak detector, designed primarily for checking telephone-cable leaks. Capable of locating air, gas, and corona discharge leaks, the self-contained 0.7 kg unit (Fig. 2-25) detects ultrasound in the frequency range 38-42 kHz. Sensitivity is 6 dB signal-to-noise ratio at 1.8-2.4 m from a leak source consisting of a 0.13 mm orifice subject to a 2 psi (13.8 kNm^{-2}) pressure head. Similar equipment is available from other sources, e.g., UE Systems.

Expansion waves generated by *sudden* leaks in gas pipelines lead to AE-type signals that can be timed by two spaced-apart transducers to

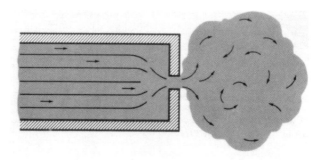

FIG. 2-24.   (Top) Ultrasonic noise produced by turbulence of leaking gas. (Middle) Examples of probes responsive in the range 36 to 44 kHz. (Bottom) Reflector increases sensitivity and distance range. Illustrations courtesy Delcon Div./Hewlett Packard and Communications Technology Corp.

FIG. 2-25. Leak detector with headset, model 42A15, made by Dukane Corp.

locate the source of the leak (Rocha and Finch, 1986; Chapter 3, Fig. 3-62). Even a continuous leak can be located by timing if the leak's noise signature can be picked up at two separated transducers and be cross-correlated (Kupperman et al., 1987; Chapter 4).

2.2 Thermometry

Ultrasonic thermometers are usually designed to respond to the temperature-dependence of sound speed c. In special cases where only one particular temperature is of interest, such as the temperature of a phase change or the recrystallization temperature, the temperature-dependence of reflectivity or attenuation may be utilized.

Ultrasonic thermometers have found applications in the temperature range -80 to $+250°C$, where the "quartz thermometer" offers resolution of 0.1 millidegree and linearity superior to that of platinum resistance thermometers. Fluidic thermometers, usually oscillating below 10 kHz, are used for aircraft, aerospace and other engine applications. Fluidic sensor temperatures up to 1200°C are fairly common; materials limitations have restricted applications at 2200°C to short-life experiments. At higher temperatures, approaching 3000°C, thin-wire probe-type ultrasonic sensors offer accuracy, simpler profiling, faster response and/or longevity advantages over thermocouples. In 1963–1967 studies of the transport properties of gases, ultrasonic pulse techniques were used to measure temperature in the 10,000 to 20,000°C range, using the gas as its own sensor. Some 20 years later, related techniques were used to measure boiler and stack gas temperature.

The number of quartz thermometers sold to date is estimated to be in the thousands, of which ~1600 were sold by 1975. In contrast, only a small number of thin-wire thermometer probes and corresponding instruments had been sold by 1985; applications include nuclear fuel pin centerline temperature measurement, critical heat-flux experiments, chemical retort (oil shale) experiments and the temperature compensation of torsional density sensors such as those described in Section 2.3.

Most applications for the preceding devices have been developed since about 1960, although the origin of the techniques themselves can be traced to earlier decades. The thin-wire technique, for example, is traceable historically to the pioneering work of Bell, 1957, and in some respects, to the notched-bar dissertation of Frederick, 1947. Quartz crystal studies date back to the Curie brothers' discovery of piezoelectricity in 1880, but the linear coefficient (LC) cut underlying one form of the quartz thermometer is of a more recent vintage (1964). Surface acoustic wave (SAW) resonators became available as inexpensive temperature sensors around 1985.

2.2.1 Physical Principles

The temperature-dependence of sound speed c may be computed for ideal gases:

$$c = (\gamma RT/M)^{1/2}, \tag{2-3}$$

where γ = specific heat ratio C_p/C_v, R = gas constant, T = absolute temperature and M = average molecular weight. Uncertainties in composition, e.g., incompletely burned fuel entering a turbine section of a jet engine, place one limit on the ultimate accuracy obtainable. Impedance mismatch, attenuation in low-density and/or turbulent gases and ambient noise place other limits in potential applications where the use of Eq. (2-3) has been proposed. To overcome noise and attenuation over long paths, tens of meters, Nuspl et al., 1986, found it expedient to reduce frequency to the 0.5 to 3 kHz range and employ correlation detection techniques. In this way the average temperature was measured in the exit plane of large utility boilers at temperatures well over 1000°C, and over path lengths of ~10 to ~20 m. See also Green, 1985. Equation (2-3) is applicable at cryogenic temperatures too (Cataland et al., 1962).

Generally speaking, sound speed c equals the square root of a stiffness term divided by a density or inertial term. For liquids and solids, the temperature-dependence of c may be computed (i.e., predicted) in cases where the temperature dependence of the appropriate terms have been measured previously. The temperature-dependence of c is illustrated in Fig. 2-26 for some gases, liquids and solids. For specific media c exhibits

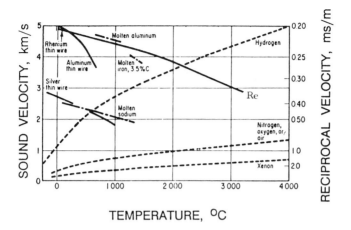

TEMPERATURE, °C

FIG. 2-26.   Sound velocity in gas, liquid and solid media as a function of temperature.

a square root dependence on temperature (for gases), on the reciprocal of compressibility (for liquids) and on elastic moduli (for solids).

Two basic measurement principles have been used in instruments responding to c: resonance and pulse (nonresonance). The quartz thermometer senses temperature T by the 1000 Hz per °C change in resonant frequency of a small 28 MHz quartz disk. Fluidic devices oscillate at a frequency determined by sensor geometry and c of the gas passing through the device. A resonant probe due to Bell, 1972, may be described as a small tuning fork whose frequency changes with elasticity and density changes. Bell previously introduced nonresonant thin-wire T-sensing probes. A hybrid technique wherein a broadband pulse impulsively drives a short sensor into resonance was developed by Fowler, 1971. A variety of pulse techniques, multizone profile thin-wire sensors and right-angle wire sensors have been investigated during the decade 1977–1987. At Sandia, Carlson et al., 1977, and Field, 1986, developed five-zone sensors with zone lengths as short as 1 cm, for temperatures up to nearly 3000°C.

2.2.2   Equipment

Figure 2-27 shows the Hewlett-Packard quartz thermometer; Fig. 2-28, a fluidic thermometer; Fig. 2-29, a nonresonant single-zone wire sensor; and Fig. 2-30, a nonresonant multizone wire sensor. Figure 2-31 illustrates multipath concepts proposed in 1966 for profiling the temperature in a gas (Lynnworth and Carnevale, 1966). Figure 2-32 shows one of the meth-

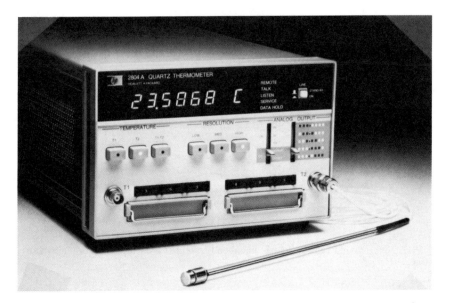

FIG. 2-27. Quartz thermometer and oceanographic temperature sensor assembly. This probe operates from -40 to +120°C, and is rated for 10,000 psi (equivalent to ocean depth of over 4 miles, or nearly 7 km). Nominal operating frequency is 28 MHz; sensitivity to temperature change is approximately 1000 Hz per °C.

ods proposed in 1970 for steel billet thermometry (Lynnworth, 1970). By the mid-1980s multipath T profiling had been demonstrated in hot gas (Green, 1985) and in hot steel (Wadley et al., 1986). For temperature measurements inside the human body the method of Davis and Lele, 1985, shown in Fig. 2-33, is of interest. Figure 2-34 illustrates the noninvasive

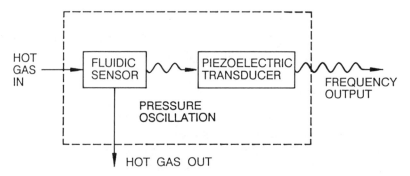

FIG. 2-28. Fluidic thermometer element and transducer, operational schematic, after McMillan and Pamperin, 1972.

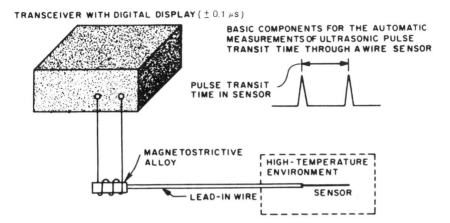

FIG. 2-29.   Temperature sensing probe, with nonresonant single-zone sensor.

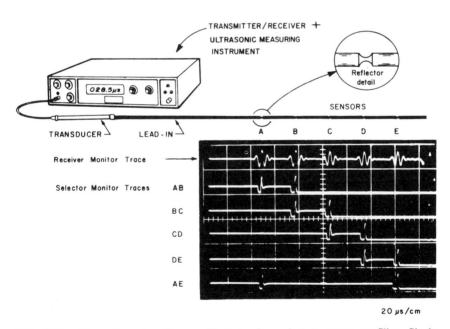

FIG. 2-30.   Schematic and oscillogram illustrate ultrasonic temperature profiling. Single line containing series sensors is scanned by selecting echoes according to sensor position along the line. Transit time between selected pair of echoes corresponds to temperature between reflection points. Echo pairs AB, BC, CD, DE yield profile; pair AE yields average temperature. After Lynnworth and Patch, 1970. © 1970 ASTM.

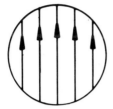 

FIG. 2-31.   Tomographic thermometry proposed for gases by Lynnworth and Carnevale, 1966, and for hot steel billets by Lynnworth and Patch, 1970.

(clamp-on) measurement of temperature of a liquid within a conduit, such as liquid sodium in a stainless steel pipe. A table comparing several *resonator* designs proposed or used as ultrasonic thermometers appears in Chapter 5.

2.2.3   Fluid Temperature and Flow Profiles in a Pipe

Figure 2-4 shows a four-chord quadrature method developed to measure the area-averaged flow velocity in a round pipe. Data from the four individual paths can also be interpreted in terms of the flow velocity profile and the sound speed profile. To the extent that sound speed c is related to fluid temperature T the c profile yields the T profile. If there are substantial T gradients across the pipe, T profile information could be important in calculating volumetric flow rate at standard conditions, or mass flow rate.

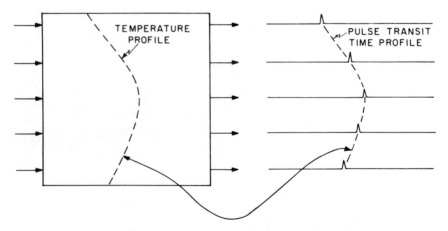

FIG. 2-32.   Ultrasonic determination of temperature distribution inside large metal billets, proposed by Lynnworth, 1970b. Measuring the transit time versus elapsed (real) time was also proposed, to monitor the approach to equilibrium. See also, Carnevale et al., 1964.

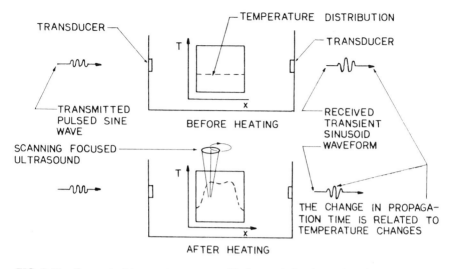

FIG. 2-33. Concept of temperature measured before and after hyperthermia therapy. The development of this noninvasive diagnostic technique was directed towards simultaneous operation with scanning focused ultrasound as the heating modality. This necessitated independent operation of both ultrasonic fields. To limit interference between the ultrasonic fields, narrow-band quartz transducers are operated at harmonically unrelated frequencies. Signal-averaging of the received diagnostic beam minimizes interference from the heating field. After Davis and Lele, 1985. © 1985 IEEE.

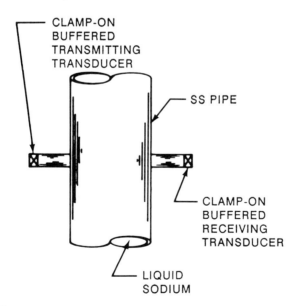

FIG. 2-34. Clamp-on ultrasonic thermometer. The temperature of liquid sodium was determined in 1968 by measuring the transit time of an ultrasonic pulse transmitted across the diameter of a ~25-cm OD stainless steel pipe. Clamped-on probes were noninvasive and were removed after the test. Test site: Argonne National Laboratory, oscillator rod facility, May 1968.

2.3   Density and Porosity

Some ultrasonic measurements of density ρ and porosity q may initially appear analogous to well-known gamma-ray measurements based on backscatter or on absorption principles. However, the ultrasonic energy/ matter interaction is quite different from that for gamma rays. The differences may be clarified with reference to the following applications of ρ measurement in gases and liquids, and ρ and q measurements in solids.

Consider probe-type resonant and nonresonant densitometers, which are ordinarily wetted or immersed partly or totally in the fluid under test, and also transmission measurements wherein c or α is a function of ρ or q. The former type may be understood by analyzing sound propagation in the probe (which is "loaded" by the fluid). The latter type is understood by analyzing propagation in the fluid or solid medium itself.

2.3.1   Resonant Probe Principles

As an example of a vibrating element (resonant vane/hoop) density sensor, consider the design of ITT Barton in Fig. 2-35. Introduced in 1971, it consists of a sensing vane symmetrically positioned across a supporting cylinder. In operation, the vane oscillates to move with a simple harmonic motion, causing an acceleration of the surrounding fluid.

The vane oscillates at a resonant frequency determined by the density of the surrounding fluid. As the fluid density increases, the frequency of vibration decreases substantially in accordance with the following relationship:

$$\rho = A/f^2 + C, \qquad (2\text{-}4)$$

where A and C are constants and f = frequency.

The probe is installed in a line containing fluid (one type for liquids, another for gases). A detector within the probe senses the frequency of the vibrating vane. This signal is amplified in the transmitter and energizes a driver within the probe with a minimum force to sustain oscillation at the system resonant frequency.

Probes of this type can operate with fluids in static or dynamic situations, and from cryogenic temperatures to over 100°C. Precision is 0.1% of full scale. Applications include aerospace, chemical, petrochemical, refinery, pipeline, cryogenic, and food-processing. Liquid applications are limited to viscosities less than ~100 centipoise. (See remarks in Section 4.2.2.1 on the improvement in ρ accuracy for natural gases, if the sound speed is known, based on a correlation discovered by Jaeschke and Hinze, 1987.)

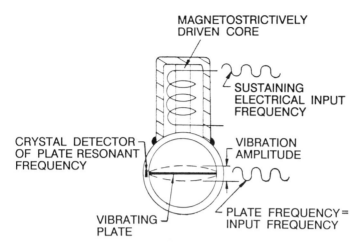

FIG. 2-35.   ITT-Barton vibrating plate densitometer, showing magnetostrictive driver and piezoelectric detector. After November, 1972.

In another resonant probe design, Fig. 2-36, liquid (even of high viscosity) passes through a U-tube, and the resonant frequency of the filled tube indicates ρ.

Hoop mode vibrations are utilized in fluid densitometers (e.g., Abbotts, 1972; see Fig. 2-37). (Both the vane and U-tube probes normally resonate in the audible, not ultrasonic, range.)

Since about 1955, several investigators have considered measuring the electrical impedance of a crystal radiating into a liquid to determine the liquid's characteristic impedance ρc. This approach has not yet proven practical.

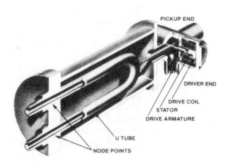

FIG. 2-36. Dynatrol® U-tube resonator measures density/specific gravity/%solids concentration. Illustration courtesy Automation Products.

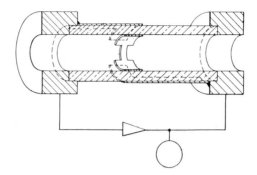

FIG. 2-37.   Densitometer probe consisting of hollow piezoelectric cylinder. After Abbotts,
1972.

Relative to an orbiting densitometer, gases at densities as low as air at
140 to 280 km have been tested with a resonant quartz crystal used as a
microbalance. Response was reported as linear for pressures from 10^{-4}
to 2×10^{-6} Torr, with sensitivity of 1.7×10^{8} Hz/g, and accuracy of 6%
(Roder, 1974).

2.3.2   Nonresonant Probe Principles

A pulse-echo sound velocity probe, Fig. 2-38, developed by NUSonics
primarily for liquid composition and concentration applications, has also
been used for ρ determinations in cases where ρ is a known function of
c. Probes may be mounted recessed or in-stream. In either case the mea-

FIG. 2-38.   Sound velocity probe consisting of transducers and a corner reflector, and
normally used for composition and interface measurements, can be used in some cases to
determine liquid density. Illustration courtesy NUSonics.

surement is based on propagating a pulse between the transducer and the reflector over a defined liquid path.

A waveguide sensor of noncircular cross section (Fig. 2-39) uses torsional waves to measure ρ of fluids, especially liquids. Analogous to the multizone thermometer, the torsional sensor can be built as a multizone density profiler and can also sense liquid level and, in some cases, viscosity (Lynnworth, 1977; Bau, 1986; Kim and Bau, 1989; Chapter 9).

Transmission Principles

Gas Density.   The relative amplitude of a received pulse after transmission across a gas (Fig. 2-40) depends mainly on the impedance mismatch at the buffer rod/gas interfaces, and on the attenuation losses in the gas. For the symmetrical case illustrated, and neglecting beam spread, it may be shown that, in terms of the reference echo amplitude A, and the first and second amplitudes B and C corresponding to single and triple transmissions across the gas path L, the gas density ρ_1 is given by

$$\rho_1 = (Z_0/4c_1) \, |B/A| \, (B/C)^{1/2} \tag{2-5}$$

where c_1 = sound speed in the gas and Z_0 = buffer rod characteristic

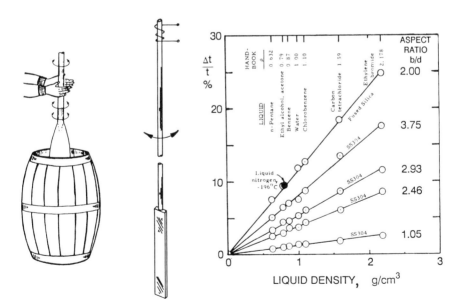

FIG. 2-39.   Torsional densitometer. After Lynnworth, 1977, 1980. Analyzed by Bau, 1986 and Kim and Bau, 1989.

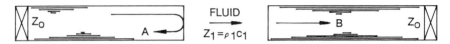

FIG. 2-40. Transmission measurement of fluid properties using buffer rods.

impedance. (Mismatch and attenuation losses may be separated alternatively by using two different paths L_1 and L_2 or by transmitting two different frequencies, since $\alpha \sim f^2$, but mismatch losses are independent of f when dispersion is absent.)

In gas volumes containing temperature gradients, there exist ρ gradients even though pressures may be substantially uniform. Measuring the ρ distribution is important, for example, in aerodynamic tests in ducts where total gas mass flow rate is to be computed as the area integral of local ρV products. One noninvasive ultrasonic approach to this problem is to measure the c distribution by a multipath transmission method. Now, if the c distribution can be translated to a T distribution (see Fig. 2-26), and if ρ is known at one location, say, at the inside wall of the duct, then in principle T and ρ can be combined to yield ρ as a function of duct coordinates. In effect, this exploits the expression for c in an ideal gas, $c = (\gamma p/\rho)^{1/2}$.

Liquid Density.   For liquids which can be interrogated as in Fig. 2-40 at a frequency where αL is very small compared to mismatch losses, the density is given by

$$\rho_1 = (Z_0/c_1) \, |B/4A|, \qquad\qquad (2\text{-}6)$$

where the subscript 1 now refers to the liquid.

On the other hand, in slurries, sludge, or similar two-phase media, αL may be very large compared to mismatch losses. In some such cases the fluid density may be empirically correlated with attenuation.

Solid Density.   The density of a nonabsorbent solid member, at least one surface of which is immersed in a liquid such as water, can be measured by reflection coefficient principles.

The density of solid sheets of known thickness l very thin compared to wavelength λ may be determined based on reflection and transmission coefficient principles. Density is computed using an approximation of the form

$$\rho = AZ_0/\pi Bfl, \qquad\qquad (2\text{-}7)$$

where A and B are amplitudes of reflected and transmitted rf (radio frequency) bursts of center frequency f, and Z_0 = characteristic impedance of the medium on both sides of the sheet.

Porosity.   Analogous to liquid densitometry based on proportionality or correlation of ρ and c, useful determinations of porosity q in some ceramics, particularly porcelain, have been obtained from c measurements. See Fig. 2-41. In relation to dielectric properties of porcelain insulators, Filipczynski et al., 1966, reported that optimum sintering occurs at the maximum wave velocity corresponding to minimum porosity (example: 1380°C firing temperature). Within limits, porosity is reduced by higher firing temperatures, the effective wave path being reduced as gas inclusions are filled in by melting. However, too high a temperature (1410°C) causes outgassing, creating new pores, and reducing c. To avoid the q ambiguity, one would follow c during the firing process, or one limits the furnace temperature.

Spriggs, 1962, proposed exponential equations of the form $E = E_o^{-b_o q_o - b_c q_c}$ (where E = modulus of porous material, E_o = modulus of pore-free matrix, b_o and b_c are empirical constants, and q_o, q_c are volume fractions of open and closed pores) to fit elastic moduli data versus porosity. This attempts to take into account open and closed pores, for low porosities, e.g., q up to ~0.16 for cold-pressed and sintered alumina (q = void volume/total specimen volume). Martin and Haynes, 1971, proposed an equation of the form $E = E_o - k_2 E_o q^{2/3}$ where k_2 = empirical constant depending on average void properties. This expression appears valid for a wide range of porous solids, for q up to a "critical porosity" where $E < 10\% E_o$. Hasselman, 1962, considering the voids as a dispersed phase within a continuous phase matrix, proposed an equation for the porous composite of the form

$$E/E_o = bq/[1 - (b + 1)q] \qquad (2\text{-}8)$$

but pointed out that in many cases, an increase in q would be accompanied

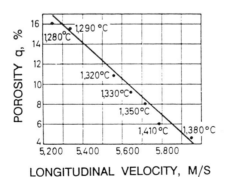

FIG. 2-41.   Variation of absolute porosity versus longitudinal wave velocity for one type of electrically insulating porcelain, after Filipczynski et al., 1966, p. 255. © 1966 Butterworths.

by a change in microstructure or type of porosity. The q-dependence of elastic properties such as E, G or σ cannot then be expressed by one single equation valid for $0 < q < 1$.

Handley et al., 1987, characterized porosity in graphite epoxy composite laminates with polar backscatter and frequency-dependent attenuation. The goal of their research was to:

> examine ultrasonic methods for the characterization of porosity in fiber-reinforced composites. Five uniaxial graphite/epoxy composites with 1% to 8% volume fraction of solid glass inclusions to model "porosity" were investigated using two complementary ultrasonic techniques. For the polar backscatter method, samples were insonified at a polar angle of 30 degrees and an azimuthal angle centered at 45 degrees with respect to the fiber orientation. For each specimen data were acquired at 121 sites by translating the sample over an 11 by 11 grid in 2 mm steps. At each site the azimuthal angle was varied in 5 degree steps from 35 to 55 degrees and the resulting spectra were averaged in order to remove background variations not attributable to porosity. The electromechanical response of the measurement system was deconvolved using the method of log spectral subtraction with a water-only path calibration trace taken at normal incidence from a stainless steel reflector. Polar integrated backscatter was obtained by averaging over the useful bandwidth. Polar integrated backscatter exhibited a good correlation with the volume fraction of "porosity," with correlation coefficient $R = 0.9$. For the frequency-dependent attenuation method, data were acquired at 441 sites on a 21 by 21 grid in 1 mm steps. Signal loss relative to a water-only path was obtained as a function of frequency using log spectral subtraction. Signal loss was expanded about the center frequency of the useful bandwidth and the excess attenuation was expressed relative to a "pore"-free region. The rate of increase with frequency of excess attenuation exhibited a good correlation with the volume fraction of "porosity," with correlation coefficient $R = 0.8$. Attenuation-based techniques are most readily applied under circumstances in which access to both sides of the specimen is feasible, whereas backscatter techniques lend themselves to measurements limited to single surface access.

Theory and experiments relating to cast aluminum containing up to 6% porosity and average pore radii from 10 to 150 μm are reported by Lewis et al., 1985.

Correlation of reflectance and porosity of sea floor sediment has been studied by several investigators since at least 1965, when Breslau measured ocean bottom returns in the Atlantic, at 12 kHz (see, for example, studies by Faas, 1969). Correlation coefficients cited in these porosity/reflectance studies range from 0.706 to 0.97.

PAR Scientific (Denmark) developed an ultrasonic system to measure the density of polyethylene sheet. As reported in *Ultrasonics* (July 1987,

p. 251), the system uses a rectangular piece of the material under test (no special surface treatment is required) and provides a digital readout of the density in a matter of seconds. The time required for the whole measuring process is less than 1 minute.

The plastic sheet is placed in a water tank between an ultrasonic transducer and a reflecting plate. A microprocessor initiates a pulse of ultrasound that travels through the plastic sheet and reflects back to the transducer from the end plate. An analysis of the various reflected pulses allows the velocity of sound in the plastic to be measured. In these materials, the velocity of sound is linearly related to the density, and the microprocessor presents the density as a digital readout. For polyethylene in the range 0.92-0.96 g/cm^3, an accuracy of ± 0.0005 g/cm^3 or better is claimed.

2.3.3   Fluid Density and Mass Flow Rate in a Pipe

For well–defined liquids and sometimes even for ill–defined gases (e.g., flare gases) the sound speed c correlates with liquid density or gas molecular weight, respectively, if temperature is known. In the liquid cases where c is interpretable as ρ, the contrapropagation method of flow velocity (V) measurement in a pipe (Section 2.1) can also yield c, ρ and then the mass flow rate M. Here, $M = K\rho VA$, where K = meter factor and A = cross sectional area of the pipe. In flare gases (Smalling et al., 1984; Chapter 4), c has been related empirically to average molecular weight. Therefore, if pressure is known, M can be calculated.

Whether ρ is measured directly by a sensor, or inferred from c or T, the same $K\rho VA$ expression for M applies. Configurations for measuring ρ and M appear in Chapter 5.

2.4   Interface Sensing

2.4.1   Object Present or Absent

Moving objects can be counted as they interrupt a beam of ultrasound. In one system, an ultrasonic beam with a frequency of approximately 40 kHz passes between transmitter and receiver. The beam is interrupted as an object comes between transmitter and receiver, producing a signal which is used to de-energize a relay in a remote power supply. The transmitter and receiver each employ piezoelectric crystals. As an ultrasonic beams hits the receiver membrane the built-in piezoelectric crystal converts the mechanical vibration into an electric oscillation. This oscillation is fed back into the circuitry. As the vibration ceases, the circuitry is detuned to de-energize the relay in the power supply. The receiver has a built-in adjustment potentiometer to adjust the system to different distances. The maximum distance between transmitter and receiver is ~3 m. The ultrasonic beam diameter can be reduced by means of cones and similar focusing devices to achieve higher sensing power to detect objects as small as 1.5 mm, up to an object the size of a railroad freight car.

Typical applications are illustrated in Fig. 2-42. Proven applications include: counting or signaling on production lines, conveyor belt monitoring, breakage warning in paper mills, and edge or width control. See also: Frederick, 1965, p. 215.

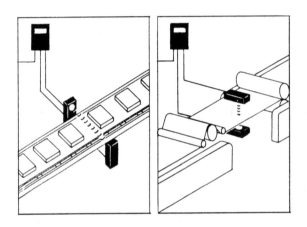

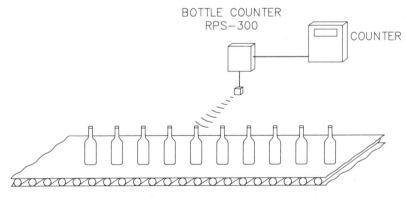

FIG. 2-42. Top: Counting and presence/absence (broken web) applications of interrupted sound beam, courtesy Endress and Hauser. Bottom: Ranging and counting with polymer electret transducer, courtesy Migatron.

2.4.2   Thickness by Reflection and Transmission Techniques

Industrial ultrasonic measurements of metal, plastic or glass thickness are commonly made by timing the round-trip interval for a longitudinal wave pulse to traverse a sheet, plate or conduit wall and converting the measured time to thickness, for materials of known c. Tens of thousands of ultrasonic thickness gages have been sold which operate on this principle. Digital or analog readouts are sometimes augmented by oscilloscope or equivalent displays, as required in some corrosion studies, gaging of inhomogeneous materials like reinforced truck tires or gaging of very thin layers in a multilayer sandwich at 50 or 100 MHz (Fig. 2-43). Rangeability is typically from ~0.02 mm to >1 m, but not necessarily in one instrument. Accuracy typically is better than 1%, and response time in portable instruments is typically ~1 second. Resolution depends in part on the material and geometry under test and the ultrasonic frequency. As a numerical example, resolution of the thickness of hollow metal turbine blades used in aircraft engines is typically about 0.15 mm when tested at 20 MHz, in blade regions where multiple backwall echoes can be detected. The thickness of some plastic coatings on glass lamps can be resolved down to 0.025 mm using a 100 MHz delay line transducer.

Thickness is sometimes measured by resonance techniques. For example, resonance instruments used to be available to measure the thickness of metal sheets down to <1 mm. The instruments operated on the principle that as frequency is swept, standing waves are set up in the

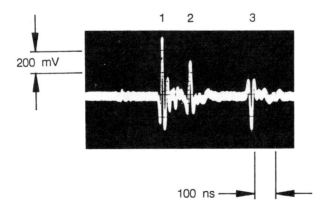

FIG. 2-43.   Thickness measurement of teflon coating, 75 μm (3 mils, approximately the thickness of a human hair) on top of glass. Equipment: Panametrics pulser/receiver model 5601 and 50-MHz transducer V215-BC.

sheet, causing discrete indications of loading on the crystal, at a sequence of harmonic frequencies corresponding to sheet resonances. Somewhat analogously, spectral analysis of a broad-band pulse has been used to measure layer thicknesses down to ~0.1 mm (Fig. 2-44). Vibro-Meter has developed an ice thickness sensor for airplanes, based on resonance principles. The detection element is a diaphragm vibrating between 20 and 50 kHz. Its resonant frequency is increased by a layer of ice but decreased by water, oil or dirt. (See Lustenberger, 1986.)

Resonance thickness gages are also in wide use to monitor the film thickness of vapor-deposited or sputtered coatings. Here, a quartz crystal's resonant frequency is monitored, as that crystal intercepts a sample of the coating. As the coating builds up, the frequency is reduced (Fig. 2-45).

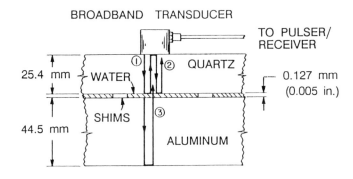

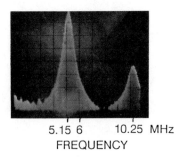

FIG. 2-44. Thickness measurement of intermediate layer of relatively low characteristic impedance. (Top) Experimental arrangement. (Bottom) Spectrum of trailing end of echo 1 from fused silica/water interface, showing impulse-induced resonance. After Papadakis and Fowler, 1971. © 1971 AIP.

FIG. 2-45.   Sloan model 2000 digital deposition thickness monitors, and sensor head containing 1.27 cm diameter, AT cut, 5-MHz quartz crystal.

In industrial applications of pulse echo thickness gages, special cases of unusual difficulty arise when one or more of the following conditions prevail: rough surface; curved surface; c unknown; T gradients present; thickness bounded by interface of low reflectivity for L or S waves; high α; dispersive material; material in motion at high speed; high background noise level; material surface at high temperature; thickness inaccessible (e.g., non-contactable), irregular, poorly defined, or outside the range of standard transducers or electronics.

For such special problems, less-routine supplemental approaches are sometimes considered. These include use of: waves other than longitudinal, oblique incidence, focused waves, transmission and reflection coefficients, multiple transducers, special couplants, dry coupling at high pressure, buffer rods or bubblers, frequencies outside the 2- to 10-MHz range normally employed.

The location of interfaces is obviously related to other categories in Table 2-1 such as level, position, and thickness, in that a distance is to be determined. In the present section, we consider interfaces between two media whose acoustic impedances are sufficiently close so that the reflection coefficient $R \ll 1$, or whose interface is so rough that reflections are too weak to be detected reliably. Some petroleum products flowing in pipelines illustrate the former case, while a solidifying steel ingot illustrates the latter case. By stating the problem in terms of low reflectivity, we intend to rule out the approach of timing the reflection from the interface. The most obvious remaining approach is to use transmission, to measure c, α, or terms proportional to their reciprocals.

Transmission approaches may be nonintrusive or intrusive. Nonintrusive transducers may be mounted outside a pipe through which liquid is flowing, or they may momentarily contact the hot exterior of a solidifying ingot. Intrusive probes such as the c-probe of Fig. 2-38 are typically mounted in a flange, penetrating the pipeline. Most commonly, the interface between two similar liquids flowing in a pipe is detected by an intrusive c-probe, as the interface passes the probe. For example, Zacharias, 1972, reported that in tests on 22 gasolines, there was enough difference in c so that interfaces between any two of them could be detected using a probe such as that in Fig. 2-38.

A different interface location problem prevails in Fig. 2-46, which is not amenable to an intrusive approach. Here we consider a layered two-phase medium. For the symmetrical sandwich or layered model depicted, suppose that sound speeds c_x and c_y are known, as is the total distance z between transducers: $z = x + y$. From a measurement of the transit time t from one transducer to another, the interface location can be determined as $x/2 = c_x(c_y t - z)/2(c_y - c_x)$. This transmission method has been used by Jeskey et al., 1977 in experimental studies of steel ingot solidification, for ingots where $z \approx 20$ cm. Surprisingly, despite coarse approximations such as taking c_y as a constant for molten steel, and c_z corresponding to an average of the outside wall temperature and the solidus temperature, computations of the solid skin thickness $x/2$ based on ultrasonic data were reasonably close to values determined independently by calipering the dumped ingot wall during solidification, and determined by computations using solidification rate theory.

The scatter observed in the transit time data corresponds to a thickness uncertainty of several mm. This uncertainty is comparable to the calculated distance between liquidus and solidus isotherms in the solidifying ingot. See Sections 7.3, 7.4.

2.4.3 *Level of Liquids and Solids*

Level measurement has become one of the most commonly-encountered application areas for ultrasonics. As in other areas, success may be anticipated provided sufficient care is executed in screening each application. A survey article in *Instrumentation Technology* (September 1974) summarized the principal ultrasonic level detectors for liquid, solid, and multiphase materials (foam, material A on B, etc.) available at that time. Both invasive and noninvasive devices were already widely used. Outputs were responsive to specific levels or could indicate levels continuously. Improvements that occurred after that survey include: higher-temperature transducers; buffer rod gap-type probes; sensors using previously-neglected modes such as torsion, Lamb or shear; microprocessor-based

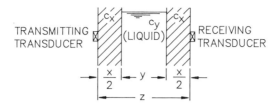

FIG. 2-46. Model of two-phase layered medium through which transit time is measured, after Jeskey et al., 1977.

electronics with improved programmability for new applications and self-test features; and low-cost equipment, due in some cases to mass produced polymer transducers (Section 9.7).

Most ultrasonic liquid level sensing methods may be categorized according to the breakdown of Table 2-4. When trying to understand the differences between one type versus another, parameters such as access route, number of transducers and wave types such as those included in Table 2-5 may be helpful. Probably the most common ultrasonic level devices function much like relays or switches which are normally open or normally closed. (NO: acoustic waves not conducted across the path; NC: acoustic waves readily transmitted across the path.)

Many clamp-on liquid level gages respond to presence or absence of echoes from the opposite wall, indicating the presence or absence of liquid at the transducer height. Clamp-on bottom-mounted designs are also available. Sometimes the dampening effect of a liquid in contact with the wall is exploited. This is most sensitive for plastic-walled vessels, whose impedance is nearly matched by the liquid inside. For continuous level gages in which travel time in a fluid is interpreted as distance, compensation for c changes or gradients can be achieved by a series of equidistant reflectors each positioned to reflect a small part of the incident beam (Frederick, 1965, p. 213); by measurements across a known path such as the diameter; or for known liquids, from their temperature.

Most liquid level gages are probe types. Resonant types are dampened when contacted by the product. Pulse types can be arranged to transmit across a gap when immersed in a liquid. Alternatively, transmission across an air gap can be interrupted by attenuating media such as granular solids. Pulse travel time is sometimes used to measure the distance to a float adjacent to or connected to a magnetostrictive waveguide, as illustrated in Section 7.9.3.

Noncontact types usually measure transit times from above the level in question. These are usually T-compensated [see Eq. (2-1)] and operate in silos, over conveyors, etc.

Table 2-4. Ultrasonic Methods Used to Measure Liquid Level Categorized According to Clamp-on or Non-clamp-on Approaches

	Clamp-on Transducer(s)			Wetted Transducer(s) or Foreign Sensor(s)			Waveguide(s) or Buffer(s)		
	Reflections Within Wall (Ringdown) Dampened Locally by Liquid Load	Transmission Along Wall of Container Senses Coupling into Liquid Inside Container	Transmission in Fluid: NO/NC Acoustic Switch or Time of Flight	Loading or Dampening	Short Fixed Gap = NO/NC Acoustic Switch	Fixed or Variable Path in Fluid; NO/NC Switch or Time of Flight	Internal Propagation Senses Level	External Propagation Senses Level	Passive Noise Receiver Senses Level
Point or Multipoint	X		X	X	X	X	X	X	X
Continuous		X	X			X	X	X	

60

Table 2-5. Examples of Some Design Considerations that May Influence the Selection of One Ultrasonic Method Over Another

Wetted Materials	Accuracy	Sound Speed Compensation	Wave Type	Number of Transducers	Required Maintenance	Calibration	Interfering Variables
Container itself	1% of reading	Not necessary	Long'l.	1	Daily	Req'd. at factory	Dust
Stainless steel	1% of range	Based on temperature	Shear	2	Weekly	Req'd. in field	Foam
Teflon	± 1 mm	Based on echo(es)	Lamb	>2	Monthly	Not required	Bubbles
Other	± sensor radius	from reflection(s) at	Rayleigh		Annually		Scale
		known distance(s)	Ext'l.		Never		Residue
			Tors'l.				Change in temperature or composition
			Flex'l.				Bio build-up
							Other sources of attenuation
							Mixers or other mechanical structures in path
							Target surface unflat, tilted, rough, unsettled
							Acoustic short circuit
							Wall echoes
							Ambient noise (fluid-borne) acoustical, structure-borne, vibration, electrical)
							Little difference between acoustical properties in liquid and vapor

Feedthrough or Seal	Response Time	Frequency, MHz	Hazardous Zones and Safety/Code Approvals	Consequences of Erroneous Measurement or Faulty Alarm	Other Factors	Cost per Point	Degree of Invasiveness	Active/Passive	Access
Not applicable	Milliseconds	0.02	Zone 0, 1 or 2	Nuisance, no hazard	Availability of spares	<$10.	Noncontact	Active	Top
All metal	0.1–1 s	.04	UL, FM, CSA, PTB, Cenelec. . . .	Potentially polluting or hazardous to health	History of duplicate or similar equipment in duplicate or similar applications	<$100.	Momentary contact	Passive	Side
Gasket	1–10 s	.1	Intrinsically safe	Loss of life likely	Training required to operate equipment properly	<$1000.	Permanent clamp-on	Both	Bottom
O-ring	Other	.2	Earthquake proof		Retrofit or designed into container	>$1000.	Penetration		Other
Pipe thread compound		.5	Explosion proof						
Glass-to-metal		1.							
		2.							
		5.							

61

In torsional waveguides of noncircular cross section, the transit time increases in proportion to the depth of immersion in the liquid (Lynnworth, 1977, 1980; Bau, 1986; Chapter 7). Interpretation requires knowledge of the density. If ρ is unknown or variable, an alternative possibility is to measure transit time in the unwetted portion of the waveguide, down to the small echo generated at the vapor/liquid interface.

In this section it may be appropriate to mention depth sounders or ocean-bottom profilers, since these operate on pulse-echo principles similar to the noncontact types used in silos.

Figures 2-47 through 2-49 illustrate some of the variety of approaches, modes and probes for level measurement. Despite the widespread use of ultrasonic liquid level sensors, and their apparent simplicity, readers are nevertheless cautioned that application problems such as buildup, adhesion, condensation, bubbles or other factors can frustrate an otherwise straightforward installation. Measurements in highly viscous media, or of stirred material where the top surface is sloping or curved, and measurements in the presence of moving machinery such as agitators, represent difficult or impractical situations. Continuous measurements in cryogenic liquids, and in high temperature liquids such as molten sodium, have been difficult. Cryogenic level has been measured using thermal oscillations, i.e., the sounds of boiling when a warm solid contacts liquids

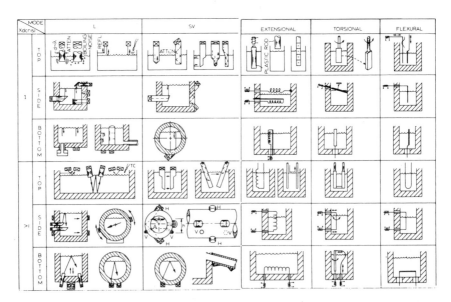

FIG. 2-47. Liquid level ultrasonic approaches categorized by wave type, number of transducers and location of transducers. After Lynnworth, 1979. © 1979 Academic Press.

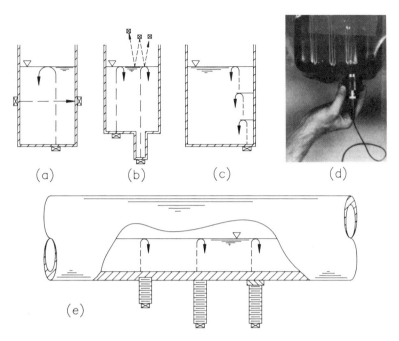

FIG. 2-48.   Liquid level measurements using transducers not in contact with the liquid.
(a)–(c) Sound speed compensation by transmission and reflection methods. (d) Hand-held
transducer senses level of automobile transmission fluid. (e) Ringdown in thick wall creates
deadbands, making it difficult to measure liquid level continuously, especially if c varies in
liquid because of a wide temperature range. Remedy proposed here is the use of two or
more additions to the pipe to create unequal time delays, thereby staggering the deadbands
of different transducers.

such as helium or nitrogen (Laplant and Flood, 1972). Day and Smith,
1973, developed transducers that wet rapidly in liquid sodium in the 150
to 260°C range, as part of their work on under-sodium viewing and ranging.
One application has been the measurement of sodium level using an immersed transducer and a 19-mm ID tube (Smith et al., 1974). External
bottom-mounted and side-mounted longitudinal mode sensors have been
used by a number of investigators (e.g., designs of Fowler and Elfbaum
reported in Lynnworth, 1986; Jacobson et al., 1987a; Dam, 1987; Gates,
1987).

2.4.4   Ranging, Position

Ranging or position sensing usually consists of pulse-echo time measurements in a medium of known or compensated sound speed, between two
or more points. The method has been used for some time in the animal

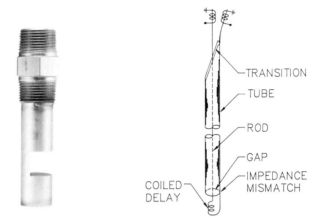

FIG. 2-49. Left: Example of a longitudinal wave gap-type liquid level probe, courtesy Introtek. Right: Experimental extensional wave remote gap probe with concentric rod and tube.

kingdom (Fig. 1-1) and in recent decades in military, industrial and bioengineering (aids to the blind) applications. The medium is usually air or water. Figures 2-50 through 2-55 illustrate several ranging and position applications.

Figure 2-50 illustrates the principle of echo location of a solid object such as a stainless steel member in a nuclear reactor tank filled with liquid sodium, or a stirrer or agitator in a chemical reactor vessel. See also, Section 2.6.6, and Day and Smith, 1973.

Figure 2-51 shows oscillograms corresponding to the position of a movable magnetostrictive wire through a coil bobbin. If the wire is isopaustic, e.g., the type used in T-insensitive delay lines, then the time interval between selected echoes is directly proportional to position.

Figure 2-52 illustrates a linear displacement device introduced by Tempo Instrument in the early 1970s. The device used a nickel-cadmium wire enclosed in a measuring spar. A permanent magnet, placed on the bar at the location to be measured, magnetostrictively launched an ultrasonic pulse at the location of the magnet. The time required for the ultrasonic pulse to travel back down the wire was measured by a quartz-crystal-controlled clock and transformed into a dc output ranging from 0 to 10 volts. Ranges were available to 1.5 m, and accuracy was reported as ±0.15%. The temperature coefficient of scale factor was less than 0.02% per °C. Improved versions of this type of equipment, available from MTS Systems, are shown in Section 7.9.2.

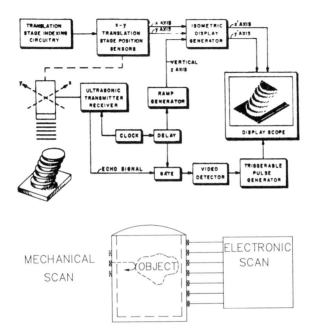

FIG. 2-50.   (Top) Under-sodium viewing system block diagram, after Day and Smith, 1973. © 1973 IEEE. (Bottom) Noninvasive determination of position of submerged object. Box labeled "electronic scan" represents a steerable array, a synthetic aperture focusing technique (SAFT), a pixel-driven search method, etc. The choice depends in part on whether or not the shape of the object is known, and if known, whether the shape is spherical or more complicated than spherical.

Figure 2-53 illustrates the SAC graf/pen®. This device included a data tablet, a stylus or cursor, and a control unit. Strip magnetostrictive sensors of precisely selected sound speed, installed on the x and y sides of the tablet, received signals from the stylus or cursor. The stylus combined a ball-point pen with a tiny spark gap which generated an ultrasonic pulse. A later model with cursor used a piezoelectric crystal to generate ~ 100 kHz pulses. The x and y travel times were digitized at rates up to 200 coordinate pairs per second and a resolution of 2000×2000 line pairs. Introduced in 1969, over 1000 graf/pens were sold by the end of 1974.

More recent (post-1985) approaches to xy position sensing using airborne sound or Rayleigh waves on the curved glass surface of a terminal are illustrated in Figs. 2-54 and 2-55. A coordinate resolver developed in France for maps employs Lamb waves, as described in Chapter 7, Section 7.9.5.

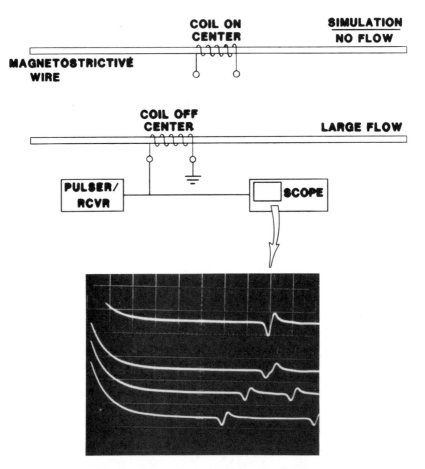

FIG. 2-51.   Time interval between separated pulse pair is proportional to distance that coil is off-center. Sweep approximately 10 μs/div. After Lynnworth, 1987.

2.4.5   Bubbles and Particulates

Detection and measurement of bubbles and particles is important in industrial application areas such as heat transfer, fuel lines, liquid transfer, instrumentation lines, etc., and in biomedical applications (Dam, 1986). Discontinuities generally influence c, α, and the phase of the reflected wave. Discontinuities having an acoustic impedance *lower* than that of the liquid matrix cause the phase of the reflected pressure wave to *reverse*; high-impedance discontinuities do *not* cause phase reversals. Lossy media (complex impedance) lead to phase shifts other than 0 or 180°. In principle, discontinuities can be detected by transmission or reflection methods, and sometimes by passive methods. (Compare with the analogous NDT problem of detecting a large number of small inclusions or defects in a solid, e.g., porosity or graphite flakes in castings, or disbonds in a laminate.)

Detection of water boiling is readily demonstrated with simple pulse-

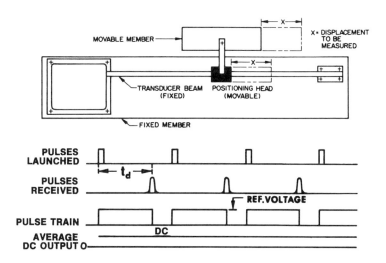

FIG. 2-52. Temposonics™ linear displacement transducer is based on measuring time interval for an ultrasonic pulse to travel between two points on a nickel-cadmium magnetostrictive wire waveguide. The wire is enclosed in a nonmagnetic tube (transducer beam). Current pulses are applied to the wire at a quartz crystal-controlled repetition rate. In this version of the Weidemann effect, ultrasonic pulses are generated at the position where a magnetic field intercepts the wire. These pulses arrive at a fixed reference a time interval later. (See pulse diagram.) The magnetic field is generated by a movable permanent magnet positioning head. A pulsewidth-modulated pulse train is obtained, with the pulsewidth modulated by the launching pulses and the received pulses, respectively. The pulse height is controlled by reference zeners and the pulse train is filtered to provide a DC output signal. (See also, Chapter 7, Section 7.9.2.) Illustration courtesy MTS Sensor Division.

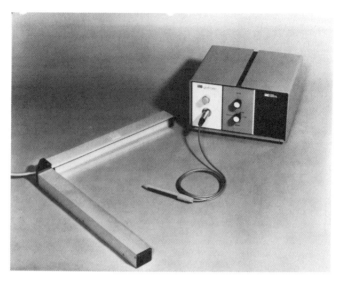

FIG. 2-53. SAC "graf-pen" GP-3 digitizer. Photo courtesy Science Accessories Corp.

FIG. 2-54. Ultrasonic digitizer measures airborne times of flight of 70-kHz pulses emitted from the tip of the stylus to microphones in the Grafbar™. Unit shown has 0.1-mm resolution, and 100 points per second slant digitizing rate. Illustration courtesy Science Accessories Corp.

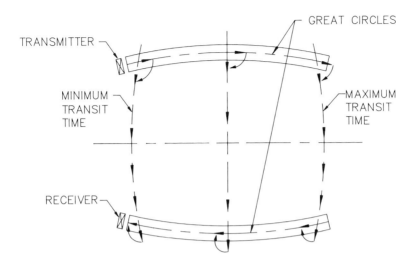

FIG. 2-55. Finger location on a curved glass screen is located by the dip in the Rayleigh (surface) wave signal transmitted between reflective arrays. The time of the dip indicates position. After Adler and Desmares, 1986. © 1986 IEEE.

echo equipment (Lynnworth, 1971). Detection of steam bubbles in water under high pressure is reported by Hulshof and Schurink, 1985. Where *incipient* boiling is to be detected, the problem is much more difficult. Acoustic detection of boiling in liquid sodium at temperatures up to 650°C has been approached by using lithium niobate transducers operated as acoustic emission (AE) receivers (see also Section 2.6.5), as reported by Anderson et al., 1972.

Response to *particles in motion* underlies most Doppler blood flow-meters, the particles being the red corpuscles. (Satomura, 1957; Azimi and Kak, 1985.) An industrial application of particle monitoring is reported by Whitesel et al., 1986. Details on these applications appear in Section 7.8. Industrial flowmeters now use both Doppler and non-Doppler methods to measure the flow of particulates, or the flow of fluids containing bubbles and/or particulates.

2.5 Elastic Properties

Equations relating sound speeds for appropriate modes and polarizations to corresponding elastic constants, elastic moduli, and Poisson's ratio are available in the literature. For isotropic bulk or slender solids the equations take on the simplified forms given in Section 3.7. For industrial applications, ultrasonics provides a well-established way of obtaining the required data, for temperatures from cryogenic levels up to the melting point or destruction temperature of the specimen. For isotropic *slender* specimens, the nomogram in Section 3.7 may be used. Measurements of c in *loaded* specimens provide data on higher order elastic constants (Krause et al., 1975–1985; see also Section 2.7.3).

Measurement techniques include resonance and pulse. Magnaflux marketed the "Elastomat" resonance equipment for many years. Cylindrical test specimens, typically ≲1 cm diameter × ~10 cm long are supported at nodal positions. The resonant frequency typically is measured as a function of T, up to ~1000°C. With appropriate furnace, coupling members, and refractory specimen, resonance techniques have been used to over 2500°C. Resonant frequency of carbon sheet steels is indicated by Control Products Company's Modul-Γ drawability tester.

A variety of pulser/receiver combinations have been devised for elastic property studies, and also a variety of specimen configurations, especially for high temperature studies. The principal configurations or experimental approaches for high temperature moduli studies in solids using pulse techniques are the notched bar (Frederick, 1947); the notched wire (Bell, 1957); use of mode conversion to obtain shear wave data (Reynolds, 1953); and momentary contact (Carnevale et al., 1964). Frederick's method has been used at Sandia in studies of graphite to nearly 3400°C (Gieske, 1975). The author in 1968 combined Bell's method with electrical self-heating to obtain moduli data in Mo, Re, and W at the melting points of these metals, 2610, 3180, and 3410°C, respectively. These experiments used the first Panatherm®. This was a pulser/receiver/time intervalometer instrument developed by B.M. Gordon and colleagues at the Gordon Engineering Company to measure automatically the time interval between a pair of selected echoes to ±100 ns. (See Section 3.5.3.2.) In the 1970s a later version of the Panatherm was used in a number of laboratories for moduli determinations (Lynnworth, 1973). Echoes from which Young's and shear moduli can be obtained simultaneously are illustrated in Fig. 2-56.

In general, routine ultrasonic methods are applicable to measuring moduli in a large number of materials and geometries. However, specimens which are dispersive by virtue of their inhomogeneous construction or bounded geometry are not easily evaluated by routine procedures. To measure dispersion, i.e., variation in phase velocity versus frequency, variable path and/or variable frequency techniques may be required. Illustrations of phase and group velocity measurements, and of the potential interpretive errors attending the uncompensated use of pulse techniques in dispersive specimens, are contained in Thompson, 1973, and in Lynn-

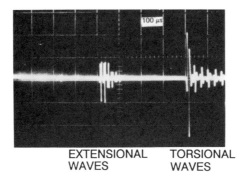

EXTENSIONAL   TORSIONAL
WAVES   WAVES

FIG. 2-56. Simultaneous magnetostrictive generation of extensional and torsional waves is used to measure elastic moduli and Poisson's ratio in slender rod or wire specimens. After Lynnworth, 1972.

worth et al., 1973b. The former work relates to Lamb waves in steel plate; the latter to L and S waves in woven reinforced composites.

For laboratory and also for on-line measurements of elastic modulus in fibers, filaments, films and papers, H.M. Morgan Company's dynamic modulus tester (Fig. 2-57) has been widely used for over twenty years. A bibliography relating to this tester, containing over twenty references, is available from that manufacturer. About 300 of these instruments were estimated to be in use in 1975. For lossy materials a Viscoelasticity Rheometer made by Nametre measures complex moduli. (See Chapter 8.)

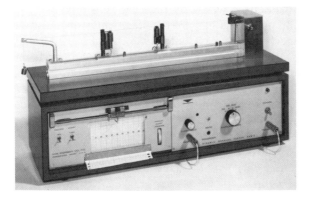

FIG. 2-57. Dynamic Modulus Tester PPM-5R shown with Fiber Scanner Option. The Fiber Scanner Option is used for testing natural and synthetic fibers such as polyester, nylon, cotton, spandex, hair, etc. Photo courtesy H.M. Morgan Co.

Transit time or sound speed measurements can monitor the setting of concrete (Carleton, 1986). One can also use data on c to match graphite electrode sets to control their burning rate. Useful frequencies may be as low as ~50 to 150 kHz for small specimens in the lab, and ~25 to 50 kHz for large structures encountered in the field.

2.6   Selected Nondestructive Testing Topics

Nondestructive testing (NDT) includes a broad range of activities such as inspection, materials evaluation, quality control and process control, both open loop and closed loop. See, for example, McMaster, 1959, Krautkrämer and Krautkrämer, 1983 or Cielo, 1988. Topics included in ultrasonic NDT represent the activity of thousands of workers and a period of time of some fifty years. A number of books, journals, and professional societies are devoted exclusively to these topics. Therefore, it is clear that any attempt to summarize briefly the scope of NDT technology and equipment would be inadequate.

For those readers unfamiliar with NDT practice, it may be of interest to note that ultrasonic tests are most often confined to the 1 to 10 MHz decade. Longitudinal wave pulses are principally used. The transducer is usually housed in a case, such that pulses can be coupled manually either through a thin film of liquid (contact test) or through a substantial distance of water (immersion test) such that the test piece is in the far field. (Laser and electromagnetic transducers are examples of ongoing approaches to avoid the need for physical contact to the object or medium under test.) The same transducer serves as transmitter and receiver in most cases. Focused waves at frequencies as high as 8 GHz were in use by 1985, and prospects had already been envisioned for increasing to 100 GHz in superfluid helium (Quate, 1985). Angle beam and buffer rod techniques are also used. Electronic equipment usually performs the functions of a pulser/receiver and oscilloscope, plus various additional or optional functions such as gates, alarms, analog or digital displays, recorders for scans, etc. An example of a digital flaw detector introduced around 1985 is shown in Fig. 2-58.

One aspect of open loop process control to which NDT contributes is Statistical Process Control (Cielo, 1988). SPC looks at the output periodically after the output is finalized to ascertain *whether the process had been under control.* If the answer is "yes," then the process is allowed to go on. If the answer is "no," then the process is stopped and repaired. SPC can utilize many ultrasonic methods which are normally thought of as being in the realm of NDT for the measurement of the properties of the output on a periodic basis. For instance, Fig. 2-59 portrays an elastic

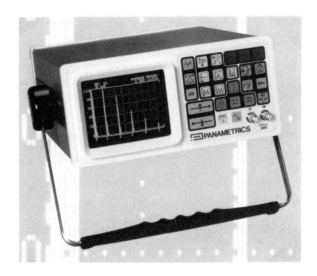

FIG. 2-58.   Digital flaw detector introduced by Panametrics about 1985 offered a number of features new at that time for that type product: completely digital, μP-controlled; portable (5.7 kg/7.7 kg with battery); large bright screen; sealed touch panel, color-coded and functionally labeled, to control all instrument functions and settings; status display shows the value of all test parameters so that an operator can subsequently reproduce any test setup (even that of another operator), without test blocks. (Fowler and colleagues, 1987, priv. comm.)

modulus measurement combined with a flaw detection measurement. The elastic modulus correlates with strength, and both properties arise from a casting process which may go out of control. SPC using the ultrasonic velocity could be used to check on the casting process while 100% NDT would still be used on the particular part shown in Fig. 2-57 because of safety requirements (Papadakis, 1976).

The following subsections indicate some of the ways that active and passive ultrasonic NDT methods have been applied in the laboratory and in industry. The topics are selected to show the use of c and α measurements, and their interpretation relative to materials properties, characteristics and departures from specifications.

2.6.1   *Cracks, Flaws, Inclusions, Inhomogeneities, Disbonds*

The defects represented by the above labels are usually detected by pulse-echo single transducer techniques, but pitch-catch and through-transmission two transducer techniques are also used. Surface cracks, for ex-

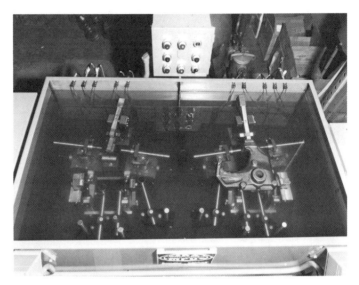

FIG. 2-59.   Tank, transducers, and jigging for ultrasonic immersion tests of cast iron spindle supports for the front suspension of trucks. A right-hand piece is shown, while the left-hand piece is removed to display the three transducers aimed up at critical areas to look for porosity by pulse-echo, and the two other transducers mounted horizontally for velocity measurements by through-transmission. These two transducers are aimed at a test cube for calibration. The cube pivots out of the way when the part to be tested is inserted. Ultrasonic velocity is correlated with yield strength in nodular iron so cast iron parts can be sorted for strength. (Illustration courtesy Ford Motor Co.) After Papadakis, 1976. © 1976 Academic Press.

ample, reflect surface waves. Internal cracks and inclusions reflect longitudinal or shear waves. Disbonds, historically one of the more difficult conditions to quantify, are often tested by reflection and/or transmission techniques with the record displayed as a C-scan. This is a form of plan view of the inside of the part. In some circumstances resonance tests may indicate bond quality.

2.6.2   Anisotropy and Texture

Anisotropy influences c and α. For example, the elastic constants along different axes of single crystals have been determined from c measurements (McSkimin, 1953). Attenuation in polycrystalline solids is partly due to discontinuities in acoustic impedance at each grain boundary.

Determination of texture, particularly in rolled materials, has been demonstrated in metals and nonmetals. Various wave types and polarizations have been used to measure c versus direction in the specimen. Fowler,

1969, showed that extensional waves could be used to determine the variation in modulus in rolled Ti sheet, in specimen strips cut with axes at 0°, 45°, and 90° to the rolling direction. Martin, 1968, suggested that a quartz shear wave crystal and a Q-meter could be used to show that, as the crystal was rotated, the resonant frequency extrema for rolled sheet metal would correspond to polarizations parallel and perpendicular to the rolling direction. Papadakis, 1973, demonstrated line-contact coupling for determining texture in paper, and later (1974) demonstrated ultrasonic spectroscopy as applied to the calculation of the texture parameter $\Delta c/c$ for rolled aluminum. (See also, MacDonald, 1981.)

The reflectivity of roughened surfaces was investigated by Blessing et al., 1984. They found, using 10 MHz tone bursts in water, and 2.3 MHz tone bursts in air (wavelengths comparable), that the scattering increased with surface roughness, but not exactly as predicted by the theory of Haines and Langston, 1980.

2.6.3   Grain Size in Metals

The dependence of the attenuation and scattering of ultrasound on the grain size of metals and alloys has been studied over a wide range of frequencies. The different structures and correspondingly different grains in test samples are often obtained by using different heat treatments. The experimental data confirm the theoretical conclusions relative to the increase in attenuation with increasing frequency and grain size.

As Babikov, 1960, noted, in polycrystalline bodies the sound-scattering mechanism at relatively low ultrasonic frequencies (of the order of a few MHz) is similar to the scattering of sound by fine-grained particles, and at high frequencies it becomes similar to the process of diffusion. In the determination of grain size in metals by ultrasound, one should utilize those frequencies at which the scattering or absorption of ultrasound is strongly dependent on the grain size.

For the practical application of the ultrasonic pulse method of measuring the grain size in metals and alloys, Babikov recommended the following frequency range: the upper limit of the applied frequencies is suitably bounded by a wavelength equal to or somewhat greater (by a factor of two or three) than the grain diameter, while at the lower limit, the wavelength should exceed the grain diameter by a factor of 15-20.

The study of ultrasonic absorption and scattering in steels is of considerable practical interest. For construction and instrument steels, heat treatment has the greatest influence on the ultrasonic attenuation. It follows from Fig. 2-60 that quenched steel has the lowest absorption. On the other hand, for an overheated structure, which has the distinguishing feature of a large grain size, the absorption of sound is much greater.

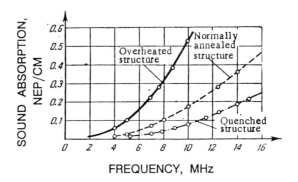

FIG. 2-60. Dependence of the absorption of ultrasound on frequency for steel subjected to various heat-treatments, after Babikov, 1960. © 1960 Plenum Press.

Therefore, ultrasonic methods could be utilized in quality control of the heat treating of a metal. A literature survey according to Saniie and Bilgutay, 1986, indicated that ultrasonic evaluation of grain size is a practical and widely used NDT technique in current manufacturing applications. Note that in Babikov's 1960 book attenuation was emphasized, whereas by 1986, as Saniie and Bilgutay point out, scattering techniques, despite their need for considerably more signal processing, "are fast becoming the trend in ultrasonic grain size estimation." Saniie and Bilgutay, 1986, used homomorphic signal processing (homomorphic deconvolution or cepstrum method) to interpret backscattered echoes in terms of grain size. This nonlinear processing utilizes Fourier transform and logarithmic operations to convert convolutional signals to additive form, which are then more easily decomposed. Their use of a homomorphic wavelet recovery system to transform grain signals in the time domain to cepstrally smoothed functions in the frequency domain is illustrated in Fig. 2-61. This may be viewed as one of a number of techniques aimed at removing randomness in the backscattered signals and extracting parameters related to the frequency-dependent attenuation coefficient and hence to the grain size.

In subsequent work, Saniie et al., 1988, determined the variation of attenuation with depth in the specimen, i.e., the attenuation coefficient profile.

2.6.4 Phase, Microstructure and Nodularity

Ultrasonic measurements of c and α during *phase changes* in metals have been reported by Bell, 1963 and his co-workers Hub, 1963 and Thorne, 1963. Figure 2-62 is an example of Hub's 1963 work on iron,

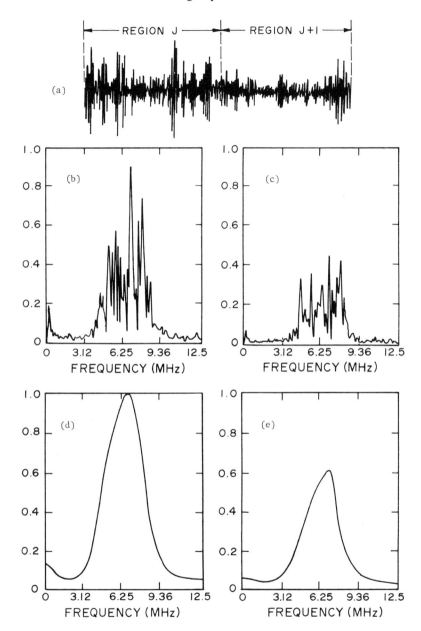

FIG. 2-61. (a) A typical grain signal. (b, c) Magnitude spectra of regions J and J + 1, respectively. (d, e) Cepstral smoothing of magnitude spectra of regions J and J + 1, respectively. (f) Dependence of the attenuation coefficient on frequency in stainless steel (SS) having average grain diameters of 50, 172 and 320 μm. After Saniie and Bilgutay, 1986. © 1986 AIP.

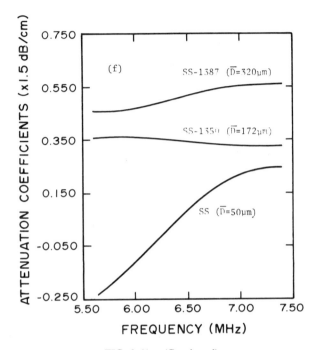

FIG. 2-61.   (Continued)

showing the influence of alpha, gamma, and delta regions on sound propagation as temperature was increased.

A decade later ultrasonic studies of *microstructure* in steel, particularly
type SAE 52100 yielded the data in Fig. 2-63, where c and α are plotted
versus temperature T for shear waves (Papadakis et al., 1972). It was
found that c for both longitudinal (L) and shear (S) waves dropped steeply
on heating, due to the solution of carbides between the A_1 and A_{CM} temperatures. On cooling, c hysteresis is observed. This appears related to
the solution and precipitation of carbides, changing the carbon concentration in the iron lattice. Below 900°C, the behavior of α_S and α_L differed,
indicating different responses to possible absorption mechanisms. Above
900°C, α_S and α_L were both attributable mainly to grain scattering. This
conclusion was supported by their ratio $\alpha_S/\alpha_L = 6.7$ at 900°C, in good
agreement with the theoretical ratio $\alpha_S/\alpha_L = (\frac{3}{4})(c_L/c_S)^3$ for Rayleigh scattering by crystallites of cubic symmetry (see also, Section 2.6.3).

In the SAE 52100 alloy work just cited, α_S remained high on cooling,
indicating irreversible grain growth. But in many pure polycrystalline metals, it is found that while α increases rapidly on heating near $T_M/2$, where
T_M = absolute melting point (K), α reversibly reduces to its prior value

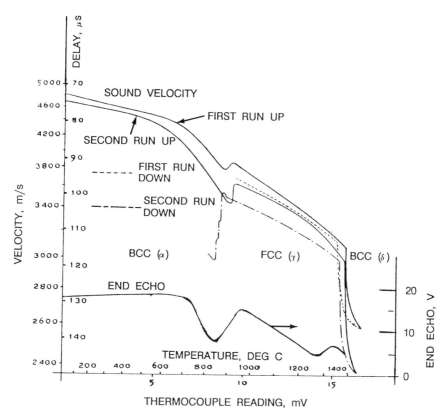

FIG. 2-62.   Variation of sound velocity and end echo amplitude versus temperature measured using pulse technique up to 1450°C by Hub, 1963, in iron specimen 2-mm diameter × 17-cm long.

on cooling. The observed absorption apparently correlates with a *recrystallization* mechanism, the most likely mechanism being absorption in the grain boundaries ($f < 0.2$ MHz). Further, the temperature at which the modulus versus T curve shows a break in slope correlates with the temperature at which α versus T shows a break, Fig. 2-64 (Papadakis et al., 1974). A potential application of such observations would be to measure c, α, or both, to indicate when a particular polycrystalline metal was above its recrystallization temperature. If one controlled T based on such a measurement, the energy necessary to hot-work a metal could be reduced in some processes. It may be of interest to note that in the laboratory, if one electrically self-heats wire specimens, the "recrystallization break points" may be identified in terms of simultaneous c and α measurements, even if T is not measured *per se*. That is to say, by increasing

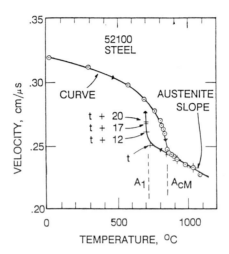

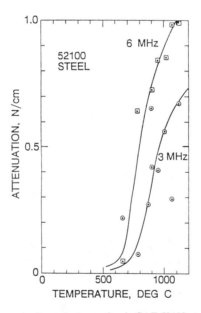

FIG. 2-63.   Shear wave velocity and attenuation in SAE 52100 steel versus temperature on heating and cooling. After Papadakis et al., 1972. © 1972 AIP.

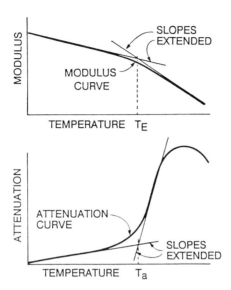

FIG. 2-64.   Method of defining the break points T_E and T_α in the modulus and attenuation curves. After Papadakis et al., 1974. © 1974 AIP.

the heating current gradually, one can readily measure c at some (unknown or known) T near $T_M/2$, where α has suddenly increased by about an order of magnitude over its value at room temperature.

Carbon in the form of graphite is often used as an additive in the production of cast iron, amounting to 2 to 4 percent by weight in typical castings. The microstructure of the graphite within the cast iron has major effects on the casting's mechanical properties. When the graphite arranges itself as thin flakes the result is gray iron, which is hard and brittle. When the graphite takes the form of spherical nodules the result is *nodular iron*, which is soft and malleable.

Foundries need to check nodular iron for uniformity. It is important both that the distribution of graphite in the casting be uniform, and that the graphite inclusions be of the right form (nodules rather than flakes).

There is a consistent difference in sound velocity between pure iron, nodular cast iron, and gray cast iron. Typically, pure elemental iron will have a velocity of approximately 5900 m/s, nodular iron will have a velocity of approximately 5600 m/s, and gray iron will have a velocity of approximately 4800 m/s. Exact velocities for a given application will vary depending on alloy composition, grain structure, and other process variables, and should be verified on calibration standards made from the material to be tested. (Diamond and Lutch, 1971.)

A pre-1975 example of one of several dozen *nodularity*-testing installations in automobile engine manufacturing plants is indicated in Fig. 2-
59. This illustrates the complicated shapes in which high precision, high
accuracy sound speed measurements are sometimes required. A *c*-based
iron nodularity tester introduced in 1987 is shown in Fig. 2-65. It combines
a mechanically-calipered thickness with transit time. The transducers are
fitted into the modified jaws of the calipers.

The field of composites offers many challenges and opportunities for
process monitoring and control by ultrasonics. *Microstructural* aspects
may be taken to include periodicity of reinforcements, fiber orientation,
bonds between components, resin/reinforcement ratio, porosity, degree
of cure, departures from homogeneity, especially out-of-spec departures,
etc. The application of ultrasound for evaluating these and other microstructural parameters may be found in the NDT and composites literature.

2.6.5 Acoustic Emission

About twenty years ago, in the mid-1960s, it was predicted that acoustic
emission (AE) would be the fastest growing field in acoustics during the
seventies. Whether AE or SAW (surface acoustic wave) disciplines turned
out to "grow more" is not quickly ascertained. AE symposia, books,

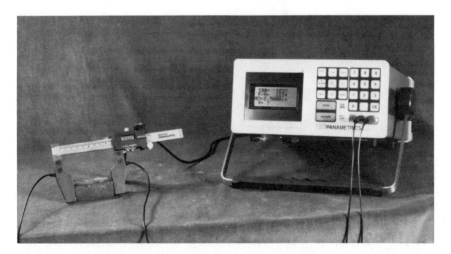

FIG. 2-65.   Ultrasonic velocimeter designed for cast iron nodularity testing and other applications combines a distance measurement by conventional calipers having a digital output,
but modified by installation of small ultrasonic transducers in the jaws. In this way transit
time is measured over the path between the jaws simultaneously with the mechanical distance
measurement (Elfbaum, 1987, priv. comm.).

over 300 journal articles by 1975, special issues, newsletters, an acoustic emission working group, an ASTM subcommittee E07.04, several organizations entering the field, existing companies enlarging their facilities, introduction of equipment of increasing sophistication, calibration standards—all these indicate increasing interest and industrial acceptance of acoustic emission as a means of monitoring potential failure mechanisms in critical parts or structures.

Acoustic emission (AE) systems "listen" to leaks and to material and structural phenomena by amplifying, filtering, recording, displaying and analyzing signals emitted by the phenomena. Welds have received considerable AE attention, both after and even during the welding process. Well-known applications are found in the nuclear, petrochemical, and aircraft industries, and also in monitoring bridges, buildings, reinforced composites, and even wooden beams.

Well documented AE studies include other areas such as: martensitic phase transformations, dislocation processes, glass fiber and whisker reinforced composites, fracture mechanics, plastic deformation, crack growth during hydrogen embrittlement, stress corrosion cracking and low cycle fatigue, and detection of unwanted particles inside sealed transistors. Sonic signature analysis has been used in monitoring bearings (see Liptai et al., 1972).

The so-called shock pulse method for monitoring bearings detects the development of a mechanical shock wave caused by the compression between rolling elements and raceways. At the instantaneous moment of impact a compression (shock) wave is generated and radiates through the bearing housing. At impact a large local acceleration of the material is initiated. The magnitude of this acceleration is solely dependent upon the impact velocity and is not influenced by any mechanical vibration (Hazen, 1987).

The compression wavefront (shock pulse) caused by a mechanical impact sets up a dampened oscillation in the instrument's piezoelectric transducer at its resonant frequency.

Unlike vibration analysis equipment, which normally picks up a signal in the 20-20,000 Hz range, the shock pulse method detects an ultrasonic signal within a narrow band near 32 kHz. The 32 kHz band corresponds to strong emissions from a running bearing. As the bearing damage grows the signal increases in intensity.

In the SPM Instrument Company's equipment (Fig. 2-66) coupling of the transducer to the structure utilizes a stainless steel stud threaded into a tapped hole, or bonded to the structure. The transducer may be quickconnected to the stud or coupled permanently. (Compare with "hybrid" flowmeter transducers, Section 4.1.2.4.)

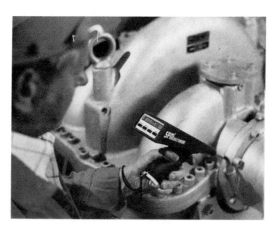

FIG. 2-66.   Shock pulse method of detecting bearing wear. Illustration courtesy SPM.

Generally speaking, acoustic emission instruments are available to locate cracks by triangulation, to measure count rates, to discriminate
against noise, and to analyze the amplitude distribution of signals.

Calibration devices and test sets to verify triangulation programs have
utilized phase changes, extensional waves coupled via a sewing needle
point, breaking a pencil point, and sparking.

Details on specific applications are given in the bibliography and in
literature available from AE system vendors, and in EPRI Conference
Proceedings on Incipient Failure Detection (e.g., March 1987). (See also,
Sections 2.1.3 and 2.4.5).

AE "dosimeters," or ultrasound level meters, may be required as safety
equipment in areas falling within the provinces of government or consumer agencies such as BRH, FCC, FDA, OSHA, or CPSA.

Waveguide techniques were used in an extensive high temperature leak
detection program reported by Famiglietti and Ghia, 1987. See Fig. 2-67.

2.6.6   Imaging, Holography and Microscopy

An introduction to the status of these topics in the mid-1980s is provided
by references such as the following special issues published by the IEEE:
Trans. Sonics and Ultrasonics, July 1984 and March 1985; *Trans. Ultrasonics, Ferroelectrics, and Frequency Control*, May 1987. Appropriate
sessions in the proceedings of the annual IEEE Ultrasonics Symposium
may also be cited, e.g., 1986. The journal *Acoustic Imaging* is another
archival source of information on current theories, techniques, equipment
and applications. The January 1987 issue of this journal, for example,

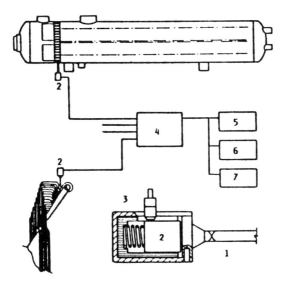

FIG. 2-67.   Block diagram of the RAF leak detection system for low pressure preheaters and steam headers. 1–Waveguide. 2–Acoustic emission transducer. 3–Clamping device. 4–Signal conditioning unit. 5–Multipoint chart recorder for continuous monitoring. 6–RMS voltmeter. 7–Spectrum analyzer. After Famiglietti and Ghia, 1987.

contains some 50 abstracts from the twelfth in a series of international symposia on Ultrasonic Imaging and Tissue Characterization.

Fundamentally, it is not essential to convert information about a test specimen from acoustical waves to light waves, in order to make decisions concerning the location, size, shape, orientation, or motion of detected inclusions. Bats have proven this. However, for most operators of ultrasonic test equipment, a visual image is desirable. The various oscilloscope or scan-recording modes, including those which give an appearance of perspective, provide some degree of imaging, in effect. Sokolov's (1934, 1950) image converter is described in various texts (Blitz, 1967, p. 109). Displays now can be on an oscilloscope, computer terminal, or TV screen. The possibility of using color on a TV display to represent acoustical information has been demonstrated for blood flow in and out of the heart. Schlieren methods have been used to visualize sound waves. Powder methods, which demonstrate standing wave patterns, are familiar to many students of acoustics.

In holography, both amplitude and phase information of the wavefronts emerging from the flaw can be recorded in terms of light intensity on film. The hologram or coded interference pattern contains the diffraction information necessary to reconstruct the original flaw wavefronts and thus produce a true visible image of the flaw. As Christensen, 1988, notes, the phase information of ultrasonic waves at $f < 10$ MHz can be digitized and stored, permitting reconstruction of images without recourse to film-recorded interference patterns similar to those required for optical frequencies, $f \approx 10^{14}$ to 10^{15} Hz.

Within the 1965–1975 decade, a number of reports appeared on the use of holography for inspecting thick-walled pressure vessels, locating and identifying submerged objects and for other NDT applications (Collins, 1973). Biomedical applications, although generally beyond the scope of this review, are important and well documented. Several organizations have built holographic or other imaging systems, which may in time or during appropriate times become standard products. Holosonics introduced their holographic equipment in 1971, and about 30 NDT systems were sold in the first four years since their introduction.

Sokolov imaging equipment was manufactured by James Electronics around 1970, but after some seven systems were sold, it was decided to withdraw this product from the commercial market.

One can categorize imaging equipment performance in terms of the resolution being on a macro scale or micro scale. If we define microresolution to mean resolution on the order of 0.1 mm or less, then most imaging and holographic equipment can be said to resolve in the macro scale. Holosonics' system 200, for example, provided a 3D focused image in real time, essentially achieving the theoretical half-wave resolution limit. For common metals, tested below 5 MHz (S waves) or 10 MHz (L waves), $\lambda/2$ exceeds 0.1 mm.

RCA developed a system whose resolution approached the micron border, namely "Ultrasonovision" (Vilkomerson, 1974). Ultrasonovision was a large aperture high resolution system for the visualization and quantitative analysis of acoustic wavefronts. It was based on a Michelson interferometer in which one mirror is a thin metallized pellicle submerged in water and through which passed the acoustic wave (see Fig. 2-68). A scanned laser beam was used to rapidly measure and display the displacement amplitude of the acoustic wave.

On the micro scale, Zenith developed a concept for an acoustic microscope (Korpel et al., 1971). By producing images of microstructural detail, it opened up a new dimension in ultrasonic visualization. One form of this acoustic microscope is manufactured by SonoScan (Fig. 2-69). In order to achieve resolution in the micron size range, high acoustic frequencies, typically 100 MHz and above, are employed. Furthermore, high resolution detection is accomplished in "SLAM" models by a noncontacting focused laser beam probe which is rapidly scanned to produce real time acoustic micrographs (Kessler, 1985, 1988; Lin et al., 1987).

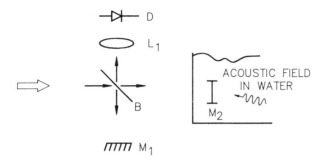

FIG. 2-68.   Basic arrangement of Ultrasonovision system. B is a beam splitter. M_1 is the rigid external mirror, M_2 is the flexible mirror or pellicle. L_1 collects the light from the interferometer onto the photodiode D. After Vilkomerson, 1974.

Applied to both the material and biological sciences, acoustic microscopes reveal surface and subsurface information on structure which directly pertains to variations in density and elasticity on a microscopic scale. Grain boundaries may be observed in the bulk of a material, for example, as well as microfractures, inclusions and disbonds. Furthermore, by employing the interference mode of operation localized stresses may be revealed.

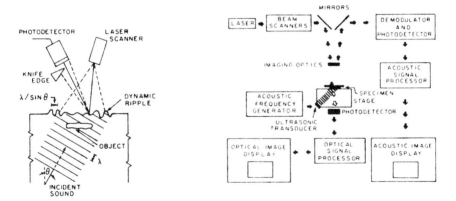

FIG. 2-69.   Left: Schematic diagram of the scanning laser acoustic microscope ("SLAM"). Right: Laser detection of acoustic energy at an interface. Optical sensors for acoustic waves and photoacoustic effects are used because no mechanical contact needs to be made to the solid material being examined. They are easy to scan and they can have good definition. The major disadvantage is that they are extremely insensitive compared to piezoelectric detectors. In photoacoustic applications, they are employed to detect modulation of the surface temperature, surface displacement, and surface slope of a solid, as well as the temperature variation of transparent solids, liquids and gases. After Kino et al., 1986. © 1986 IEEE.

These and other industrial imaging applications and problems may be understood or at least anticipated by referring to analogous studies on biomedical applications aimed at tissue characterization. In some of these studies, analyses include the estimation of the attenuation coefficient, mean spacing of scatterers, and general texture as revealed by an amplitude histogram. These properties have been demonstrated clinically to indicate disease in the heart, liver, and spleen, in cases where the lesions are not apparent on the image.

The most serious degradation to ultrasonic images is coherent speckle, which is comparable to that encountered using laser illumination. Subtle lesions are often missed because of this noise source. It arises from the signals from adjacent scatterers adding and cancelling in a random fashion, depending on their relative phase. One possible remedy combines images with independent speckle patterns to reduce the noise.

Unlike x-rays, ultrasonic waves are significantly affected by materials having different propagation velocities. Thus, the inhomogeneities in the human body corrupt the various focusing methods and distort the resulting image. A remedy developed at Stanford by the mid-1980s involves a correction system that introduces delay corrections to the received signals on each element of an ultrasonic transducer array. These delays are varied based on a criterion of performance within the image.

Returning to industrial applications, an example of a C-scan image ob-

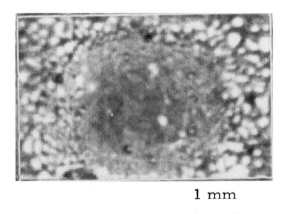

1 mm

├────────┤

FIG. 2-70. Example of bond evaluation of an integrated circuit (IC) assembly (Thermcon substrate with Si chip). White areas indicate disbond in die attach. Radiographically undetectable, these results were obtained at 100 MHz using a Panametrics HyScan instrument. After Clark, 1986.

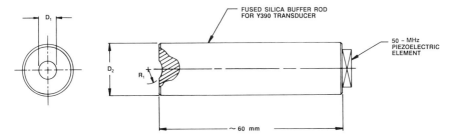

FIG. 2-71.   Fused silica focusing buffer rod, used by Fowler in 1971 at 50 MHz, and later at 100 MHz.

tained with a Panametrics Hy-Scan operating at 100 MHz is given in Fig. 2-70. This image of subsurface bond quality in a semiconductor chip was obtained using a 100-MHz transducer in its fundamental mode, bonded to a fused quartz focusing element. The 100-MHz design is similar to the 50-MHz 1971 design in Fig. 2-71, due to Fowler.

2.7 Other Measurements, Other Applications, Special Topics

2.7.1 Transport Properties: Viscosity, Thermal Conductivity

According to classical theory, sound absorption in gases is due to the sum of viscous and thermal conduction effects (Chapter 3, Section 3.7). The α ratio due to these effects lies between about 1 and 3, for common gases:

$$\alpha_{vis}/\alpha_{th} = 4\eta C_v \gamma/3 K_{th} (\gamma - 1), \tag{2-9}$$

where η = coefficient of viscosity, C_v = specific heat at constant volume, γ = specific heat ratio, and K_{th} = coefficient of thermal conductivity, which equals $2.5 C_p \eta/\gamma$ for monatomic gases. Experiments have established that classical theory accounts for absorption in monatomic gases without ionization. In the equation for classical sound absorption in a monatomic gas, substitution of $2.5 C_p \eta/\gamma$ for K_{th} leads to the equation relating η to α given in the next section. In diatomic and polyatomic gases, in binary mixtures, and in reacting gases, attenuation is observed to be higher than the classically predicted value, due to diffusion, relaxation, or internal effects.

2.7.1.1   Viscosity

During the 1960s, Carnevale et al. reported η determinations in argon to $\sim 10^4 K$, where η was calculated from the measured attenuation coefficient α and sound speed c, at a gas pressure $p = 1$ atm:

$$\eta = \alpha \gamma c p / 2\pi^2 f^2 [(4/3) + 5(\gamma - 1)/2\gamma]. \tag{2-10}$$

In nonmetallic liquids the classical absorption is essentially due to η since K_{th} effects are negligibly small in comparison. Excess attenuation is observed in nonmonatomic liquids, which, as for some gases, can dwarf the classical effects. See, for example, Herzfeld and Litovitz, 1959. Non-invasive monitoring of viscosity changes which are characteristic of one polymerization process was reported by Lynnworth et al., 1971. The attenuation of pulses transmitted across the diameter of 75 mm stainless steel pipes containing liquids having viscosities of ~ 50 and $\sim 10^4$ centipoise responded differently to these large differences in viscosity. But reproducible resolution of small differences of η is another matter.

Presently available equipment for η measurements generally is not based on the above transmission concepts using the medium as its own sensor, but rather, on the damping effects of viscous liquids upon an immersed probe.

Viscosity Probes for Liquids

Ultrasonic η probes utilize several different geometries. Figure 2-72 shows an immersed torsional wave piezoelectric transducer (Mason, 1947) with which η was measured from 0.01 to 6 poises. Liquids of higher viscosity exhibit a shear stiffness, complicating the analysis of the mea-

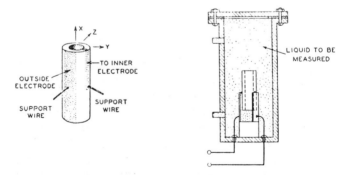

FIG. 2-72. Experimental arrangement for measuring η in light liquids, using a torsional transducer, after Mason, 1947. The torsional ADP crystal, as shown at the left, is nodally suspended by three wires, and immersed in the liquid under test.

sured crystal resonance frequency and resistance. Mason et al., 1949 also utilized shear wave reflectance at oblique incidence (Fig. 2-73). Shear wave reflectance at normal incidence (Fig. 2-74) is reviewed by Moore and McSkimin, 1970.

In the Bendix/Combustion Engineering Model 1800 viscometer, for which a variety of metal probes are available, the sensor generally consists of a protected resonant blade including a magnetostrictive portion at one end, and stainless steel at the other (Fig. 2-75). This instrument was introduced in December 1968. Nineteen years later, it was estimated (S. Bostic, 1987, priv. comm.) that between 500 and 800 of this type equipment had been sold, mainly for in-line use. It operated as follows.

The probe is inserted into the liquid and the degree of "shearing" or "viscous drag" is translated electronically into a value for that particular liquid, for $\eta\rho$ values ranging in four scales up to 50000 cp $\times$ g/cm^3. Probes are available to accommodate numerous liquids at temperature ranges up to $+200°C$ and a pressure range from 0 to 750 psig ($\sim$5 MPa). If desired, a number of probes can be connected via a multipoint unit to a single electronic subassembly.

The electronic subassembly produces a short pulse of current to a coil situated inside the probe and around the thin blade sensing element. The resulting magnetic field excites the magnetostrictive member and causes the blade to vibrate longitudinally. When the vibration amplitude is attenuated to a predetermined level, another pulse is triggered. Thus the pulse repetition rate is a measure of $\eta\rho$. To the extent that ρ is known, η can be determined. Accuracy is $\sim$2% for Newtonian fluids. For non-Newtonian fluids, reproducibility is $\sim$1% of full scale.

In Poland, an ultrasonic viscometer was developed wherein a mechanical resonator was excited into the torsional mode. The decay curve was viewed on a screen, calibrated in units of η for liquids of known ρ (Filipczynski et al., 1966, p. 124). In 1977 the author observed that the *ve-*

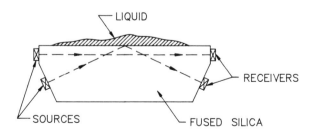

FIG. 2-73.   Viscosity determined from shear wave reflection coefficient at *oblique* incidence, after Mason et al., 1949. See also, McSkimin, 1960.

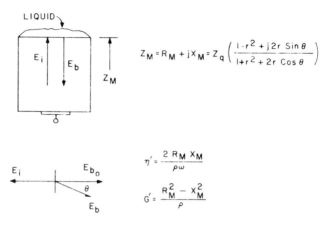

$$Z_M = R_M + jX_M = Z_q \left(\frac{1 - r^2 + j2r\,\sin\theta}{1 + r^2 + 2r\,\cos\theta} \right)$$

$$\eta' = \frac{2\,R_M\,X_M}{\rho\omega}$$

$$G' = \frac{R_M^2 - X_M^2}{\rho}$$

FIG. 2-74. Basis of shear reflectance method at *normal* incidence, after Moore and McSkimin, 1970. © 1970 Academic Press.

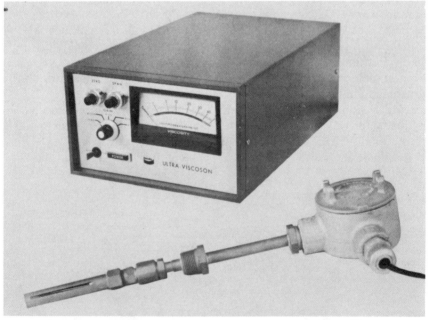

FIG. 2-75. Ultra-Viscoson model 1800 viscometer senses damping effect on resonant probe, thereby determining viscosity × density product $\eta\rho$. Photo courtesy Combustion Engineering.

locity of torsional waves decreased in a noncircular sensor immersed in cold glycerine by an amount greater than could be attributed to density. Ten years later Bau and Kim (1987, priv. comm.) found a useful relation between the torsional velocity in a threaded metal tube and the square root of $\eta\rho$ for glycerine/water mixtures. See Chapter 9, Section 9.1.

In the past ten years or so, the damping effect of a viscous fluid upon a vibrating sphere has been utilized in an oscillation viscometer due to Nametre. This device utilizes acoustic but not ultrasonic vibration frequencies. See Fig. 2-76 and Section 9.1 for probe illustrations and examples of in-line applications.

Measurements of the increasing η (along with increasing moduli) were reported by Papadakis, 1974 for epoxy during polymerization. He utilized reflection coefficient (R) measurements of longitudinal and shear bulk waves at 5 MHz, and R and α measurements in immersed wires guiding extensional and torsional waves at $\sim$100 kHz. The standard wire probe

FIG. 2-76.   Standard oscillating rod sensor is part of an in-line "Viscoliner" viscometer, and is one of several probes available from Nametre. See Fitzgerald and Matusik, 1986; Fitzgerald et al., 1988a.

and readout equipment used in this work, Fig. 2-77, was originally intended for thermometry and moduli applications.

In a series of papers, Harrold and Sanjana, 1986–1987, reported on their use of embedded 1.5 mm diameter polyester fiberglass waveguides as sensors to monitor polymerization. The expendable extensional mode waveguides ultimately became part of the reinforcing network within the composite material. See Chapter 9, Section 9.1.

A rheology sensor consisting of a buffered 1 MHz transducer was used by Cohen-Tenoudji et al., 1987, to monitor the dynamic viscosity of a resin curing thermally at 145°C. The real and imaginary components of the shear modulus were determined throughout the cure cycle.

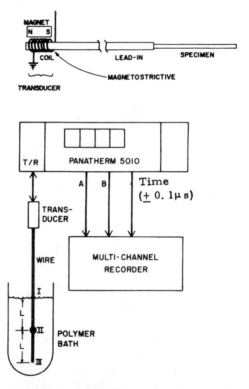

FIG. 2-77.   Detail of thin wire experimental arrangement to monitor polymerization automatically. The echo amplitude outputs A and B were from the weld (II) and the specimen end (III), respectively, at first. When the meniscus echo (I) grew to a preset trigger level, channel A began monitoring the meniscus (I) and Channel B switched to the weld (II). After Papadakis, 1974.

2.7.1.2   Thermal Conductivity and Other Transport Properties

Besides η, measurements of α may be interpreted in terms of the thermal conductivity coefficient K_{th}, the diffusion coefficient D_f, activation energy E, relaxation strengths S, rotational collision number Z_{rot}, etc. Direct measurements of these basic parameters are usually not considered routine industrial practice, and standard instruments dedicated to any one of these measurements apparently are not yet available. Indirect measurements of a particular transport parameter, e.g., K_{th} are sometimes approached by empirical correlation. Such correlations can be established when c and/or α respond in a predictable manner to the same material variables that contribute to K_{th} variations in a part such as an ablative heat shield. In still other cases, mechanisms such as transport of carbon into rhenium at $\sim 2200°C$ can be identified as a diffusion process, based on the time-dependence of changes in c. In one such experiment (Fig. 2-78), the process was identified as diffusion by monitoring c versus exposure time, when a self-heated rhenium wire was surrounded by graphite felt, in vacuum. Bell, 1963, has shown that the temperature T_D above which diffusion proceeds rapidly, e.g., nitrogen into titanium, at $T_D =$ 881°C (the "beta-transus") can also be determined by the dramatic changes in c and/or α when T_D is exceeded.

Bradshaw, 1972, separated η and K_{th} contributions to α by simultaneously measuring the Q for different modes of a gas-filled cavity (radial, longitudinal; see Fig. 2-79). Measurements can be made for a variety of fluids, and over a wide range of temperature and pressure.

Blessing and Flynn, 1987, at NBS found that a good correlation existed between the insertion loss of 2.5 to 25 kHz pulses and the R-value of glass fiber insulating materials. Their acoustic test yielded results in minutes that ordinarily takes several hours to obtain using thermal measurement procedures. In the NBS acoustic test, the samples of insulation are contained in individual baskets having acoustically-transparent bottoms. Extension of this acoustic test method from its laboratory stage to quality control during manufacture and to field use (e.g., characterizing attic insulation) was reported to be underway in cooperation with ASTM Committee C16 on Thermal Insulation.

Thermal Diffusivity

Thermal diffusivity, α_{td} (which equals $K_{th}/\rho C_p$ where ρ = density and C_p = specific heat at constant pressure) was determined in plexiglass, polyethylene and other media by measuring with thermocouples the temperature rise produced by focused ultrasound (Newman and Lele, 1985).

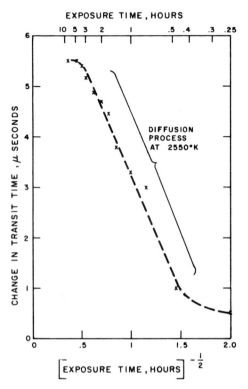

FIG. 2-78. Diffusion of carbon into rhenium at ~2250 K is indicated by linear portion of plot. Data were obtained using self-heated rhenium wire, ½-mm diameter × 50-mm long, surrounded by graphite felt. Source: M.S. McDonough, in Panametrics Final Report, Contract NONR-3918(00), 1967.

Peak focal intensities ranged from 200 to 800 W/cm². Single pulse, repeated pulses or cw were used. In subsequent work the temperature rise was measured ultrasonically (Davis and Lele, 1985; Fig. 2-33).

2.7.2 Pressure, Force, Vibration, Acceleration

Pressure influences sound propagation in solids, liquids, and gases, but in different ways. In solids, applied pressure leads to so-called stress-induced anisotropy, as shown by Crecraft, 1964, Hsu, 1974, and others (see Section 2.7.3). In liquids, the effects of pressure p upon c and α are usually small (relative to effects in gases), but the frequency of relaxation peaks can be shifted significantly. However, sound speed in the sea increases significantly with depth D due to hydrostatic pressure. The depth

LONGITUDINAL DISTRIBUTION   RADIAL DISTRIBUTION

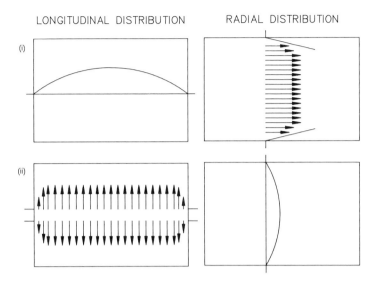

FIG. 2-79.   Particle velocity distributions in a cavity excited (top) in the first longitudinal mode, and (bottom) in the first radial mode. After Bradshaw, 1972.

coefficient (pressure effect) is $\Delta c/\Delta D = 0.017$ m/s per m; $(\Delta c/c)\Delta D = 1.1 \times 10^{-5}$/m. In real gases, c increases as p increases, while α is inversely proportional to p. For a nonideal gas, the equation of state may be written $pV = RT + Bp$, where V = volume and B = second virial coefficient. From this, $c = [\gamma(RT + 2Bp)/M]^{1/2}$.

In a liquid containing bubbles, it is possible to determine pressure from their resonance behavior. Bubbles of air with radius >100 μm, if compressed and rarefied adiabatically, resonate at a frequency proportional approximately to the $\frac{3}{4}$ power of the ambient pressure. In water, an air bubble of 100 μm radius resonates at 32 kHz at atmospheric pressure, increasing to 53 kHz at 2 atmospheres. In a double frequency method due to Shankar et al., 1986, it was demonstrated that changes in bubble size could be tracked as pressure increased from 20 to 100 mm Hg above ambient, to within 15 mm. See Fig. 2-80.

Ultrasonic equipment developed for measuring p includes a Hewlett-Packard instrument, utilizing another special cut in quartz, and a Wallace and Tiernan sonar manometer. In the former device, Fig. 2-81, the quartz crystal responds almost linearly to p, but requires temperature compensation. The probe assembly contains the quartz crystal p-sensing oscillator and a reference oscillator. The p-dependent difference frequency is transmitted up the cable to the signal processor. The p-sensing oscillator typically varies from ~0.5 to ~1 MHz. Range of the quartz pressure gage is

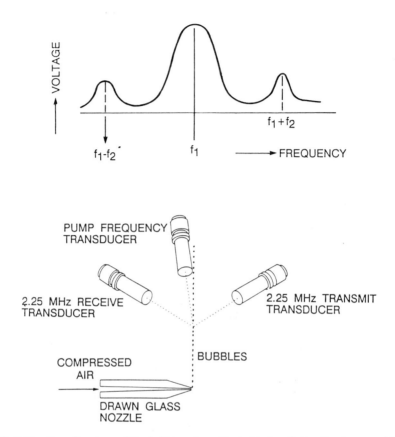

FIG. 2-80. Top: Spectrum of the bubble echo subjected to two fields. One is a swept low frequency field and the other is a high frequency imaging field. The side bands appear at the sum and difference frequencies. Bottom: Geometry of the transducer arrangement. The low frequency transducer is arranged at an angle so as to reduce the effects of the generation of sum/difference frequencies due to direct impact of the pump field on the high frequency transducers. After Shankar et al., 1986. (See also, Chapelon et al., 1988). © 1986 Butterworths.

0 to 844 kg/cm^2 (0 to 12000 psia), 0°C to 150°C, and resolution is <0.7 g/ cm^2 (0.01 psi), at a 1 second sampling period. Introduced in 1971, uses include: oil well logging, oceanographic research, pulse tests in wells, and monitoring underground detonations. See Section 9.2 for additional details.

In the latter device, Fig. 2-82, ultrasonic pulses measured the relative heights of two mercury columns. One piezoelectric transducer mounted at the bottom of each leg of the manometer transmitted an ultrasonic pulse through the mercury to the surface and received the echo from that sur-

FIG. 2-81.   The 2811B Quartz Pressure Gage consists of a 2813B Quartz Pressure Probe and a 2816A Pressure Signal Processor. The 2813B Quartz Pressure Probe converts bottom hole pressure to an electrical signal whose frequency is proportional to pressure. This signal is transmitted from the downhole probe to the surface through a single conductor, armored electric line. The 2816A pressure signal processor connects to the electric line on the surface and conditions the pressure-related signal. The signal then drives a frequency counter and is converted to a pressure reading in the Surface Recording Package. The 2820A Surface Recording Package displays and records pressure test data. The 2816A pressure signal processor is mounted in the 2820A. Photo courtesy Hewlett-Packard.

face. By transmitting the ultrasonic pulse simultaneously in both legs, the difference in time between reception of the echoes of the two legs was related to the height difference.

By holding c constant, the time difference was related to a height difference. A time reference (master-clock oscillator), was chosen to yield different calibration units. Inches of Hg, psi, mm Hg, mb, psf, N/m^2, and Pa were the standard units available.

Maintaining a constant speed of sound and constant mercury density required careful temperature control. A mercury-in-glass thermostat was the thermal reference. Control was better than $\pm$ .028°C for accuracy to ± 13 μm Hg at 81 mm Hg.

Reports on the use of SAW oscillators as pressure sensors have appeared from time to time since about 1975. Some are cited or reviewed by Hauden, 1987.

Piezoelectric transducers have been widely used for many years to measure force, vibration, acceleration, torque, etc. Such applications are to be expected from the very definition of a piezoelectric material. In this

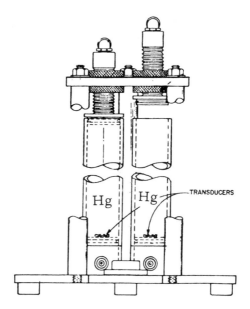

FIG. 2-82.   In digital U-tube sonar manometer that was commercially available in the 1970s, ultrasonic pulses were used to measure the difference in heights of two columns of mercury. Illustration courtesy Wallace & Tiernan Div., Pennwalt Corp.

section, we wish to merely note that quartz and other materials are used in applications from cryogenic temperatures up to ~800°C, in nuclear environments, up to at least 100 kHz, and provide up to 3 axes of component resolution in packaged units. Application details are readily available from a number of vendors. See Figs. 2-83 and 2-84. With the intro-

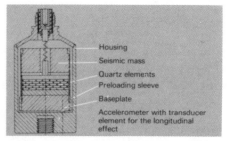

FIG. 2-83.   Piezoelectric instrumentation includes two- and three-component force-measuring quartz transducers, multicomponent force-measuring dynamometers, high-temperature pressure transducers, and engine pressure gages for measuring pressure, force, acceleration, torque, cutting forces, etc. Photo courtesy Kristal Instruments.

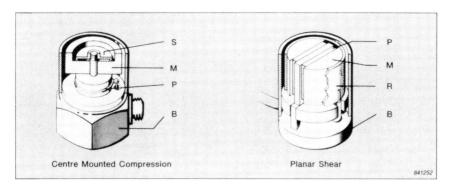

FIG. 2-84.   Compression and planar shear accelerometer designs from Brüel & Kjaer. M = seismic mass, P = piezoelectric element, B = base, R = clamping ring and S = spring.

duction of piezoelectric films, sensors such as those in Fig. 2-85 have emerged.

2.7.3   Stress and Strain

A number of ultrasonic studies have been aimed at evaluating residual stress in metals, based on changes in c with stress. Unfortunately, factors other than stress (e.g., inhomogeneities, texture) influence c too, and unless the effects of these other variables can be eliminated or compensated for, the satisfactory determination of internal stresses remains elusive. One approach to separating stress and texture anisotropy is due to MacDonald, 1981.

Measurements of horizontal shear or Rayleigh velocity in two orthogonal directions, as in the emat experiments of Clark and Moulder, 1985, of NBS-Boulder, or of Alers, 1988, respectively, may provide solutions to the problem of eliminating the effect of all variables other than stress,

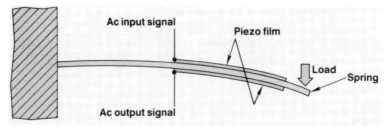

FIG. 2-85.   Force or force-related parameters can be sensed using PVDF piezoelectric polymer film transducer. Illustration courtesy Pennwalt.

and might eventually lead to a clamp-on (noninvasive) pressure gage for pressurized vessels. It is also interesting to note that the acoustic birefringence, B, which equals the normalized difference in velocities, is calculated from the ratio of twice the difference in orthogonal transit times $2\Delta t$, to their sum, $t_1 + t_2$. The formula $B = -2\Delta t/(t_1 + t_2)$, except for the denominator's exponent, is almost identical to that used in calculating flow velocity V in some transit time flowmeters. Accordingly, instruments developed for one purpose might be reprogrammed to suit the other.

Another way to eliminate factors other than stress is to interrogate the specimen over a given path before and after the stress is applied. In some cases, stress can be resolved to better than ~ 100 psi. The limit depends on material, wave type (L or S), propagation direction, and polarization relative to the stress axis, path length, and on the resolution limit of the electronics. Sing-around and pulse-echo-overlap methods have been used to resolve c changes of 10 to 20 ppm.

It is interesting to compare ultrasonic sensitivity to stress with strain. Hsu, 1974, presents data for 1018 steel and 2024-T4 aluminum. For steel, using 10 MHz shear waves propagated perpendicular to the compressive load direction, but with particle motion parallel to load direction, Hsu measured c increases of 0.1% per 20,000 psi. The strain equals the stress divided by Young's modulus, or about 0.07% for the 20,000 psi (1.38×10^8 N/m^2) stress. Thus the change in c due to stress is about 1.5 times the strain.

Can one apply the above type of data to measure the tension in a bolt, with access to the head? Not easily. If the bolt is faced off square to the axis at both ends, and is available for measurement before and during tightening, and there are no significant T changes, the task becomes more reasonable.

In the 1970s, Heyman, 1976, devised marginal oscillator techniques for high-resolution transit time measurements in bolts. Steblay, 1986, introduced a small hole as a reference so that strain could be measured, Fig. 2-86.

A rather specialized means for recognizing torsional stress in magnetostrictive wires or rods was discovered accidentally in the Fe-Co-Va-Mn alloy Remendur which had been straightened by a rotating-jaw machine (Lynnworth, 1972). It was unexpectedly observed that in wire thus straightened, torsional wave magnetostrictive transduction was strong, and relatively uniform and permanent, compared to the Wiedemann effect normally induced by electrically magnetizing a wire circumferentially. Heating the straightened wire red-hot for a few seconds relieved the torsional stress, as indicated by the observation that such heat treatment eliminated the torsional transduction effect in the heated segment of the wire.

FIG. 2-86.   Hole(s) in threaded member provide reference reflector(s) for strain measurement. After Steblay, 1986.

2.7.4   Composition

Given a mixture of two materials A and B, it appears intuitively reasonable to expect sound propagation (c and α) to bear a relationship to the ratio a/b, where a and b represent the relative amounts of A and B. For simplicity we will not deal with α in this section, nor with determinations based on dispersion in c.

Given an *ideal gas mixture*, from Eq. (2-1), it is seen that the average molecular weight M and average specific heat ratio γ determine c at a given T. For binary mixtures, expressions for the equivalent molecular weight and equivalent ratio of specific heats are given by Noble et al., 1964, as well as approximations appropriate to their operation of an ultrasonic gas chromatograph for binary mixtures. Response is generally linear with concentration up to ~10% sample in carrier on a weight basis.

For flare gases it has been found that the average molecular weight M can be computed from the sound speed and temperature, usually to an accuracy on the order of 2%. This unexpected high accuracy is obtained without measuring γ, the specific heat ratio. Apparently M and γ are not independent; γ appears to be a simple monotonic function of M, for the flare gases tested in the study reported by Smalling et al., 1984. See also, Section 4.2.2.2.

For binary mixtures of *liquids*, c has been measured for ~100 combinations. Empirical rules relating c to concentration or mole fraction of the added component involve Rao's constant and also Wada's constant (called the molecular compressibility). Examples are plotted in Beyer and Letcher, 1969. See Fig. 2-87.

In the nuclear industry, it is necessary to carefully document and control the status of heavy water. This fluid must be registered and tracked under international regulations to prevent unauthorized use. Heavy water producers and reactor operators frequently measure the concentration of heavy water in "normal" water to determine the mole fraction of D_2O.

Mole fraction has traditionally been measured with expensive, time consuming laboratory-based techniques, such as infrared absorption or mass spectrometry, which require opening sealed containers to remove samples for assay.

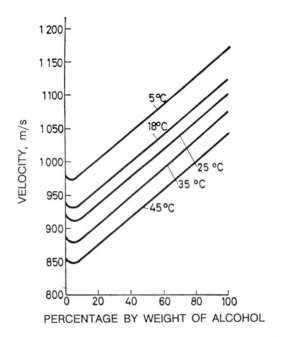

FIG. 2-87.   Relationship between velocity and concentration at different temperatures for an ethyl alcohol-carbon tetrachloride mixture (after Derenzini and Giacomini, 1942). CCl₄ is unassociated but ethyl alcohol is associated. This accounts for the nonlinearity. See, for example, Blitz, 1967, p. 126.

In cooperation with the Brookhaven National Laboratory and under the auspices of the United States Program for Technical Assistance to IAE Safeguards, a noninvasive clamp-on concept was developed into an instrument for the assay of the mole fraction of heavy water in sealed containers. The resulting instrument used the pulse–echo method in conjunction with a correlation between mole fraction, temperature and the velocity of sound in the D_2O/H_2O mixture. The instrument is portable, field operable, and it can be used with a printer or computer to provide a permanent record of all measurement parameters and test results. (Elfbaum, 1987, priv. comm.)

Composition measurements in many liquids are routinely conducted using NUSonics' sonic solution analyzers. Normally both c and T are measured (Zacharias, 1970). The sing-around technique is used to measure c with errors typically as small as 0.01% of reading. With careful T control or measurement, concentration can be determined in some cases to 0.02% (sulfuric acid, 85 to 100% region). Errors can increase when the liquid includes bubbles, solid particles larger than 50 μm, emulsions with large liquid droplets, and slurries. The sonic solution analyzer has also been used to determine yeast slurry consistency (Feil and Zacharias, 1971) and percent solids in foods (Zacharias and Parnell, 1972).

Composition determination in solids may be illustrated by an application in the glass industry (Hagy, 1973). Ultrasonic techniques for determining absolute and differential thermal expansion of titania-silica glasses have been applied to a fused sandwich seal, composed of two glasses with slight composition differences. For the titania-silica system, it was shown that for any two compositions with small titania differences the expansion coefficient differential remains constant from $-195°C$ to $925°C$. A direct correlation between ultrasonic velocity and thermal expansion was established for this glass system and led to successful nondestructive measurements. With an experimentally defined relationship, the measurement of the ultrasonic velocities yields absolute or differential expansions. Excellent agreement with seal testing data was shown to exist with differential data taken by photoelastic and ultrasonic methods.

It is interesting to note that by using a modified cell for a substitution method in the through-transmission mode, a NUSonics sound velocimeter normally used for liquid velocimetry can be used for solid specimens too (Zacharias et al., 1974).

Hygrometry applications may be mentioned here. In one case, sound speed in paper and various organic materials was measured with extensional waves. As moisture levels increased, c decreased (Stungis and Merker, 1976). In another case, which represents a rather large number of commercial applications, a coated resonant quartz crystal served as the water vapor detector (Crawford et al., 1964). In this sorption hygrometer, a 1 μm coating of the hygroscopic polymer is applied to the crystal. The change in resonant frequency due to the sorbed water is given by $\Delta f = \Delta f_o (W/W_o)$ where Δf_o is the frequency change due to the coating, W is the weight of the water sorbed, and W_o is the weight of the coating. To measure other vapors, different selective sorbing substances are used. A detector for UDMH, for example, is reported to respond "instantly," but requires ~ 1 minute for quantitative determination of concentration (Varga, 1974). Piezoelectric sorption hygrometers have been available from DuPont. In the mid-1980s, several investigators studied higher-frequency surface acoustic wave (SAW) devices as moisture sensors, e.g., Joshi and Brace, 1985. See Fig. 2-88.

2.7.5 SAW Sensors and Membrane Transducer Applications

The preceding sections on applications were mostly organized in terms of the measuring method. However, when certain methods come into exceptionally wide use, the methods and their applications may become

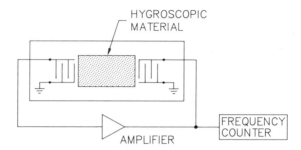

FIG. 2-88.   Block diagram of a SAW humidity sensor. After Joshi and Brace, 1985. © 1985 IEEE. If the hygroscopic material is replaced by a thin-film meander line heater, a flow sensor for gases is obtained (Joshi, 1988).

nearly synonymous. Stated another way, it may be more instructive to group a variety of applications which share a common solution than to discuss these applications in separate parts of the text. Two such areas involve SAW sensors and membrane transducers. In both cases it has been reported that government or commercial support for developing the resulting devices exceeded 10^6 dollars. Because of the large efforts in the past few decades to understand and manufacture SAW devices such as oscillators, convolvers, delay lines and filters, many workers have encountered practical problems where the performance originally sought was frustrated by undesired effects of temperature, humidity, stress, etc. Recognizing that these "undesired" effects might lead the way to a useful sensor for temperature, humidity, etc., some workers redefined their goal, aiming for sensors responsive to a particular variable instead of a device immune to all environmental parameters. In other cases, however, SAW sensor development was the goal right from the start.

The March 1987 *IEEE Trans. UFFC* was a special issue devoted to acoustic sensors. Half of the twenty papers in this issue dealt with surface acoustic wave (SAW) sensors, for which the measurands included: biological components; concentration of gases such as organic vapors, NO_2 and H_2S; magnetic field; position; acceleration; force; pressure; flow velocity and temperature. Other applications of Rayleigh or surface waves include characterization of surface quality, stress, thickness of layer a on b, etc. (Viktorov, 1967). Recognizing that the Rayleigh wave may be thought of as a special case of a plate wave or Lamb wave, one can anticipate that more general Lamb wave sensors will have advantages over SAW devices in some situations. See, for example, Wenzel and White, 1988.

The reference path idea illustrated in Fig. 2-89 is employed in a number

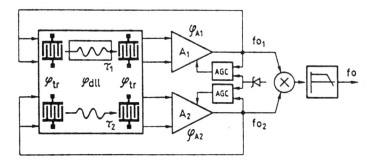

FIG. 2-89.   Sensor principle using dual delay-line oscillator. After Venema et al., 1987. ©
1987 IEEE.

of SAW sensors. Note the use of dual paths, one of which is intended to
respond selectively to one measurand only. Common-mode interferences
(temperature, humidity and aging, for example, in an ideal chemosensor)
tend to be cancelled. [The use of dual paths in SAW devices has ante-
cedents or analogs in other differential path systems, e.g., phase contrast
(Section 3.6.1.4), upstream and downstream oscillating loops in early ultrasonic flowmeters (Kritz, 1955), and in a sense, in a Wheatstone bridge.]

Further details on the SAW sensors grouped in this introductory section
may be found in the IEEE special issue cited above, or in the case of
temperature in Chapter 5.

Polymer electret transducers are used in mega-unit quantities in camera
echo ranging, and in an increasing number of industrial applications. Low
cost and availability of the transducers and associated circuitry for transmitting, receiving and signal processing underlie such applications. Fields

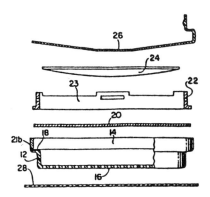

FIG. 2-90.   Polaroid electret transducer. After Kirby and Paglia, 1985.

Table 2-6. Sensor Classification Scheme Due to White, 1987. © 1987 IEEE

A. Measurands
  A1. Acoustic
      A1.1 Wave amplitude, phase, polarization, spectrum
      A1.2 Wave velocity
      A1.3 Other (specify)
  A2. Biological
      A2.1 Biomass (identities, concentrations, states)
      A2.2 Other (specify)
  A3. Chemical
      A3.1 Components (identities, concentrations, states)
      A3.2 Other (specify)
  A4. Electric
      A4.1 Charge, current
      A4.2 Potential, potential difference
      A4.3 Electric field (amplitude, phase, polarization, spectrum)
      A4.4 Conductivity
      A4.5 Permittivity
      A4.6 Other (specify)
  A5. Magnetic
      A5.1 Magnetic field (amplitude, phase, polarization, spectrum)
      A5.2 Magnetic flux
      A5.3 Permeability
      A5.4 Other (specify)
  A6. Mechanical
      A6.1 Position (linear, angular)
      A6.2 Velocity
      A6.3 Acceleration
      A6.4 Force
      A6.5 Stress, pressure
      A6.6 Strain
      A6.7 Mass, density
      A6.8 Moment, torque
      A6.9 Speed of flow, rate of mass transport
      A6.10 Shape, roughness, orientation
      A6.11 Stiffness, compliance
      A6.12 Viscosity
      A6.13 Crystallinity, structural integrity
      A6.14 Other (specify)

  A7. Optical
      A7.1 Wave amplitude, phase, polarization, spectrum
      A7.2 Wave velocity
      A7.3 Other (specify)
  A8. Radiation
      A8.1 Type
      A8.2 Energy
      A8.3 Intensity
      A8.4 Other (specify)
  A9. Thermal
      A9.1 Temperature
      A9.2 Flux
      A9.3 Specific heat
      A9.4 Thermal conductivity
      A9.5 Other (specify)
  A10. Other (specify)

B. Technological Aspects of Sensors
  B1. Sensitivity
  B2. Measurand range
  B3. Stability (short-term, long-term)
  B4. Resolution
  B5. Selectivity
  B6. Speed of response
  B7. Ambient conditions allowed
  B8. Overload characteristics
  B9. Operating life
  B10. Output format
  B11. Cost, size, weight
  B12. Other (specify)

C. Detection Means Used in Sensors
  C1. Biological
  C2. Chemical
  C3. Electric, Magnetic, or Electromagnetic Wave
  C4. Heat, Temperature
  C5. Mechanical Displacement or Wave
  C6. Radioactivity, Radiation
  C7. Other (specify)

D. Sensor Conversion Phenomena
  D1. Biological
      D1.1 Biochemical transformation
      D1.2 Physical transformation
      D1.3 Effect on test organism
      D1.4 Spectroscopy
      D1.5 Other (specify)

Table 2-6.   (*continued*)

D2. Chemical	E5. Semiconductor
D2.1 Chemical transformation	E6. Liquid, gas or plasma
D2.2 Physical transformation	E7. Biological substance
D2.3 Electrochemical process	E8. Other (specify)
D2.4 Spectroscopy	
D2.5 Other (specify)	**F. Fields of Application**
D3. Physical	F1. Agriculture
D3.1 Thermoelectric	F2. Automotive
D3.2 Photoelectric	F3. Civil engineering, construction
D3.3 Photomagnetic	F4. Distribution, commerce, finance
D3.4 Magnetoelectric	F5. Domestic appliances
D3.5 Elastomagnetic	F6. Energy, power
D3.6 Thermoelastic	F7. Environment, meteorology,
D3.7 Elastoelectric	security
D3.8 Thermomagnetic	F8. Health, medicine
D3.9 Thermooptic	F9. Information, telecommunications
D3.10 Photoelastic	F10. Manufacturing
D3.11 Other (specify)	F11. Marine
	F12. Military
E. Sensor Materials	F13. Scientific measurement
E1. Inorganic	F14. Space
E2. Organic	F15. Transportation (excluding
E3. Conductor	automotive)
E4. Insulator	F16. Other (specify)

of application of polymer transducers are probably in most or all of the 15 specific areas in White's 1987 classification, plus music, recreation and others. (See Table 2-6.) An illustration of an electret transducer appears in Fig. 2-90. Examples of applications of polymer transducers are collected in Section 9.7, grouped according to the transducer being of the piezoelectric polymer (or copolymer) type, or an electret (e.g., Polaroid).

Table 2-7.   Examples of Annually-Revised Books and Periodicals for Vendor and Product Information

1. *Chemical Processing* magazine, mid-November issue, even years (Putman)
2. *Control Equipment Master* [CEM] (Chilton)
3. I & CS (Chilton)
4. ISA *Directory of Instrumentation 1986–1987* [or subsequent] (ISA)
5. Proc. Symposia of National Engineering Societies
6. *Research & Development* magazine, October issue
7. *Thomas Register* (Thomas Publishing Co.)
8. 1986 [or subsequent] *Sensor and Transducer Directory* (North American Technology)
9. Measurement and Control *Handbook and Buyer's Guide* (Measurements and Data Corp.)

3. Theory and Measurement Techniques

Theories on the generation, propagation and detection of ultrasonic waves underlie ultrasonic measurement techniques. Given the existence of a number of excellent books covering the theory of electro-acoustic transduction, wave propagation and other fundamental aspects of physical acoustics, such as interactions between sound waves and the host medium, it seems unnecessary to repeat such material here. Accordingly, in this chapter, the emphasis is on measurement techniques. Some theory and general principles are presented in support of these techniques but no attempt is made to be as rigorous as in texts whose main purpose is to present the theory of generation, propagation and detection of ultrasound.

3.1 Waves, Media, Boundary Conditions

The principal waves that are used for measuring physical properties and process control parameters are the first six entries listed in Table 3-1. These waves may be used alone (to "ask one question at a time") or in combination with one another. The "other" waves listed include some

Table 3-1.   Principal Wave Types Used (or Encountered) in Industrial Measurements of Physical Properties and Process-Control Parameters

	Media	
Wave	Solid	Fluid
Longitudinal/Compressional	•	•
Transverse shear	•	
Extensional	•	
Torsional	•	
Rayleigh and Rayleigh-like	•	
Lamb and Lamb-like	•	
Other, e.g., head, lateral, Love, Sezawa-like, Stoneley, Franz-type creeping wave, . . .	•	• •

circumferential types that may be encountered unintentionally, such as fluid-borne slow waves with speeds comparable to that in the fluid adjacent to an immersed elastic body. (Luppe et al., 1985; Überall, 1973; Überall and Huang, 1976; Stoneley, 1924; Franz, 1954.)

Only the first two types can propagate in unbounded media. Depending on the dimensions of the medium, its properties and the boundary conditions, any one of the above wave types may propagate with some degree of "guiding."

The questions one might ask upon being introduced to these waves include:

- How fast do they travel?

- At what rate are they attenuated?

- How do these waves interact with the host medium?

- What happens at interfaces?

Detailed knowledge of the *particle motion* is not essential to our present task. Illustrations of particle motions for the principal wave types appear in Krautkrämer and Krautkrämer, 1983, and in some of the other books in the list of references. A few remarks summarizing the trajectories for the main wave types appear at the end of Section 3.1.1.

Readers interested in *nonlinear acoustics*, a topic not covered in this book, are referred to the work of Breazeale, 1987, Beyer, 1965, and the book by Novikov et al., 1981 (trans. 1987) on nonlinear underwater acoustics. Nonlinear acoustics and macrosonics comprise topic [25] in *The Journal of the Acoustical Society of America*, issued monthly.

3.1.1 Phase and Group Velocity; Dispersion; λ Ambiguity

The first question listed above, dealing with wave speeds, has occupied the attention of acousticians for several centuries. Phase and group velocities c and U, respectively, are most precisely interpreted for continuous waves:

$$c = f\lambda = \omega/k \qquad (3\text{-}1)$$

and

$$U = d\omega/dk, \qquad (3\text{-}2)$$

where $\omega = 2\pi f$, f = frequency, $k = 2\pi/\lambda$, and λ = wavelength (Stratton, 1941). Velocity dispersion, wherein the speed of sound depends on the frequency, is observed in certain structures including plates, rods and coatings, wherein at least one of the dimensions perpendicular to the propagation direction is comparable to λ, and also in certain materials including plastics, elastomers, fiber-reinforced composites, porous elastic solids and relaxing fluids. Dispersion due to *geometry* is familiar in the propagation of Lamb waves in sheets. Dispersion intrinsically due to the *material itself* is a familiar characteristic of "Silly Putty," in which the elasticity of the material depends on the loading rate, or test frequency. Therefore, sound velocity depends on frequency.

Dispersive materials are sometimes tested with a short broadband pulse whose peak is taken to propagate at the group velocity. However, for pulses as short as ~1 to 3 cycles, the received signal can be quite smeared by dispersion and distorted by frequency-dependent attenuation. Here, unless measurements are transformed into the frequency domain, it is hard to justify the use of short pulses, because c and U generally are not definable for such broadband pulses propagating in dispersive media. For narrower band tone bursts containing, say, ~10 to 30 cycles, and enveloped by a rectangle, truncated Gaussian bell curve or a $\cos^2$ function, a center frequency can be defined. Then c and U can be associated with that frequency, and one can measure the phase of the received signal— or at least the "apparent" phase. Unfortunately, even for narrowband tone bursts, if the pathlength and center frequency are fixed, measurements in a dispersive specimen may contain an ambiguity in the number of wavelengths in the path ("λ ambiguity"), preventing a reliable determination of c or U.

On the other hand, c and U may be determined accurately, using either variable path or variable frequency methods, to avoid or resolve the λ ambiguity problem. It can be shown that by increasing the frequency (fitting more wavelengths into the specimen) the slope of the curve of phase (in units of wavelengths N) versus frequency is the group delay t_g. That is, $t_g = dN/df$. Between f_1 and f_2, if the phase-versus-frequency function turns out to be a straight line passing through the origin this means that phase velocity equals group velocity in that f_1 to f_2 region. Complications arise, however, in small specimens due to echoes, multipaths and mode conversions. (Lynnworth et al., 1973b.)

Even if the curve of phase versus frequency is not a straight line, c can still be determined in cases where the number of wavelengths in the specimen (of length L) is known unambiguously. For example, the measurement can be conducted down to a sufficiently low frequency, or there may be known limits on c, such that at some frequency, N is known. Thus, from $\lambda = L/N$ one obtains the phase velocity as $c = f\lambda$.

Instead of continuously varying the frequency to generate a curve of N versus f, one can continuously vary the path to generate N versus x. The slope is $dN/dx = 1/\lambda$. This means $\Delta x = \lambda$ for $\Delta N = 1$. Once λ is known, c is obtained as λf. Repeating this procedure yields $c(f)$ and dc/df. This concept is illustrated by Section 3.6.

Once c and dc/df are known, the group velocity U can also be obtained from, or cross checked by,

$$U = c/[1 - (f/c)\,(dc/df)].\qquad(3\text{-}3)$$

The measurement of c and U does not require that one use *plane* waves. In their studies of thin films (<1 μm), Crean et al., 1987, found a correlation between microstructure and *surface* wave dispersion.

Sachse and Kim, 1987, utilized a *spreading transient* propagating between a well-characterized point source and point receiver to obtain phase and group velocities and the attenuation coefficient versus frequency for both P and S waves, including direction-dependence, in composites. They credit the fields of seismology and acoustic emission (AE) for providing analogies and helpful background technology. Earlier use of point contacts may also be found in elastic moduli resonance measurements (e.g., Spinner, 1956).

Particle trajectories are summarized in Table 3-2 for the principal ultrasonic waves used in research and industry.

Phase velocities of some of the principal waves may be summarized as follows (Tu et al., 1955; Kolsky, 1963 and 1964; Bau, 1986; Kim and Bau, 1986; others):

Longitudinal:

$$c_L = (E/\rho)^{1/2}/[(1 - \sigma)/(1 + \sigma)(1 - 2\sigma)]^{1/2} \quad \text{for} \quad \lambda/a > 2.5;\qquad(3\text{-}4)$$

$$c_{LP} = (E/\rho)^{1/2}/(1 - \sigma^2)^{1/2} \quad \text{for} \quad \lambda \geqslant \text{plate thickness.}\qquad(3\text{-}5)$$

Table 3-2.   Particle Trajectories for the Principal Waves Used in Ultrasonic Research, Measurement, Devices, Nondestructive Testing and Process Control

Wave	Particle Trajectory
Longitudinal and extensional	To and fro in direction of propagation; transverse motion in solid rods depends on Poisson's ratio σ and diameter/wavelength ratio $2a/\lambda$.
Transverse shear	Orthogonal to direction of propagation, and in simplest case, polarized linearly.
Torsion	Circular path in plane $\perp$ axis if waveguide is round; for *non*circular waveguide, warping complicates trajectory (and also reduces the wave speed).
Rayleigh	Ellipse in plane $\perp$ surface and $\parallel$ to the direction of propagation, provided Rayleigh wave is sinusoidal (Kolsky, 1963, p. 85).
Lamb, symmetrical, lowest-order mode	Edge view of plate resembles a straight snake swallowing a series of equally spaced eggs.
Lamb, asymmetrical, lowest-order mode	Plate flexes like a flag in a steady wind, with periodic waviness.
Love	$\parallel$ surface, $\perp$ direction of propagation, in layered material in which speed in the coating is less than in substrate ("slow coat").

Extensional:

$$c_e = (E/\rho)^{1/2} [1 - \sigma^2\pi^2 (a/\lambda)^2] \tag{3-6}$$

for radius a small but not negligibly small compared to wavelength λ. This approximation reduces to

$$c_e = (E/\rho)^{1/2} \quad \text{as} \quad a/\lambda \to 0. \tag{3-7}$$

Transverse shear:

$$c_t = (G/\rho)^{1/2}; \tag{3-8}$$

Torsion, lowest-order mode, in round waveguide:

$$c_{t_\bullet} = (G/\rho)^{1/2}; \tag{3-9}$$

in elliptical waveguide of aspect ratio $= 3$:

$$c_{t_\bullet} = 0.6 \, (G/\rho)^{1/2}; \tag{3-10}$$

in square waveguide:

$$c_{t\blacksquare} = 0.9184 \ (G/\rho)^{1/2}; \tag{3-11}$$

in rectangular waveguide of aspect ratio $= 3$:

$$c_{t\blacksquare} = 0.56 \ (G/\rho)^{1/2}; \tag{3-12}$$

in diamond waveguide of aspect ratio $= 3$:

$$c_{t\blacklozenge} = 0.57 \ (G/\rho)^{1/2}. \tag{3-13}$$

Rayleigh: c_R is determined from the root between 0 and 1 of the Rayleigh equation (Viktorov, 1967)

$$\eta^6 - 8\eta^4 + 8(3 - 2\xi^2)\eta^2 - 16(1 - \xi^2) = 0, \tag{3-14}$$

where $\eta = c_R/c_t$ and $\xi = c_t/c_L = [(1 - 2\sigma)/2(1 - \sigma)]^{1/2}$. Thus c_R/c_t is a function only of Poisson's ratio σ, a graph of which appears in Knopoff, 1952, and Brekhovsikh, 1980. A less formidable equation yielding c_R is the well-known Bergmann approximation, from which the Rayleigh velocity in an isotropic solid is given by

$$c_R = (G/\rho)^{1/2} \ (0.87 + 1.12\sigma)/(1 + \sigma). \tag{3-15}$$

A linear approximation that matches Knopoff's calculations at $\sigma = 0$ and $\frac{1}{2}$ is

$$c_R = c_t \ (0.0813\sigma + 0.8744), \tag{3-16}$$

while the best-fit straight-line approximation calculated on the basis of Knopoff's c_R/c_t ratios at $\sigma = 0$, $\frac{1}{4}$ and $\frac{1}{2}$ is

$$c_R = c_t \ (0.1618\sigma + 0.8759), \tag{3-17a}$$

or

$$\sigma = 6.1545 \ (c_R/c_t) - 5.3898. \tag{3-17b}$$

These approximations are compared in Table 3-3 with Knopoff's calculations at $\sigma = 0$, $\frac{1}{4}$ and $\frac{1}{2}$ [and essentially repeated in Brekhovskikh, 1980 (except for an erroneous value for Poisson's ratio of $\frac{1}{3}$, which should read $\frac{1}{4}$)]. The principal velocity ratios are plotted as a function of Poisson's ratio in McMaster, 1959.

As a numerical example, c_R in steel is about 3000 m/s. On the free surface of an elastic half-space, c_R is not dispersive if the elastic constants do not vary with depth. The Rayleigh wave is thus a special case of a plate wave, special in the sense of being nondispersive. The Rayleigh velocity c_R is the phase and group velocity asymptote for zero-order symmetrical and antisymmetrical Lamb waves (Viktorov, 1967, pp. 73, 74)

Table 3-3. Comparison of Velocity Ratios Computed from Rayleigh Equation by Knopoff, 1952, and from Bergmann and Linear Approximations

				c_R/c_t			
Poisson's Ratio σ	Knopoff Calculation	Bergmann Approx.	% Difference	Linear Approx. Matched at End Points	% Difference	Linear Approx.[a] Best Fit for $\sigma =$ 0, $\frac{1}{4}$ and $\frac{1}{2}$	% Difference
0	.8744	.8700	− .50	.8744	0	.8759	+ .17
$\frac{1}{4}$	.9194	.9200	+ .07	.8946	− 2.7	.9164	− .33
$\frac{1}{2}$	.9553	.9533	− .21	.9553	0	.9568	+ .16
RMS error, %	—	—	.55	—	2.7	—	.40

[a] $c_R/c_t = 0.1618\sigma + 0.8759$; $\sigma = 6.1545(c_R/c_t) - 5.3898$

and, according to experimental observations of Tu et al., 1955, c_R *seems to be the limiting (minimum) velocity for compressional waves in a rod of radius* $a \approx \lambda$. To remember this rod result, it may be helpful to visualize the Rayleigh particle orbits as ellipses that just fit neatly in the rod when their major axis $= \lambda_R = a$. (Individual particle displacements, however, are much less than λ; otherwise the small-amplitude, linear elastic model would be violated.) A number of dimensionless ratios are plotted as a function of the Rayleigh to longitudinal velocity ratio in Lees, 1975.

The summary of phase velocities continues below.

Flexural waves in a round bar, calculated for $\sigma = 0.29$ (Kolsky, 1963, p. 71):

$$
\begin{array}{ccccccc}
a/\lambda & 0 & .2 & .4 & .6 & .8 & \geq 1 \\
c_f/(E/\rho)^{1/2} & 0 & .40 & .51 & .56 & .57 & .58
\end{array}
$$

Flexural waves in a thin plate. The zero-order antisymmetric plate or Lamb wave a_o exhibits a phase velocity proportional to the square root of frequency (Viktorov, 1967):

$$c_F = [E/12 \rho (1 - \sigma^2)]^{1/4} (2\pi f d)^{1/2} \tag{3-18}$$

In steel, for $df > 10$ mm·MHz, for the a_o wave, c_F essentially equals c_R.

The group velocity of extensional waves derived from Eq. (3-6) and accordingly valid only for small a/λ is given by

$$c_{eg} = (E/\rho)^{1/2} [1 - 3\sigma^2\pi^2 (a/\lambda)^2] \tag{3-19}$$

It is seen that $c_{eg} \to c_e$ as $a/\lambda \to 0$. This is equivalent to saying the extensional wave is nondispersive in the long-wavelength limit: $dc_e/df \to 0$ as $a/\lambda \to 0$.

Notation. In Eqs. (3-4) to (3-19), the following symbols are used:

E, G	Young's, shear moduli of elasticity
σ	Poisson's ratio ($\sigma = E/2G - 1$)
a	Bar radius
λ	Wavelength
c	Wave velocity
d	Plate thickness
f	Frequency

3.1.2   Attenuation Coefficient α

The second question listed above, on attenuation, is generally more difficult to answer, at least in the sense that while c can be measured with an absolute accuracy approaching 10^{-6} it is difficult to measure the attenuation coefficient α to much better than 10^{-2} or 10^{-3}. This is largely due to the relative difficulty in measuring amplitudes or Q compared to time intervals or resonant frequencies. The difficulty is compounded by α being a function of frequency, f, whereas c often is not a function of f. Furthermore, beam spread (diffraction) usually influences the measurement of α much more than the measurement of c. In addition, for lossless media, adjacent impedances can influence the amplitudes of echoes while producing no effects on phase shift other than 0 or π.

As background for the following discussion, let us define $\alpha = (1/x) \ln (A_o/A)$ where x = path increment, A_o = amplitude of a plane pressure wave at an initial point, and A = amplitude at a second point where the plane wave has propagated the additional distance x.

Also, R = reflected pressure/incident pressure. (Differential-path methods of measurement will be explained in Section 3.6.)

As an example of a fixed-path α measurement, consider the pulse-echo arrangement of Fig. 3-1 showing transducer, buffer and specimen, and echoes A, B and C, and normally incident sound pressure amplitude A_i.

Pulse-echo determinations of the attenuation coefficient α generally require knowledge of, or compensation for, the sound pressure reflection coefficient R at the surface where the sound enters the specimen. Both R and α can be determined by measuring the amplitudes of echoes A and B (originating at the front and rear surfaces of a specimen of thickness or length L) and a multiple reverberation echo C. For this case the equations for R and α, derived by Papadakis, 1968, are

$$R = -\left[A_o C_o/(A_o C_o - B_o^2)\right]^{1/2}; \qquad (3\text{-}20)$$

$$\alpha = \left[\ln R\, B_o/C_o\right]/2L, \qquad (3\text{-}21)$$

where subscript zero means echo amplitudes A, B and C have been corrected for beam spread and transducer and amplifier nonlinearities. (The *plane wave* model does not include beam spread.)

There are cases, however, where R and α are preferably determined from A, B and an echo other than the specimen reverberation echo C. R

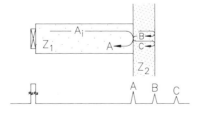

FIG. 3-1. Notation for α and R determinations for pulse-echo buffer methods. Z_1 and Z_2 are characteristic wave impedances in the first and second medium, respectively.

can sometimes be determined theoretically from impedance and geometry considerations or determined experimentally from the ratio of buffer rod end echo amplitudes with and without the specimen coupled. Given A_o, B_o and R, one finds $\alpha = (1/2L)\ln[(A_o/B_o)(R^2 - 1)/R]$ Np/cm. (Lynnworth, 1973a). For the special cases $R = \pm\ 0.618$, this equation simplifies to $\alpha = (1/2L) \ln |A_o/B_o|$ Np/cm. Note that when $\alpha = 0$, if $|R| = 0.618$ then $|A_o| = |B_o|$ and $C_o = R$. R is positive when $Z_2 > Z_1$, as is seen from the usual expression for R at normal incidence: $R = $ reflected/incident pressure $= (Z_2 - Z_1)/(Z_2 + Z_1)$ where the Z's are characteristic impedances for the longitudinal or shear wave. Sometimes a notch, shoulder, or mass added temporarily to a waveguide provides a reference echo for computing R. See Fig. 3-2.

Since α is usually a sensitive function of frequency, and R is sometimes a function of frequency too, the above determination should use echo amplitudes measured at a particular frequency, e.g., the center frequency of a tone burst. If broadband pulses are used, then A, B and C should be spectrum-analyzed, with calculations again based on diffraction-corrected amplitudes at each frequency of interest. In other words, α and R may be determined from echo amplitudes measured in the time or frequency domain. See Papadakis, 1976; Tang et al., 1988.

Damping.   Attenuative damping materials are normally and often necessarily installed *immediately behind* the transducer element [as in most NDT transducer assemblies, e.g., Fig. 3-3(i)]. But other locations are possible. In slender waveguide experimental systems, a mildly-damping material such as adhesive tape is often attached over a 10 to 20 cm length of lead-in between transducer and sensor (or sometimes after the sensor) to reduce ringing [Fig. 3-3(ii, iii)]. In the extensional/torsional mode converting waveguide system of Fig. 3-3(iv), by locating bonded damping materials *forward* of the magnetostrictive transducer's attachment point,

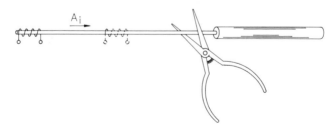

FIG. 3-2.   Pliers apply massive termination temporarily, providing a reference echo for computing R at a joint. Alternatives include sampling the incident wave amplitude A_i as the reference for computing R.

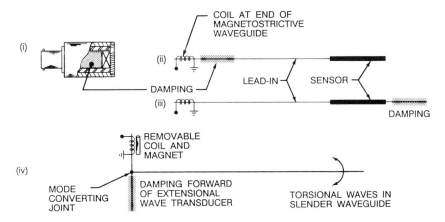

FIG. 3-3. Damping material located (i) behind a piezoelectric NDT transducer element, and (ii-iv) elsewhere in magnetostrictive waveguide systems.

one retains the option of removing the coil, in addition to achieving some degree of separation of the dual problems of shaping the wave and damping it.

3.1.3 Acoustic Impedance

Acoustic impedance, usually explained by analogy to electrical impedance, may be taken for longitudinal waves in lossless media to be the ratio of sound pressure P to particle velocity u. (See, however, Greenspan, 1972.) Acoustic impedance largely determines the reflection and transmission of energy at interfaces between adjacent media. It therefore influences resonator bandwidth and ringing. Another reason for examining impedance is that in a medium of characteristic acoustic impedance $Z = \rho c$, the intensity of a plane wave is $I = P^2/2\rho c$.

In homogeneous bulk media the sound pressure reflection and transmission coefficients at normal incidence are simple functions of the characteristic acoustic impedances, or ρc products, on each side of the interface:

$$R = (Z_2 - Z_1)/(Z_2 + Z_1) = (r - 1)/(r + 1), \qquad (3\text{-}22)$$

where $r = Z_2/Z_1$, and

$$T = 2Z_2/(Z_2 + Z_1) = 2r/(r + 1). \qquad (3\text{-}23)$$

If $r < 1$, then $R < 0$ (phase reversal). If $r = 1$, then $R = 0$ and $T = 1$. For large r, $T \rightarrow 2$.

For guided extensional waves the reflection and transmission coefficients are calculated from mechanical impedance terms that take the cross sectional area A in each waveguide into account:

$$Z_e = \rho c_e A, \tag{3-24}$$

while for torsional waves the polar moment of inertia about the axis, J, becomes important:

$$Z_t = \rho c_t J, \tag{3-25}$$

where $J_\bullet = \pi a^4/2$ for round wires of radius a, and $J_\blacksquare = (1/12)(b^3 d + bd^3)$ for rectangular cross sections of aspect ratio b/d.

Media of dissimilar characteristic impedances Z_O and Z_L can be matched at one frequency f_1 and its odd harmonics $3f$, $5f$, etc., by introducing a quarterwave impedance matcher of intermediate impedance $Z_m = (Z_O Z_L)^{1/2}$, where Z_L = load impedance.

Broader bandwidth is achieved by using two or more layers of intermediate impedances. Current practice seldom uses more than two layers. According to the optimization criteria of Desilets et al., 1978, the two intermediate quarterwave layers should have characteristic impedances so that the Z steps vary monotonically from Z_O to Z_L:

$$Z_{ma} = Z_L^{3/7} Z_o^{4/7}, \tag{3-26}$$

$$Z_{mb} = Z_L^{6/7} Z_o^{1/7}. \tag{3-27}$$

For narrowband matching, the neglected nonmonotonic multilayer proposal of Fry and Dunn, 1962, may be of interest. Silk, 1984, draws attention to the "brass-faced" nonmonotonic nonresonant (and therefore potentially broadband) method due to Aranovich and Prudov, 1970, in which the matching thicknesses may be as small as $\lambda/20$ to $\lambda/30$.

Another broadband impedance matching alternative is to use nonresonant thicknesses large compared to the pulse length. The "matching" benefit, however, is less than that obtained if the same material were used in a quarterwave resonant length (Lynnworth, 1965). To cite a calculation from that work, if water is used to match from fused silica to air, a transmission gain of ~5 dB is predicted for a nonresonant layer, compared to ~20 dB at quarterwave resonance.

In Eq. (3-25) c_t includes a shape factor K (Spinner and Valore, 1968), that is, $c_t = K(G/\rho)^{1/2}$, where $K < 1$ unless the cross section is circular. In 1977, the author noticed that for rectangular cross sections of constant area A, for aspect ratios increasing from 1 to 10, the fractional increase in J is almost exactly the same as the decrease in K.

This means that at a joint between circular and rectangular cross section materials of about equal torsional wave characteristic impedances, by matching areas, one essentially matches torsional impedances. Further, if both materials have about the same σ, then, by matching areas, one will essentially match impedances for both torsional and extensional modes. As a corollary, a desired degree of impedance mismatch is achievable by means of controlled area mismatching. As another corollary, the value of $K_{\blacksquare}$ can be estimated as the reciprocal of $J_{\blacksquare}$ /J$_\bullet$ for waveguides of constant A: $K_{\blacksquare} = J_\bullet$ /$J_{\blacksquare}$ or $J_{\blacksquare}$ $K_{\blacksquare} = J_\bullet$, where the solid subscripts indicate the cross sectional shape.

3.1.4   *Impedance Matching*

Impedance matching is familiar in optics (coated lenses), electrical engineering, musical wind instruments and speakers, and in ultrasonic horn welders, transducers and waveguide joints. In ultrasonic devices the motivation to match impedances sometimes stems from the need to maximize energy transfer into a medium. Often, however, the main requirement is to eliminate echoes, dampen the ringing of a device whose resonances are troublesome, or broaden the bandwidth of a transducer.

With respect to damping the ringing of a transducer disk, Silk, 1984, proposes as a figure of merit for the trapping of energy in the disk,

$$W = (Z_D - Z_A)(Z_D - Z_B)/(Z_D + Z_A)(Z_D + Z_B) \qquad (3\text{-}28)$$

where Z_D is the acoustic impedance of the disk and Z_A and Z_B are the acoustic impedances of the surrounding materials. W represents the square root of the trapped energy after one oscillation and, in a simple case, would represent the amplitude ratio of cycle n to cycle $n - 1$. A high value of W (>0.75) represents a situation in which the disk rings for a long time, a medium value ($W\sim0.3$) would be acceptable for much ultrasonic work, according to Silk, while values of $W < 0.1$ imply very short (broadband) probe response.

The containment of Z_A and Z_B needs to be considered in practical devices. Z_B is normally a backing sealed within a transducer housing. Z_A would appear to be more complicated, because this member, in the simplest model, is exposed to the load that could be a corrosive fluid, a gas containing solvents, etc.

One solution to this problem, used in making quarterwave impedance-matched transducers for ultrasonic flowmeters for natural gas and flare gas, is to sacrifice some of the gain achieved by quarterwave matching, in exchange for an impervious metal seal (Lynnworth et al., 1981, 1984). Examples of transducers constructed in this manner are given in Fig. 3-

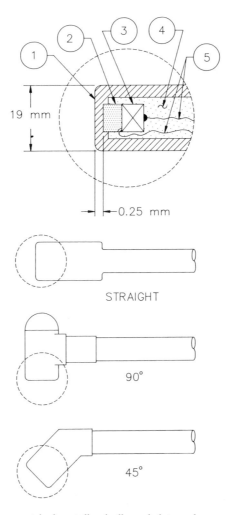

FIG. 3-4. Impedance-matched metallurgically-sealed transducers. 1–Thin window. 2–Quarter wave impedance matcher. 3–Piezoelement. 4–Potting/Backing material. 5–Lead wires.

4. The "window" thickness is typically 50 μm for SS and 250 μm for Ti. At 100 kHz these windows are very thin compared to wavelength, so the loss due to them is small.

The first pair of Ti-housed piezoelectric transducers of this type to be installed in a flare line as part of an operational flowmeter were removed and examined after six months of exposure to flare gases. (Flare gas ordinarily would be expected to foul flow sensors.) A thin deposition was observed. As this did not appear to degrade performance, the transducers were reinstalled and resumed continuous operation without maintenance (Smalling et al., 1984). Five years after the initial installation, operation was continuing, with no maintenance and apparently no degradation. In this type of industrial service, despite the hostile conditions of the flare gas, the Ti seal apparently maintains the integrity of the quarterwave impedance-matched interior. Wilson et al., 1979, used a similar metal encapsulation method to house polymer piezoelectric transducers, but without the impedance matching layer (see Section 9.7). In the presence of background acoustic noise, a *narrowband* matched device may have advantages with respect to filtering as well as directivity. On the other hand, enhancement of signal/noise by using short-duration phase-coded or broadband chirp waveforms requires *broadband* transducers.

It is noted in passing that when the dry flat face of a 100 kHz transducer of the types in Fig. 3-4 is covered with a thin taut plastic film, ~ 10 μm, entraining an air film, a 10-dB gain is typically observed, without sacrifice of bandwidth.

Impedance adjustment within the piezoelement by means of composite mixing, use of coral-like structures, various connectivity patterns, or other material modifications, are discussed in Silk, 1984. Carnevale et al., 1967c, described how co-author Carey added μballoons to epoxy to create a low-Z matching material on the front of a transducer. Carey then faced off the layer until 1-MHz pulse shapes and amplitudes transmitted across an air gap were optimized. See Section 5.1.1.4.

In 1983 work at Stanford, a quarterwave rubber layer overcame some of the transducer/air mismatch (Fox et al., 1983). Five years later Khuri-Yakub et al., 1988, presented a two-layer matching approach for air transducers wherein the best second layer is to be found when the first matching layer is not optimal. As a numerical example, if the first layer (GE RTV615 + 50-μm air bubbles) exhibits a $Z_1 = 0.3$ MRayls the desired $Z_2 = 5.8$ MRayls. A 1-MHz air transducer built according to this design yielded a round-trip insertion loss of 45 dB.

If water is the medium into which broadband radiation is desired, the preferred impedance for a *single* quarterwave plate, according to the optimization criteria selected by Thorne, 1987, is $Z_m = (Z_L^2 Z_c)^{1/3}$, where Z_L = impedance of the water load and Z_c = impedance of the piezoelectric ceramic. Thorne used acrylic, the impedance of which was 78% of the optimum Z_m for his case. For a *two-layer* match the impedances should be tailored as prescribed by Eqs. (3-26) and (3-27), according to the optimizing criteria of Desilets et al., 1978, or Silk, 1984.

Deka, 1987, used multiple layers of readily available materials to match 450-kHz transducers into air for non-contact NDT of steel, and presence, position and level applications. Deka's matched noncontact NDT solution may be contrasted with Van Valkenburg's earlier (1973) order-of-magnitude-lower-frequency noncontact NDT of lower-impedance media like plywood and rubber tires, accomplished without matching the transducers to air.

Geometry. Just as experimental horns match musical instruments or speakers to open air, so too can conical or stepped solid transformers accomplish matching of guided elastic waves. As an example, in Chapter 8 the tapered glass rod (due to Fowler and Bradshaw, and reported in Lynnworth et al., 1977) is drawn down to a diameter thinner than a hair, to approximately match the extensional wave impedance of SiC modulus test specimens of even smaller diameter, under 25 μm.

Wave Type. While it may be convenient to associate ρc with a material, it is important to remember that, when dealing with wave impedances, the wave type is important. Analysis of spherical waves, for instance (Kinsler and Frey, 1962) shows that at distances far from a point source ($kr \gg 1$), the complex ratio of pressure to particle velocity approaches ρc, the value for plane waves. Nearer to the source, however, reactive as well as resistive components are present: $Z = [(kr) (\rho c)/(1 + k^2 r^2)] [kr + j]$, where $k = 2\pi/\lambda$, r = distance from source and $j = (-1)^{1/2}$.

3.1.5 Reflection and Transmission Coefficients at Normal and Oblique Incidence

At *normal* incidence the R and T coefficients applicable to adjacent homogeneous isotropic elastic media can be calculated from their impedances. At *oblique* incidence, sound speeds come into play too, and the partition of energy among the several modes becomes more complicated. At the interface between two immiscible liquids, R and T need only be calculated for longitudinal waves, as a function of the angle of incidence θ_1 (Officer, 1958; Krautkrämer and Krautkrämer, 1983):

$$R = \frac{\cos \theta_1 - (\rho_1 c_1/\rho_2 c_2) [1 - (c_2/c_1)^2 \sin^2 \theta_1]^{1/2}}{\cos \theta_1 + (\rho_1 c_1/\rho_2 c_2) [1 - (c_2/c_1)^2 \sin^2 \theta_1]^{1/2}}, \tag{3-29}$$

and

$$T = \frac{2 \cos \theta_1}{\cos \theta_1 + (\rho_1 c_1/\rho_2 c_2) [1 - (c_2/c_1)^2 \sin^2 \theta_1]^{1/2}}. \tag{3-30}$$

One of the simplest oblique cases would be that represented by two immiscible inviscid liquids of equal sound speeds but differing in ρ (e.g., oil on water at an appropriate temperature). Only longitudinal waves are supported. There is no ray bending and no mode conversion. The expressions for R and T, derived from the more general liquid/liquid equations (3-29) and (3-30) are

$$R = \frac{(1 - \rho_1/\rho_2)}{(1 + \rho_1/\rho_2)} \tag{3-31}$$

and

$$T = \frac{2}{(1 + \rho_1/\rho_2)} \tag{3-32}$$

These R and T coefficients, being independent of the angle of incidence because of the equality of sound speeds, could just as well have been calculated by the earlier equations (3-22) and (3-23) for R and T at normal incidence.

Another special case occurs when both liquids have equal impedances but unequal sound speeds. From Eq. (3-29), or from the sound power reflection and transmission coefficient α_r and α_t as given in Kinsler and Frey, 1962, or Blitz, 1967, it is readily shown that for *Z-matched liquids* R, T, α_r and α_t depend on c_2/c_1 and the angle of incidence:

$$R = \frac{\cos \theta_1 - [1 - (c_2/c_1)^2 \sin^2 \theta_1]^{1/2}}{\cos \theta_1 + [1 - (c_2/c_1)^2 \sin^2 \theta_1]^{1/2}}, \tag{3-33}$$

$$T = \frac{2 \cos \theta_1}{\cos \theta_1 + [1 - (c_2/c_1)^2 \sin^2 \theta_1]^{1/2}}, \tag{3-34}$$

$$\alpha_r = \left[\frac{\cos \theta_1 - \cos \theta_2}{\cos \theta_1 + \cos \theta_2} \right]^2, \tag{3-35}$$

and

$$\alpha_t = \frac{\cos \theta_1 \cos \theta_2}{(\cos \theta_1 + \cos \theta_2)^2}, \tag{3-36}$$

where the angle of transmission θ_2 is related to θ_1 through Snell's Law (see Section 3.1.7).

If $c_1 = 2 c_2$ and if $\theta_1 = 45°$, then $\theta_2 = 20.7°$, $R \approx 0.14$ and $\alpha_r \approx 0.02$. R and α_r are small but nonzero even though $Z_2 = Z_1$.

The last special case of Eqs. (3-29) and (3-30) to be considered might correspond to a "pipeline gasoline interface detector" configured as an oblique-incidence interrogation between clamped-on transducers, where the two gasolines in a steel pipe shall be presumed to be separated by a planar interface and have equal densities but unequal sound speeds. Accordingly,

$$R = \frac{\cos\theta_1 - (c_1/c_2)\,[1 - (c_2/c_1)^2\sin^2\theta_1]^{1/2}}{\cos\theta_1 + (c_1/c_2)\,[1 - (c_2/c_1)^2\sin^2\theta_1]^{1/2}}\,, \qquad (3\text{-}37)$$

and

$$T = \frac{2\cos\theta_1}{\cos\theta_1 + (c_1/c_2)\,[1 - (c_2/c_1)^2\sin^2\theta_1]^{1/2}}\,, \qquad (3\text{-}38)$$

which are close to zero and one, respectively, if c_1 and c_2 differ only by a few percent.

Within a given isotropic elastic solid, vacuum bounded, mode conversion from longitudinal to shear occurs for all angles of incidence between 0 and 90°, but possibilities for complete conversion occur only for $\sigma \leq 0.26$ and incident angles $\theta_1 > 38°$.

For obliquely incident shear, mode conversion does not occur for incident angles $> 45°$. Again, complete conversion occurs only for $\sigma \leq 0.26$ and for $45° > \theta > 26°$.

These relations are most easily seen in graphs due to Arenberg, 1948, reproduced in Section 3.7. The 30° beveled buffer rods in Fig. 3-5, made of a material having $\sigma = \frac{1}{4}$, illustrate the only case where mode conversion is 100% complete *and* where the incident and reflected mode converted waves are perpendicular.

For *solid/liquid* planar interfaces between two semi-infinite lossless media, the equations for longitudinal and shear reflection and transmission coefficients may be found in Officer, 1958, Krautkrämer and Krautkrämer, 1983, and other texts. Illustrative graphs appear there and in McMaster, 1959, Kinsler and Frey, 1962, in several papers by Mayer and

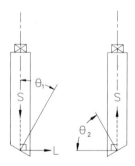

FIG. 3-5. Beveled buffer rods, $\sigma = \frac{1}{4}$, $\theta_1 = 30°$, $\theta_2 = 60°$, axes parallel. S to L and L to S mode conversions are 100% complete and orthogonal. After Lynnworth, 1969b.

co-workers, 1965 to 1979, in Lynnworth and Chen, 1975, and in references cited therein. (For *viscous* liquids, the *complex* reflection coefficient can be interpreted in terms of viscosity—see Chapter 2. McSkimin, 1959, includes sample calculations of the ratio of input to output amplitudes, for a silicone oil/fused silica interface, for the special case where the shear wave in the solid propagates at 45° to the normal. Studies of lossy systems include those of de Billy and Quentin, 1984, and McSkimin, 1960 and 1964.

Transmission at the interface between a homogeneous isotropic elastic half-space and a homogeneous, transversely isotropic half-space is analyzed by Roberts, 1988.

At an appropriate angle beyond the second critical angle, Rayleigh waves can be launched in the solid (strictly, *leaky* Rayleigh waves), but only if the shear wave velocity exceeds the longitudinal velocity in the liquid, *and furthermore*, if the liquid to solid density ratio, ρ_L/ρ_S, does not exceed limits that depend on the ratios of the liquid velocity to the longitudinal and shear velocities in the solid (Brower et al., 1979). These authors cite ice/water as one of several combinations for which leaky Rayleigh waves can *not* exist, because the liquid density is too high compared to that for ice.

For homogeneous *solid/solid* combinations, R and T depend on the boundary conditions (slip-free or liquid layer, for example). The interested reader is referred to Krautkrämer and Krautkrämer, 1983, Brekhovskikh, 1980, Kühn and Lutsch, 1961, or to the works of Mayer and co-workers, 1965–1979. For *inhomogeneous* media, see Ewing et al., 1957, and Brekhovskikh, 1980.

For discussions or applications of transmission (leakage) from solids into water, the following may be of interest: shear (Van Valkenburg, 1962; Lynnworth et al., 1982; Lynnworth, 1979); Lamb (Deighton et al., 1981; Gillespie et al., 1982); Rayleigh or other leaky surface waves (Van Valkenburg, 1957; Deighton et al., 1981; Gillespie et al., 1982; Lynnworth and Nguyen, 1985; Matson et al., 1987).

The leakage of energy from a homogeneous elastic solid into an adjacent lossless liquid has been calculated for a number of specific cases, with results depending on the wave types, angles and materials involved, and, for Lamb waves, on the mode (a_0, a_1, . . .), frequency-thickness product and number of wetted surfaces (one or two). Table 3-4 lists calculated numerical results for vertically polarized (SV) shear waves incident at 45° (Lynnworth, 1979) and for Rayleigh waves (Deighton et al., 1981), where the adjacent liquid is water. One application of the leakage calculation of Deighton et al., 1981, to the problem of liquid level (Gillespie et al., 1982) includes a configuration which is of interest partly because of its use of

Table 3-4.   Reflection Loss in dB Per Bounce
For 45° SV Waves and Attenuation
in dB Per Wavelength for Rayleigh
Waves Compared For Aluminum and
Steel Immersed in Water

Wave	Aluminum	Steel
SV	2.3	0.8
Rayleigh	1.5	0.5
Ratio	1.5	1.6

immersed reflectors, and partly because of its similarity to a subsequent
analogous measurement using fiberoptics (Morris et al., 1987).

These ultrasonic and fiberoptic analogies, both directed toward liquid
level measurement where the leakage of energy needs to be redirected
upwards, are compared in Fig. 3-6.

3.1.6   Beam Spread, Diffraction Loss, Seki Parameter

Diffraction or beam spread losses are usually estimated on the basis of
the analysis of a rigid piston cw radiator. In the near field, where the Seki
parameter $S_k = \lambda z/a^2 < 4$, the diffraction loss L_d, in dB, is roughly equal

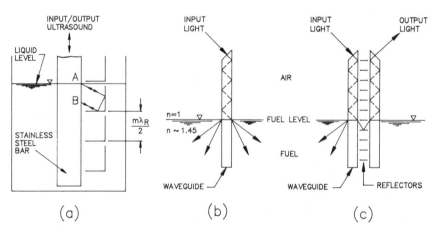

FIG. 3-6.   Comparison of ultrasonic and fiberoptic approaches to liquid level, using leaky
waves and reflectors in the liquid to direct the leaked energy back into a receiving waveguide.
(a) Arrangement of corner reflecting structure, after Gillespie et al., 1982. (b) Light coupling
out of slab waveguide at liquid surface. (c) Coupling between waveguides with reflectors
present, after Morris et al., 1987.

to S_k. Here, z = axial distance from transducer and a = transducer radius. Beyond S_k = 4, in the far field, the sound intensity is inversely proportional to the square of the distance. Thus, L_d increases 6 dB per double distance (Fig. 3-7). Diffraction losses occur in transducer windows, wedges, pipe walls and buffer rods, not just in the fluid.

Under cw excitation, a circular disc of radius a acting as a piston generates a major lobe with a null at θ = $\sin^{-1} 0.61\lambda/a$ (i.e., width of major lobe = 2θ = $1.22\lambda/a$) beyond which there are ~$2a/\lambda$ side lobes. The 3-dB beam width is given by a total included angle of approximately $\lambda/2a$.

Kinsler and Frey, 1962, p. 173, give a few numerical examples of "beam widths" corresponding to where the far field intensity falls to some fraction of its axial value I_o, for λ = $a/4$. Their examples are:

I/I_o	-3 dB	-6 dB	-10 dB	First null
Beam width (2θ)	7.4°	10.1°	12.9°	17.3°

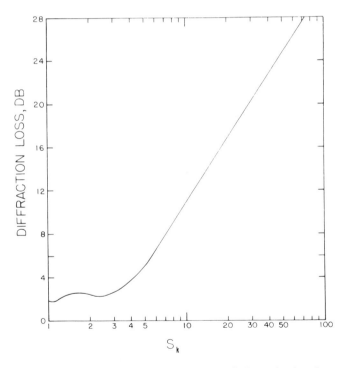

FIG. 3-7.   Diffraction loss in dB versus Seki parameter S_k for a circular piston source of radius a radiating at the wavelength λ in an isotropic medium to a circular receiver, distance z away, also of radius a and coaxial with the source. Inverse square relation in far field (S_k > 4) means loss increases by 6 dB per double distance. Adapted from Papadakis, 1966.

Corresponding to the first of these examples, the approximation that the 3–dB beamwidth $\approx \lambda/2a = (a/4)/2a$ yields $2\theta = 7.2°$, fairly close to the exact calculation, $7.4°$.

Rschevkin, 1963, McMaster, 1959, and other authors discuss the cw acoustic field on the axis of a circular piston in varying degrees of detail.

Papadakis and Fowler, 1971, point out that one of the consequences of *broadband* [as opposed to narrowband (cw)] operation, is that nulls are smeared.

In the next section, "Refraction," a diagram including the energy reflection coefficient α_r indicates that $\alpha_r = 1$ at the first and second critical angles. But, this diagram is based on calculations that neglect beam spread. Thus, in practice, where beam spread occurs, *total* reflection is not observed even though the central ray itself may be 100% reflected.

In that same diagram, as well as in the mode conversion diagrams due to Arenberg, 1948, no mode conversion is predicted at normal incidence. Again, in practice, beam spread leads to a small but nonzero conversion. See, for example, Guyott and Cawley, 1988.

Another example showing how the unjustified disregard for beam spread leads to false conclusions is pointed out by Censor and Newhouse, 1986. In their example, a transducer whose axis is perpendicular to the axis of flow in a pipe yields useful pulsed Doppler returns, even though the central ray would not. The explanation is simple. Because of beam spread (or focusing—see Fig. 2-8) some of the interrogating energy interacts obliquely with the flow. The flow velocity is derived from the limits of the relevant Doppler spectrum.

The *finite width* and *finite duration* of practical sound beams should also be borne in mind when applying to practice the results calculated for cw wavefront models of infinite extent. For the analogous electromagnetic or optical case, Stratton, 1941, p. 499, points out that, according to Picht, 1929, a *finite*-width light beam cannot be totally reflected, even when the indices and incident angle would lead one to expect total internal reflection.

3.1.7 Refraction

Refraction, or ray bending, occurs when a sound ray crosses interfaces between media of different sound speed. (In optics the analogy would be a light ray encountering an index change, e.g., air/water, air/glass.) If the index, or sound speed, varies continuously, then the ray path is curved so as to go from A to B in the shortest possible time. With respect to curved acoustic paths, see Officer, 1958, Brekhovskikh, 1980, Camp,

1970, and articles on underwater acoustics (ray tracing) appearing, for example, in *The Journal of the Acoustical Society of America.*

As a curved-path example, if the sound velocity profile in a pond linearly decreases with depth d from the surface, $c = c_o - gd$, then the sound path of a ray initially horizontal is a circular arc of radius c_o/g.

If the sound speeds are constant in two adjacent media, but the interface is curved, focusing or defocusing occurs (lens action).

Snell's Law,

$$c_1/\sin \theta_1 = c_2/\sin \theta_2, \tag{3-39}$$

can usually be used to calculate the refracted angle. However, if one of the media is in rapid motion (liquid at high velocity in a pipe), then the refracted angles deviate from the value predicted in Eq. (3-39). Another set of limits, perhaps not immediately obvious from Eq. (3-39), has to do with mode conversion and total reflection at or beyond critical angles. This may be illustrated with a simple numerical example corresponding to a planar interface between water and steel (Fig. 3-8). If a longitudinal wave in the water is incident at the first critical angle near $\theta_i = 14°$, Snell's Law predicts that the refracted longitudinal wave will be transmitted in the steel at $\theta_L = 90°$ and the refracted shear at $\theta_S = 33°$, approximately. But, in fact, no energy is transmitted into the steel at the critical angle; the incident energy is totally reflected, at least to the extent that the plane-wave (no beam spread) model adequately depicts the true situation.

Focusing.   Papadakis, 1972, points out the advantages with respect to consistency in physics and optics, if one defines the acoustic index n so that in solids likely to be used as lenses with water, $n > 1$. This is obtained if $n = c_{solid}/c_{liquid}$, analogous to the optical index $n = c_{free space}/c_{dielectric medium}$. Using Papadakis' notation, the optics lens equation can be used in acoustics, e.g., $1/F = (n - 1) [(1/R_1 + 1/R_2)]$, relating focal length F and lens radii R_1 and R_2. For a plano-curved lens the radius of curvature $R_c = F (n - 1)$.

In acoustic microscopy a common method of focusing uses a buffer with a curved end radiating into water (Lemons and Quate, 1973; Chou et al., 1988; Fig. 2-71). Other methods of focusing include cylindrical-shell and spherical-shell transducers radiating into solid or liquid media, phased arrays, and Fresnel lenses and Fresnel focusing using interdigital electrodes unequally spaced (Nomura et al., 1983; Nomura and Yasuda, 1985). Atalar et al., 1987, used a round acrylic rod as a solid cylindrical mirror to focus shear waves at oblique incidence, resulting in focused SAW in aluminum [Fig. 3-9c (iv)]. Shattuck and Nouhi, 1988, describe a focused

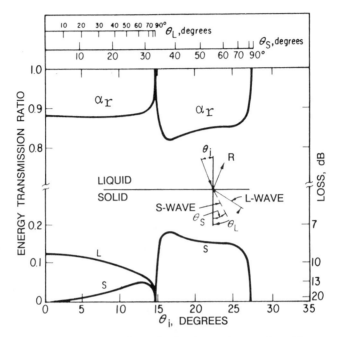

FIG. 3-8. Energy transmission ratios and refraction angles are plotted here as a function of incident angle θ_i for both longitudinal waves L and shear waves S at a water/steel interface. The S-wave has about twice the energy transmission ratio of the L-wave mode (reflection losses lower by 3 dB), at angles that are often used in clamp-on flowmeters. α_r is the energy reflection coefficient. These calculated results neglect beam spread and finite beam width.

optoacoustic transducer that provides an ultrasonic pattern spatially similar to the optical pattern used to illuminate it. In water, their 2.8-MHz focusing prototype yielded a 3–dB beamwidth of 3.5 mm measured at a range of 92 mm.

Many ranging and noncontact presence sensing applications in air, especially at distances measured in tens of meters, do not require focusing (Garwood, 1988). But, to accurately measure the perpendicular distance to a surface, say ±0.1 mm at a range of 250 mm, Hickling and Marin, 1987, found it necessary to focus. For a spherically focused transducer, the spot size diameter at the focus is $d_s = 1.4\lambda F/d$ where F = focal length and d = transducer diameter. Because of practical limits on λ due to attenuation in air above ~1 MHz and on F/d due to reasons analogous to those which limit the minimum optical f-number (ratio of focal distance to lens diameter) the smallest practical spot size in air at atmospheric pressure is about 0.8 mm at 1 MHz (Hickling and Marin, 1986). In the

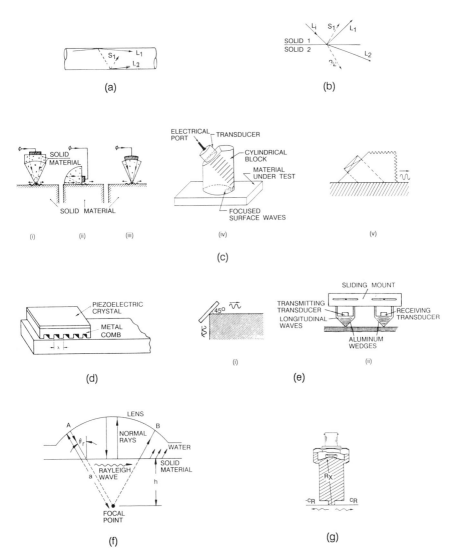

FIG. 3-9. Examples of transducers, waveguides, wedges, knife-edged chisels and corner interactions for generating by mode conversion, the following waves: longitudinal, shear, Lamb, Rayleigh, Love. (a) Reflection within a buffer rod. (b) Reflection and refraction at solid/solid interface. (c) Oblique incidence, wedge transducers: Line focus (after Latuszek, 1986, © 1986 IEEE); (i) cylindrical lens; (ii) cylindrical mirror; (iii) cylindrical transducer; (iv) point focus using shear crystal and cylindrical wedge, after Atalar and Kömen, 1987, © 1987 IEEE; (v) conventional wedge. (d) Comb radiator (Viktorov, 1967, © 1967 Plenum). (e) Corner converter (Viktorov, 1967); knife-edged chisel for longitudinal/Rayleigh conversion (Elkind et al., 1986). (f) Focused longitudinal (after Clarke et al., 1985, © 1985 Butterworths) and (g) Shear waves focused at a "point" contact of diameter $\approx \lambda$.

laboratory, at still shorter ranges, e.g. 0 to 0.5 mm, distance resolution of 0.2 µm in air was reported by Fox et al., 1984, *without* focusing.

3.1.8   Mode Conversion

The commonly used techniques for mode converting may be divided according to the wave types, and nearly equivalently, according to how the medium is bounded. We shall first treat mode conversions between pairs of longitudinal, shear, Lamb, Rayleigh and Love waves. In order to propagate in an elastic solid, each requires that one or more transverse dimensions be unlimited. Next we treat mode conversions between pairs of extensional, torsional and flexural waves. Each of these can propagate in an elastic waveguide, all of whose transverse dimensions are small compared to λ.

3.1.8.1   Longitudinal, Shear, Lamb, Rayleigh, Love

Each of these waves can be generated directly without mode conversion or transmitted between coupled media without conversion. For example, longitudinal and shear are usually generated with appropriately cut crystals (e.g., X- or Y-cut quartz), and Lamb and Rayleigh, by interdigital electrodes on a piezoelectric substrate. Love waves, which are horizontally polarized (SH) shear waves in a plate, can be launched by a shear wave transducer coupled to one end of the plate (Brekhovskikh, 1980).

With respect to *mode conversions* between any pair, Table 3-5 summarizes the possibilities proposed or demonstrated to date. In this table, " = " means "same as in symmetrically–corresponding off–diagonal term," justified by reciprocity. The sign "—" means either "probably impractical" or "no existing mode conversion." The terms "refraction" and "wedge" are admittedly redundant. Comb methods are discussed in Viktorov, 1967.

Examples of transducers and elastic systems for effecting these tabulated conversions appear in Fig. 3-9. In some cases, the efficiency of coupling and mode converting is assessed by monitoring an echo having the same mode as the incident mode, e.g., the reflection coefficient R.

3.1.8.2   Extensional, Torsional, Flexural

Techniques for mode converting between extensional and torsional or flexural, or between torsional and flexural, are illustrated in Fig. 3-10. The contacts (forces) are primarily tangential or orthogonal. (See May, 1964; see also, flextensional transmitter of Inoue et al., 1987.)

To monitor the *efficiency* of mode conversion, transmission or reflec-

Table 3-5. Techniques for Mode Conversion

Incident Wave	Longitudinal	Shear	Mode Converted Wave		
			Lamb	Rayleigh	Love
Longitudinal	—	Reflection, refraction, wedge	Refr., wedge, comb, waveguide	Point, knife edge, corner, refr., wedge, comb, waveguide	—
Shear	=	—	Refr., wedge, comb, waveguide	Point, knife edge, corner, refr., wedge, comb	Refr., wedge (SH)
Lamb	=	=	—	Overlapped surfaces	—
Rayleigh	=	=	=	—	—
Love	=	=	=	=	—

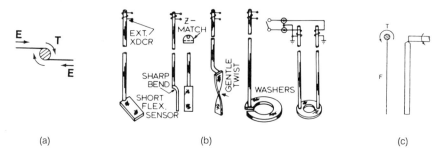

FIG. 3-10.   Mode conversion techniques for extensional, torsional and flexural waves. (a) Extensional/torsional. (b) Extensional/ flexural, after Lynnworth, 1978. © 1978 IEEE. (c) Torsional/ flexural.

tion coefficient measurements may be used (Fig. 3-11). The *purity* of conversion to torsion can be tested quickly in waveguides of noncircular cross section as follows. One immerses the torsional member in water, advancing axially deeper and deeper, while observing the waveform to see if the waveshape is preserved. If the lowest-order torsional mode is dominant, all cycles are delayed equally and no energy leaks into the water. If another mode is present, e.g., the flexural mode, it will be observed to "run through" the torsional packet as the noncircular section becomes fully immersed. (The effect resembles differential path measurements of a pulse in a dispersive medium.) The wetted waveform differs from the dry waveform in amplitude and shape. An example of a mode converting

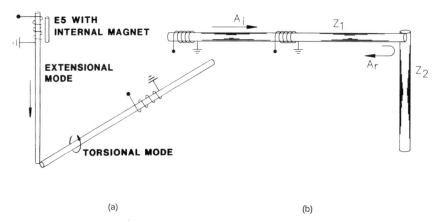

FIG. 3-11.   Monitoring mode conversion efficiency by (a) transmission and (b) reflection coefficient methods. After Lynnworth, 1987.

sensor in which the torsional mode appears to be pure is illustrated in Chapter 6, Section 6.2.2, Fig. 6-19 (inset).

3.1.9 High-Temperature Waveguide Techniques for Nondispersive Waves—Extensional and Torsional Modes, $f \approx$ 100 kHz

Since elastic solids transmit ultrasonic waves right up to their melting point, it is natural to consider their use as high-temperature sonic transmission members. To minimize heat transfer in the waveguide's axial direction, the use of waveguides of small diameter, ~1 mm, allows the use of materials at temperature extremes almost without regard for their thermal conductivity. In the *passive* mode, acoustic emission waveguide applications occur in connection with monitoring pressure vessels for leaks or crack propagation (Chapter 2, Section 2.6.5). Waveguides may be welded, brazed, screwed to or clamped onto the structure in question. At the end of the waveguide remote from the high temperature, one attaches a piezoelectric transducer, or, according to Lynnworth and Spencer, 1972, a magnetostrictive transducer. In some cases the entire waveguide is magnetostrictive, avoiding the transducer joint, or even contact (Kaule, 1964). Further, this allows the waveguide to be treated as an expendable material in cases where it may be necessary to feed it continuously through the excitation/detection coil into a remote hostile region.

In the *active* mode, usually pulse-echo, initial applications were mainly related to studying elastic moduli and metallurgical/acoustical correlations at phase changes (Bell, 1957). A few years later, *thermometry* applications for single zones emerged (Bell, 1963, Thorne, 1963, and Hub, 1963), and soon after, for multiple zones (Lynnworth and Patch, 1970). See Chapter 5. Bundles of thin rods have also been proposed as alternatives to bulk wave buffer rods (Section 3.1.10) in the hope of achieving efficient broadband transmission without mode conversion and without dispersion. Apart from high temperature applications, Bau, 1986, Jen, 1987, and their colleagues are among investigators who extended earlier analyses of bare and clad waveguide sensors, respectively. See, for example, Thurston, 1978, and May, 1964.

The "techniques" of generating, conveying and detecting guided extensional and torsional waves at a nominal frequency $f \approx$ 100 kHz may be divided into functional categories, the main ones being those in Table 3-6. Most of the "high-temperature" techniques of this section can be generalized to techniques for temperature extremes and remote sensing.

Table 3-6.   Guided Waveguide Techniques and Tips. (See Also Chapter 5, Section 5.2.1 and Chapter 8, Section 8.2.)

Transduction	If using magnetostrictive materials, choose 49 Fe/49 Co or similar alloy; Ni; ferritic stainless steel (type 430); or isopaustic alloy. Bias with magnet for extensional waves, or with current pulse through transducer or through auxiliary coil for torsional waves. Use ferrite washers to limit axial flux leakage. Use coil length $\approx \lambda/2$. Couple electromagnetically through thin-wall tube (SS, $\sim\frac{1}{4}$ to $\frac{1}{2}$ mm wall) if all-metal pressure boundary is required. To produce microstrains $>10^3$ consider rare earth magnetostrictive alloys.
Supports/feedthroughs	Threads; rubber stopper; diaphragm; cotton-like fibrous refractory strands; graphite cloth; teflon O-rings or swaged ferrules; very fine wire; low-density packing foam; magnet; low-reflectivity fibrous washers spaced inside a sheath. Use motion to avoid sticking or echoes at contacts. Use surface wave *inside* a tube to avoid reflections from external supports.
Damping	Soft rubber; W-loaded epoxy; steel wool compressed; masking tape; shrink tubing.
Reflectors	Diameter or material change (ΔZ); notch; hole $\sim$1 mm dia; kink; sharp bend; clamp or other added mass.
Materials for long waveguide, $\geq$3 m	Choose material compatible with thermal and chemical environment, e.g., SS (stainless steel); refractory metal or other alloys, preferably with m. pt. > twice the absolute service temperature, to avoid attenuation usually occurring above recrystallization temperatures; single crystals; sapphire fiber; strands for flexibility; electroplate or otherwise coat for corrosion resistance; shape and coat for required thermal response, e.g., thin black Ti ribbon for fast response. Use high Z to minimize echo amplitude from supports, e.g., choose dia $\geq$ 3 mm instead of $\leq$ 1 mm for lead-in.
Sensor	Material, cross section and size appropriate to application, e.g., 1 mm diameter for temperature sensor; 1 $\times$ 3 mm envelope for diamond cross section torsional density sensor; $\rightarrow$ 10 mm dia or envelope dimensions for rugged tubular sensor.
Joints	Mechanical, chemical or metallurgical bond, i.e., threaded or press fit; weld, braze, solder, epoxy; thermosetting cement. Honey or pressure for temporary connection at room temperature.

Summary of Interactions Between Guided Waves and Adjacent Liquid

Examples may be cited where the presence (hydrodynamic mass, inertia or viscosity) and motion of the liquid adjacent to a waveguide influences the propagation of ultrasound within that waveguide:

- Damping of extensional waves by viscous liquid (Roth and Rich, 1953)

- Reduction in flexural mode phase velocity in Al strip immersed in water (Ageeva, 1960).

- Damping of waves in waveguide with disks, by adjacent liquid (Mongan, 1961).

- Reduction in extensional phase velocity in threaded rod immersed in water (Lynnworth, 1975; Hyseresis, Section 3.1.13).

- Attenuation of Lamb modes due to leakage into water (Deighton et al., 1981; Gillespie et al., 1982).

- Attenuation of Rayleigh and shear waves due to leakage into water (Van Valkenburg, 1957, 1962).

- Reduction in torsional mode phase velocity in waveguide of non-circular cross section immersed in a liquid (Lynnworth, 1977; Bau, 1986; Section 6.2.2).

- Damping and reduction in phase velocity of torsional waves in threaded metal tube immersed in viscous liquid (Kim and Bau, priv. comm., 1987; 1989; Section 9.1).

- Echo generated at liquid/vapor interface, for torsional waves in noncircular waveguide (Smith and Junger, 1961; Lynnworth, 1977; Sections 6.2.2 and 7.5.2.2).

- Echo generated at liquid/vapor interface, for flexural waves in round tube (Dieulesaint et al., 1987; Section 7.5.2.3).

- Increase or decrease in Rayleigh velocity depending on direction of water flow adjacent to surface (Solie and Tonning, unpubl., 1972, reported in Lynnworth, 1979a, p. 435, Fig. 17).

- Rayeigh and plate wave sensors (*IEEE Trans. UFFC*, March, 1987).

- Increase or decrease in guided-wave velocity depending on direction of water flowing in a metal tube (Jacobson and Lynnworth, unpubl., 1988).

3.1.10   High-Temperature Buffer Rods for Longitudinal and Shear Waves

Ultrasonic flowmetering, thermometry, liquid level measurements for molten metals, nondestructive reflectometry, elastic moduli determinations and other applications frequently require bulk waves (longitudinal, shear) to be conveyed to or from a medium or structure under environmental conditions that are so hostile chemically or thermally that it would be impractical to subject a transducer to these conditions for the time required to accomplish the measurement. Sometimes momentary contact reduces the time constraint, but, even here, contact durations for red hot steel may be as long as 10 seconds. Variations of the steel poker used to tend a fire while protecting the operator's hands will be recognized in modern (1947–1987) buffer rods for longitudinal and shear (bulk) waves.

Frederick, 1947, 1948, seems to have been the first to use a bulk wave solid buffer with a discontinuity near the hot end to measure by ultrasonics the properties of a solid specimen at high temperature, ~1000°C. McSkimin, 1959, used a threaded buffer of fused silica to about 500°C. Lehtinen, 1960, measured the sound speed in molten Fe at 2 MHz by using a water cooled SS buffer rod, ~35 mm dia × ~500 mm long, as a step towards cast iron quality control. (See Section 3.6.3.1.) In this sense Lehtinen demonstrated a precursor to measurements a quarter-century later in molten Al made by Mansfield, 1984. Mo buffer rods were used by Macedo and Litovitz, 1965, to measure viscous relaxation in molten boron trioxide and zinc chloride to ~1400°C. Fused silica buffers have been used to measure sound speed and attenuation in gases and plasmas, where momentary contact permitted fused silica to be used to ~10 times its melting point (Carnevale and coworkers, 1960–1967). Fused silica rods were typically smooth cylinders 12.7 mm OD × 200 mm long. [At the opposite temperature extreme, following suggestions of Kytomaa (1987, priv. comm.), fused silica buffer rods of 2.54 mm dia × ~200 mm long were used in flowmeter experiments at MIT where the liquid was nitrogen at −200°C (Lynnworth, 1988b).] Fused silica buffer rods, dimensioned nearly the same as in this cryogenic flowmeter work, were used by Carnevale et al., 1964, to measure elastic properties in solids to 1000°C.

If the buffer rod conveys *longitudinal* waves, sidewall reflections cause mode conversion to shear, resulting in energy lost from the initial pulse and generation of many spurious delayed echoes that become extremely troublesome in pulse-echo measurements. If the buffer is metallic, threading is the most commonly-used technique for reducing contributions from the sidewalls. Sather, 1968, found a second use for threading. He pressure-coupled elastic moduli specimens to the buffer using a threaded coupler. Details on threaded buffers used by Youngdahl and Ellingson in the past decade or so are given in Section 7.3.2.4. Knurling and grooving are alternatives to threading. The use of noncircular cross sections such as triangular and hexagonal prisms has not been successful in suppressing sidewall echoes and mode conversions. In principle, for long monochromatic tone bursts, a rectangular cross section can delay the diagonal shear so that, after reconversion to longitudinal, its phase at the rod end lags that of the original (unconverted) acoustic burst by an odd number of half periods. This suppresses some of the mode converted noise. But the tone burst's leading and trailing edges remain uncancelled.

Longitudinal wave buffers of diameter $\gg \lambda$ were proposed in the late 1960s as a possible remedy for sidewall problems. In 1988 the opposite was proposed; namely, the use of acoustically *slender* buffer, for 100 kHz waves intended for surface and internal temperature measurements in red hot steel billets. (See also, Gelles, 1966.) Another proposal in the 1960s was to use *shear* waves instead of longitudinal, because shear waves do not mode convert near grazing incidence, in fact, not beyond 45°. The higher attenuation coefficient at a given wavelength, however, along with perceived or actually greater difficulty with coupling, have propably discouraged most investigators from using shear to test polycrystalline materials, especially if the information that is sought can be obtained by using longitudinal waves.

Apart from buffer sidewall echoes, buffer noise due to granularity in polycrystalline buffer material, and due to mode conversion to shear at off-normal incidence at the buffer's hot end (an inevitable consequence of beam spread) limit the resolution of defects, compared to predictions for an ideal buffer. For some temperature or thickness measurements in hot steel, where the required signal to buffer noise ratio need not be as high as for flaw detection, buffer techniques, albeit imperfect, suffice. See, for example, the high temperature buffer work of Youngdahl and Ellingson, 1986, Krupski, 1979, or Jeskey, (1987, priv. comm.) reviewed in Section 7.3.2.4.

Pitch-and-catch constructions have also been investigated, but here the limit with respect to flaw detection is sometimes imposed by unwanted leakage across the specimen's surface between the transmitter and receiver buffer. For some applications other than flaw detection, e.g., near-surface temperature measurement, the "leakage" between a pair of parallel-axis buffers might indeed provide the signal whose travel time yields the temperature. (In the high-temperature rolling transducer of Section 3.3.2, a pitch-catch arrangement is used to detect internal defects in hot steel slabs.)

The possibility of a fiberacoustic bundle, suggested by Gelles, 1966, some day may provide the best high temperature solution for a contacting buffer rod probe, because it could avoid dispersion, sidewall conversions and grain noise. Current interest concentrates elsewhere, however, namely, on noncontact emat and laser methods, e.g., Boyd and Sperline, 1988; Alers et al., 1988; Burns et al., 1988; Wadley et al., 1986; Boyd et al., 1989. The merits of noncontact testing, however, were recognized and demonstrated with magnetostrictive transduction by Kaule as early as 1964. In some potential noncontact applications of lasers the following problems need to be solved: attenuation by fog or mist due to vaporization of spray on continuous caster; safety related to high-energy laser; cost (Droney, 1988, priv. comm.).

Lastly, regarding gas temperature measurements, we mention longitudinal wave buffers such as air-filled steel tubes used by Traina, 1985, and Nuspl et al., 1985, 1986. These are traceable in principle to He-filled SS tubing used earlier by Apfel, 1962, to ~1000°C.

In the through-transmission applications where the fused silica buffers were used, their smooth mode converting exterior posed no particular acoustical problem. However, when the radiating end of a buffer is exposed to high pressure, as when pressure coupling is the method of choice for coupling into a pipe at high temperature, the smooth wall may be abandoned in favor of a wall at least threaded in part, to provide a grip, even if not for suppression of mode converted echoes. Oblique-incidence buffer rod designs used in a clamp-on flowmeter at 260°C are shown in Fig. 3-12. The buffer's wall is threaded near the hot end only. This allows the buffer material to be selected for acoustical transmission and thermal insulation properties, while allowing a strong stainless steel or other pipe jacket of dissimilar expansion coefficient to be used as a protective shield and heat transfer device. (Lynnworth, 1988b,c; Matson et al., 1987).

For cryogenic applications where the fluid is liquid oxygen or liquid hydrogen, the number of allowable gasket or sealant materials is quite limited. Examples of all-metal swaged ferrule or ring-sealing methods for removable buffer rods (e.g., low-carbon 300-series stainless steel) are shown in Fig. 3-13. An example of an extensional buffer similar to one currently under investigation for cryogenic applications (NASA SBIR feasibility study) is shown in Chapter 6, Fig. 6–17.

Apart from cryogenic or high-temperature motivations, another use of buffer rods is to generate *circularly polarized transverse waves*. In this technique, due to Einspruch, 1964, the buffer material must be anisotropic (cubic crystal) and of length $l = (1/4f) [c_1 c_2/(c_2 - c_1)]$ in order to convert a linearly polarized transverse wave to a circularly polarized one (Fig. 3-14). Here, c_1 and c_2 are the velocities of linear polarized transverse waves with displacements parallel to [110] and [1$\bar{1}$0].

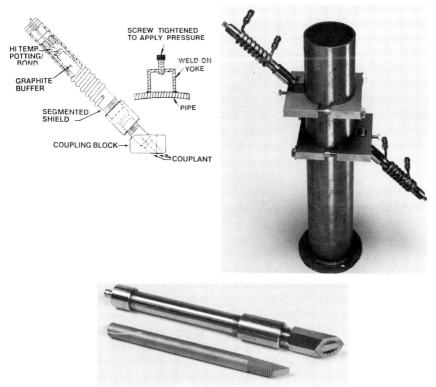

FIG. 3-12. Oblique-incidence buffer rod designs for high-temperature clamp-on applications.

3.1.11 Nonreciprocity

According to Kinsler and Frey, 1962, Helmholz' acoustical reciprocity theorem may be rephrased into the following electroacoustical reciprocity theorem:

> If a simple source of strength Q_1 at point A produces a sound pressure P_2 at point B, then a simple source of strength Q_2 at point B will produce a sound pressure P_1 at point A such that $Q_1/P_2 = Q_2/P_1$.

Kinsler and Frey point out that, in this form, the theorem is analogous to a commonly used reciprocity theorem for linear passive electrical networks.

The term "simple source" refers to a source whose dimensions are small compared to the wavelength, which means that at distances far from

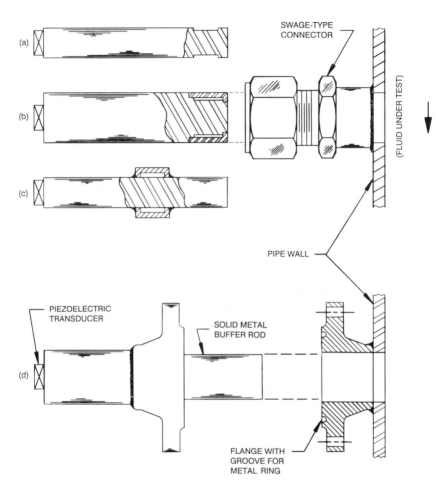

FIG. 3-13.   Methods of securing removable wetted metallic buffer rods using swaged metal ferrules or metal rings, thereby avoiding the use of *non*metallic gaskets or other *non*metallic sealants. (a) Circumferential groove captures swaged ferrule. (b,c) Tube welded to solid buffer provides a region at tip or elsewhere which can be properly deformed by swaged ferrules. (d) Ring-type joint obtained with flange grooved for metal ring.

the source compared to the source dimensions, the pressure is independent of the shape of the source.

Censor and Newhouse, 1986, refer to a different formulation of the reciprocity principle, which, according to Morse and Ingard, 1968, states that the spatial properties (radiation pattern) of a transducer are the same whether it acts as a transmitter or as a receiver. The reciprocity theorem underlies several procedures for calibrating transducers (Sachse and Hsu,

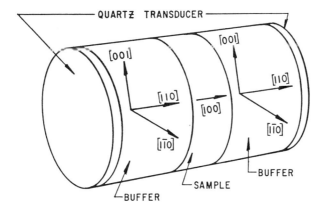

FIG. 3-14. Buffer rod arrangement for producing circularly polarized elastic waves. The length of one buffer is adjusted such that the two shear waves arrive at the buffer/sample interface out of phase by $\pi/2$. The second buffer, which is optional, may be the same as the first. After Einspruch, 1964. © 1964 AIP.

1979) and it is also a useful basis for explaining, to first order, the operation of certain ultrasonic instruments, for example, a contrapropagation transit time flowmeter. One could even say that it is so useful a theorem that one may suffer for a long time trying to account for errors in a particular measurement before realizing that the theorem may not be applicable in a given set of circumstances. The rest of this section is devoted to examining apparent violations of the reciprocity theorem of concern in contrapropagation flowmetering, attributable to electrical or electroacoustical nonlinearities, to flow, or to flow patterns which have a different effect on the received pressure according to whether said patterns are near the transmitter or receiver.

In the contrapropagation flowmeter the flow velocity V of a fluid moving between the transducers is determined from upstream and downstream transit times. If the sound speed c is a known constant, then V can be determined from the difference in upstream and downstream transit times, $\Delta t = t_2 - t_1$. In practice, accurate measurements of V to say $\pm 1\%$ often require Δt to be measured in liquids to about ± 1 ns. Thus our present concern with nonreciprocity lies not so much in amplitude differences between upstream and downstream pressures P_2 and P_1 but in differences in their phase or transit times. It will also be understood that while an ideal contrapropagation measurement would actually interchange the positions of the transducers, it is not practical to do so, particularly not at a high rate of change of the direction of interrogation. The circuits for transmitting and for receiving likewise may not be identical for both directions, e.g., mismatched diodes. The two transducers, even if identical, behave in a nonlinear, one might say mischievous, manner, if driven hard enough or if examined closely enough. From these considerations it seems plausible that even at zero flow, any of the nonlinear or nonidentical conditions represented in Fig. 3-15 could lead to nonzero Δt by more ns than would be considered negligible.

Even if upstream and downstream transducers and electronics were identical and linear, flow disturbances near transducer A or B can have

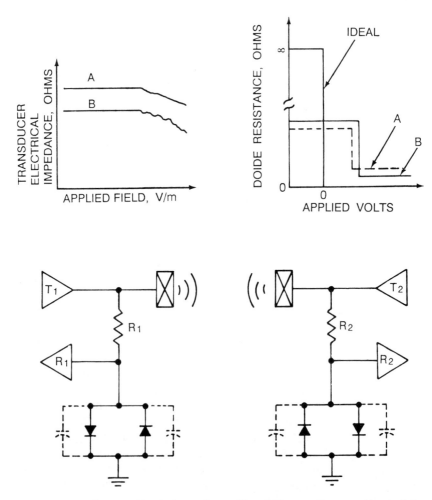

FIG. 3-15. Nonlinear nonidentical transducers (A and B) or nonlinear nonidentical electronic components associated with nonreciprocal behavior in a contrapropagation flowmeter.

differing effects on the observed transit time, according to the direction
of sound propagation. This may be explained in a qualitative manner by
considering that a rotational effect that interacts with a sound beam near
the launch site will affect the beam over the rest of its trajectory, whereas
a similar interaction near the receive site affects the beam only for a short
distance. An analogy might be of a child who suffers a misfortune early
in life versus a much older person who suffers the same misfortune but
much later in life and so is affected for a much shorter time and, perhaps,
is not affected in exactly the same way. The above explanation or analogy
may help account for the difference in waveshape, amplitude and, most
important, erroneous transit time as measured to a particular zero crossing, associated with unequal hydrodynamic effects in the sound path near
the inlet and outlet regions of an axial-path offset-style flowcell as represented in Fig. 3-16.

It turns out that even in a flowcell configuration as in Fig. 3-17, where
the disturbances to the flow are minimal, as V/c increases from 0 to 1,
deviations from reciprocity increase, not just with respect to transit time,
but also with respect to pressure (Ingard and Singhal, 1973). It seems that
the greater the flow, the less the reciprocity.

3.1.12   Synthetic Apertures and Pseudo-Arrays

In radar, a synthetic aperture antenna refers to the effect of generating a
long antenna by signal processing means rather than by the use of an
actual long physical antenna. The long radar antenna is synthesized by
moving a small antenna relative to the target, with either the antenna or
the target, or both, moving. The received signal is processed coherently
over an integration time. The synthesized antenna length is given by the
trajectory of the small antenna relative to the target during the coherent
integration time.

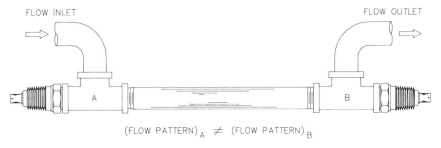

FIG. 3-16.   Asymmetrical hydrodynamic interaction with contrapropagating sound beams
is associated with nonreciprocal transmissions and erroneous transit time measurements.

FIG. 3-17. Reciprocity at zero flow but not at Mach 1.

In ultrasonic devices, the initial electroacoustic source of ultrasound is often small. But, if the sound is conveyed by a leaky waveguide or structure, then the source may be effectively extended over a region much larger than the transducer. This provides a coherent source of ultrasound, as if the small "electroacoustic antenna," commonly called the transducer, were swept through the leaky element at the speed of sound.

Ultrasonic investigators have utilized, in effect, the synthetic-aperture idea to explain transmission through sheets immersed in water, in liquid flowmeters and liquid level sensors, in radiating transducers and in shear modulus measurements in solid cylindrical buffer rods at high temperature. Examples of such work include Sanders, 1939, Hughes et al., 1949, Swengel, 1955, Van Valkenburg, 1957, 1962, Dunegan, 1964, Sunthankar, 1977, Karplus and Raptis, 1978, Brazhnikov, 1978; Raptis and Sheen, 1982, Gillespie et al., 1982, Baumoel, 1984, Lynnworth, 1979, 1982, 1988(unpubl.), 1988a, Lynnworth and Nguyen, 1985, and Motegi et al., 1987. Illustrations selected from some of these investigators' works appear in Fig. 3-18. Christensen, 1988, refers to the use of synthetic apertures in holography. Burch and Burton, 1984, by using synthetic aperture focusing techniques (SAFT), were able to achieve the theoretical lateral resolution of half the transducer diameter, $D/2$. This may be compared to the corresponding value for the 6-dB beamwidth of an unfocused pulse-echo transducer, $\lambda d/D$, where d is the range to the target and λ is the wavelength. Thus, SAFT yields improvement if $d > D^2/2\lambda$. *Real-time* synthetic aperture imaging is discussed by Ozaki et al., 1988.

Transducer arrays typically consist of either a multiplicity of physically distinct transducer elements (Chilowsky and Langevin, 1923) or a single piezo element with an electrode pattern distributed so that the piezo element can be excited nonuniformly or sampled at discrete points over its surface. Arrays are particularly useful for generating a prescribed field pattern, including the steering of that pattern through a region of interest, or along a particular direction such as 45° to the surface of the radiator, as in the ultrasonic heat meter described by Mágori, 1985. (Chapter 4, Section 4.6.3).

Johnston and Miller, 1986, mimicked a two-dimensional array of piezoelectric receivers by positioning a point-like receiving element sequen-

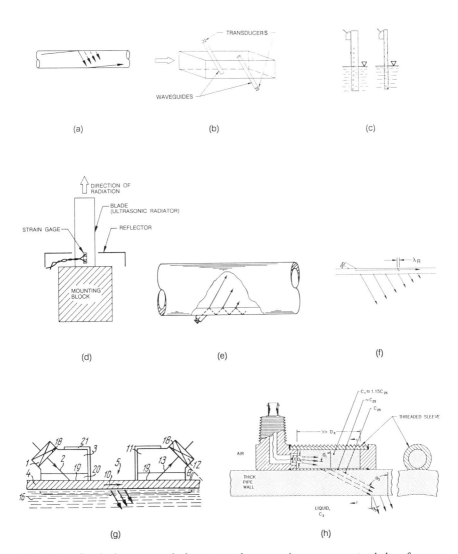

FIG. 3-18. Synthetic aperture, leakage or mode conversion over an extended surface as utilized by (a) Hughes et al., 1949, (b) Swengel, 1955, (c) Van Valkenburg, 1957, 1962, (d) Sunthankar, 1973, (e) Karplus and Raptis, 1978, (f) Lynnworth and Nguyen, 1985; Lynnworth, 1988a, (g) Brazhnikov, 1978, and (h) Lynnworth, 1988, unpubl.

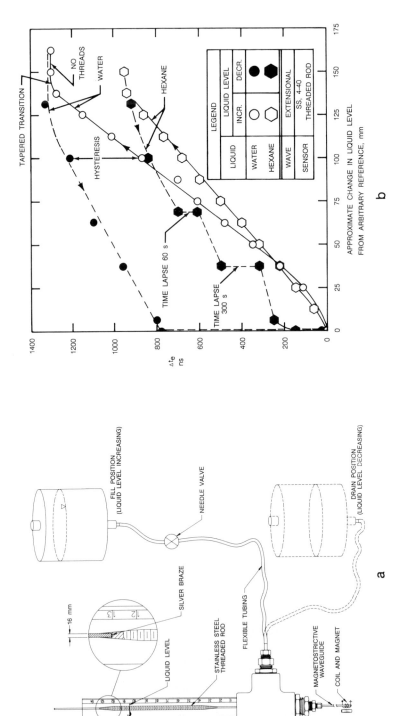

FIG. 3-19. (a) Hysteresis experiment. (b) Laboratory test results.

152

tially at a number of locations representing the elements of the array, and refer to an array mimicked in this way as a *pseudo-array*.

3.1.13 Hysteresis

Hysteresis refers to the difference in readings when the measurand (same input) is increasing versus decreasing, during full-range traverses. In some mechanical sensors hysteresis might be expected due to the difference between static and dynamic friction. On the other hand, in an ultrasonic sensor or measuring system having no apparent moving parts (except for the microstrains of the transducer or sensor elements connected thereto) one might not expect hysteresis.

As an example of an ultrasonic sensor in which hysteresis is present or absent, depending on the application, consider a liquid level sensor made of a vertical threaded rod, Fig. 3-19a, where the interrogating wave is the extensional mode. When tested in water at room temperature under conditions where the level h increased and then decreased, all in a matter of minutes, the initially-dry sensor appears to behave linearly when h increases (Fig. 3-19b), except for the end portions of the threaded sensor, which are tapered to match impedances. The increase in extensional wave transit time, Δt_e, is attributed to the hydrodynamic mass sensed by the threaded portion of the waveguide (Wambsganss, 1977). But when h decreases, the response differs, apparently due to unevaporated water sticking to the threads. (Similar hysteresis effects can occur at low g despite smooth sensors—see Jacobson et al., 1987a.) What if the liquid does *not* coat or otherwise remain attached to the threads? The data in Fig. 3-19b for hexane disclose less hysteresis, since hexane at room temperature is more volatile than water. In a well ventilated situation hysteresis would be still smaller, approaching zero for a highly volatile liquid.

3.2 Transduction

Transduction for our purposes is an electroacoustical (or electromechanical) energy conversion process to or from acoustical (or mechanical) energy. Usually some special property of a material is involved, e.g., anisotropy in the case of piezoelectricity, magnetism or permeability in the case of magnetostriction, electrical conductivity for emats, etc. A brief description, mainly intended to suggest the scope of transduction and references for further elaboration, is contained in Table 3-7.

Some types of transduction inherently include coupling to a material, i.e., emat, esat, laser. However, even if an ultrasonic wave has been

Table 3-7.   Principal Electroacoustic Transduction Means Used in Ultrasonic
             Measurements and Process Control and in Exploratory Investigations or
             Special Applications

Piezoelectric.   "Pressure electricity"—Curie brothers, 1880. Coupling between elastic and dielectric phenomena in materials having no center of symmetry. Occurs in natural crystals (quartz, tourmaline, etc.) and in manufactured ferroelectrics, polymers and copolymers.

Magnetostrictive.   Magnetically induced constriction or expansion in ferromagnetic media; Joule, 1847. Reverse effect attributed to Villari, 1865. Torsion mode due to Wiedemann, 1862.

Electrostatic (esat).   As transmitter, Coulomb force acts on alternating polarization. As receiver, change in gap dimension generates change in voltage (capacitor action).

Electromagnetic (emat).   As transmitter: Lorentz force acts on an eddy-current-carrying conductor in a magnetic field. As receiver: Velocity detector. Analogous to motor/generator action. (Frost, 1979; Hutchins et al., 1986; Hu et al., 1988; van den Burg et al., 1988).

Laser and Optical.   Thermoelastic, ablative or evaporative generation, interferometric detection. (Monchalin and Héon, 1986; Hutchins, 1988; Addison et al., 1987).

Other.   X-ray, thermal, spark, mechanical, explosive, magnetostatic, Wiegand/Barkhausen effect, MHD etc. (Sachse and Hsu, 1979; Swift, 1988).

---

generated in Material A, it still is important to consider ways of transmitting (coupling) efficiently the energy in that wave into a second Material B.

3.3   Coupling Methods for Ultrasonic Waves

A number of topics in ultrasonics are frequently introduced with the help of an analogy from electrical circuit theory (e.g., impedance) or optics (focusing, refraction or other wave phenomena). But the problem of defeating the air film when coupling ultrasound (Fig. 3-20) while perhaps analogous to the problem of eliminating films as barriers to heat transfer from one solid to another, does not seem to have a well-known counterpart in the electromagnetic domain. So let us create one. Imagine that we want to transmit light between two adjacent glass lenses of similar indexes whose curvatures mate exactly, but one of the lenses happens to be reflectively metalized on its curved face so as to appear opaque. What we need is some sort of safe solvent to eliminate the reflective barrier to light transmission without deteriorating the lenses.

   In the acoustic or ultrasonic longitudinal wave case, if a thin air film exists between two solids, the transmission of ultrasound is almost com-

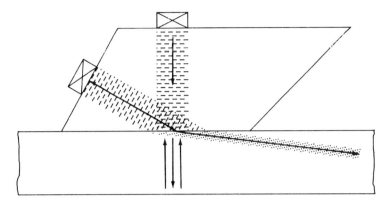

FIG. 3-20.   Ultrasonic inspection device includes beam for monitoring coupling between two solids. After Van Valkenburg, 1954.

pletely blocked by the film, due to impedance mismatch. [For the same reason, a thin solid sheet, even paper, almost completely blocks transmission across an air path. (That's why acoustic echo ranging cameras don't automatically focus properly through windows.)] The effectiveness of either blockage (air gap or sheet) is readily calculated from the "transmission line" equations for energy transmission. These calculations predict that, even for very thin interruptions, if the impedance ratio is large enough (e.g., $Z_{steel}/Z_{air} > 10^4$) the transmission coefficient is $\ll 1$. (See, however, Szilard's 1964 discussion of finite-amplitude transmission of ultrasound across ~0.1-μm air gaps in metals.)

In water or other common liquids the longitudinal wave characteristic impedance contrast between the liquid and the solid transducer or solid medium to be tested is not as large as in air, so it is no longer as important for the liquid gap to be eliminated. In fact, it is often convenient to couple ultrasound from transducer to part by using a water bath or liquid column. However, in order to introduce ultrasound as from a solid buffer into a liquid metal, *wetting* of the buffer is essential (Lehtinen, 1960; Smith et al., 1974; Mansfield, 1984). A gas film between the two would act as a major barrier, blocking transmission. (Even when nondestructively testing with water, as in an immersion tank, wetting agents are often added to improve the coupling.) In passing we note that if ultrasound were to be generated or detected in a liquid metal using emats, wetting would be of no concern, provided the effective gap were $\leq \frac{1}{4}$ mm.

Returning to the air gap obstacle to coupling, one remedy is to replace the air with a sonically conductive medium of impedance closer to that of the solids, e.g., water, oil, propylene glycol, gels, wallpaper paste with glycerine, silicone grease, epoxy, rubber, urethane, Teflon or other resilient sheet, molten plastic, glass, salt or flux, soft metal foil, solder, low melting point glass, glass frit, eutectic formed by contributions from appropriate platings on the adjacent surfaces (Krause, 1968); ceramic bond (Butler et al., 1979), diffusion bond, weld. High pressure, $\sim 10^3$ bar, and/or the displacement (Szilard, 1964) can bring adjacent solids of routinely machined surface finish (~ 3 μm rms roughness) into sufficiently intimate contact to overcome the gap impedance. Optically flat surfaces (gage blocks) can be coupled at least in the low-MHz range by wringing them together without any extraordinary coupling pressure. Optical contacts, if unspoiled by dust, are usable even into the GHz range (Smith, 1965; Carr, 1965).

At temperatures remote from ambient by more than about ± 200°C the choice of temporary or field-installable couplants is limited. Teflon tape has been used to couple clamp-on transducers to a steel pipe for over one year at 260°C (Matson et al., 1987). Gold foil has been used to over 500°C by Karplus for several years (priv. comm., 1987), and Al foil down to − 196°C (N_2 b. pt.) by Nguyen et al., 1982, both in clamp-on flowmeters. Krautkrämer and Krautkrämer, 1983, review couplants and coupling methods for high-temperature NDT applications, as does Andrews in Szilard's 1982 book. To couple *shear* waves at *normal* incidence, the couplant must have sufficient rigidity at the frequency in use. For *temporary* shear-wave coupling, honey, thermosetting cements, phenol salicylate, quick-setting epoxy, optically flat surfaces, or high pressure (>100 bar) may be used. Epoxies or solder are typically used for *permanent* bonds.

At *oblique* incidence the coupling methods for vertically polarized shear (SV) can be identical to those for longitudinal waves, i.e., water, oil, grease, cements, etc. Note that high viscosity or bonding is *not* required for *oblique* shear (SV). To couple horizontally polarized shear (SH), use the same methods as for shear at normal incidence (Lynnworth, 1967a,b).

Among the authors who have analyzed transmission across gaps filled with air or couplants, who have investigated transmission across optically flat or pressure-coupled surfaces, or who have proposed novel coupling methods or couplants, especially for high-temperature applications, the following abbreviated list may be of interest as a starting point for readers concerned about this important topic: Arenberg (liquid-coupled SV at 45°, 1948); Farrow (molten glass film, 1954); Erdman (water jet, 1956); Knopoff, 1957; McDade et al., diphenyldibenzyl silane to 287°C, 1959; LoPilato, in Hastings et al., 1961 (neoprene sheet, cut from a glove, used as dry couplant with porous composites); Sproule (dry-coupled spring-loaded rounded needlepoint transducer, 1961); Crecraft (pressure applied at room temperature to $>10^3$ bar, 1964); Carnevale et al. [pressure coupling at high temperature, to 1000°C, 1964 (abstr., 1963)]; Szilard, 1964; Smith (optical flats), 1965; with Gussenhoven, 1965; bonds made for GHz reflectivity experiments of Carr, 1965; with Holt and Gussenhoven, 1966; with Smith, 1968; Carey, in Carnevale et al., 1967c (used $\lambda/4$ layer of silica μballoon/epoxy composite to impedance-match into air and other gases); Sather (screw pressure for high-temperature coupling of L and S waves, 1968); Krause (Au-In bond, 1968); Fowler and Lynnworth (Al, Zn foil, for high-temperature high pressure couplant, 1970), Gavin and Anderson (Pt foil, 1974); Karplus and Tupper (Au and Pt foil, 1975); Araki and Matsunaga (Au foil between threaded buffers and a hot pipe, patent filed 1975; issued 1977). Smith et al., 1974, obtained rapid wetting of SS in liquid Na. They etched SS at 1000°C in vacuum; a layer of Au is deposited to passivate the face until the transducer is immersed, whereupon the Au dissolves, leaving a clean surface exposed to Na. Other sources are Butler et al. (400-day tests at 400°C, 1979); Forster and Karplus (Au foil, 10 MPa/1500 psi coupling pressure to 467°C, 1980; Karplus et al., 1981); Goldstein, Teflon tape sliding contact, 1982; Molina (gel, 1982); Szilard, 1982; Foster et al. (silver-platinum paste for fluorides to 577°C, 1983); Krautkrämer and Krautkrämer, 1983; Lynnworth and Nguyen (screw pressure for extensional/torsional waves, 1984); Kupperman and Lanham (dry ceramic powder couples S waves, 1985); Djordjevic and Traugott (liquid jet, 1985); Drescher-Krasicka et al. (Teflon tape pressure coupling from 4 to 725 K, or from the temperature of liquid He up to 452°C, 1985); Buynak and Crane (permeable membrane of $Z \approx Z_{water}$, which allows water column to leak slowly through it, 1987); Sheen et al. (high-temperature lubricant containing Ni particles, 1987); Matson et al. (Teflon tape used in clamp-on flowmeter at 260°C, 1987); Matthews et al., 1987 (fusing glass fiber waveguides after twisting them together, 1987); Chleck (double-sided adhesive tape, 1988, priv. comm.)

These coupling concepts are summarized in Table 3-8. This table retains electroacoustic noncontact coupling entries to provide certain perspectives and comparisons, even though the "coupling" mechanism in such cases is distinct from that used to overcome gap impedance mismatch.

3.3.1 Categories in Tabular Form

The following table categorizes couplants according to the coupling medium, followed by examples within each category.

Table 3-8. Couplant Categories, Couplants and Coupling Means Used in Ultrasonic
Measurements, Testing and Process Control

---

1. Solid Couplant
    Bond or joint
        Weld, braze, solder, high remelt eutectic, cement, epoxy, fused glass, crystal, wax,
        threaded joint
    No bond, no joint
        Powder, soft metal foil (Au, Pt, Al, etc.), Teflon tape, urethane and other plastic films,
        rubber, lubricant, grease, flux or buffer that melts on contacting hot object, threaded
        buffer, wedge, expendable plating (e.g., Au) preserves clean wettable radiating trans-
        ducer or housing surface until plating is dissolved by liquid metal
2. Liquid Couplant
    Low viscosity/easy flow
        Water, lubricant (oil, glycol), liquid jet
    High viscosity/restricted flow
        Pastes, gels, glycerine, liquid in permeable membrane
3. Gas, Air or Vacuum (Noncontact)
    Magnetic
        Magnetostrictive, magnetostatic, Wiegand
    Electromagnetic
        X-ray, laser, Lorentz force, optical, thermal
    Electrostatic
        Spark, capacitive, electrostrictive, fringe field
4. Contact Without Couplant
    High pressure, optical contact, particle displacement bridges the gap

---

3.3.2 Momentary Contact at High Temperature

Confronted at Avco in the late 1950s with the problem of measuring
plasma temperature on the order of 10^4K, Carnevale recognized the possibility of adapting the acoustic thermometry approach of Mayer, 1873,
by contacting the plasma only momentarily, but long enough to measure
sound speed c. Carnevale swept a pair of fused silica buffer rods through
the plasma and synchronized the ultrasonic ($\sim$1 MHz) pulse transmission
to obtain a measurement when the plasma essentially filled the 6.35 mm
gap between the buffers (Carnevale et al., 1961, 1962).

Shortly after co-founding Panametrics in 1960 (Parametrics, at that
time), Carnevale proposed to NASA to extend the momentary contact
method in gases by varying the gap, thereby obtaining simultaneously
both the sound speed c and the attenuation coefficient α. Thus the sum
of the transport properties η and K_{th} were measured in hot gases and
plasmas up to 10^4K and beyond, with the temperature being determined
from c (Carnevale et al., 1967b). Carnevale also proposed to ONR (Office
of Naval Research) the adapting of the momentary contact technique to

the problem of determining the elastic moduli in bulk solids at elevated temperature. This proposal included the idea of obtaining shear-wave velocity c_t by mode conversion, but the author's experiments at Panametrics in 1963 up to ~1000°C showed that it was easier to measure c_t without mode conversion, instead using pressure-coupling at normal incidence. This method coupled longitudinal waves up to ~1000°C too (Carnevale et al., 1964; abstract, 1963). See Fig. 3-21. (Pressure-coupling at *room* temperature had been studied independently and approximately simultaneously in the United Kingdom by Crecraft, 1964.) High-temperature momentary-contact dry-pressure-coupling experiments on solids continued for a number of years at Panametrics, with considerable effort devoted to developing the furnaces for heating refractory metals in a suitable atmosphere, while providing valved access ports for the momentary contact interrogation. Some preliminary results were obtained in a bar of Pt to 1470°C (m.pt. 1610°C) and in a bar of W up to 2500°C (m.pt. = 3410°C) using a high-gas-pressure graphite heater furnace very similar to that designed by Pears et al. of the Southern Research Institute (Fig. 3-22). Data obtained in these Pt and W experiments appear only in an ONR report (Lynnworth and Papadakis, 1974, pp. 7-31 to 7-36). The highest temper-

FIG. 3-21.   Manually operated levers and toggle clamps used in the early 1960s for high-pressure momentary-contact coupling at normal incidence of shear waves at elevated temperature. Shear modulus was measured to 1000°C in fused silica, steel, and refractory metals and alloys. After Carnevale et al., 1964; (Abstr., 1963).

FIG. 3-22. Cam-actuated pneumatic cylinders automatically drive tungsten probes into high-pressure momentary-contact with heated specimen. Using this apparatus, c_T was measured at Panametrics in the mid-1960s to 2000°C in Ta, and c_L to 2500°C in W. Basic oven was designed by C.D. Pears of the Southern Research Institute, 1963.

ature at which *reliable* momentary contact data on sound speed and attenuation were obtained in solids in that ONR program was 1200°C (Papadakis et al., 1972).

It was suggested in Carnevale et al., 1964, that the momentary-contact pressure-coupling technique might be useful in the field of NDT for testing steel or other materials early in the production stages. For such potential applications of momentary-contact pressure-coupling, both reciprocating and rolling contact designs emerged (Carnevale and Lynnworth, 1967; Lynnworth, 1969; Figs. 3-23 and 3-24). In the 1960s, one of the problems investigated was the detection of the extent of an axial cavity in billets rolled from ingots and known as "piping." Despite encouragement and financial support from several steel companies in the U.S., and subsequent tests at one of Bethlehem Steel's plants (Droney and Pfeiffer, 1980), a "piping" success was not forthcoming in that era. Reciprocating momentary contact probes were, however, later adapted successfully to measuring the *thickness* of red hot steel, as reported by Krupski, 1979, and Jeskey (1987, priv. comm.), summarized in Section 7.3.2.4.

The idea of making high-pressure rolling contact by using modified

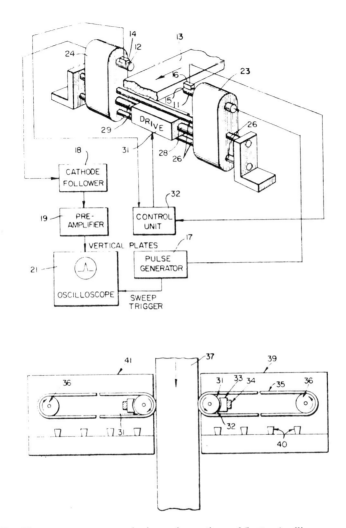

FIG. 3-23.   Momentary contact employing reciprocating and (bottom) rolling pressure coupling. After Carnevale and Lynnworth, 1967.

rollers like those already in existence in steel mills that produce billets and blooms occurred to several investigators. One apparently promising design of a pressure-coupled wheel search unit that includes two transducers emerged recently as a result of a cooperative program among AISI, NBS, Argonne National Laboratory and Magnaflux (Kupperman, 1987, priv. comm; see Fig. 3-25). Unfortunately, sufficient details on the design and performance of this transducer are not available, from which one

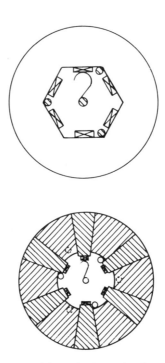

FIG. 3-24.   Buffer rods incorporated into wheel search unit proposed for high temperature testing. After Lynnworth, 1969a.

could make comparisons with earlier high-pressure rolling transducer designs such as that due to Uozumi, 1971. The introduction of this pressure-coupled rolling transducer into a steel mill in Canada in 1988 may provide data that demonstrate the extent to which the latest improvements solve the practical problems associated with detecting the extent of the oxidized piping defect in red-hot steel billets and blooms, and thereby provide a reliable control signal for hot cropping in the rolling mill. In China, pressure coupling at high temperature has been investigated by Ganlin et al., 1988.

Roller-pressure transducers for testing hot steel that emerged in the 1970s in Japan and the United Kingdom are represented by the works of Uozumi, 1971 and Andrews et al., 1974.

Reminiscent of the blacksmith's craft (Fig. 3-26), Goode and Lewis, 1975, developed a hammer-like transducer for the ultrasonic inspection of steel blooms and billets at elevated temperatures to detect piping and segregation by using a momentary-*impact* pressure-coupling technique. Their apparatus comprises a steel head assembly mounted for reciprocating movement. Two ultrasonic transducers, one for transmission and the other for reception, are mounted on a split head separated by an air gap. The head is reciprocated in such a manner as to cause impact upon the sample with an impulsive blow of short dwell time ($<$400 μs), once during each reciprocation of the head. Tests were reported on a variety of mild steel specimens at surface temperatures up to about 1100°C, but

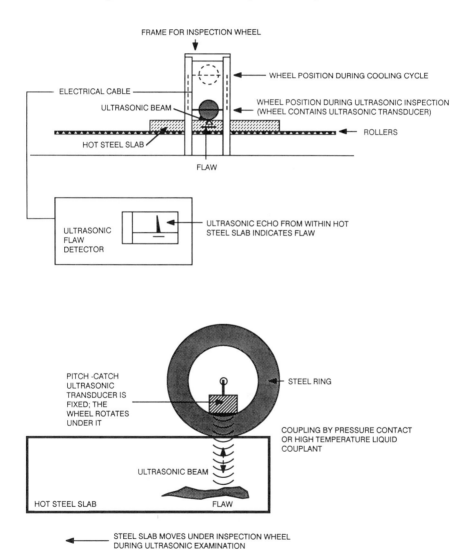

FIG. 3-25.   Pressure-coupled rolling transducer due to AISI, NBS, ANL and Magnaflux. Illustration supplied by ANL.

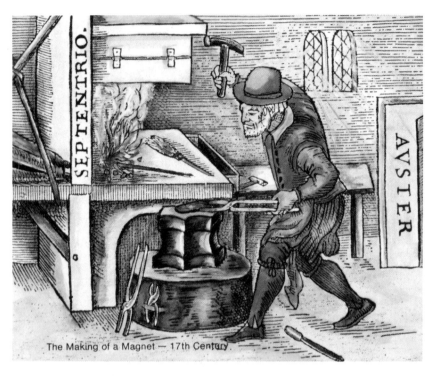

FIG. 3-26.   Early use of momentary contact in a high-temperature application. After Gilbert, 1600. Illustration courtesy *Measurements & Control*, IEE and Peregrinus.

only for descaled surfaces. See Fig. 3-27, which illustrates several uses of impact, not only at high temperature.

Apart from high-temperature applications, *momentary impact* is useful in vibration studies and acoustic resonance testing. Kistler and PCB Piezotronics, for example, manufacture quartz impulse force hammers so that impact force and structural response can be spectrum analyzed (Morse et al., 1972; PCB Piezotronics, 1974). In Fig. 3-28, acoustic res-

$\longrightarrow$

FIG. 3-27.   Examples of hammer head strikes to provide impulses or a momentary coupling for ultrasound. (a) Instrumented hammer is used to impulse the structures whose response is to be determined, courtesy Nicolet Scientific Corporation, 1975. (b) "Receiver" accelerometer(s) are mounted at critical locations on structure, after PCB Piezotronics, 1974. (c) Special hammer, upon striking the hot steel billet, provides a momentary coupling for ultrasound, after Goode and Lewis, 1975. (d) Block diagram of momentary impactor, and impact recording and analyzing apparatus to study resonance in catalytic converter substrates, after Papadakis, 1976.

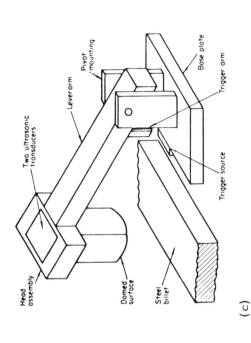

(c)

Two ultrasonic transducers

Head assembly

Lever arm

Pivot mounting

Base plate

Trigger arm

Trigger source

Domed surface

Steel billet

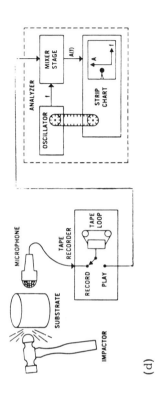

ANALYZER

OSCILLATOR

MIXER STAGE

A(f)

STRIP CHART

MICROPHONE

TAPE RECORDER

TAPE LOOP

RECORD

PLAY

SUBSTRATE

IMPACTOR

(d)

(a)

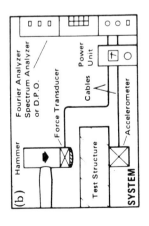

(b)

Hammer

Force Transducer

Fourier Analyzer
Spectrum Analyzer
or D.P.O.

Power Unit

Cables

Test Structure

Accelerometer

SYSTEM

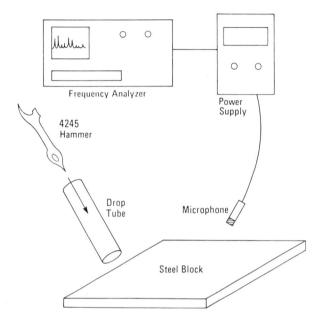

FIG. 3-28.   Schematic illustration of the hardware set-up used for acoustic resonance testing employing impact excitation. After Munce et al., 1986. © 1986 IEEE.

onance testing employing impact excitation (controlled drop test) was used to determine case depth in small sintered parts, e.g., a print hammer of approximate dimensions 4 × 30 × 1.5 mm (Munce et al., 1986). The percentage change in resonant frequency between two hammers is approximately proportional to the difference in case depth.

3.3.3   Transducer Clamps, Shoes and Ports; Precision Port Plugs; Minimizing Any Storm in a Port by Wedge, Screen or Membrane; Layout and Hot Tapping Tips

The theory and techniques in this section are discussed with respect to contrapropagation flowmeter applications but may be adapted to other situations.

Clamps vary from simple canvas or metal straps, tracks, guides, magnetic hold-down devices, and yokes, to more elaborate structures. Designs vary according to the accuracy required, time allowable for installation, permanent versus temporary use, temperature, clamping pressure required, type of transducer, odd or even number of traverses, pipe size, range of pipe size, pipe material, environment, manner of controlling the spacing (e.g., measurement by a ruler, by transit time, or control by a template or spacer bar), etc. Examples are illustrated in Chapter 2, Figs. 2-17 to 2-20. Some designers prefer that the clamp and transducer be separable. But economic considerations sometimes favor a combined unit.

In the early to mid-1980s it appeared that a pair of externally contoured square-holed stainless steel shoes, Fig. 3-29, might be a clever way to adapt a single pair of flat-faced clamp-on transducers to a rather wide variety of pipe sizes (Lynnworth and Adsmond, 1984). However, the shoe solution seems to have been quickly outstepped by transducer developments in Japan and in the U.S. whereby, by 1985, a single pair of flat-faced clamp-on transducers could generate in the pipe wall, shear or other waves, respectively, independent of pipe size, and over large ranges in pipe size. For example, one clamp-on 1-MHz transducer available from Tokyo Keiki could be used on most pipes between ~50 to 500 mm nominal diameter, while one 500-kHz transducer due to the author was used on pipes from ~33-mm to 4.7-mm outside diameter (Lynnworth and Nguyen, 1985). In some 1988 applications, however, the pendulum swung back to shoes. In one group of applications, shoes were fabricated not just to provide a smooth, flat coupling surface, but also to make the sum of the shoe thickness Y plus the pipe wall thickness W equal to a constant, e.g., 10 mm. The main purpose here was to create pipe conditions conducive to obtaining reproducible phase velocity and reproducible signal spectrum. Since the phase velocity of Lamb waves, for example, is a function of the frequency-thickness product their use in pipes of different W ordinarily requires different values of f to obtain a particular phase velocity.

FIG. 3-29.   Square-holed shoes, bonded to curved surface of pipe, provide flat face for repeatable coupling to flat-faced clamp-in transducer. After Lynnworth and Adsmond, 1984.

But if one can externally transform W to $W + Y = $ const. then Lamb waves of but one frequency suffice for obtaining that phase velocity.

To avoid uncertainties in the interrogated path due to refraction uncertainties, or to define a path beyond the limits imposed by refraction, it is common practice to fabricate a spoolpiece with holes or ports for the transducers or to drill such holes in existing piping. The ports are normally fitted with standard couplings or flanges to accommodate the transducers. The ports can also be plugged with special precision plugs that accommodate removable transducers (Fig. 3-30).

Somewhat offsetting the potential high-accuracy advantage of ports over the undisturbed boundary associated with clamp-on are the risks that the ports may clog with undissolved solids or deposits like paraffin, or become a trap for entrained gases, or, at high flow, contribute to excess turbulence right in the path of the interrogating sound beam. *Remedies*

FIG. 3-30.   Precision plugs accommodate removable transducers, while providing the controlled paths associated with wetted transducers.

for port-induced problems include filling the port with epoxy or other suitable material (Lake, 1962); covering the port with a fine-mesh screen that is nearly transparent to ultrasound in liquids *or* gases (Lynnworth, 1975b), or covering the port with a membrane such as a plastic film (Pfau, 1970, 1973) or metal shim. In liquids, a thin metal shim, ~50 μm, even at a 45° angle of incidence, is relatively transparent to low-MHz pulses. (For an explanation see Fig. 2-6 or one of the following references: Lynnworth, 1979, p. 474; Sanders, 1939; Deighton et al., 1981; Bertoni and Tamir, 1973; Bar-Cohen and Chimenti, 1985, 1986; Bar-Cohen, 1987.)

Layout problems for contrapropagation flowmeters typically require that the location and spacing of a pair of transducer locations or ports be determined accurately in the field. (Shop methods are not included here.) Accuracy is somewhat limited by, or determined by, pipe sizes, materials and access and environmental constraints (high temperature, noxious

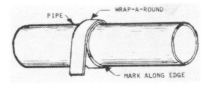

FIG. 3-31.   Flexible layout aid for pipe, courtesy Contour Sales Corp.

FIG. 3-32. "Sidefinder" centerpunch tool uses vee, fixed level and punch to locate holes at discrete angles around horizontal pipe at 45° increments. This includes the most common case, 3 and 9 o'clock positions.

FIG. 3-33. Commercially-available centering heads have a V- or Y-shaped head, adjustable dial bubble level and (left) manually-operated or (right) spring-loaded hardened centering pin. Illustration courtesy Contour Sales Corp.

fumes). For pipes larger than about 300-mm diameter, dimensional tolerances of ±1 mm and angular tolerances of ±1° are usually achievable with templates or fixtures. Spacing can be controlled more easily if both transducers lie on the same side of the pipe.

Templates can be made of paper or, for high-temperature layouts, of Kapton or similar polyimide plastic sheet, flexible asbestos-free gasket material or metal. Errors may accrue because of changes in sheet size between the layout room and pipe environment or because of the sheet

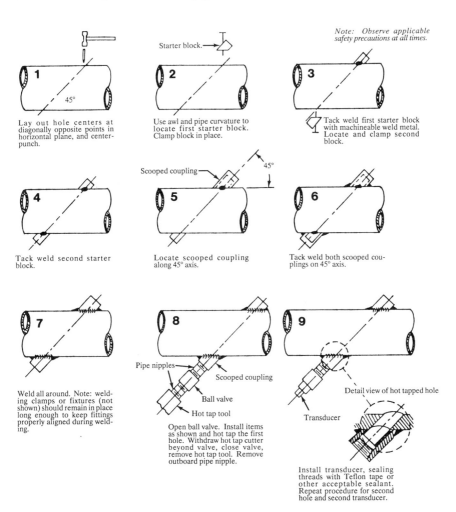

FIG. 3-34.   Nine-step guide to hot tapping ultrasonic ports along 45° axis. After Warburton and Lynnworth, 1983. © 1983 ISA.

thickness. Therefore layouts should be checked by independent mechanical or laser inspection means or accomplished by laser in the first place if such equipment is available. Standard Wrap-A-Round® or similar "runaround" templates are available from welder suppliers (Fig. 3-31) but if these are used with a soapstone marker or chalk, the resulting tolerances which are normally adequate for pipefitters may not be adequate for an ultrasonic measurement where accuracy is sought.

For the case where the two transducers must be located on *opposite* sides of the pipe, say at the 3 and 9 o'clock positions of a horizontal pipe,

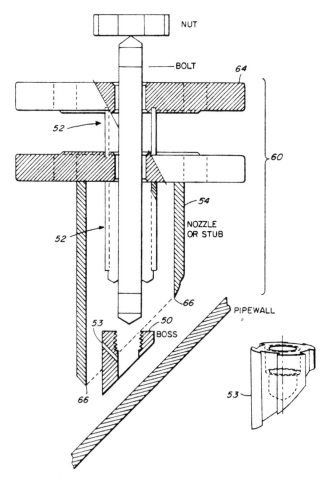

FIG. 3-35. Jigs and fixtures for hot tapping ultrasonic angled ports, due to Smalling et al., 1986.

a "Sidefinder" centerpunch tool composed of a vee, punch and level may be useful, Fig. 3-32. The commercially available tool shown in Fig. 3-33, termed a centering head, provides a more general solution to center-punching at any selectable angle around a horizontal pipe. For round pipe, opposite sides are also locatable by folding the template or encircling sheet

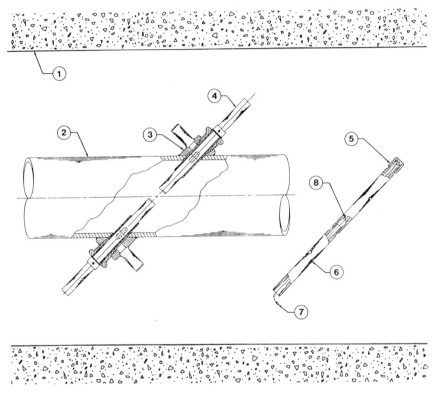

FIG. 3-36. *Segmented* alignment bar aligns couplings for welding on opposite sides of a pipe, when concrete vault is too small to allow one-piece bar to be installed and removed. (Compare with straight-bar one-piece method reported by Warburton and Lynnworth, 1983.) The design of the proposed segmented extended-well transducer shown at right was similarly motivated by the dual constraints of a small concrete vault and the desire to measure by the contrapropagation method, over the longest path possible in a steel pipe over 1 m in diameter, the flow of water plus entrained air. In this application the low pressure two-phase fluid had *a priori* unknown (but presumably repeatable) attenuation characteristics. The threaded segments, added one at a time, allow the path in this low-pressure fluid to be decreased in steps up to 30 cm, so that $\alpha(f)$ can be determined experimentally. Legend: 1-Concrete vault. 2-Steel pipe. 3-Scooped (contoured) or beveled pipe coupling with optional bleed or flush port. 4-Segmented alignment bar. 5-Transducer potted in pipe cap and first segment. 6-Add-on segment. 7-Electrical connector. 8-Potting.

in half. During layout, out-of-roundness and wall-thickness data should be recorded, since such data will normally be needed to compute volumetric flowrates from the measured transit times.

Having scribed or centerpunched the pipe at locations that are to be drilled, the question of how the drill can be prevented from walking arises. This is particularly evident if the hole is oblique to the pipe, say 45°. The solution is to guide the drill (or other cutting tool such as a piloted hole saw). A starter block or equivalent jig allows the hole to be started normal to a flat surface (Warburton and Lynnworth, 1983; Smalling et al., 1986). This solution can also be adapted to cases where the pipe is pressurized, requiring "hot tapping." See Figs. 3-34 and 3-35, which illustrate a few of the steps, tools and techniques developed in the early 1980s for hot tapping ports for liquid and flare gas flowmeters. Practice on scrap parts generally proves beneficial until adequate experience is gained so the procedures can be executed safely and accurately, with a minimum of chip or coupon contamination inside the pipeline. To align couplings or flanges on opposite sides of an empty (unpressurized) pipe, a one-piece or segmented alignment bar is used (Fig. 3-36.)

3.4   Decoupling Methods

One approach to decoupling is to try to do the opposite of whatever was recommended in the previous section on coupling. In contrast to the objective of coupling, which is to maximize the transmission of the ultrasonic signal, the objective in decoupling is to minimize the transmission of unwanted acoustic energy. In many cases the source of the unwanted energy is the ultrasonic transmitter, which means the noise is coherent, not random.

3.4.1   Early Experiments on Sound Propagation in Vacuum

In his "story of acoustics," Lindsay, 1966, recounts von Guericke's mid-17th century experiment. Von Guericke rang a bell in a jar evacuated by means of his air pump and claimed that he could still hear the sound. Actually, the first to try the bell in vacuo experiment was apparently Kircher, 1650, who concluded from his observation that air is not necessary for the transmission of sound. Undoubtedly, the trouble with these investigations of von Guericke and Kircher was the failure to avoid transmission through the walls of the vessel, together with the inadequate vacuum that they were able to obtain. In 1660, Boyle, 1682, repeated the experiment with a much improved air pump and more careful arrangements and finally observed the now well-known decrease in the intensity of the sound as the air is pumped out, due mainly to increased impedance mismatch at low pressure.

3.4.2   Impedance-Mismatched Sandwich

Livengood et al., 1954, used gaskets to decouple sections of a combustion chamber, and in addition used alternating layers of gaskets and steel washers to isolate the transducers in their measurement of the temperature of an internal combustion engine.

Carnevale introduced paper gaskets between sections of a shock tube in order to interrupt the high-frequency noise generated by a bursting metal diaphragm and heading towards the ultrasonic 1-MHz transducers near the end wall. (See Section 5.1.1.4; Carnevale et al., 1967c.)

Lynnworth, 1977c, separated transmitter and receiver sections of a flowcell, interposing sealing materials of dissimilar acoustic impedance to improve the isolation. Flange bolts and nuts can similarly be isolated acoustically with fibrous or plastic sleeves and washers (sometimes used for *electrical* isolation too).

In a flowmeter for medical gases, Kou et al.,1984, decoupled each of two 100-kHz PZT crystals from their aluminum holders by using a three-point suspension of soft silicone rubber elements cut from O-rings. Silicone rubber was also used as an isolating pad behind the crystals. To test the decoupling, foam rubber was temporarily stuffed into the flowcell to block the airborne signal. No acoustic short circuit was observed, demonstrating the adequacy of their solution.

In the flare gas ultrasonic flowmeter described by Smalling et al., 1984, 1986, low impedance packing gland materials are used to provide a gas seal between steel or stainless steel members while not coupling 100-kHz ultrasound efficiently. See Fig. 3-37.

In a flowmeter for water, using wetted transducers, isolation was improved by 6 dB by interposing threaded plastic bushings between each 1-MHz stainless steel transducer body and the stainless steel flowcell.

3.4.3   Small Contact Areas: Points, Knife Edges, . Membrane

In the Tracor ultrasonic gas chromatograph in Chapter 9, Section 9.4, the transducers are held as delicately as practical by using mechanical supports of *minimum contact area* (Noble et al., 1964). (If pressure is applied to the point contact, then pressure-coupling occurs, rather than isolation.

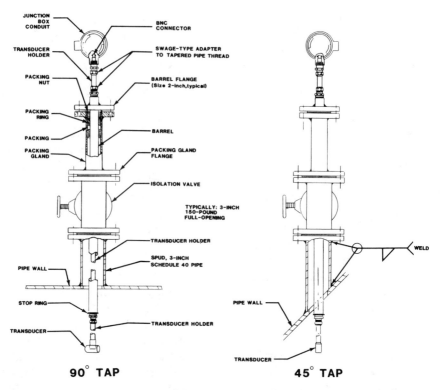

FIG. 3-37. Low-impedance (and high attenuation) of packing gland sealant contribute to acoustic isolation in a 100-kHz transducer holder used in flare gas flowmeters. After Smalling et al., 1984, 1986.

See, for example, Fowler, 1971a, who pressure-coupled 100-kHz pulses from the point of a hardened needle into structures to simulate acoustic emissions.)

In the previously mentioned shock tube work of Carnevale et al., 1967c, *knife edge* support is provided by a Teflon spool. See Chapter 5, Fig. 5-4.

In experiments conducted at ORNL by Rogers and Miller, 1982, to develop and evaluate a feedthrough for a magnetostrictive waveguide sensor, the pressure seal was made of a *stainless steel membrane*, 130 µm at its thinnest section (Fig. 3-38). Because of mechanical impedance mismatch, it provided a low-reflectivity seal, and coupled little energy into the adjacent structure.

A *thin plastic membrane* of the type commonly used to wrap foods, ~18-µm thick, and stretched taut over a metal container, isolated 100-kHz transducers from one another and allowed clamp-on interrogation

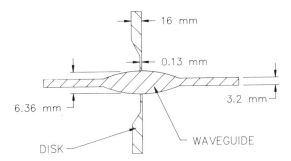

FIG. 3-38. Pressure seal designed by Rogers and Miller, 1982, and the author.

through the membrane of the level of water in a styrofoam cup (Fig. 3-39a,b; Lynnworth, 1986). In another experiment, the same transducers were gel-coupled to diametrically-opposed points on the sides of the cup, with coupling pressure increased to deform the cup until the transducer faces were parallel. Short circuit was absent. The clamp-on airborne transmission is shown in Fig. 3-40.

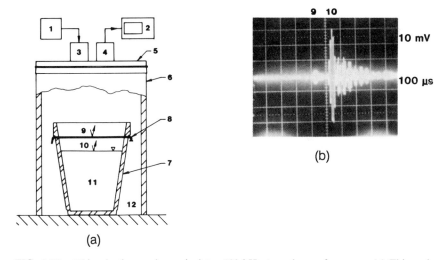

(a)

(b)

FIG. 3-39. Thin plastic membrane isolates 100-kHz transducers from cup. (a) Thin rod, (8, bicycle spoke) used as reference reflector, in laboratory experiment with transducers (3,4) outside top-sealed container (6). (b) Oscillogram shows reference echo (9) and water echo (10). 1-Panametrics pulser/receiver 5055. 2-Oscilloscope. 3-Transmitter transducer. 4-Receiver transducer. 5-Thin plastic film, ~18 μm, stretched taut over top of container (6) and sealed with elastic band. 6-SS pipe, 100–mm ID × 150–mm long. 7-Styrofoam coffee cup. 8-Bicycle spoke, 2.4–mm ø. 9-Spoke echo. 10-Water echo. 11-Water. 12-Air. After Lynnworth, 1986.

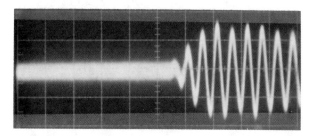

FIG. 3-40.   Clamp-on transmission of 100-kHz pulse across air in a styrofoam cup that was deformed by pressing the gel-coupled transducers against opposite sides until the transducer faces were parallel. In this oscillogram, no acoustic short circuit noise can be seen prior to the arrival of the 100-kHz airborne signal.

A gas-filled balloon left over from a birthday celebration became a party to what began as an isolation experiment (Fig. 3-41). The oscilloscope shows no acoustic short circuit. If this experiment is repeated under more controlled conditions, the sound speed/molecular weight relation is easily verified by filling the empty balloon with gases of known molecular weight M and known specific heat ratios, all at the same temperature.

3.4.4   High-Attenuation Coupler

Oftentimes materials of low impedance also are highly attenuating, e.g., cork, fiber-reinforced plastic, plastic foam. Hence such materials are doubly effective in achieving isolation. (Folds, 1972, observed during tests

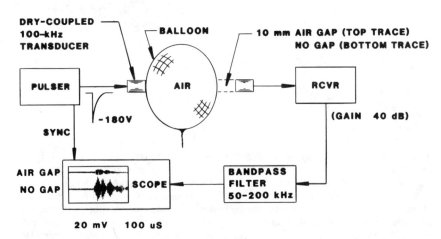

FIG. 3-41.   Balloon isolates transducers and provides an easy way to demonstrate dependence of sound speed on gas molecular weight.

of plastics, elastomers, and syntactic foam that those materials exhibiting a large temperature coefficient of sound velocity tended to be more attenuating.)

Gasket isolators, for example, were introduced by Livengood et al., 1954, into a special combustion chamber to reduce 2-MHz acoustic short circuit. Asbestos-like gasket materials may be used to isolate flanged transducer holders from flanged ports, or to isolate two halves of a flowcell from each other (Lynnworth, 1975b,c). Pipe couplers made of rubber are available in plumbing supply stores, and may be considered as an inexpensive way to interrupt metal-borne acoustic transmission from one part of a conduit to another.

3.4.5   Delay Path: Reentrant Cavity, Spiral

When short pulses or tone bursts are used (as opposed to cw), time-domain isolation can be achieved by delaying the short circuit noise until after the signal has been received. Since sound travels faster through most solids than through most fluids, the delayed-noise solution is not always practical. Gases other than H_2 or He and/or pipes of large diameter require long delay paths and, hence, are the most troublesome with respect to this method of isolation. At low temperatures, even low molecular weight gases exhibit low sound speeds.

In a laboratory feasibility study, Lynnworth and Carnevale, 1966 considered the possibility of using a reentrant delay mount and also a coiled Cu tube delay mount to support transducers, between which c was to be measured in H_2. Whether the damping means necessary to suppress vibrations would reintroduce acoustic short circuit, was not determined. (Reentrant diagram: Fig. 3-42). Numerical examples of the calculated length x_r required to delay 11-bit coded bursts of ultrasound for a 2-inch stainless pipe filled with air or water, assuming frequencies of 100 kHz and 1 MHz as commonly used in those media, are given in Table 3-9.

If the short circuit delay path is *short* compared to the travel time of interest in the fluid, and if the short circuit noise is of short duration (i.e., no long ringdown tail) then at the time the sought signal pulse finally arrives, noise may have already subsided to a value that is of no concern. (See Figs. 3-43 through 3-45).

3.4.6   Interrupted Geometry—Bumps and Grooves, Knurl, Threads, Orthogonal Shear Wave Crystals

Interruption of plate waves (Lamb, Rayleigh) by introducing discontinuities can be thought of as putting to good use scatterers that in other circumstances might be classed as rejectable defects as judged by their

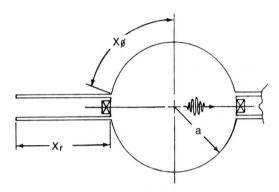

FIG. 3-42.   Example of reentrant cavity mount to delay acoustic short circuit pulse.

high reflectivity. Some of the early beam drift flowmeter patents proposed bumpy or groovy interruptions as obstacles to acoustic short circuit.

The common methods for scattering waves in buffer rods (threads, knurl) could be applied on flowcells if the cost and stress rise were tolerable.

In a high-temperature clamp-on feasibility study, chevrons were built up as weld beads on the steel between the transducers. The improvement in signal-to-noise ratio attributable to them was not determined. (Nguyen et al., 1982).

In general, other methods in this section are probably to be preferred over add-on bumps, grooves or other mechanical modifications to a pressure boundary. However, it should also be noted that in pitted steel pipe, rusty and corroded due to long exposure, the acoustic short circuit intensity is much less than in clean pipe of otherwise equal material and dimensions.

Orthogonally-polarized shear-wave 1-MHz crystals were used in a clamp-on flowmeter project in the early 1970s to increase the isolation

Table 3-9.   Calculated Length x_r in Orthogonal Reentrant SS Delay Mount for 11-bit Barker Code and 2-inch ($\sim$50 mm ID) Steel Pipe. PW = Pulse Width = $11/f$ Where f = Carrier Frequency. c_L/c_o = Ratio of Longitudinal Velocity in SS to That in the Fluid

Fluid	f, Hz	PW, μs	c_L/c_o	x_r, mm
Air	10^5	110	$\sim$20	600
Water	10^6	11	$\sim$5	127

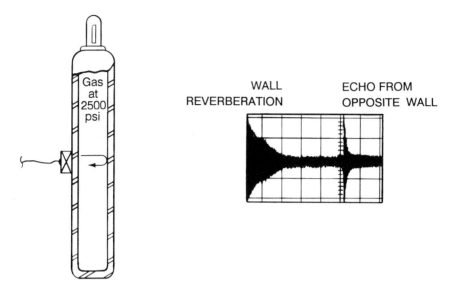

FIG. 3-43.   Noninvasive (clamp-on) measurement of pressure of a gas of known composition, temperature and pressure-dependence of c (see Carey et al., 1969, or illustration in Section 3.8), demonstrates pulse-echo detection of signal which is "isolated" in time after the ringdown has subsided. After Lynnworth et al., 1971. © 1971 IEEE.

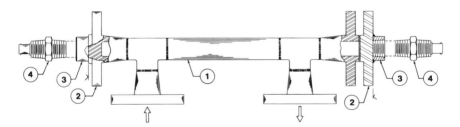

FIG. 3-44.   Flowcell concepts for high-pressure gas [e.g., steam, N_2 or O_2 at 2000 psi (136 bar) provides "time domain isolation" for short-duration pulses whose ringdown, arriving as acoustic short circuit energy via a metal path between the transmitting transducer and the receiving transducer, terminates before the gas-borne signal arrives. Legend: 1–Offset-style, weld neck flanged flowcell made of steel for steam or N_2, monel for O_2, etc. 2–Solid blind flange. 3–Half-coupling, similar to that used with precision plugs of Fig. 3-30. 4–Removable or permanent transducer, as in Fig. 3-30. Gasket, bolting, weld details and cabling not shown. In an alternate geometry (Fig. 3-45) the transducers interrogate the flow along a folded 45° zigzag path, such that the flow itself remains axial. The axial flow route avoids the pressure drop associated with the two 90° turns in the above configuration, and allows the axial conduit section to be built with plain ends suitable for welding into a pipeline, without the expense of end flanges.

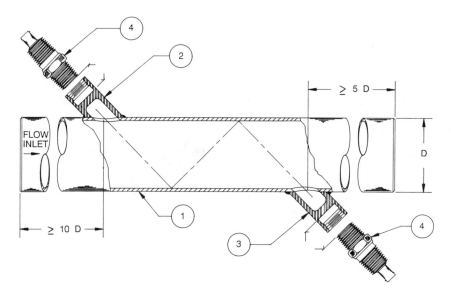

FIG. 3-45. Alternative to flanged design of Fig. 3-44, using all-welded pressure boundary, provides straight path for flow, zigzag path for ultrasound. Legend: 1–Pipe of minimum length approximately 18 diameters. 2,3–Special couplings. 4–Removable or permanent transducer, same as in Fig. 3-44.

between simultaneous cw upstream and downstream channels. Using the orthogonally-polarized arrangement in Fig. 3-46, 6-dB isolation improvement was observed over the same crystals in a parallel configuration (Pedersen and Lynnworth, 1973).

3.4.7 Dampener Between Transducers—Highly Attenuative, Impedance-Matched, Distributed or Localized

In testing a clamp-on flowmeter, if one finds excessive acoustic short-circuit noise, the intuitive reflex is to dampen it. Thus one would probably try to coat as much of the area as possible between the transducers with an absorptive material. Materials sold for vibration damping usually are designed to solve problems up to ~20 kHz, so their effectiveness at frequencies five to fifty times higher may be unknown. Of course, the ready availability of sheets of rubber, Teflon, or other flexible materials may offer an immediate and adequate solution.

In the mid 1980s, Nguyen (priv. comm) recognized that the proprietary

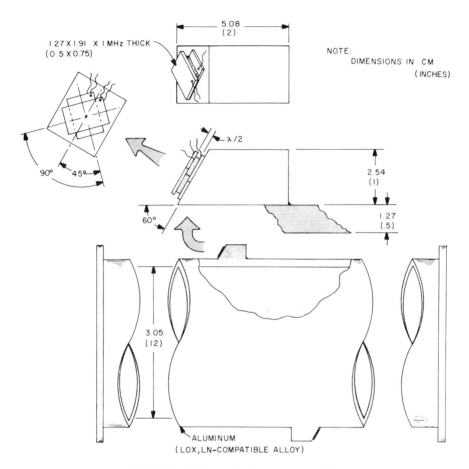

FIG. 3-46.   Orthogonal shear wave crystals.

backing materials used to damp ultrasonic vibrations in broadband NDT transducers might make excellent dampers on a pipe. In such transducers the backing impedance Z_b closely matches that of the piezoelectric crystal, and Z_b can be controlled by the ratio of, for example, tungsten to epoxy. Z-matched damping material can be applied, bonded or coupled to a pipe in different ways. In a zigzag clamp-on flowmeter, clamps providing ~6-dB improvement per clamp were found especially helpful in suppressing early pipe-borne plate or surface waves corresponding to a single traverse of the liquid, thereby making it much easier to selectively detect a triple or quintuple transit (Lynnworth and Nguyen, 1985). See Fig. 3-47.

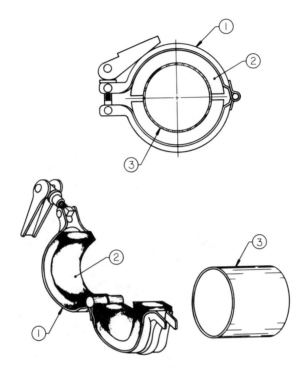

FIG. 3-47. *Z*-matched damping means built into a removable clamp. 1–Sanitary clamp. 2–Damping material contoured to outside diameter of flow conduit. 3–Flow conduit. After Nguyen, priv. comm., 1984.

3.4.8 Electrical Filtering, Quadrature Demodulation and Subtraction of Noise or Clutter

Since the attenuation coefficient α in most materials increases in proportion to the second or higher power of frequency f, the effectiveness of damping techniques diminishes at low f. At the same time, beam spread is greater for low-frequency energy. While the beam spread may tend to weaken low-f contributions, it is clear that one can not easily aim a transmitter so that low-f energy does not radiate to the receiver. The source of low-f short circuit noise in many cases is due simply to the radial mode action within a so-called thickness mode transducer.

A simple remedy, sometimes sufficient, is to introduce high pass *filtering*. If the signal and noise modes differ in frequency by an order of magnitude, a single pole filter may suffice.

In clamp-on flowmeter applications, especially if the pipe is of relatively small OD and of thick wall, the short circuit typically includes much noise at the signal carrier frequency. If the ultrasonic method of flow measurement is *cross-correlation* of *tags*, then a *quadrature demodulation* technique due to Jacobson may improve the signal to noise ratio by about 20 dB (Jacobson et al., 1985).

Sometimes short circuit noise even at the same frequency as the fluid-borne signal can be identified by its stationary character, in contrast to the jittery character of a signal that has travelled through a turbulent, flowing medium. Also, fluid-borne contrapropagating transmissions are delayed or advanced by the flow, so a time exposure of an oscilloscope display can reveal which parts of the trace have a single image versus those parts having a double image (Fig. 3-48.)

If the pipe-borne noise can be measured or predicted at the receiver transducer, in principle it can be *subtracted* electronically: digitize and/or store noise; measure signal plus noise; subtract noise. Analogous situations occur, for example, in moving target indicators (Mason, 1964) and in a torsional mode waveguide liquid level sensor containing spurious but stationary echoes (clutter). In the latter example the echo generated at

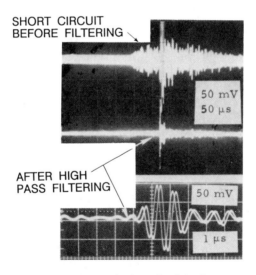

FIG. 3-48. Elimination of much of acoustic short circuit in clamp-on application by means of 1-MHz high pass filtering. Fluid: oil, $T = 8°C$, $\upsilon = 14$ cs, $\rho = 0.850$. Steel pipe OD = 324 mm, $W = 13$ mm. Couplant, propylene glycol. Top trace: pulser/receiver only. Middle trace: output of high pass filter. Bottom traces: double exposure of manually-interchanged interrogation direction yields $\Delta t \approx 200$ ns, from which V is found to be 1.52 m/s. Note that residual short circuit is stationary; only the fluid-borne portion of the signal exhibits Δt under flow. In this test, the signal to noise ratio is S/N = 22 dB. After Lynnworth, 1982a. © 1982 Inst. MC.

the liquid's surface may be not much larger than the clutter. But if the clutter can be substantially reduced by the digitize/store/ subtract routine, a usable signal emerges whose arrival time correponds to the liquid level. (See Section 3.6.2 and Figs. 3-74 and 3-75 therein.)

In NDT, synchronous grain noise often frustrates one's attempt to detect echoes from small defects within a polycrystalline specimen. One remedy is to move the transducer around a bit so that grain noise tends to appear random relative to the nearly stationary echo from the small target. In one study of clamp-on flowmeters, equivalent remedies are under investigation that aim to reduce by electronic means the timing errors caused by synchronous noise or clutter.

3.5 Electrical Excitation Waveforms, Detection, Digitization, Timing and Amplitude Sensing— Passive Techniques

This section is concerned with transmitted waveform and receiver detection techniques used in *active* systems, and receiver techniques used in *passive* systems. The signal generating and processing techniques are often analogous to their predecessors in the communications and radar/ sonar target identification and detection fields. Sometimes special methods are needed in ultrasonics to deal with degraded signals, spurious signals or noise that are characteristic of, and unique to, the bounded or unbounded media through which ultrasound propagates, or uniquely associated with ultrasound/measurand interactions.

3.5.1 Excitation Waveforms

Conceptually, perhaps the two simplest transmitter waveforms are cw (continuous wave) and delta function (spike): one frequency, and its dual in the time domain, one event. The bandwidths of these extreme cases are completely opposite, being zero in one case and infinite in the other. Between these limiting cases one finds a number of special cases that may be examined from the perspectives of modulation and bandwidth. The more complicated transmitter waveforms are usually motivated by the need to overcome a problem that could not be solved with a simpler one. Considerations of how bats or dolphins navigate, hunt and communicate, or how people talk, convinces one that at least in nature, simple waveforms seem to have evolved into waves of more complex structure, along with corresponding detection and signal processing means, until a satisfactory solution was obtained. (Graff, 1981; Lewis, 1983; Purves and Pilleri, 1983).

A list of some transmitter waveforms, along with suggested references for more complete discussion of their implementation, pros and cons, is given in Table 3-10. Waveforms are illustrated in Section 3.5.5.3, Fig. 3-60.

3.5.2 Receiver Concepts

Receivers used in ultrasonic measurements are conveniently grouped into one of these categories—broadband, narrowband or (for a coded waveform) a correlation or matched filter receiver.

Broadband and narrowband receivers are characterized by their "Q" defined by

$$Q = f_o/(f_U - f_L), \tag{3-40}$$

where f_o is the center frequency and f_U and f_L are the highest and lowest

Table 3-10. Transmitter Waveforms

Type	Example/Remarks/Reference
Cw (or quasi-cw)	Used in HP quartz thermometer to excite quartz crystal oscillator at a frequency near 28 MHz that is linearly related to temperature.
Delta (spike)	Used in NDT flaw-detection equipment and many other instruments to launch the interrogating pulse.
Square wave tone burst	Used in high voltage pulsers to drive transducers hard at controlled frequency, usually near resonance.
Sinusoidal tone burst	Used for pulse measurements within a narrow bandwidth.
Amplitude modulation within tone burst	Used to compensate for undesirable aspects of transducer's impulse function (sometimes called inverse filter technique).
Frequency modulation	Used to find maximum usable frequency; servoed to maintain fixed number of wavelengths in path. Fishbacher, 1959, 1963; Pedersen and Lynnworth, 1973; Redding, 1978; Mágori, 1985; Mylvaganam, 1989.
Phase coded modulation	Pseudo-random, M, Barker or other codes may be imprinted upon the phase, to allow correlation detection techniques to detect arrival of signal that appears buried in noise. Due to Pedersen, in Lynnworth and Pedersen, 1972; Brookner, 1977; Jacobson *et al.*, 1987b.
Truncated Gaussian	Maximizes bandwidth for time-limited pulse. Brookner, 1977; Silk, 1984, pp. 54–55.
Cos^2	Practical approximation to truncated Gaussian. Brookner, 1977.

frequencies within the passband (3–dB points). Broadband receivers typically have a Q of 4 or less and narrowband receivers a Q of 10 or greater.

Narrowband systems make use of cw or long burst transmissions that allow very good signal-to-noise ratios (S/N) for moderate peak transmitted power. The narrowband receiver also rejects noise outside the passband, giving a further improvement in S/N. One penalty of a narrowband receiver in time-interval measurement is a "phase ambiguity"—i.e., the inability to distinguish the arrival time of a particular cycle of a nominally single-frequency transmitted signal. This often limits the use of narrowband receivers to applications where the time (or time difference) measurement does not exceed one period of the transmitted signal. It may also not be possible to "gate" or "window" a narrowband signal to avoid the effects of multipath interference. Still another limit is response time, which is inversely proportional to bandwidth. Phase ambiguity is sometimes avoided electrically by using two frequencies not harmonically related or by combining the narrowband measurement with a broadband one (coarse, fine).

Broadband receivers are usually used with pulse transmission, as opposed to cw. The broadband receiver minimizes the distortion of the received pulse, simplifying the detection system and avoiding the phase ambiguity and some multipath problems. However, high-peak transmission power is required to obtain reasonable signal-to-noise ratios.

Correlation and matched filter receivers combine most of the advantages of narrowband and broadband receivers (Taub and Schilling, 1971; Brookner, 1977). The transmitted signal usually consists of a burst modulated in phase and/or amplitude by a pseudo-random code or a frequency modulated burst (chirp). Correlation and matched filtering are not independent but different techniques for yielding the same optimum filtering. These techniques provide good signal-to-noise ratios with moderate peak transmitted power and without the phase ambiguity or multipath problems of cw. Modern receivers in ultrasonic systems often use fast analog-to-digital converters and implement the correlation or matched filtering by using digital processors.

3.5.3 Timing by Arming and Zero Cross Detection

Having received a transmitted signal, the next questions are:

1. When did it arrive?
2. How strong is it?

Sections 3.5.3 to 3.5.5 deal with the first question, and Section 3.5.6, with the second.

3.5.3.1   Leading Edge Arming and Timing

If a broadband pulse is distinct from the background noise and appears as a solitary event, its arrival can be associated with any easily defined, unambiguous aspect of the pulse. For example, the arrival of the pulse in Fig. 3-49 could be taken as point A, B, C, etc. See Table 3-11.

Arguments can be made in favor of or against any of these choices. In practice, the choice would almost certainly weigh factors not included in the simplified model.

As a starting point for comparing some of the possibilities in Table 3-11, let us note that, if a zero crossing is to be used, H could be selected if the zero cross detector were armed by any part of the leading edge between A and E that exceeds a pre-set level, e.g., $V_o/2$.

If we anticipate that the received signal will be distorted from time to time, then a better way to decide when it arrived may be to evaluate more of the information contained in the pulse than merely one point such as H. The question is, how much data can we afford to handle versus how much can we afford to disregard?

Before getting into the more complicated arming and timing methods, it is appropriate to note that leading edge *timing* was used successfully in a large number of instruments, as pointed out by Greenspan and Tschiegg, 1962 (Section 3.5.4.1 on the sing-around method). Also, by way of background, let us refer to ultrasonic "Reflectoscopes" or similar NDT equipment (McMaster, 1959). With such instruments echo arrival time was usually estimated or measured to some point on the leading edge by the observer. In *dynamic* situations, where not only distance x but its variation in time, dx/dt, were of interest, various means of recording the changing arrival time of MHz pulses emerged, according to the photographic [e.g., Edler and Hertz, 1954 (Fig. 3-50); Lynnworth, 1962] or later, video technology (e.g., Smalling et al., 1984; Hansmann and Kirby, 1986) that was available to record echo patterns as seen on an oscilloscope.

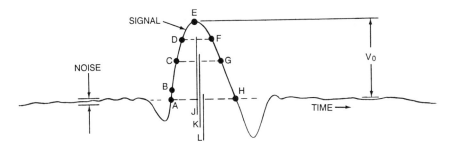

FIG. 3-49.   When did this pulse arrive?

Table 3-11.   Examples of Possible Arrival Criteria for the Broadband Pulse in Fig. 3-49

A First distinct zero crossing	E Peak	J Center of 3-dB points
B Rising Amplitude = 10%V_o	F Falling Ampl. = 70.7%V_o	K Center of 6-dB points
C Rising Amplitude = 50%$V_o(-6$ dB)	G Falling Ampl. = 50%V_o	L Center of gravity between A and H
	H First zero crossing after peak	
D Rising Amplitude = 70.7%$V_o(-3$ dB)		

Automatic leading-edge timing, telemetering and recording means that did not require use of an oscilloscope were also utilized, e.g, digital thickness gages, reentry vehicle ablation gages (McGunigle and Jennings, 1975; Gieske, 1987), flowmeters and other intervalometer-based instruments.

3.5.3.2   Timing Midway between Leading and Trailing Edges

Two data points ought to be better than one, with respect to reliable timing. The level at which the two points are selected ought to characterize the pulse and ought to occur in regions of high slew rate. To measure time of arrival in the first automatic thin-wire waveguide ultrasonic thermometer, B.M. Gordon (ca. 1965, priv. comm.) proposed a selectable threshold, settable at 50 to 90% of peak (normal setting, -3 dB points) such that, assuming the 3 dB points were the threshold levels, J would define the arrival. In Gordon's method, described in detail in Papadakis et al., 1972a, 10-MHz clock pulses are divided by 2 between D and F so that the resulting count corresponds to the midpoint, J. Thus, the time interval between echoes A and B generated at the beginning and end of a temperature sensor is measured between the "centerlines" of A and B, each centerline being the midpoint of the threshold's intersections with that echo's main lobe (leading and trailing edges).

If the pulses were narrowband instead of broadband, however, both the foregoing zero crossing or threshold centerline methods (H, J) would be jeopardized by ambiguity in determining when a particular threshold is crossed. Several ways of dealing with such ambiguity, due to Schmidt, 1975, Mitsuta, 1974, Biber et al., 1980, Laenen, 1984, Eck, 1977, and Loosemore and Mustin, 1978, are reviewed in Lynnworth et al., 1985.

3.5.3.3   Integrated Threshold Arming

Referring to Fig. 3-51, several typical 100–kHz waveforms are shown as they occur in flowmeter tests.

In order to achieve greater immunity to the amplitude fluctuations and

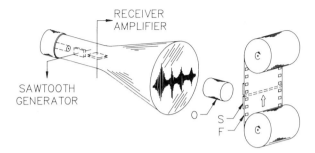

FIG. 3-50.   Apparatus for continuously recording dynamic ultrasonic cardiogram time interval data on film. In this method, due to Edler and Hertz, 1954, a horizontal slit S was mounted in the image place of the camera objective O (ordinarily used for photographing the CRT screen of a Reflectoscope), the slit width being 0.5 mm. Directly behind this slit, a 24-mm film F was continuously moved at right angles to the slit at a rate of 1 cm/s. The slit was placed parallel to, but a little above, the image of the x-axis of the CRT, so that light coming from the x-axis and the time scale on the CRT screen can not reach the film. If no echo signal appears on the CRT screen, nothing will be marked on the film except the transmitted pulse signal (main bang) which will be recorded as a straight line parallel to the direction in which the film is moved. If a constant, non-pulsating echo signal appears on the screen, even this will produce a straight line on the film parallel to the transmitted pulse line. The distance between these two lines is then proportional to the distance from the crystal to the reflecting boundary (heart wall). If the echo signal pulsates along the x-axis a curve will be recorded on the film. The distance of this curve to the transmitted pulse line will correspond at any moment to the crystal to reflecting boundary distance. In this way, the variations of the crystal to reflecting boundary distance were recorded with respect to time. After Edler and Hertz, 1954.

distorted shape of the bursts than was possible with most pre-1985 techniques, it was decided to use an integrated threshold arming method. In one form of this method, a capacitor is charged in proportion to the amount the half-wave rectified signal exceeds a threshold V_n. When the capacitor's voltage exceeds the preset arming level, a zero cross detector is armed, so that time is measured to the next positive-going or negative-going zero crossing. The contributions to the voltage being integrated by the capacitor are approximately proportional to the amplitudes of the successive cycles in the received burst. Therefore, the integration steps are unequal [Fig. 3-51(c)]. If an early, small cycle falls into the noise, i.e., below the V_n threshold, its contribution is hardly missed because the integrated threshold can be set to be tolerant of small contributions being present or absent. Previous arming techniques, despite a similar objective, generally retain an undesired requirement to not "miss" a weak cycle in order to know precisely at how many cycles into the burst one is measuring the zero crossing. In some earlier work, attenuated pulses are discarded or processed differently from nonattenuated pulses.

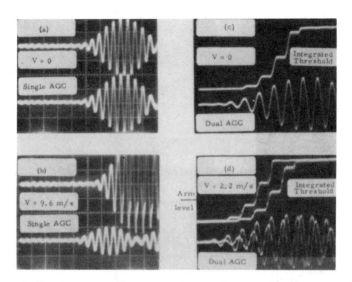

FIG. 3-51.   Oscillograms of 100-kHz bursts interrogating N_2 gas flowing through an offset-style flowcell demonstrate the difficulty of determining arrival time unambiguously. This 1985 solution used dual AGC and the integrated threshold arming method. In (a) and (b), upper trace is for the upstream direction, lower trace is for the downstream direction. After Lynnworth et al., 1985. © 1985 IEEE.

Further details appear in Lynnworth and Matson, 1985, and in Wallace et al., 1985. The reliability of this method has been demonstrated in flare gas ultrasonic flowmeters since 1982. On a much larger scale, a similar integration method was reported by Biber et al., 1980, with respect to Polaroid's echo ranging camera, now finding use in industrial non-camera applications too (Section 9.7).

3.5.3.4   Center of Energy

Consider a narrowband sine- or bell-developed tone burst. A "center of energy" arming method due to Bradshaw and Pedersen, 1984, operates as follows. A "center" is found such that the integral from the beginning of the received pulse to the center equals the integral from the center to the end. Time is measured to the zero crossing immediately following the centerline arming, Fig. 3-52.

3.5.3.5   Derivative of Envelope Defines Peak

This method was developed at Oak Ridge National Laboratory (ORNL) by D. McNeilly (ca. 1983, priv. comm.) for timing broadband pulses, as in Fig. 3-49 at point E. McNeilly's method involved differentiating the pulse, such that a zero crossing occurred in the derivative waveform at the same time that E occurred in the actual wave.

3.5.3.6 Analysis of Digitized Waveform

Digital processing oscilloscopes provide one example of equipment capable of examining more than one point in a pulse to decide when that pulse arrived. Tasman et al., 1982, reported improvements in timing broadband pulses, by processing the digitized waveform. Unfortunately, much of the potential improvement over, say, timing merely between 3-dB points, is unattainable if the waveguide sticks to a sheath and generates coherent spurious echoes. Thus, signal quality remains an important parameter in predicting the accuracy and resolution limits of signal processing techniques of increasing sophistication. (See also Section 3.5.5.3.)

3.5.3.7 Dual-Frequency Technique Resolves Phase Ambiguity

In some early contrapropagation flowmeters, the sought sub-ns resolution was indeed achieved using phase measurement techniques. However, at a single frequency, flow range is limited by phase ambiguity. Two frequencies can resolve this ambiguity (Gutterman, 1985; Mágori, 1985). The unambiguous range is increased from the half-period (or period) of the first single-frequency carrier to the half-period (or period) of the beat

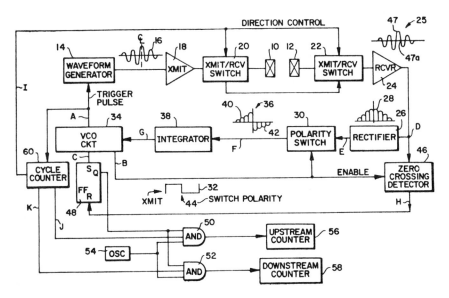

FIG. 3-52. Center of energy arming method. After Bradshaw and Pedersen, 1984.

frequency for the two-frequency system. This can be illustrated with a graph of phase versus time, Fig. 3-53. Investigations of two-tone ultrasonic rangefinding systems are represented by the work of Smith and Schoenwald, 1984; Schoenwald and Smith, 1984; and Ono et al., 1984.

3.5.4 Timing in the Frequency Domain

3.5.4.1 Sing-Around

The invention of the sing-around velocimeter, according to Greenspan and Tschiegg, 1962, occurred over fifty years ago (Shepard, 1943; patent filed 1937). From the perspective of the mid- to late-1980s it appears to have triggered a breakthrough in ultrasonics. "Sing-around," a term coined in 1948, eventually led to thousands of instruments being manufactured for high precision intervalometry, including contrapropagation flowmetry. Sing-around *flowmeters* have been in use since at least 1964 (e.g., Baba, 1964), with improvements continuing to this day (e.g., Delsing, 1987). See also, Chapter 10, Section 10.3.

Most of the remarks in this section are taken from a published lecture by Greenspan and Tschiegg, 1962. The sing-around velocimeter reviewed by Greenspan and Tschiegg is automatic, has fast response, and is easily adapted to recording. The models discussed by Greenspan and Tschiegg

are restricted to use with liquids that show no appreciable frequency dispersion, except possibly at very high frequencies. They must be designed, adjusted and calibrated for a particular class of liquids within which the total variation of the speed and attenuation of sound is not too large. These changes are commonly caused by changes in temperature, pressure or composition. The velocimeters have high stability and are therefore especially

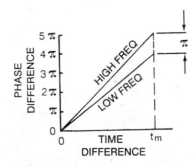

FIG. 3-53. Ambiguity exists for $t \geq t_m$ but not for $t < t_m$. At t_m, the phase difference is π.

adapted to differential measurements, such as that of sound-speed gradients in the sea. Other examples of differential measurement are the determination of the effect of dissolved air on the speed of sound in water near the turning point (approximately 74°C).

A sing-around velocimeter may be thought of as a cylindrical tank the ends of which are electroacoustic transducers, and the whole filled with the liquid under test. A voltage pulse is applied to the "sender," and a corresponding pulse of sound travels through the sample liquid and is received and converted to an electrical pulse by the receiver. In order to define uniquely the time interval between the pulses and to specify their location in time, some characteristic must be selected that will still be recognizable after the pulse has been distorted by transmission through the liquid and by the bandwidth limitations of the transducers. [In the instrument discussed in 1962] the pulse position is specified by the instant at which it begins to rise from the noise.

The choice has several important consequences. To begin with, a pulse-modulated carrier has now no advantage over the much simpler video pulse, even though the distortion of the former would be much less. Further, it becomes essential that the pulse rise rapidly. This is no problem so far as the input (transmitter) pulse is concerned, but the received signal rises relatively slowly. The fast rise is restored by amplification; nevertheless, there is introduced an unknown delay equal to the time that the output pulse spends below the noise. This delay depends on the attenuation characteristics of the liquid. It is primarily for this reason that the velocimeter must be calibrated and used on a class of liquids within which the attenuation characteristics are not too variable.

The timing is automatic. The received pulse after suitable amplification and reshaping is again applied to the sender; thus the device regenerates and the pulse repetition frequency (prf) depends upon the speed of sound in the liquid and to some extent upon electrical and other delays.

A major source of difficulty is the existence of multiple echoes between the transducers. The various sets of echoes, each set arising from a different primary pulse, are not synchronous because of the electrical time delay. Various means of eliminating the reflections have been used. In the case of a straight path, the transducers are tilted slightly out of parallel so that all pulses received other than the first are lost in the noise. In the case of a bent path, where the sound is reflected back nearly on itself, the transducers and reflector occupy their geometrically correct positions, but the reflection coefficient is purposely made rather small. The first received pulse is attenuated by reflection once and the second three times; the result is that all pulses received after the first are negligible. Hard rubber, Teflon, and perforated metal are suitable materials for the reflector.

A bent path minimizes errors that arise from mass motion of the liquid, and is preferred for field models, and in the laboratory in cases where a large volume of sample, which necessitates vigorous stirring for maintenance of thermal equilibrium, is used. In cases where the liquid is contained in a small tank immersed in a temperature-controlled bath, the straight path is satisfactory.

The advantages of a bent path are even greater for a doubly bent path with two reflectors. This arrangement was used on all near-1962 models.

Greenspan and Tschiegg described the principle of operation as follows.

A block diagram of the sing-around principle is shown in Fig. 3-54. The input transducer is energized by a trigger-type, pulse-forming circuit that produces short pulses. This circuit is adjusted to run free at a prf somewhat less than the expected minimum operating prf. The pulses of pressure produced by the input transducer travel down the sample liquid in a time P/c, where c is the speed of sound and P is the path length. The received pulses are amplified and shaped and are used to synchronize the original pulse-forming circuit. If t_e is the sum of the electrical delays and the time lost in the noise, the total time delay is

$$1/f = t_e + P/c. \tag{3-41}$$

The prf, f, is measured and perhaps recorded. Both t_e and P are obtained by direct calibration with a liquid in which the speed of sound is known. If the velocimeter is to be used in the sea, for example, a suitable calibration liquid is distilled water. Readings of f with distilled water at various tem-

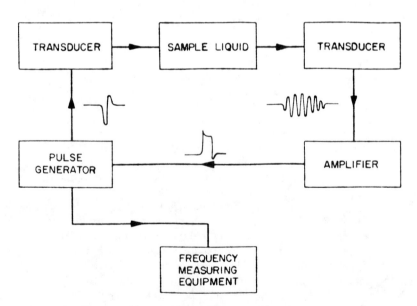

FIG. 3-54.   Block diagram showing sing-around principle. After Greenspan and Tschiegg, 1962.

peratures between 0 and 60°C cover the range which would be obtained in the sea where the extremes of temperature are 0 and 40°C and the salinity reaches perhaps 4%. Corresponding to each temperature of the distilled water is a known speed c and an observed prf, f. These determine the unknowns t_e and P in Eq. (3-41. It is also possible to determine t_e by measuring f for two different known values of P. This method is both more cumbersome and less accurate; the length P in Eq. (3-41) is only an effective length and is difficult to define in an absolute sense, especially in the case where the receiving transducer is not accurately parallel to a wavefront.

3.5.4.2 Phase Locked Loops

Oscillating loops, as he termed them, were conceived by Kritz in 1944 (Kritz, 1958; filed 1948; priv. comm., 1986), as a means of measuring flow velocity using two (or four) transducers. Filed several weeks later than the sing-around flowmeter patent of Garman et al., 1954, Kritz's patent had the technical advantage of operation with a manually or electronically switched pair of transducers, whereas Garman et al. required four transducers. Both patents derive a difference frequency Δf proportional to flow velocity V. Kritz's method required the upstream and downstream cw or pulsed carrier frequencies to be servoed such that the wavelength was the same in either direction (equivalent to saying same number of wavelengths in each direction) resulting in a Δf proportional to V. In the sing-around flowmeter of Garman et al., the difference in the prf's was proportional to V. Kritz's patent recognized that the sum of the frequencies is independent of V. Also, he proposed (but did not illustrate or demonstrate) clamp-on, leaving this notable step to Kalmus et al., 1954.

3.5.4.3 Resonant Frequency

In high-Q systems such as tuning fork sensors, SAW resonators and the like, it is often easier to achieve high precision in "timing" by obtaining the c-related data in the frequency domain. In a bounded region of sound speed c the resonant frequency f is directly proportional to c and inversely proportional to the travel time over the path, e.g., between opposite faces of a crystal. Whether time or frequency is more fundamental is a question not to be answered here, if it is answerable at all. The coexistence of counting, frequency and period measurement functions in individual instruments suggests that a choice is available.

In industrial thickness gaging, resonance principles [Branson "Vidigage," UKAEA (United Kingdom Atomic Energy Establishment-Harwell) ultrasonic micrometer (Aldridge, 1967) seem to have given way to time-of-flight techniques. But some of the most sensitive ultrasonic instruments are resonance-based (HP Quartz Thermometer and Quartz Pressure Gage, Sloan thickness or coating gage for sputtered coatings).

To measure the speed of sound in an object, one can, in principle, use pulse timing techniques or resonant frequency techniques. Suppose the object is a free thin disc of sound speed c and thickness x. At halfwave resonance, $x = \lambda/2$, and the resonant frequency $f_o = c/\lambda = c/2x$. As a numerical example, if $x = 1$ mm and $c = 5000$ m/s, $f_o = 2.5$ MHz.

Alternatively, suppose we wish to measure c by pulse-echo techniques for the same object. How high a frequency is required? Let us assume a total pulse width of 1.5 cycles. To obtain *distinct* echoes, the round-trip travel time $t_{rt} = 2x/c$ would need to be not less than 1.5 cycles and so $f = 1.5 f_o$. As the path or part gets smaller and smaller, if the measurement system is bandwidth-limited, one may be forced to use resonance. As a corollary, if one seeks or only has access to a small sensor, choose (or at least consider) resonance. (See Fig. 3-55.).

Several practical examples illustrate this conclusion: (a) The small HP Quartz temperature and pressure sensors mentioned earlier; (b) Ultrasonic spectroscopy of thin films (Papadakis and Fowler, 1971); (c) Measurements of elastic moduli in slender specimens *too short* for separate echo identification, wherein the use of impulse induced resonance provides a practical solution (Fowler, 1971b); (d) Consider a specimen 100 mm long $\times$ 10 mm diameter. If tested at halfwave resonance (e.g., Köster,

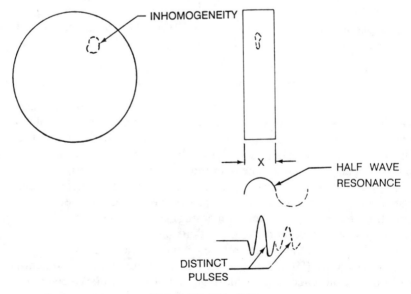

FIG. 3-55. Thin disk.

1948) the d/λ ratio would be (10 mm)/(200 mm) = 0.05. But if tested with a commonly used pulse technique at 100 kHz, d/λ = 0.2, typically (for c = 5000 m/s). In other words, compared to λ, the diameter is much smaller (as desired) in the lower-frequency resonance test.

For objects of *irregular shape*, resonance (modal analysis) testing appears to generate information more easily than do pulse methods (Munce et al., 1986). Even for objects of regular shape, their "soundness" has long been judged by the pitch and quality of the tone (Chapter 1, Fig. 1-1c). The ease of use, and economy, may be dominant factors in choosing resonance, independent of accuracy or other considerations.

Finally, consider the characterization of the elastic properties of small specimens of piezoelectric material. These, too, often lend themselves most naturally to resonance testing (e.g., Warner et al., 1967).

Depending on the equipment available, pulse techniques (as is well known in NDT) still offer advantages for locating an inhomogeneous area, for multizone sensing, or for avoiding undesirable resonances due to standing waves. Whereas resonance yields "average" properties from a single frequency result, pulse techniques can yield information on non-average properties, i.e. inhomogeneties, at the price of more measurements and perhaps more complicated equipment. The 1.5:1 frequency ratio calculated above for distinct echoes disregards the fact that by using deconvolution, thickness can be measured by pulse techniques even when the echoes interfere (Kleinschmidt and Mágori, 1985).

3.5.5   Timing by Pattern Recognition/Correlation

The *IEEE Standard Dictionary of Electrical and Electronics Terms* (1984 edition) defines pattern recognition as the identification of shapes, forms, or configurations by automatic means. In this section, however, we wish to expand the scope of pattern recognition to include nonautomatic means too, such as an operator making a frequency measurement when a particular pattern is recognized on an oscilloscope or other display means. Two examples of pulse-pattern recognition are illustrated that have been widely used in the laboratory for accurate phase velocity measurements: pulse echo overlap and pulse π point. An automatic tracking version of the former was announced in 1987 by Par Scientific of Denmark, 1987. This may be compared with the phase sensitive technique for absolute measurement of velocity reported by Toulouse, 1987, which is capable of being automated. Correlation detection using pseudo-random-noise coded cw transmissions was demonstrated by measuring upstream and downstream ultrasonic phase velocity automatically in 1972, in a contra-propagation flowmeter electronic design due to Pedersen (Lynnworth and

Pedersen, 1972). Waag et al., 1972, used correlation detection in their Doppler flowmeter. Fifteen years later, an eleven-bit Barker-coded tone burst version appeared (Jacobson et al., 1987b), followed by a thirteen-bit Barker code compression technique for NDT (Chen and Deng, 1988).

As background for timing with correlation detection, we define it for modulation systems according to the above-cited 1984 dictionary: *Correlation detection* is

> based on the averaged product of the received signal and a locally generated function possessing some known characteristic of the transmitted wave. *Notes*: (1) The averaged product can be formed, for example, by multiplying and integrating, or by the use of a matched filter whose impulse response, when reversed in time, is the locally generated function. (2) Strictly, the foregoing definition applies to detection based on cross-correlation. The term correlation detection may also apply to detection involving auto-correlation, in which case the locally generated function is merely a delayed form of the received signal.

Useful explanations and examples of coding and cross-correlation analysis appear in Brookner, 1977, Sato et al., 1978, Dixon, 1984, Beck and Plaskowski, 1987, Hayward and Gorfu, 1988 and in Chen and Deng, 1988.

3.5.5.1 Pulse Echo Overlap

This technique, when applied to a transducer/buffer/specimen system, yields a highly accurate measurement of the round-trip transit time t_{rt} for a narrowband or broadband pulse. This was accomplished in systems originating in the early 1970s by superimposing the echoes A and B from the front and back of the specimen on subsequent sweeps of an oscilloscope. The time between sweeps was derived from a highly stable oscillator whose period T_s was made exactly equal to t_{rt} by the operator adjusting for best overlap, Fig. 3-56. For details see Papadakis, 1973, 1976.

3.5.5.2 Pulse π Point

When this "overlap" technique is applied to a transducer/buffer/nondispersive specimen system, the carrier or center frequency f of a tone burst is swept until a series of constructive ($+$) and destructive ($-$) interferences are observed, corresponding to standing wave interferences in the specimen.

One uses tone bursts of duration much longer than the transit time in the specimen. One then measures frequencies at which successive interferences occur in a sensor that is several wavelengths long at the lowest frequency. If one plots the number of wavelengths in the sensor versus

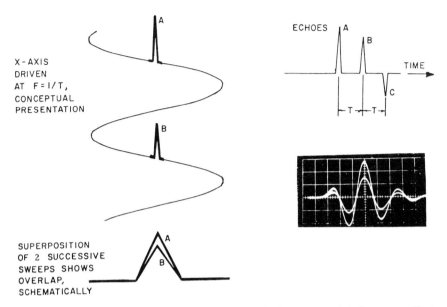

FIG. 3-56. Pulse echo overlap technique originally relied on operator's judgment to adjust an oscillator until the overlap pattern was obtained. Absolute accuracies on the order of 10 ppm were achieved. Oscillogram shows A and B echoes overlapped when oscilloscope X-axis is driven at the frequency $F = 1/T$. Echoes A and B are intensified using Z-axis modulation.

in-phase or out-of-phase frequencies, from which the slope is readily obtained, one can determine the sought time delay between echoes.

From the delay the phase velocity c is determined. From the amplitudes of the envelopes, the attenuation coefficient α also can be determined. The name "pulse π point" refers to the fact that as f is swept, successive maxima and minima correspond to additional halfwave interferences (180°, or π radians). The method may be clarified with the aid of two oscillograms (Fig. 3-57). Although not as accurate as the previous method (pulse echo overlap) the pulse-π-point method requires less equipment. For a full description, see Papadakis, 1971.

3.5.5.3 Correlation Detection by Using Coded Waveforms

Coded patterns occur in nature. For example, tree rings have been used for timing in the sense of dating an old wooden beam. To date the beam, one first needs a reference pattern. Douglass (1919, 1928 and 1936, e.g.) of the University of Arizona examined cross sections of hundreds of trees in a southwestern part of the U.S. to develop a master calendar spanning

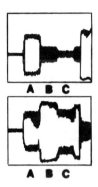

FIG. 3-57.   Destructive and constructive interference of tone burst echoes from beginning and end of specimen. After Papadakis, 1971.

some six hundred years from the 1300s to the early 1900s. The nonrecurring pattern of rings enables one to slide a pattern from a beam along the master calendar until a unique match is found. In the language of timing, one finds the delay from t = 0 at which the cross-correlation between the beam pattern and the reference is maximum. More recent accounts of crossdating and dendrochronology appear in Fritts, 1976, and in the *Smithsonian Magazine*, July 1985.

By the mid-1980s several investigators pointed out the merits of *correlation detection* as a means of enhancing ultrasonic transit time measurements in flowing media (Laenen, 1984; Marttinen and Luukkala, 1985; Nuspl et al., 1986; Jacobson et al., 1987b). In fact, as early as 1971 an M-sequence phase-modulation method reported by Okujima and Ohtsuki, 1971, enabled one to range gate within transmitted and received beams (illustrated in the author's 1979 review, p. 499). In 1972 a coherently coded cw transit-time flowmeter employing correlation detection, designed by Pedersen and reported in Lynnworth and Pedersen, 1972, had been demonstrated with a response time of 20 ms. At the same conference, Waag et al., 1972, reported their use in Doppler blood-flow measurements, of pseudorandom binary phase coded cw or tone bursts plus correlation detection. In subsequent Doppler studies, stationary noise modulation of transmitted cw or bursts was practiced by Cooper and Newhouse, 1976, Jethwa et al., 1975, and others. The advantages of a pseudorandom code, however, is that it can be delayed by a shift register and so does not require a delay line. (See Azimi and Kak, 1985.)

Correlation detection would appear to be particularly useful in determining the transit time of pulses masked by or attenuated into the noise; pulses whose amplitude and/or phase are jittery (which may be caused by high turbulence and/or multiphase media); or pulses that are distorted because of multipath phenomena such as those encountered in flow measurements across rivers or in closed conduits in which transverse gradients in composition or temperature lead to transverse gradients in sound speed, which in turn refracts some of the rays differently from others.

As is well known from radar signal detection analysis, under poor signal-to-noise conditions, correlation detection probably provides one of the most reliable ways of detecting a pulse. Correlation detection utilizes the energy and information content of the entire signal, not merely the information contained in a portion of one cycle. It is a form of pattern recognition in which random noise tends to be disregarded. The practicality of the technique is due in part to the availability of fast microprocessors and fast, low-cost A/D converters appropriate for the maximum ultrasonic frequencies of main interest in time-of-flight flowmeters (2 MHz).

The correlation detection flowmeter shown in Fig. 3-58 used the same basic flowcell configurations and wetted and clamp-on transducers previously used by its manufacturer for more conventional transit time contrapropagation measurements of flow. Correlation detection can also be applied to timing the flow of eddies or other inherent or induced tags, but the presence of tags is not *required* in contrapropagation measurements using correlation detection. If scatterers are indeed present, one can also adapt correlation detection to a range gated transit time mode (Embree and O'Brien, 1985) or to a range-gated Doppler mode. (See, for example, Azimi and Kak, 1985, and Section 4.5.2 on applications of the correlation detection flowmeter.) Compared to the arm/zero cross method of timing a pulse, the correlation method tends to be considerably slower because of digitizing, averaging and calculations involved. To some extent, faster response is obtainable if one is willing to purchase faster and correspondingly more expensive equipment. In a commercial correlation detection contrapropagation instrument available in 1987 (Panametrics Model 6068), the shortest response time was 1 s.

As an example of a transit-time flowmeter that employs correlation detection, consider the Panametrics Model 6068 flowmeter, which, in prototype form, began field tests in 1987. In the 6068, flow velocity is derived from the transit-time difference between ultrasonic signals transmitted upstream and those transmitted downstream through the fluid. However, the 6068 differs from earlier transit time flowmeters in the following respects:

i.   Instead of a delta-like pulse, square wave or single-frequency tone burst excitation of the transmitting transducer the 6068 drives the

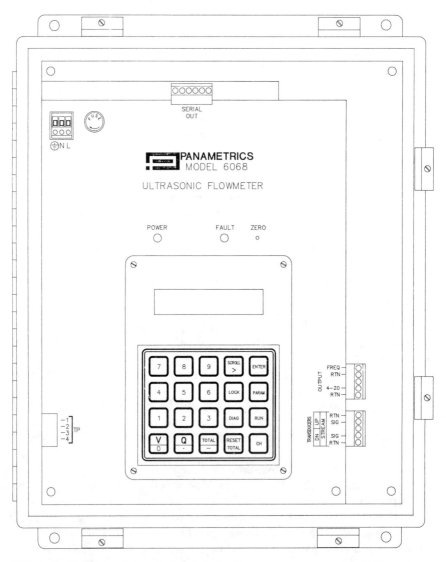

FIG. 3-58. Example of a correlation detection instrument programmed in 1987 to operate as a transit time flowmeter. In 1988, new programs extended the operating modes to reflection. In 1989 it was planned to add a third mode, tag cross-correlation.

transmitter with a binary phase encoded signal, typically 4 to 20 cycles in length.

ii. The received signal is digitized using a "flash" analog to digital converter.

iii. Rather than using threshold or zero-cross detection schemes the 6068 determines the transit time by correlating the encoded transmitted signal with the digitized received signal.

A block diagram of the 6068 is shown in Fig. 3-59. One possible transmitted code and the corresponding received signal appear in Fig. 3-60. Note that preservation of the code depends in part on transducer bandwidth. Figure 3-60 also compares the eleven-bit Barker code with other driving waveforms.

3.5.6 Amplitude Sensing

Amplitude sensing can range from signal present or absent (as in gap-type liquid level gages) to instruments that measure accurately the amplitude of echoes versus time or versus distance. Sometimes an indirect measure of signal amplitude is obtained by monitoring an AGC (automatic gain control) voltage. Reflectometers or impedometers may use reference echoes as a basis for calculating the reflectivity or impedance of a load. In the present section, we also include a proposed method for ranging to a known source without any synchronization link to that source. Regarding the above amplitude-sensing techniques, it is more important to extract the principles that are used, in contrast to specific embodiments. The latter tend to become obsolete because of equipment innovations and improvements, e.g., microprocessors, faster sample and hold circuits, digitization breakthroughs, etc.

3.5.6.1 Attenuation Comparator

In the 1950s, Matec began manufacturing ultrasonic attenuation comparators that enabled an operator to match the exponential decay of an RC circuit to a series of echoes ringing down in a specimen. At that time, this technique provided a practical way for many operators to obtain an α-related measurement in units of dB/cm or dB/μs, for example. Amplitude was "sensed" by observing the vertical deflection on a cathode ray tube. Attenuation comparators thus provided essentially an oscilloscope-type measurement that was "easy to use" with respect to α determinations (Einspruch, 1963).

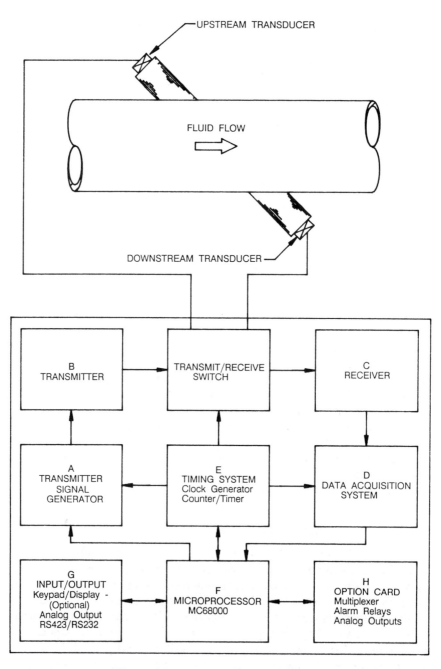

FIG. 3-59.   Model 6068 system block diagram.

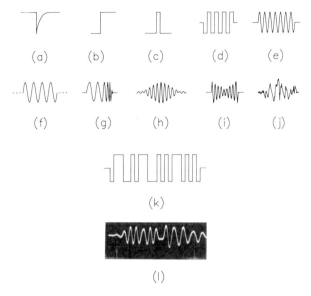

FIG. 3-60. Comparison of electrical waveforms for driving the transmitting transducer. Examples include amplitude, frequency and phase modulation. (a) Broadband spike approximates delta function. (b) Step function. (c) Half-cycle square wave. (d) Square wave burst. (e) Tone burst. (f) Continuous sine wave. (g) Chirp (fm tone burst). (h) Gaussian-enveloped or $\cos^2$-enveloped narrowband tone burst. (i) Tone burst amplitude modulated inversely, to compensate for narrowband transducer's impulse response. (j) Random noise. (k) Eleven-bit Barker phase code. (l) Signal received by broadband transducer after transmission across water path of 1-MHz ultrasonic tone burst modulated by (k).

3.5.6.2   Gated Peak Detectors

In automated flaw detectors, it is often necessary to respond rapidly to flaw echoes detected during fast scans. As an example, one might want to alarm on flaw echoes above a threshold and occurring between front- and rear-surface echoes from an immersed part undergoing a C-scan test.

In an automatic intervalometer developed around 1965 for waveguide thermometry (Chapter 5, Section 5.2.1), B.M. Gordon derived time intervals from the midpoint of threshold crossings, where said thresholds in turn were derived from the peak amplitude of selected broadband echoes. The peak amplitudes or their ratios were later applied to controlling the electrolytic etching of refractory wire ultrasonic temperature sensors and to calculating reflection coefficients and attenuation coefficients at high temperature.

These examples illustrate the need for gated peak detectors. The solution frequently has been to use a fast sample and hold device. By "fast" one means the ability to respond accurately to a high frequency pulse, say 10 or 100 MHz. If the received signal is digitized, then the peak can be extracted by digital-signal processing techniques. In some early analog solutions, comparators afforded a way to test an incoming signal against a set of reference levels $E_1, E_2, \ldots, E_n$. A diode, triode or transistor, for example, could be biased to conduct only when the incoming signal exceeded the reference (pulse height analysis).

3.5.6.3 AGC Monitors

In order to help keep track of transit-time flowmeter performance and explain certain performance irregularities, it became apparent that some form of amplitude sensing would be desirable. In some instruments, the AGC level provided a readily available monitor signal that was related, generally nonlinearly, to the acoustic signal intensity. "Low-signal" warning lights or alarms could be activated when the AGC level approached the maximum-gain limit. Preventive maintenance of transducers could be tied to the AGC level.

3.5.6.4 Reference Echoes

In some liquid level probes of Van Valkenburg, 1957, 1962, the end echo in a zigzag shear wave probe or the echoes from a series of near-surface holes in a surface-wave probe establish voltage levels for dry versus wet probes. Detailed analysis of such probes occurred long after they were invented, suggesting that the state-of-the-art was ahead of the state-of-the-science, at least in the intuitive beginnings of some aspects of ultrasonic technology. The principle of damped probes as liquid level sensors continues to find applications in the late 1980s, but now the reference echoes, if needed, can be stored digitally. Likewise, prior to the advent of digital flaw detectors for nondestructive testing, it was often necessary to "calibrate" analog flaw detectors by using flat bottom holes or side drilled holes as reference reflectors. It is still convenient to describe the sensitivity of flaw-detection systems in terms of the ability to detect an echo from a particular target in a given material such as a 1-mm diameter hole, or a 1-mm diameter spherical steel ball in water at, say, 1 m from the transducer.

3.5.6.5 Ranging by Single- or Dual-Frequency Techniques

It is a matter of common experience, at least to most dog owners, that the distance from a listener to the source of a familiar sound such as a dog's bark can be judged as "near" versus "far" from the loudness and quality of the sound. In this section we consider how the "remote-dog-bark principle'' might be converted from a qualitative near/far estimator to a quantitative ranging method.

First, in a medium without discrete reflectors, let us suppose that a known monochromatic cw source of frequency f produces a sound pressure amplitude A_o in the farfield at a distance x_o from the source. At distances more remote from the source than x_o, the pressure A will be less than A_o because of beam spread and absorption. The beam spread contribution is 6 dB/double distance. The absorption loss shall be assumed to be proportional to f^2. At sufficiently *low f*, beam spread dominates, and a farfield listing of A versus x looks like this:

$$
\begin{array}{llllll}
x/x_o & 1 & 2 & 4 & 8 & 16 \\
A/A_o & 1 & \frac{1}{2} & \frac{1}{4} & \frac{1}{8} & \frac{1}{16}
\end{array}
$$

In other words, $x = x_o A_o/A$.

At sufficiently *high frequency*, absorption dominates. Therefore $A = A_o\, e^{-\alpha x}$ or $x = (1/\alpha) \ln (A_o/A)$. As a simple numerical example, and to the extent that beam spread may be neglected, if f is chosen such that $\alpha = 1$ Np/m $= 8.686$ dB/m, then $x = \ln (A_o/A)$, in meters.

At *intermediate frequencies*, however, both losses are present. The dual-frequency technique described next provides an approach for eliminating the beam spread contribution as a source of range uncertainty, at least theoretically. [Compare concept with change in quality of a complex sound heard at different intensities (Morse, 1948, pp. 227-228).]

The *dual-frequency amplitude ranging method* may be explained as follows. In a medium without discrete reflectors, as before, let there be

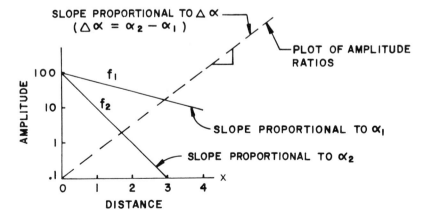

FIG. 3-61.   Graphical illustration of dual-frequency ranging method.

Table 3-12. Summary of Passive Techniques for Four Applications

Application	Passive Ultrasonic Method
1. Steady flow	Measure noise amplitude, frequency distribution. Cross-correlate noise signatures received at two stations (Kupperman et al., 1987. See Chapter 4, Section 4.5.4).
2. Sudden leak	Measure arrival time of expansion wave at two separated points (Rocha and Finch, 1986. See Fig. 3-62).
3. Crack propagation	Monitor acoustic emissions (see Chapter 2, Section 2.6.5).
4. Crack penetration	Measure noise due to pressurized fluid leaking through the crack.

two sources close together, at frequencies f_1 and f_2, each producing at the farfield distance x_o pressures of equal magnitude A_o. (One may let the aperture diameters d_1 and d_2 be inversely proportional to $\sqrt{f_1}$ and $\sqrt{f_2}$ so that beam spread losses will be the same for both waves.)

At a remote point x whose distance from x_o is to be determined, the sound pressures at the two frequencies will be $A_1 = A_o e^{-\alpha_1 x}$ and $A_2 = A_o e^{-\alpha_2 x}$. Their ratio is independent of A_o: $(A_1/A_2) = e^{-(\alpha_1 - \alpha_2)x}$, from which $x = [1/(\alpha_2 - \alpha_1)] \ln (A_1/A_2)$. A semilog plot of A_1 and A_2 versus x shows graphically how the range x is determined from the amplitude

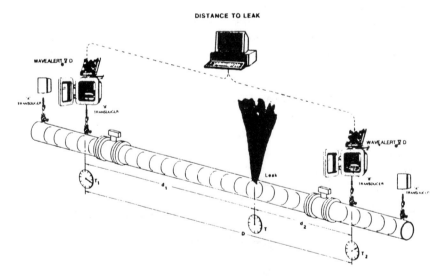

FIG. 3-62. Leak amplitude and location can be determined by using two acoustic monitoring systems, one placed at each end of the pipeline. Source: Rocha and Finch, 1986. © 1986 Helmers Publ.

ratio (Fig. 3-61). An especially simple case occurs when $f_2 = \sqrt{2}f_1$. Here, assuming α is proportional to frequency squared, $\alpha_2 = 2\alpha_1$. Therefore $x = (1/\alpha_1) \ln (A_1/A_2)$.

3.5.7   Passive Techniques for Detecting Flow, Sudden Leaks, and Crack Propagation and Penetration

Passive ultrasonic techniques associated with the applications listed in the heading are summarized in Table 3-12.

3.6   Differential Path Techniques—Continuous or Discrete Steps

With a "calibrated" instrument, it may be possible to measure the sound speed c and the attenuation coefficient α by propagating ultrasound between two fixed points A and B. But while not a differential *path* measurement, even this is a "differential" measurement of sorts—one observes the difference in transit time or amplitude when the unknown medium is substituted for one of known propagation constants (Greenspan and Tschiegg, 1962). Differential path techniques are employed to minimize bias or errors due to boundary layer effects, time delays not associated with travel time in the medium under test or energy losses other than dissipation in the medium under test. Differential path techniques reduce the reliance on assumptions about the instrument electronics or transducers.

In effect, differential path techniques seek a best fit to data such as a series of points expressing travel time versus distance, or expressing intensity versus distance, so that c and α may be determined accurately. The more points, presumably the better the accuracy. But points cost money, so sometimes two points suffice.

Probably the first use of the differential path technique in measuring sound speed accurately occurred some 250 years ago. In 1738, according to an introductory passage in Rayleigh's *The Theory of Sound*, 1877, members of the French Academy timed the sound of a cannon at different distances.

3.6.1   Phase Velocity, Attenuation Coefficient in Bulk Solids

Referring to Fig. 3-63, if a solid block of thickness W can be machined thinner and thinner, with transit time and amplitude measured as often as desired (as when thickness = X, Y, Z, . . .) then much data on prop-

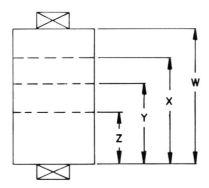

FIG. 3-63.   Block expended in differential path experiment.

agation can be accumulated before the block disappears. However, this method has one shortcoming if one wishes to repeat the experiment.

3.6.1.1   Step Wedge

The step wedge (Figs. 3-64 to 3-66) provides a sequence of different paths. If the block is homogeneous, it is a convenient specimen in which to measure c and α. Step wedges can be used for determining shear-wave propagation constants at an angle of incidence beyond about 30°, e.g., 45°, so that the longitudinal wave is not launched by mode conversion. In this oblique case it is helpful to use wedge transducers made of the same material as the step wedge to avoid ray bending. Liquid or grease couplants are usable for vertically polarized obliquely incident shear, as explained in Section 3.3.

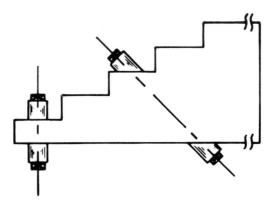

FIG. 3.64.   Step wedge for longitudinal and shear measurements at normal or oblique incidence.

FIG. 3-65.   Drill press used to pressure couple shear waves at normal incidence through a
plastic step wedge.

3.6.1.2   Double Wedge for Longitudinal and Shear Waves

Continuous incrementing of the path is possible with a pair of wedges
(Fig. 3-67). These make it easy to change the path by exactly one wavelength, $\Delta x = \lambda$, from which the phase velocity is immediately obtained
as $c = f\lambda$. Thus, c can be measured as a function of f, from which the
group velocity U can be obtained. These remarks apply to longitudinal
and to shear waves. The method gets complicated if the material is in-

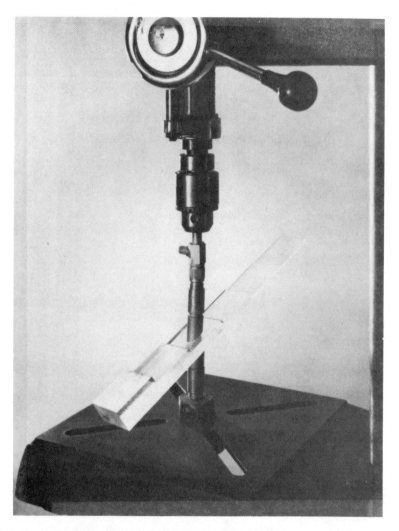

FIG. 3-66.   In this oblique incidence experiment, 2-MHz SV shear waves are coupled between acrylic probes and an acrylic step wedge specimen using a thin film of petroleum jelly. After Lynnworth, 1967b.

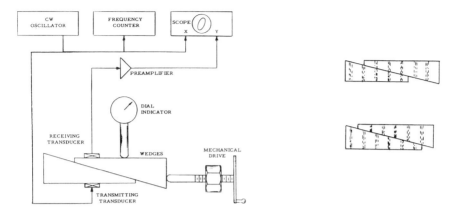

FIG. 3-67.   Sliding solid wedges allow continuous change in the path length, for longitudinal and shear waves. (a) Block diagram of system for measuring phase and group velocities, c and U, respectively. One varies path length at each fixed frequency of interest to determine c. One varies frequency while holding the path fixed, to determine U. (b) Example of difficulty encountered if material is not homogeneous. After Lynnworth et al., 1973b.

homogeneous and of limited size, but nevertheless may serve as a convenient starting point. Its application to Teflon and to a three-dimensionally-reinforced composite is discussed in Lynnworth et al., 1973b.

3.6.1.3   Sliding Wedges for Rayleigh Waves

Measurement of c_R, the phase velocity of Rayleigh waves, can utilize a pair of sliding transducers of the wedge, comb or knife edge design (Viktorov, 1967; Lakuczek, 1986). The use of appropriate low-loss or low-speed wedges, particularly where the wave in the wedge is SV shear, allowed the differential path method to be applied at high frequency (e.g., 10 MHz, Fig. 3-68) and even to solids of relatively low sound speed, e.g., graphite, Fig. 3-69, yielding the c_R data of Fig. 3-70 (Lynnworth, 1967a,b). If frequency is to be swept continuously, the comb method (assuming equidistant teeth) would be abandoned in favor of the broadband wedge or knife-edge method.

3.6.1.4   Phase Contrast Over Parallel Paths

It is fairly common in SAW sensors to attempt to null out all variables except the sought one, by means of parallel paths. Ideally, the paths differ only in the presence of a special coating in one path that is sensitive to the measurand. Examples of this technique appear frequently in the lit-

FIG. 3-68. By using an incident *shear* wave, the incident wave can be slow enough to launch Rayleigh waves in many materials, even when the wedge is made of a highly elastic (nonplastic) material. Elastic wedge can be operated at much higher frequencies than plastic wedge. View shows flint glass wedge. Wedge is coupled to the glass specimen by thin oil film, or coupled dry with a wringing or optical contact. These probes have been operated at both fundamental and overtone frequencies of the 10-MHz shear wave crystal.

erature. (See, for example, *The IEEE Transactions on Ultrasonics, Ferroelectrics and Frequency Control*, March 1987.)

3.6.2 Phase Velocity in Solid Slender Magnetostrictive Waveguides

Analogous to the sliding-on-the-surface phase velocity methods of Section 3.6.1.3 for Rayleigh wave-propagation measurements, one can launch and detect other waves in solid waveguides while retaining the advantages of a *continuously* variable path.

One particularly simple case is the solid slender *magnetostrictive* waveguide (Fig. 3-71). If a broadband pulse is launched from one end, and received at distances regularly incremented by 10 mm, a multiple exposure oscillogram with traces displaced also by 10 mm, as in Fig. 3-72, can be recorded. (Transducers for generating and detecting these pulses are shown in Fig. 3-73.) The tangent to the received signal pulses has a slope equal to c. Slender specimens, not necessarily magnetostrictive, if attached to the end of a lead-in waveguide of dissimilar mechanical impedance, generate echoes from the joint (A) and the end (B). Usually there is a reverberation echo (C). The time intervals between them, and their relative amplitudes, may be used to compute c and α as a function of, say, temperature, as explained by Papadakis et al., 1974. Waveguided echoes representing energy trapped in the specimen yield multipath data analogous to that obtained with bulk waves in a step wedge having equal steps. (Compare with Mason, 1964.)

If the waveguide system includes a member of noncircular cross section in which a torsional wave propagates (Fig. 3-74), it is possible to measure the phase velocity in that member by varying continuously (or stepwise) the depth of immersion in a liquid, typically water. Because of the difference in inertia between the wetted and dry segments (Chapter 7, Section 7.5.2.2; Lynnworth, 1977; Bau, 1986; Kim and Bau, 1988) a small echo is generated where the noncircular cross section intersects the liquid

FIG. 3-69.   Sliding wedge method measures c_R in graphite. By taking advantage of the slow shear wave velocity in acrylic, 0.5-MHz Rayleigh waves can be launched in ATJ graphite. This cannot be done using the traditional longitudinally-operated acrylic wedge transducer, nor using a water-immersed system. The Rayleigh wave velocity in this ATJ graphite specimen, 1220 m/s, is less than the speed of sound in water, ~1500 m/s. After Lynnworth, 1967b.

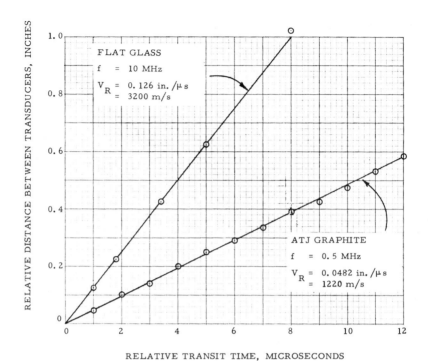

FIG. 3-70.   Rayleigh wave data measured with obliquely incident shear waves in differential path experiments.

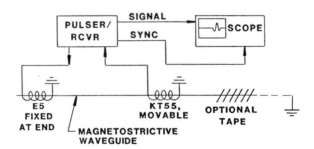

FIG. 3-71.   Block diagram shows arrangement for extensional mode phase velocity measurement using differential path technique. E5 and KT55 transducers are shown in Fig. 3-73.

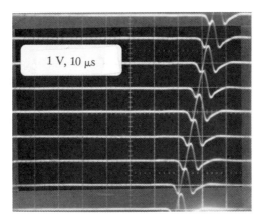

FIG. 3-72. Multiple-exposure oscillogram shows how slope of distance (y-axis) versus transit time (x-axis) yields phase velocity, 5 km/s in this example. For each trace, the path was increased by 1 cm, starting at the bottom trace.

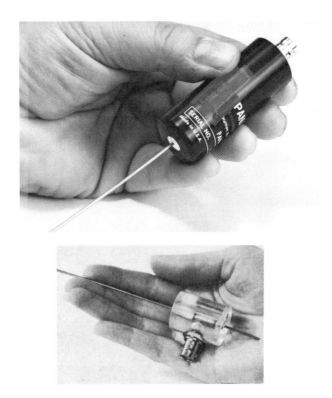

FIG. 3-73. Fixed and movable coils housed for use with magnetostrictive waveguides.

219

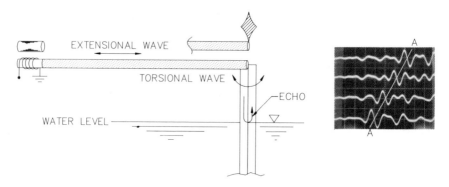

FIG. 3-74.   Left: Torsional echo generated at intersection of noncircular waveguide and water surface. Right: Phase velocity, 1875 m/s, determined from changes in transit time to echo as depth of immersion is increased in 1-cm steps. Sweep, 10 µs/div. Line AA represents graph of immersion depth versus transit times to zero crossings, the slope of which yields the phase velocity.

surface. For torsional waves in a stainless steel waveguide of diamond cross section, of aspect ratio $b/d = 3$, partly immersed in water, the reflection coefficient is small, $<10\%$. The echo is therefore difficult to detect unless spurious echoes due to ringing, contacts or supports are avoided, subtracted or suppressed by the methods of Section 3.4.8. The four-trace oscillogram in Fig. 3-75 shows the 100-kHz water/air interface echo produced by a torsional mode diamond cross section sensor, of $b/$

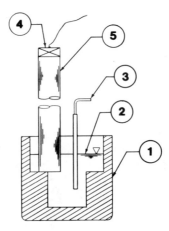

FIG. 3-75.   Sound velocity measured in molten iron. Legend: 1–Crucible. 2–Molten iron. 3–Thermocouple. 4–Transducer, 2-MHz. 5–Buffer rod, SS, water-cooled, ~34-mm ø × ~0.5-m long. After Lehtinen, 1960.

$d = 3$, immersed to depths increasing in 1 cm steps. The phase velocity in this example is $c_t = 1875$ m/s.

Similarly the phase velocity of flexural waves could be measured under conditions where reflections are produced at the liquid/vapor interface (Section 7.5.2.3; Dieulesaint et al., 1987a).

3.6.3 Propagation in Fluids

Differential paths in fluids, compared to solids, can more readily be arranged to be continuously variable. However, time constraints, as imposed by shock-heated gases, or high-temperature plasmas well above the m. pt. of the probes, may dictate that only fixed paths (discrete steps) be utilized for c and α measurements. Interferometric measurements of c usually utilize steps of $\lambda/4$, $\lambda/2$, λ, etc.

3.6.3.1 One Step in Crucible Defines Path Length in Molten Iron

Referring to Fig. 3-75, due to Lehtinen, 1960, the arrangement is self-explanatory. By measuring the temperature T of molten iron during solidification, and the travel time from the end of the stainless steel buffer to the bottom of the crucible, $c(T)$ was determined. Stirring or other means may be employed in such an experiment to minimize errors due to temperature gradients, depending on the accuracy sought.

3.6.3.2 Two or More Discrete Steps

In shock tube experiments (Carnevale et al., 1967c), use of a rectangular test section allows c and α measurements to be made in orthogonal paths of unequal lengths. The two-path arrangement was needed in order to eliminate boundary layer effects. Motegi et al., 1987, demonstrated a measurement of c in water using a pair of parallel waveguides differentially spaced up to 150 mm apart. They reported that the waveguides supported either leaky Lamb or leaky Rayleigh waves.

Four differential fixed-path ultrasonic interferometers described by Sarvazyan and Chalikian, 1988, yielded a relative error of about $10^{-3}\%$ in small samples (0.25 cm³ each) that could be tested near 7 MHz over a pressure range of 0.1 to 250 MPa.

In plasma experiments (Carnevale et al., 1967b), reciprocating probes were advanced to different positions in the plasma, yielding differential path oscillographic data similar to that in Fig. 3-76. Because of the temperature gradient in the plasma, it turns out that as the probes penetrate deeper into the plasma stream, c and α are determined at increasing tem-

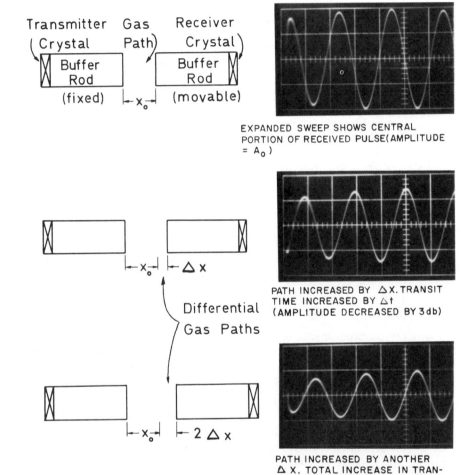

FIG. 3-76.   Schematic of probes and oscillograms, illustrating differential path technique. Simulated: Test gas, argon; T = 7000 K. After Carnevale et al., 1967b.

peratures, without changing the plasma operating conditions (gas flow, electrical power).

3.6.3.3 Continuously Variable Path: Interferometer

Examples of variable-path ultrasonic interferometers for measuring c and/or α in a high-temperature gas, a high-pressure liquid and a high-temperature liquid, respectively, are found in papers by Sherratt and Griffiths, 1934, Litovitz and Carnevale, 1955 (Fig. 3-77) and Macedo and Litovitz, 1965. The path may be changed in steps of exactly one wavelength, directly yielding $c = f\lambda$, the wave number $k = 2\pi/\lambda$ and α in units of dB/wavelength. Alternatively the steps may be units of, say, 1 mm, or time units, say 1 μs, the latter choice directly yielding c as distance/μs and α in units of dB/μs. The choice of increments is usually one of convenience or operator preference and may depend on how elements in the apparatus were calibrated.

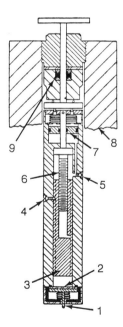

FIG. 3-77. Variable path interferometer for liquids at high pressure (2000 kg/cm²). Legend: 1–Ungrounded electrical lead-in. 2–Quartz crystal transducer. 3–Movable piston. 4–Pin to keep piston from rotating. 5–Filler plug for loading with liquid. 6–Micrometer screw. 7–Floating piston. 8–High-pressure vessel to contain acoustic system. 9–Teflon seal around rotating shaft. After Litovitz and Carnevale, 1955. © 1955 AIP.

3.7 Some Commonly Used Equations, Approximations and Selected Nomograms

Phase velocity, etc.	$c = f\lambda$, $\lambda = c/f$, $k = 2\pi/\lambda$, $\omega = 2\pi f$
Group velocity	$U = d\omega/dk = c/[1 - (f/c)(dc/df)]$
Attenuation coefficient, monatomic fluids (classical absorption, only viscous and thermal losses)	$\alpha = \dfrac{\omega^2}{2\rho_o c^3}\left(\dfrac{4\eta}{3} + \dfrac{K_{th}(\gamma - 1)}{C_p}\right)$
Amplitude (pressure) ratio	$A/A_o = (1/r)\, e^{-\alpha x}$
Intensity	$I = (\text{rms pressure})^2/\rho c$
Farfield diffraction (beam spread) loss	4 dB at d^2/λ plus 6 dB per double distance thereafter
Nearfield diffraction loss	L_d (dB) $\approx S_k$ where S_k = Seki parameter $= \lambda z/a^2$
Beamwidth between 3 dB points of piston radiator, radius a	$\theta_{3dB} \approx \lambda/d = \lambda/2a$
Focal length, plano-curve lens, where n = index	$F = (\text{radius of curvature})/(n - 1)$
Sound pressure reflection coefficient, normal incidence	$R = (Z_2 - Z_1)/(Z_2 + Z_1)$
Sound pressure transmission coefficient, normal incidence	$T = 2r/(r + 1)$ where $r = Z_2/Z_1$
Energy transmission coefficient, normal incidence	$T_e = 4Z_1 Z_2/(Z_1 + Z_2)^2 \approx 4Z_1/Z_2$ for $Z_2 \gg Z_1$
Characteristic acoustic impedance	$Z = \rho c$
Mechanical impedance, extensional waves	$Z_e = \rho c_e A$
Mechanical impedance, torsional waves	$Z_t = \rho c_t J$
Transformation by $\lambda/4$ matching layer of impedance Z_m	$Z_{in} = Z_m^2/Z_2$
Extensional velocity	$c_e = \sqrt{E/\rho}$ ($\lambda > 5d$)
Torsional velocity, round bar	$c_t = \sqrt{G/\rho}$
Torsional velocity, square bar	$c_t = .9184\sqrt{G/\rho}$
Longitudinal velocity in bulk solid (diameter $d > 5\lambda$)	$c_L = \sqrt{\dfrac{E}{\rho}\dfrac{1 - \sigma}{(1 + \sigma)(1 - 2\sigma)}}$
Shear velocity in bulk solid	$c_T = \sqrt{\dfrac{E}{\rho}\dfrac{1}{2(1 + \sigma)}} = \sqrt{\dfrac{G}{\rho}}$
Velocity ratios and Poisson's ratio	$c_e/c_L = \sqrt{\dfrac{(1 + \sigma)(1 - 2\sigma)}{1 - \sigma}}$

$$c_T/c_L = \sqrt{\frac{1 - 2\sigma}{2(1 - \sigma)}}$$

$$= \frac{1}{2} \text{ for } \sigma = \frac{1}{3}$$

Rayleigh velocity	σ	c_R/c_T
Shear velocity	0	0.8744
	$\frac{1}{4}$	.9194
	$\frac{1}{2}$	.9553

Rayleigh leakage $\quad\quad\quad\quad\quad$ ~0.5 dB/wavelength for steel/water interface

$$\sigma_{TL} = \frac{1 - 2(c_T/c_L)^2}{2 - 2(c_T/c_L)^2} = \begin{cases} 0 \text{ for } c_T/c_L = .7071 \\ .5 \text{ for } c_T/c_L = 0 \end{cases}$$

$$\sigma_{RT} = 1 - (8 - 8a^2)/(8a^2 - 8a^4 + a^6) \text{ where } a = c_R/c_T$$

$$\sigma = \left(\frac{E}{2G} - 1\right) = \frac{1}{2}\left(\frac{3K - 2G}{3K + G}\right) = \left(\frac{3K - E}{6K}\right)$$

where E is Young's modulus, G is the modulus of rigidity and K is the bulk modulus.

Snell's Law	$c_1/\sin \theta_1 = c_2/\sin \theta_2 = c_3/\sin \theta_3$
Mode conversion, solid/air boundary	Longitudinal converts to ($\Rightarrow$) shear at all incident angles except $\theta_i = 0$ and 90°; necessary but not sufficient conditions for 100% conversion: $\sigma < 0.26$ and $\theta_i > 38°$.
	Shear $\Rightarrow$ longitudinal at all angles except $\theta_i = 0$ and $\theta_i \geq 45°$; necessary but not sufficient conditions for 100% conversion: $\sigma < 0.26$ and θ_1 between 26° and 45°.
Orthogonal 100% mode conversion	$L \Rightarrow 100\%S$ at $\sigma = \frac{1}{4}$ and $\theta_i = 60°$ $S \Rightarrow 100\%L$ at $\sigma = \frac{1}{4}$ and $\theta_i = 30°$.
Doppler frequency shift, beam drift	Proportional to Mach number V/c
Contrapropagation transit time	$\Delta t = 2LV/c^2$ (approx. for $V^2 \ll c^2$)
Vortex shedding frequency	$f = 0.2V/$(strut diameter)
Cross-correlation function between $x(t)$ and $y(t)$	$R_{xy}(\tau) = \lim\limits_{T \to \infty} \frac{1}{T} \int_0^T x(t)\, y\,(t + \tau)\, dt$

3.8   Numerical and Graphical Databases— Acoustical Properties of Selected Media

Examples of data useful in the design and analysis of ultrasonic instrumentation for process control, as well as for research and experimentation, are given in Tables 3-13 to 3-16, and in the graphs and nomograms contained in Figs. 3-78 to 3-90.

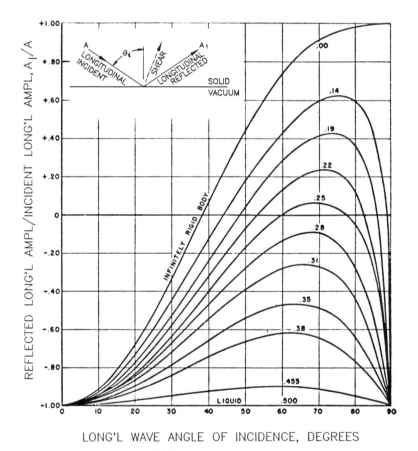

LONG'L WAVE ANGLE OF INCIDENCE, DEGREES

FIG. 3-78.   Variation of amplitude of reflected compressional wave at solid-air interface, for varying Poisson's ratios. After Arenberg, 1948. © 1948 AIP.

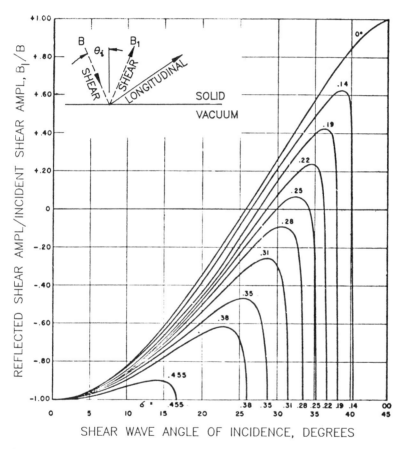

FIG. 3-79. Amplitude of reflected transverse wave at a solid-air boundary as a function of angle of incidence for varying Poisson's ratios. After Arenberg, 1948. © 1948 AIP.

Table 3-13. Elastic Constants, Wave Velocities and Characteristic Impedances of Metals, Glasses, and Plastics. (Reproduced by Permission from *American Institute of Physics Handbook*, © 1957 McGraw-Hill Book Co.)

Materials	Young's Modulus Y_0 (Newtons/m² × 10⁻¹⁰)	Shear Modulus μ (Newtons/m² × 10⁻¹⁰)	Lamé Constant λ (Newtons/m² × 10⁻¹⁰)	Poisson's Ratio σ	Longitudinal Velocity $V_l = \sqrt{\frac{\lambda + 2\mu}{\rho}}$ (m/sec)	Shear Velocity $V_s = \sqrt{\frac{\mu}{\rho}}$ (m/sec)	Extensional Velocity $V_{ext} = \sqrt{\frac{Y_0}{\rho}}$ (m/sec)	Longitudinal Impedance $Z_l = \sqrt{\rho(\lambda + 2\mu)}$ (kg/sec-m² × 10⁻⁶)	Shear Impedance $Z_s = \sqrt{\rho\mu}$ (kg/sec-m² × 10⁻⁶)
Aluminum, rolled	6.8–7.1	2.4–2.6	6.1	0.355	6,420	3,040	5,000	17.3	8.2
Beryllium	30.8	14.7	1.6	0.05	12,890	8,880	12,870	24.1	16.6
Brass, yellow 70 Cu, 30 Zn	10.4	3.8	11.3	0.374	4,700	2,110	3,480	40.6	18.3
Copper, rolled	12.1–12.8	4.6	13.1	0.37	5,010	2,270	3,750	44.6	20.2
Duraluminum 17 S	7.15	2.67	5.44	0.335	6,320	3,130	5,150	17.1	8.5
Gold, hard drawn	8.12	2.85	15.0	0.42	3,240	1,200	2,030	62.5	23.2
Iron, electrolytic	20.6	8.2	11.3	0.29	5,950	3,240	5,120	46.4	25.3
Iron, Armco	21.2	8.24	11.35	0.29	5,960	3,240	5,200	46.5	25.3
Lead, rolled	1.5–1.7	0.54	3.3	0.43	1,960	690	1,210	22.4	7.85
Magnesium, drawn, annealed	4.24	1.62	2.56	0.306	5,770	3,050	4,940	10.0	5.3
Monel metal	16.5–18	6.18–6.86	12.4	0.327	5,350	2,720	4,400	47.5	24.2
Nickel	21.4	8.0	16.4	0.336	6,040	3,000	4,900	53.5	26.6
Platinum	16.7	6.4	9.9	0.303	3,260	1,730	2,800	69.7	37.0
Silver	7.5	2.7	8.55	0.38	3,650	1,610	2,680	38.0	16.7
Stainless steel, #347	19.6	7.57	11.3	0.30	5,790	3,100	5,000	45.7	24.5
Tin, rolled	5.5	2.08	4.04	0.34	3,320	1,670	2,730	24.6	11.8
Tungsten, drawn	36.2	13.4	31.3	0.35	5,410	2,640	4,320	103.0	50.5
Zinc, rolled	10.5	4.2	4.2	0.25	4,210	2,440	3,850	30.0	17.3
Fused silica	7.29	3.12	1.61	0.17	5,968	3,764	5,760	13.1	8.29
Pyrex glass	6.2	2.5	2.3	0.24	5,640	3,280	5,170	13.1	7.6
Heavy silicate, flint glass	5.35	2.18	1.77	0.224	3,980	2,380	3,720	15.4	9.22
Light borate crown glass	4.61	1.81	2.2	0.274	5,100	2,840	4,540	11.4	6.35
Lucite	0.40	0.143	0.562	0.4	2,680	1,100	1,840	3.16	1.3
Nylon 6–6	0.355	0.122	0.511	0.4	2,620	1,070	1,800	2.86	1.18
Polyethylene	0.076	0.026	0.288	0.458	1,950	540	920	1.75	0.48
Polystyrene	0.528	0.12	0.34	0.405	2,350	1,120	2,240	2.48	1.18

A database for metals and alloys, designed for personal computers, is available from the American Society for Metals (ASM), Metals Park, Ohio. A 1987 version is called MetSel2.

Table 3-14. Sound Speed Versus Temperature in Pure Water for Temperatures on T_{68} Scale. For Finer Detail, See Del Grosso and Mader, 1972. © 1972 AIP

T_{68} °C	m/s	T_{68} °C	m/s	T_{68} °C	m/s
0	1402.388	35	1519.808	68	1554.381
1	1407.367	36	1521.745	69	1554.611
2	1412.232	37	1523.618	70	1554.799
3	1416.985	38	1525.428	71	1554.947
4	1421.628	39	1527.176	72	1555.053
5	1426.162	40	1528.863	73	1555.120
6	1430.589	41	1530.489	**74**	**1555.146**
7	1434.912	42	1532.056	75	1555.133
8	1439.132	43	1533.564	76	1555.081
9	1443.251	44	1535.015	77	1554.991
10	1447.270	45	1536.409	78	1554.862
11	1451.191	46	1537.746	79	1554.696
12	1455.016	47	1539.028	80	1554.492
13	1458.747	48	1540.256	81	1554.251
14	1462.384	49	1541.430	82	1553.974
15	1465.931	50	1542.551	83	1553.660
16	1469.387	51	1543.619	84	1553.310
17	1472.755	52	1544.636	85	1552.924
18	1476.036	53	1545.601	86	1552.504
19	1479.231	54	1546.517	87	1552.048
20	1482.343	55	1547.382	88	1551.558
21	1485.372	56	1548.199	89	1551.034
22	1488.319	57	1548.967	90	1550.476
23	1491.187	58	1549.687	91	1549.884
24	1493.976	59	1550.360	92	1549.259
25	1496.687	60	1550.986	93	1548.602
26	1499.323	61	1551.566	94	1547.912
27	1501.883	62	1552.101	95	1547.190
28	1504.370	63	1552.590	96	1546.436
29	1506.784	64	1553.035	97	1545.651
30	1509.127	65	1553.437	98	1544.834
31	1511.399	66	1553.794	99	1543.987
32	1513.603	67	1554.109	100	1543.109
33	1515.738				
34	1517.806				

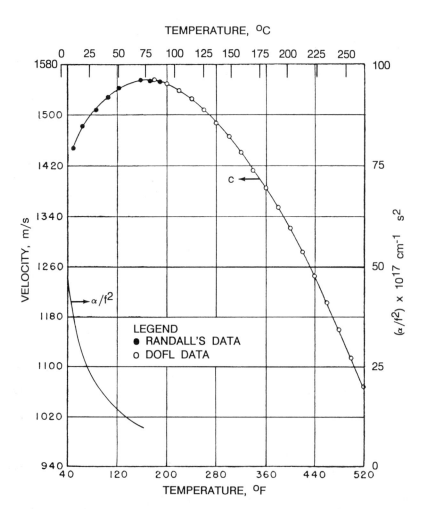

FIG. 3-80.   Velocity of sound in water as a function of temperature. After McDade et al., 1959. © 1959 AIP. Absorption versus temperature, based on data compiled and plotted in Markham et al., 1951. © 1951 AIP.

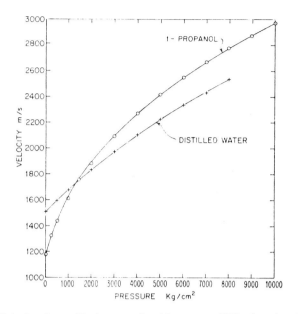

FIG. 3-81. Velocity of sound in 1-propanol and in water at 30°C, plotted against pressure. The sound frequency is 14.3 MHz. After Hagelberg, 1970. © 1970 AIP.

Table 3-15. Speed of Sound and Density for Water at High Temperatures. Sound Speed Measured by McDade *et al.*, 1959. Density Derived from Data in Crane, 1980

Temperature		Velocity	Density
°F	°C	m/s	g/cm^3
39.2	4	1422	1.000
200	93.3	1548	.963
240	116	1524	.946
300	149	1465	.918
350	177	1398	.890
400	204	1320	.859
450	232	1220	.824
500	260	1110	.784
550	288	980	.736

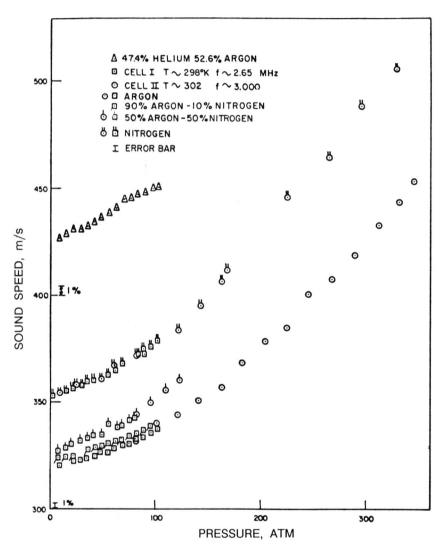

FIG. 3-82.   Sound velocity versus gas pressure in nitrogen-argon and helium-argon mixtures. After Carey et al., 1969.

Table 3-16.   Acoustic Characteristics of Liquids and Gases at Normal Pressure. Many Entries after Sutilov, 1984. Most Entries at 0°C are from the *AIP Handbook* (Gray, 1957). For a Number of These Fluids, c, dc/dt, ρc, Viscosity, and Absorption, all at 25°C, are Given in Mason, 1950, pp. 335–337. In the Last Column $Z = \rho c$

Fluid	Chemical Formula	T °C	ρ kg m^{-3}	c m s^{-1}	$Z \times 10^4$ kg/m^2 s
1	2	3	4	5	6
Nitrogen	N$_2$	-197	815	869	71
		20	1.17	351	0.041
Oxygen	O$_2$	-184	1143	911	104
		20	1.33	328	0.044
Nitric Oxide	NO	0	1.338	325	0.044
Argon	Ar	-189	1424	863	123
Neon	Ne	0	0.885	435	0.039
Carbon monoxide	CO	0	1.249	337	0.042
Carbon dioxide	CO$_2$ (high freq.)	20	1.85	268	0.052
Water	H$_2$O	20	998	1482	148
Steam	H$_2$O (1 bar)	100	0.60	405	24
Hydrogen	H$_2$	-253	355	1127	40
		20	0.10	1284	0.013
Air	—	0	1.293	331	0.04[a]
	—	20	1.29	343	0.045
Helium	He	-269	125	180	2.3
		0	0.18	965	0.017
Methane	CH$_4$	0	0.716	432	0.031
Acetone	CH$_3$CHCH$_3$	20	792	1192	94
Benzene	C$_6$H$_6$	20	878	1326	116
Bromoform	CHBr$_3$	20	2890	928	268
Bromobenzene	C$_6$H$_5$Br	50	1454	1074	156
Hexane	C$_6$H$_{11}$	20	654	1083	71
Glycerine	C$_3$H$_8$O$_3$	20	1260	1923	242
Diacetyl	C$_4$H$_6$O$_2$	25	990	1236	122
Dioxane	C$_4$H$_8$O$_2$	20	1033	1389	143
Dichlorethane	C$_2$H$_4$Cl$_2$	20	1250	1240	156
Diethylphthalate	C$_6$H$_4$(C$_3$O$_2$H$_5$)$_2$	25	1121	1470	165
Isopentane	C$_5$H$_{12}$	0	641	950	61
Aluminum	Al	965	—	4552	—
		1494	—	4262	—
Mercury	Hg	20	13590	1451	1972
Indium	In	156	7033	2215	1558
Potassium	K	75	824	1882	155
Sodium	Na	100	930	2524	235
		600	813	2255	183
Tin	Sn	230	6960	2462	1720
		495	—	2410	—
Zinc	Zn	450	6540	2700	1750
Kerosene	—	34	825	1295	107

Table 3-16. (*continued*)

Fluid 1	Chemical Formula 2	T °C 3	ρ kg m^{-3} 4	c m s^{-1} 5	$Z \times 10^4$ kg/m^2 s 6
Xylol	C_8H_{10}	20	860	1330	114
Sulfuric acid	H_2SO_4	15	1840	1440	257
Amino acid	HCOOH	20	1216	1287	156
Acetic acid	CH_3COOH	20	1050	1150	121
Spindle oil	—	25	866	1431	124
Linseed oil	—	31	932	1772	163
Olive oil	—	32	904	1381	125
Transformer oil	—	25	865	1415	122
Silicone oil DC-703	—	100	1013	1035	105
Nitrobenzene	$C_6H_5NO_2$	20	1207	1473	178
Octane	C_8H_{18}	20	703	1197	84
Paraldehyde	$C_6H_{12}O_3$	20	994	1204	120
Pentane	C_5H_{12}	20	1263	1158	146
Pyridine	$C_6H_{15}N$	20	982	1445	142
Carbon disulfide	CS_2	20	1263	1158	146
Amyl alcohol	$C_5H_{11}OH$	20	816	1294	106
Benzyl alcohol	C_7H_7OH	20	1045	1540	161
Butyl alcohol	C_4H_9OH	20	810	1268	103
Methyl alcohol	CH_3OH	20	792	1123	89
Propyl alcohol	C_3H_7OH	20	804	1223	98
Ethyl alcohol	C_2H_5OH	20	789	1180	93
Toluol	C_7H_8	20	866	1328	115
Acetic anhydride	$(CH_3CO)_2O$	24	1075	1384	149
Formamid	$HCONH_2$	20	1139	1550	177
Chlorobenzene	C_6H_5Cl	20	1107	1291	143
Chloroform	$CHCl_3$	20	1489	1005	149
Cyclohexane	C_6H_{12}	20	779	1284	100
Carbon tetrachloride	CCl_4	20	1595	938	150
Ethyl acetate	$CH_3COOC_2H_5$	20	900	1176	106
Ethyl bromide	C_2H_5Br	25	1430	890	127
Sulfur	S	110	—	1340	—
		450	—	950	—
Selenium	Se	215	—	1100	—
		560	—	870	—
Fluorine, P = 1 MN/m^2	Fl	−213	—	1039	—
		−173	—	720	—

(*a*) For air at pressure P (mm Hg) and absolute temperature T (K), the variation of ρc with T is given by $\rho c = 0.04286 (273.16/T)^{1/2} (P/760)$. Near room temperature and atmospheric pressure, $\rho_{air} \approx 1.3$ mg/cm^3 = 1.3 kg/m^3. The ratio $\rho_{water}/\rho_{air} \approx 777$, an easy-to-remember approximation.

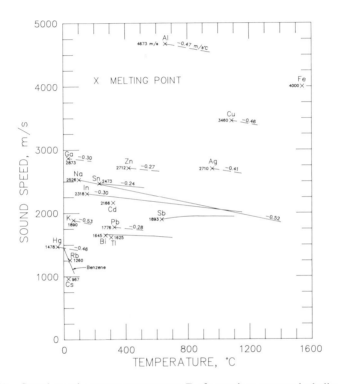

FIG. 3-83.   Sound speed c versus temperature T, after various sources including Hill and Ruoff, 1965, Webber and Stephens, 1967, and references cited therein.

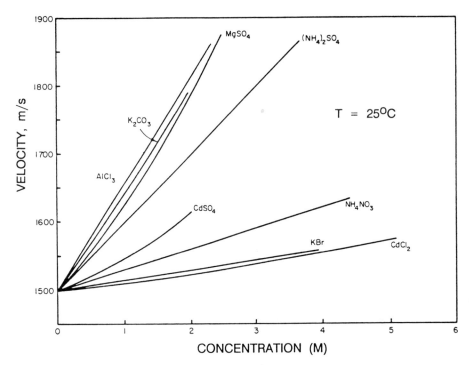

FIG. 3-84. Typical velocity data for electrolytic solutions at 25°C. After Stuehr and Yeager, 1965. © 1965 Academic Press.

FIG. 3-85. Impedance nomogram for longitudinal and shear waves. Fluids are located according to their ρ and c_L, and solids, according to ρ, c_L and c_T. Quarter-wave impedance transformation is also illustrated. For example, a $\lambda/4$ layer of fused silica matches tungsten to water (dashed semicircle). The single-layer, single-frequency matching impedance is $Z_m = (Z_1 Z_2)^{1/2}$, of thickness $\lambda/4$, where Z_1 and Z_2 are the impedances of the media to be matched. Z_m, the geometric mean, is located on an iso-Z line midway between Z_1 and Z_2. For a given solid, the vertical separation between c_L (○) and c_T (□) data points depends on Poisson's ratio σ, since $c_L/c_T = [2(1 - \sigma)/(1 - 2\sigma)]^{1/2}$. Likewise, the vertical separation between c_T (□) and, if plotted, the Rayleigh wave velocity c_R (▽) would depend on Poisson's ratio. See inset, upper left.

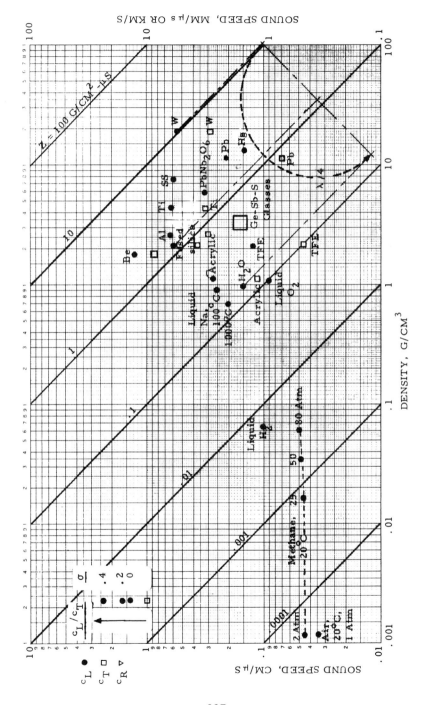

237

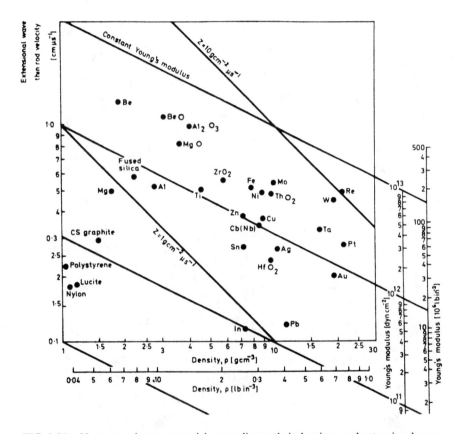

FIG. 3-86.   Nomogram locates materials according to their density ρ and extensional wave velocity c_{ext}. Lines having slope $= -\frac{1}{2}$ represent equations of the form: Young's modulus $=$ constant, $E = \rho c_{ext}^2$. Lines having slope $= -1$ are lines of constant characteristic acoustic impedance, $Z = \rho c_{ext}$. Example: fused silica, read $c_{ext} \approx 0.6$ cm/μs, $\rho \approx 2.2$ g/cm^3, $Z \approx$ 1.3 g/cm^2-μs and $E \approx 10^7$ psi. Observe that fused silica, Al, Ag and Au all have nearly the same Young's modulus. Horizontal lines, if drawn, would pass through materials of equal stiffness-to-density ratios ($c_{ext} =$ constant). After Lynnworth, 1969, © 1969 Butterworths.

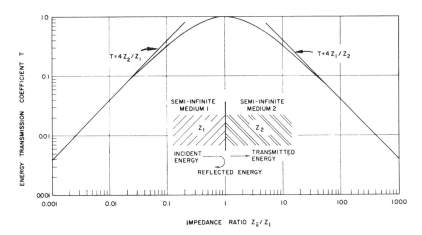

FIG. 3-87. Acoustic energy transmission across interface between two semi-infinite media, calculated at normal incidence. After Lynnworth, 1966. © 1966 IEEE.

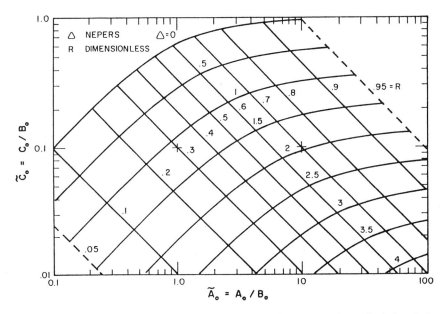

FIG. 3-88. Nomogram for sound pressure reflection coefficient R and round-trip loss $2\alpha L$. The sign of R is the same as the sign of C. $2\alpha L$ is in nepers; R is dimensionless. After Lynnworth, 1974. © 1974 Butterworths.

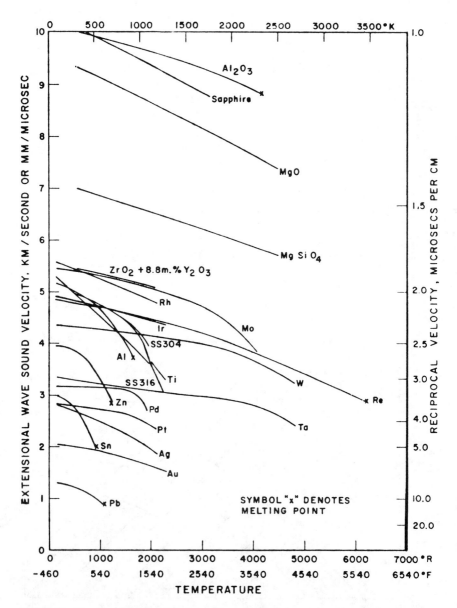

FIG. 3-89. Sound velocity versus temperature in various metals and ceramics, based on data in the literature. Note the potentially wide choice of sensor materials for use in different environments and for different temperature ranges. Source: Panametrics. After Lynnworth, 1969. See also, Papadakis et al., 1974.

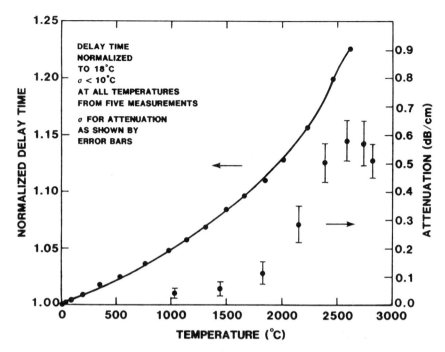

FIG. 3-90. Normalized delay time calibration and attenuation (~1-MHz) curves for K-Al-Si doped tungsten versus temperature. After Field, 1986.

Table 3-17. Examples Where Two (or More) Measurement Modes, Approaches or Disciplines Have Been Utilized to Extract More Information Than Would Have Been Possible with Any One Mode Alone

Approach	Application	Reference
Two Wave Modes		
Extensional and torsional resonance	Elastic moduli, Poisson's ratio at high temperature	Köster, 1948
Extensional and torsional pulses	Elastic moduli, Poisson's ratio at high temperature	Bell, 1957–1960; Lynnworth, 1973
Longitudinal and shear pulses, tone bursts or cw	Elastic moduli, Poisson's ratio at high temperature; materials characterization by sound speeds	Frederick, 1947, 1948; Jen *et al.*, 1988
Axial, radial cavity modes	Separation of viscous from thermal losses	Bradshaw, 1972
Two (or More) Directions/Paths		
Differential path-movable reflector	Specific heat in CO_2 to 2000K	Sherratt and Griffiths, 1934
Differential phase contrast, parallel paths	Stress regions around chords; biological tissue with *c* variations	Leung *et al.*, 1976
Before and after hyperthermia therapy, same path	Heating of tissue	Davis and Lele, 1985

241

Table 3-17. (*continued*)

Approach	Application	Reference
Reference SAW path parallel to measurand-selective path	Chemosensor, NO_2 gas sensor	Venema et al., 1987
0, 45° and 90° to rolling direction	Anisotropy of Ti sheet	Fowler, 1969
Measurements along crystal axes	Elastic constants	Mason, 1950
Measurements along different paths in a gas or in a red-hot metal billet	Temperature profile	Lynnworth and Carnevale, 1966; Wadley et al., 1986
Upstream, downstream	Flow in a pipe	Rütten, 1928
Bidirectional interrogation over parallel chords	Flow profile in a pipe (Gaussian quadrature)	Knapp, 1964; Malone and Whirlow, 1971; Nolan et al., 1986
Two (or More) Segments of Waveguide		
Extensional or torsional pulses	Profiling of temperature or density	Lynnworth and Carnevale, 1972; Lynnworth, 1977, 1980
Two Propagation Constants		
Sound speed, attenuation coefficient	Resolve ambiguity in concentration of aqueous solutions of methyl and ethyl alcohols	Babikov, 1960, pp. 162–163
Sound speed, attenuation coefficient	Transport properties at high temperature	Carnevale et al., 1967a,b
Torsional sound speed in wet and dry portions of noncircular waveguide	Density and liquid level, for fuel mass gaging in aircraft tanks (proposed)	Lynnworth, 1977; Findlay et al., 1988 (priv. comm.)
Ultrasonic + Nonultrasonic Data		
Ultrasonic + x-ray absorptions	Integrity of resin-ceramic composite	Hastings et al., 1961
Ultrasonic ranging + radiography	Measures distance to patient, optimizes x-ray technique	Kleinman, 1983
Sound speed + temperature	Molecular weight of flare gases	Smalling et al., 1984
Upstream and downstream sound speed + temperature + pressure	Mass flow rate of flare gases	Smalling et al., 1984
Sound speed + dielectric constant	Mass flow rate of aviation fuels	Lynnworth et al., 1982b
Upstream and downstream sound speed + specific gravity + composition + calorimetric data	Energy content of flowing natural gas	Pedersen et al., 1976; Munk, 1982; Scelzo and Munk, 1987

3.9 Complementary Methods

The adage, "Two heads are better than one" has several interpretations when applied to ultrasonics. In one set of interpretations, two wave modes provide more information than either mode alone (Table 3-17). In measurements utilizing ultrasonic + nonultrasonic technologies the data derived from two different disciplines complement one another.

4. Flow Applications

4.1 General Considerations

When one begins to compare *ultrasonic* methods of measuring flow with other "traditional" or nonultrasonic methods like venturi tubes, orifice plates, pitot tubes, magnetic, coriolis, thermal or target meters, etc., it becomes clear that some objective evaluation method is needed. Such a method would probably include a breakdown of applications into broad categories such as

- Closed conduit versus open channel

- Duct flow (full pipe) versus partly-full conduits

- Steady flow versus unsteady flow

- Type of fluid: compressible versus incompressible
  Newtonian versus non-Newtonian
  single-phase versus multi-phase

- Flow regimes: laminar, transitional, turbulent

- Hydrodynamic conditions: profile developed or unpredictable

- Requirements on accuracy, linearity, repeatability, reliability, cost, etc.

The comparison gets fairly complicated because there are evidently:

- Many nonultrasonic alternatives
- About a dozen different ultrasonic flow measuring approaches, at least two of which might appear equally suitable (Fig. 2-1, p. 20)
- Multiple vendors for nearly each type of equipment
- Many different kinds of applications
- Different weighting factors associated with the relative importance of, say, accuracy, size, weight and cost, according to the particular application
- Well-established methods with known limitations versus new methods promising "better" performance but perhaps with new limits or problems not yet identified.

Given a limited number of pages in which to deal with this complexity, some arbitrary decisions must be made, to delete certain topics, while retaining those which are expected to be of more lasting significance. The first general topic to be retained is *Flow Profile*. The second is labeled *Flowcells and Transducers*. In *two-phase* or *multiphase flow*, it will be understood that more than one profile exists. For example, one may be interested to know the distributions of velocity for gas, liquid and solid particulates. In many cases the *mass* flow rate profile is of interest.

To obtain a broader perspective on *flowmeters in general*, see Miller, 1983 and 1989. Within the past ten years, *ultrasonic* flowmeters have been reviewed by McCullough and Graeper, 1979, Sanderson and Hemp, 1981, and by the author, 1979, 1981, 1982. For a historical perspective the earlier reviews of Knapp, 1964, McShane, 1974, Herrick, 1977, and White, 1982, will be of interest.

4.1.1   Flow Profile

Flow profile may be important per se, or when one needs to convert a flow "reading" at a point or along a particular path to the velocity averaged over the entire cross section of the flowing medium. Flow profile depends on the fluid, the Reynolds number Re, the relative roughness and shape of the conduit, upstream and downstream disturbances, and other factors. Perhaps one of the earliest and yet still useful illustrations of flow profiles are those measured by Nikuradse, 1926, over sixty years ago. He used water flowing in smooth-walled pipe. See Fig. 4-1. To the extent that this data represents actual conditions, it is possible in principle

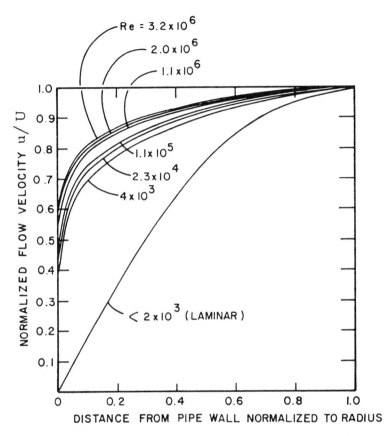

FIG. 4-1. Turbulent flow velocity distributions measured in a smooth-walled circular pipe by Nikuradse (1926a,b), compared with theoretical parabolic profile for laminar flow.

to compute the meter factor K, and thereby obtain the area-averaged flow velocity from localized samples of the velocity. Flow profiles for rough pipes are discussed in Miller, 1983, Chaps. 5 and 14. (See also, Halttunen and Luntta, 1988, Ginesi, 1987 and Fig. 4-2 regarding profile distortion and swirl due to elbows.) It will be understood that by sampling over a greater fraction of the cross section, one can lessen the risk that the result is erroneous due to the actual profile departing from that predicted. Examples of some preferred paths are illustrated in the next section, Flowcells. In the present section, however, it is convenient to list meter factors for several special cases, derived by various investigators for the smooth-wall profiles in Fig. 4-1:

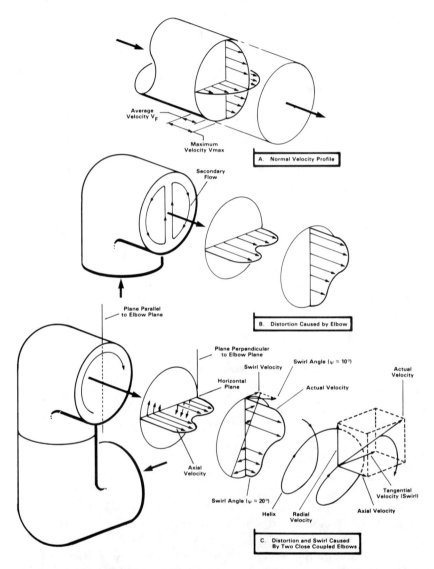

FIG. 4-2.   Profile distortion and swirl caused by an elbow alone or two out-of-plane coupled elbows. After Ginesi, 1987. © 1987 ISA.

Diametral path, laminar flow $K = 0.7500$
                     turbulent flow $K = 1/(1.119 - 0.011 \log Re)$
Midradius chord, laminar or tur- $K = 0.998 \pm 0.002$
                     bulent flow
Symmetrical segment of a midradius chord . . . see Fig. 4-3.

Again, the meter factor $K = \mathbb{V}/V_u$ where $\mathbb{V}$ = area-average velocity and V_u = ultrasonic velocity (e.g., line-average over a path such as a chord, chord segment or diameter). In principle K can be derived if the velocity profile is known (Table 4-1).

Some tilted-diameter transit time flowmeters that appeared around 1980 included profile compensation programs where Re was calculated as VD/v, and K was stored in an electronic look-up table as a function of Re (Lynnworth, 1982a). In a program called Re 1, K is tabulated according to the diametral path values that would be computed by the above equations. But when the pipe diameter D is not greater than the beam width d by a sufficiently large factor, or because of flow disturbances, other

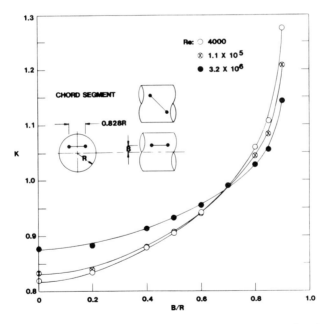

FIG. 4-3.   Calculated meter factor K for symmetrical chord segment of projected length $0.828R$ representing projection of ultrasonic interrogation path that is skewed with respect to pipe axis, and located in a plane at the normalized distance from the axis B/R, for Reynolds numbers $Re = 4000$, 1.1×10^5 and 3.2×10^6. Calculation is based on Nikuradse's 1926 data for smooth-wall pipe. After Lynnworth, 1986.

Table 4-1. Relations Among Point Velocity V_p, Maximum Velocity V_o and Average Velocity $\mathbb{V}$, for Smooth and Rough Pipes. After Schlichting, 1955 and Miller, 1983

Smooth-Wall Profile, Based on Power-Law Equation	Rough-Wall Profile, Based on Pai's Equation

Smooth-Wall Profile, Based on Power-Law Equation

$$V_p = V_o(1 - r/R)^{1/n} = (y/R)^{1/n}$$

$$n = 1.66 \log Re$$

$$\mathbb{V} = V_o \frac{2n^2}{(n+1)(2n+1)}$$

$$V_p/\mathbb{V} = [1 - (r/R)^{1/n}]/[2n^2/(n+1)(2n+1)]$$

where V_p = point velocity at radial distance r
  $\mathbb{V}$ = mean flow velocity
  R = pipe radius
  Re = Reynolds number
  y = distance from pipe wall

Rough-Wall Profile, Based on Pai's Equation

$$V_p = V_o\left[1 + \frac{s-n}{n-1}\left(\frac{r}{R}\right)^2 + \frac{1-s}{n-1}\left(\frac{r}{R}\right)^{2n}\right]$$

where $s = fRe/(32 + 46.08f^{1/2})$
  $n = (2 - fRe/32)/(1.44f^{1/2} - 1)$
and f = friction factor.

$$\frac{V_p}{\mathbb{V}} = 1 + \frac{0.715 + 1.075 \log(y/R)}{\log(0.2703\epsilon/D + 5.74/Re^{0.9})}$$

where ϵ/D = relative roughness
and $D = 2R$

values of K are required. These other values are typically determined during calibration. The program name Re 2 designates their use. In the computation of Re, if v is not treated as constant, but rather is determined from external temperature, pressure or viscosity data, or computed from sound speed (as an indicator of temperature or viscosity), the program is called Re 3 in the program library of one manufacturer.

For path segments, and to the extent that the profile obeys a known velocity distribution law, K may be computed according to the method reported by Lynnworth and Lynnworth, 1985.

The effect of the pipe roughness on ultrasonic flowmeter readings is treated theoretically by Lobachev and Myasnikov, 1980. Discussions appear in Miller, 1983, 1989. Calibration data appear in a graph presented by Zacharias, 1984.

Two interrogation methods which do *not* depend on *a priori* knowledge of the flow profile, are the following:

- 100% of cross section interrogated, with equal weight given to equal areas. For this sampling, $K = 1$.

- Multipath (usually 4-path) quadrature technique for which the area-averaged flow velocity is computed by weighting the chord samples according to their normalized distance from the pipe centerline (Chapter 2, Fig. 2-4).

Two-Phase and Multiphase Flow

To illustrate how bimodal and multimodal flow profiles arise, and to indicate why no one ultrasonic measuring mode is likely to best solve all flow measurement problems, Figs. 4-4 and 4-5 are introduced. The first of these proposes a transmission mode to measure water velocity V_1 and a reflection mode to measure the air bubble velocity V_2. By using range gating, the reflection mode would measure V_2 as a function of position along the axis, for the geometry shown. In the second illustration a third phase (solid particulates) is added. (See also, Section 4.5.8.)

4.1.2 Flowcells and Transducers

The term "flowcell" suggests a spoolpiece or other section of pipe into which transducers have been installed. Such a flowcell could be installed in a pipeline just as a venturi section or full-bore turbine meter would be installed. But since *ultrasonic* flowmeters often employ clamp-on transducers, the term "flowcell" also embraces an existing section of conduit in which flow is measured.

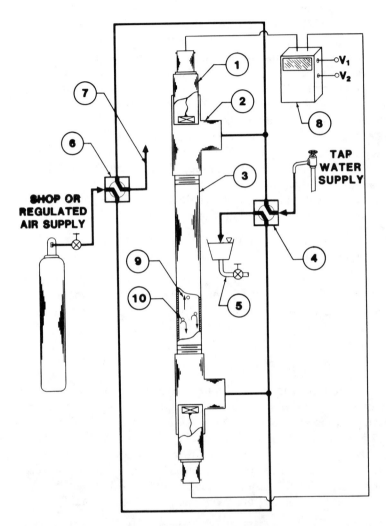

FIG. 4-4. Bimodal flow problem produced in offset-style flowcell in which air and water can be independently introduced into top or bottom tees. Combinations include: air and water flow up; air and water flow down; air up, water down. Legend: 1–Extended well transducer. 2–Tee. 3–Pipe nipple. 4–Four-port water switching valve. 5–Drain. 6–Four-port air switching valve. 7–Vent. 8–Electronic instrument uses transmission mode (9) and range-gated reflection mode (10) to measure flow velocity of water and air, respectively, and generates bimodal velocity outputs V_1, V_2.

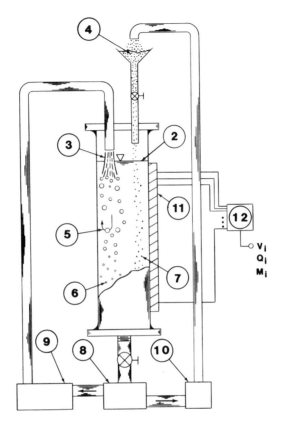

FIG. 4-5. Multimedia event in which the flow velocities for different phases are unequal. Legend: 1–Pipe. 2–Water level, exposed to atmosphere. 3–Recycled water with entrained air. 4–Sand, gravel. 5–Large air bubble rising against the downward flow of water. 6–Small air bubble carried down by water. 7–Solid particle falling faster than gravity-fed water. 8–Phase separator. 9–Water pump. 10–Solids pump. 11–Ultrasonic transducer array. 12–Ideal electronic instrument uses transmission, reflection, tag correlation, noise or other measuring modes to determine flow rates (V_i, Q_i, M_i) of the several phases as a function of position and time. Medium $i = 1$, solid; $i = 2$, liquid; $i = 3$, small gas bubbles, etc.

The exact details of the number, type and placement of transducers on a flowcell varies with installation requirements, the *type* of ultrasonic meter and manufacturer. In this section we illustrate three basic flowcell and transducer configurations.

4.1.2.1 Paths Parallel, Perpendicular or Oblique to the Flow Axis

In Table 4-2 four types of ultrasonic flowmeters are associated with the orientation of their customary interrogation paths. Examples of the flowcell and transducer configurations are given in Chapter 2, Fig. 2-1.

Table 4-2. Paths Usually Associated with Ultrasonic Meter Types

| | Orientation of Path Relative to Flow Axis | | |
Type of Meter	Parallel	Perpendicular	Oblique
Contrapropagation	X		X
Doppler	X		X
Vortex shedding		X	
Correlation of tag time of flight		X	

4.1.2.2 Wetted Transducers

A *wetted* transducer is a housed assembly which radiates sound into, or receives sound from, the fluid directly. No pipe wall or other barrier lies between the assembly and the fluid. The electroacoustic transducer element is normally protected from direct exposure to the fluid under test, as well as from contact with the atmosphere outside the conduit, particularly if that atmosphere is corrosive or flammable.

In a wetted transducer assembly, between the electroacoustic element and the fluid, there may be interposed either a plastic window, a thin metal window, or a thick metal window to withstand high pressure and abrasion. In some designs used for measurements in gases, an impedance-matching layer is sandwiched between the element and the thin window which seals the assembly (Chapter 3, Fig. 3-4).

For applications at temperature extremes the element may be isolated thermally by 10 cm or more of a wetted buffer rod.

Examples of wetted transducers are given in Fig. 4-6, where the window is "thin," "thick," or "very thick."

4.1.2.3 Clamp-On (External) Transducers

Clamp-on transducers provide one of the main incentives for choosing an ultrasonic flowmeter over competing technologies. These transducers may be temporarily or permanently coupled to the conduit. They can launch or detect the following waves in the conduit: longitudinal, shear, plate waves such as Lamb or Rayleigh-like surface waves, and sometimes other waves too (lateral, creeping, interface waves, etc.) These waves radiate into the adjacent fluid to sense flow. If the fluid is of low acoustic impedance relative to the conduit then radiation into and/or out of the fluid is inefficient, perhaps impractical. Thus, clamp-on measurements for air or steam in pipes generally remains as an on-going challenge. For water in a plastic pipe, however, clamp-on results accurate to 2% were claimed as early as 1954 by Kalmus.

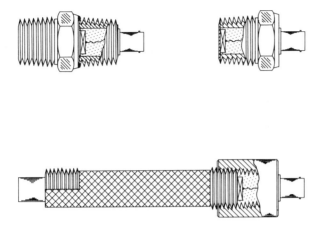

FIG. 4-6.   Wetted transducers having windows that are thin, thick or very thick compared to wavelength.

Several clamp-on transducer designs and clamps are illustrated in Fig. 4-7 and in Chapter 2, Figs. 2-9, 10, 18, 19 and 20.

Clamp-On Area-Averaging

In transit time, range-gated reflection or correlation tag clamp-on flow-meters several opportunities exist for reducing the uncertainty due to flow profile. As suggested in Fig. 4-8 the uncertainty can be reduced by sampling flow in a greater fraction of the cross sectional area, or in off-center regions or along paths other than the diameter. Depending in part on sound speeds in the pipe and liquid, however, Snell's Law often prevents one from interrogating over preferred paths such as midradii chords or Gaussian-positioned chords because of the restricted range of refracted angles.

Noncontact external transducers for flowmeters (see Section 5.1.3) are desirable and in time may become practical in cases where pipe temperatures, remoteness, or other conditions preclude physical contact between the transducer and the conduit.

4.1.2.4   Hybrid Transducers

The potential convenience of clamp-on transducers that are simply coupled to an existing pipe is sometimes offset by coupling difficulties, uncertainties in the refracted path and in pipe dimensions, pipe curvature, or inability to interrogate along chords other than those allowed by Snell's Law. Thus, accuracy may be compromised. In process control, it is usually sufficient to achieve good repeatability. Accuracy is often unimpor-

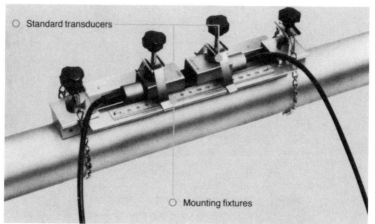

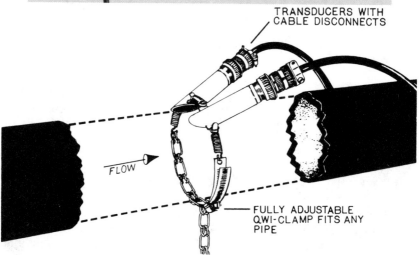

TRANSDUCERS WITH
CABLE DISCONNECTS

FLOW

FULLY ADJUSTABLE
QWI-CLAMP FITS ANY
PIPE

FIG. 4-7. Examples of clamps developed (top) for a transit time survey flowmeter, courtesy Tokyo Keiki, and (bottom) for a Doppler flowmeter applicable to different pipe sizes, courtesy K-Flow.

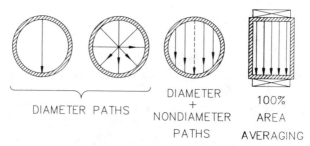

DIAMETER PATHS

DIAMETER
+
NONDIAMETER
PATHS

100%
AREA
AVERAGING

FIG. 4-8. Possible methods of improving clamp-on accuracy by reducing uncertainty in flow profile through better area-averaging. Above paths are projections in a plane perpendicular to the duct axis. Actual paths may be inclined to the duct.

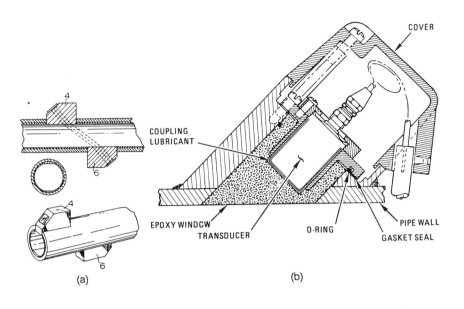

COVER

COUPLING
LUBRICANT

4

6

4

6

(a)

EPOXY WINDOW
TRANSDUCER

O-RING

PIPE WALL

GASKET SEAL

(b)

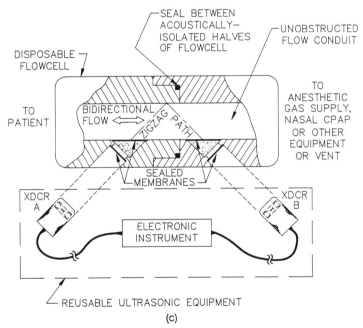

SEAL BETWEEN
ACOUSTICALLY–
ISOLATED HALVES
OF FLOWCELL

UNOBSTRUCTED
FLOW CONDUIT

DISPOSABLE
FLOWCELL

TO
PATIENT

BIDIRECTIONAL
FLOW

ZIGZAG PATH

SEALED
MEMBRANES

TO
ANESTHETIC
GAS SUPPLY,
NASAL CPAP
OR OTHER
EQUIPMENT
OR VENT

XDCR
A

XDCR
B

ELECTRONIC
INSTRUMENT

REUSABLE ULTRASONIC EQUIPMENT

(c)

FIG. 4-9. Flush-mounted windows or wedges in the wall. Designs include (a) 45°-path achieved with SV in plastic wedges (4,6) due to Lake, 1962, (b) removable L-mode transducer design with wedge-shaped epoxy window to prevent clogging of port by debris (courtesy Sparling Envirotech), and (c) disposable flowcell concept proposed for biomedical air and gas flow applications, where the windows consist of sealed-in-place membranes that are very thin compared to wavelength so as to not excessively block transmission of ultrasound into air or other gases.

tant. Where accuracy *is* important, however, the user should install the transducers strictly in accordance with the manufacturer's instructions, and take whatever additional steps are practical in order to reduce uncertainties. For example, the user may supply measured pipe diameter and wall thickness data to the instrument instead of merely assuming that the pipe is round and of dimensions and properties as tabulated in handbooks.

Another option available to some users who require accuracy is to use a *hybrid* transducer. The term *hybrid* transducer is proposed for the combination of a nonwetted transducer assembly which can be easily mounted into or dismounted from a specially prepared pressure-bounding element. This element is wetted and may be a pipe plug, blind flange or special insert forming a permanent part of the pressure boundary in a piping system. In some designs manufactured by Sparling the wetted element takes the form of an epoxy wedge which seals a port in a pipe flush with the interior wall (Fig. 4-9; see also, Lake, 1962). The wetted element can also take the form of a precision pipe plug as in Panametrics' Pan-Adapta™ plug (Chapter 3, Fig. 3-30). (Compare with quick-connect glue-on transducer mount referred to in Chapter 2, Section 2.6.5, in connection with Fig. 2-66.) In flanged piping systems, the pressure-bounding element

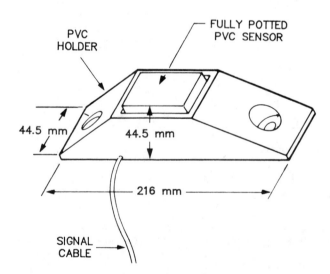

FIG. 4-10. Instream sensor for use in open conduits over 300-mm wide or for use in large concrete pipes where the outside of the pipe is not accessible. After mechanical installation the housing is grouted to the sidewall, producing a streamlined profile in the conduit. (Compare with sensor mounting considerations for boat speed, Section 4.6.4.) Reference: Kyser et al., 1988. Illustration courtesy Badger Meter.

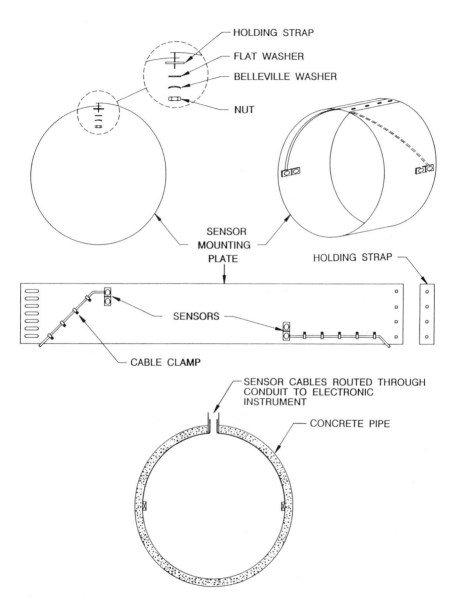

FIG. 4-11.   Internal strap design that is manufactured with the instream sensor (Fig. 4-10) accurately positioned and mounted. This mounting arrangement is used for attaching transducers to the inside wall of pipe. Reference: Kyser et al., 1988. Illustration courtesy Badger Meter.

would be a blind flange against whose unwetted face a transducer may be coupled temporarily or permanently.

Clamp-In Designs

In some cases, it is necessary to attach transducers to the *inside* wall of the conduit without penetrating the conduit near the points of installation. Designs for this purpose (as well as for open-channel flow measurements) have been available from Badger Meter for a number of years (Section 4.6.1; Fig. 4-75; Kyser and Heyden, 1984). One clamp-in design, employing a slip-in metal hoop that controls the spacing of the instream transducers, was used in conjunction with a solar-powered battery-powered transit time meter (Kyser et al., 1988; Figs. 4-10 and 4-11). Transducer assemblies containing spare crystals and designed to be lag-bolted to the inside walls of rectangular concrete conduits are shown in Lynnworth et al., 1986, p. 586.

4.2   Contrapropagating Transmission Applications

Contrapropagating transmission measurements of flow velocity V typically involve propagation in the upstream and downstream directions over the same path. The path may be resolved into two components, one of which is parallel to the sought component of flow. For duct flow, this is usually in the axial direction. Typically, upstream and downstream directions are interrogated by waves launched simultaneously or in an alternating sequence at rates determined independently by the electronics, or by the flow itself. Almost always, the two *transit times* (or terms related thereto, such as their sums, differences or reciprocals, or certain ratios) provide the desired measure of V. Sometimes, *amplitude* measurements of the upstream and downstream waves may be related to V.

While configurations within this category originated as early as 1928, significant industrial progress did not occur until 1947, when Swengel, 1955, succeeded in electronically switching the direction of interrogation alternately upstream and downstream over the same path. Swengel also addressed the profile question, using line source transducers to average flow in large rectangular ducts.

Angle-beam transducers which *could* have been clamped on the outside of pipes, as is common nowadays, were reported in 1951 in a different ultrasonic discipline, nondestructive testing. However, their widespread industrial use by Yamamoto and co-workers in clamp-on contrapropagating flowmetry did not begin until 1963, in Japan, for pipes on the order of 1–m diameter (Baba, 1964). Kritz, 1955, reported significant progress using wetted transducers installed along tilted diameters in pipes of 0.1 m diameter, and presented graphical corrections for flow profile in terms of the Reynolds number. He also showed how, in principle, the mass flow rate of a liquid could be obtained ultrasonically by responding to the liquid's characteristic acoustic impedance Z as well as to its flow velocity V or Mach number V/c. Kritz further suggested that some off-diameter path might be less sensitive to profile. But nearly twenty years passed before that special path was found by others to lie on or near the midradius chord, and its near-immunity to laminar and turbulent profiles was demonstrated at least analytically for axisymmetric conditions.

In the meantime, Knapp, 1964, suggested a multi-chord Gaussian quadrature method for sampling and properly weighting flow along more than one chord. This idea evolved independently in the late 1960s and 1970s in multichord contrapropagating equipment manufactured by Westinghouse and ORE. Another area-averaging contrapropagation method was investigated by the author and co-workers in the 1970s, wherein 100% of the cross section of small square or rectangular ducts was obliquely interrogated, often in zigzag fashion, using a square or rectangular-enveloped beam. Oblique interrogation of 100% of the cross section of circular ducts has been reported for biomedical applications such as blood flow in flexible tubes only a few mm in diameter (Drost, 1980) or air flow as encountered in respiratory studies (Buess et al., 1986; Lechner, 1983).

Transit time flowmeter equipment is currently available from about a dozen manufacturers, with applications mainly in liquids, and with temperature extremes from cryogenic to over 300°C being investigated. Accuracy in some cases is 0.5%, and in other cases, response time can be as short as 1 ms. At present, the most obvious limits on performance are due to excess attenuation [as caused by gas bubbles or other scatterers, classical plus internal or molecular absorption effects, impedance mismatch (especially for gases at low pressure), beam spread and turbulence], acoustic interference, transducer curie point, jitter, and profile uncertainties. The efforts of numerous investigators, only a few of whom are identified here, are aimed at overcoming these and other limits. Applications occur in media which are gaseous, liquid or solid, and their mixtures.

Regarding the origin of the contrapropagation method of measuring flow, some writers like to draw parallels between it and the classic ether drift experiment conducted just over a century ago by Michelson and Morley. Although the upstream-downstream mathematics appears similar for interrogations by light and sound, the physical interactions are notably

different. Rayleigh (1877, p. 2) cautions his readers about the influence
of wind on a measurement of sound speed in air. See also, Beyer, 1966.

4.2.1 *Liquids*

4.2.1.1 Conduit ID from 1 mm to Several Meters

One *lower limit* on the conduit ID which allows ultrasonic waves to be
transmitted axially upstream and downstream is imposed by cut-off conditions (Beatty, 1950). Cut-off conditions are expressed in terms of di-
ameter/wavelength: the shorter the wavelength λ, the smaller the allowable ID. In practice it has been found that 2-MHz pulses can be
transmitted through water paths at least 10 cm long in thin wall stainless
steel tubing of 1-mm ID. Here, ID/λ = 1.33 (Lynnworth and Nguyen,
1986. See also, Noble, 1968). Aside from the cut-off limit, the small ID
flowcell is subject to blockage by entrained air or vapors in the end cavities
or in the tube itself. Bleed valves sometimes remedy this problem. Bleed
valves can be connected to crosses which replace tees at the ends of an
axial path offset style flowcell. (See also, Shields and Faughn, 1969, regarding propagation of sound waves in tubes.)

It is evident that as the ID ranges down to 1 mm or below, volumetric
flowrate Q and mass flow rate M become quite small relative to their
values in more common piping. Thus, a potentially important application
area for ultrasonics is the measurement of liquid flow rates near or below
$Q \approx 100$ mm^3/s or below $M \approx 100$ mg/s. In 2-MHz laboratory tests on
hypodermic-size (1- to 2-mm ID) flowcells conveying water, it was found
that transmission was maintained as long as back pressure was high
enough, and provided bleed valves could be opened to release air when
it collected in the acoustic path. At an ID = 3 mm, water path lengths
of 300 mm have been interrogated at 2 MHz.

Measurement problems associated with low flow rates, steady as well
as pulsatile, may be solved using long axial path offset style flowcells.
The tubing sections of $\frac{1}{4}$- and $\frac{1}{2}$-inch OD ($\sim$6.3- and 12.7-mm OD), Fig. 4-
12 (Jacobson et al., 1987a), are $\sim$60-cm long. For the trapezoidal pulse
flow bursts, calibration data for water, obtained with the $\frac{1}{4}$-inch flowcell,
indicate satisfactory performance over a 10:1 flow range. Lengths up to
1 m have been used in flowcells of 19 to 25 mm ID.

The square-holed pipe design that originated as a means of sampling
the entire cross section of flowing fluid in order to minimize the effects
of flow profile on the accuracy (Lynnworth 1975b) was selected in one
high pressure, high temperature application for a different reason. In this
application it was necessary to accurately measure the flow rate (nomi-

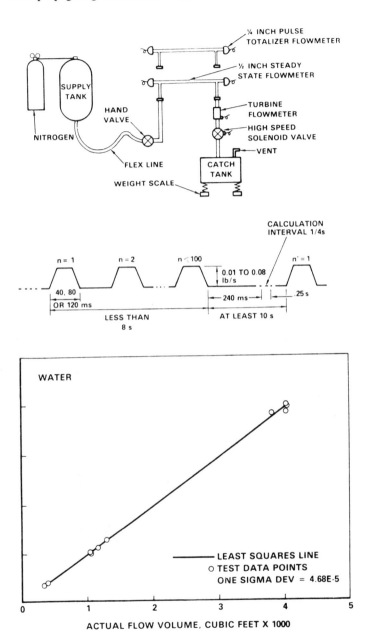

FIG. 4-12.   Unsteady flowmeter application. Top: Test arrangement at McDonnell Douglas STL ground test facility. Middle: Trapezoidal flow pulses. Bottom: Calibration results using water, measured using the flowcell made from ¼-inch stainless steel tubing in the axial path offset style configuration. After Jacobson et al., 1987a. © 1987 IEEE.

nally 7000 kg/hr) of condensate water at 297°C and >80 bar (>8 MPa). Pressure drop (ΔP) had to be minimized, $\leq 10^4$ Pa (<2 psi) to prevent flashing. A square insert, approximately 25 × 25 mm, was welded inside heavy-walled pipe fittings (Fig. 4-13). Threaded buffer rods separated the transducers from the high temperature. Designed by Nguyen generally as described in Nguyen and Lynnworth, 1983, the assembly was welded and hydrostatically tested at 2110 psi (14 MPa) in accordance with applicable Canadian and ANSI codes for pressure vessels.

The ΔP < 2 psig requirement ruled out orifice plates or other pressure-dropping sensors, and the nonconducting nature of the liquid ruled out magmeters. Otherwise either of these "conventional" flowmeters might have been preferred, or at least have been considered.

The area-averaging and reflective properties of the 100%-interrogated rectangular duct are utilized in an ultrasonic heat meter made by Siemens (Mágori, 1985). See Section 4.6.3. Cross sectional duct dimensions are ~1 or 2 cm, providing 3% of reading (or better) accuracy for water flows from 0.05 to 1.5 m³/hr. Presumably thousands of these meters were installed in Europe within a year or so of their introduction.

Given the foregoing square, rectangular and round solutions for area averaging, can we find a general solution? Imagine that flow consists of an interrogation of all the flow streamlines within the conduit. Imagine further that, for 100% area averaging, each streamline is sampled by a corresponding ray. Let the fluid's sound speed be constant and uniform. From this picture it seems reasonable to assert: the flow profile of a fluid in a conduit can be weighted properly by plane waves propagating be-

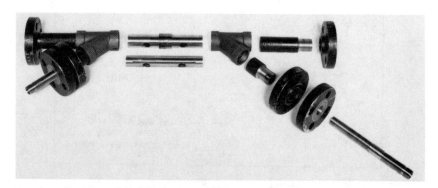

FIG. 4-13. Photograph of an ultrasonic flowmeter selected by a Canadian plant because of its low pressure drop, and its ability to measure a liquid at high temperature and high pressure (297°C, >80 bar). The square-holed insert promotes area-averaging and simplifies zigzag transmission over an axially-extended path L which exceeds the hydraulic diameter D by a factor equal to the number of traverses N_t. $N_t = 4$ in this example.

tween a source transducer and a receiver transducer, each transducer having the same size, shape and orientation as the conduit, with the waves intersecting the fluid such that no streamline is neglected. As a corollary: proper weighting is preserved if, for a fluid flowing in the z direction, the width W_z of the envelope of the wave, as measured (projected) in the z direction, is multiplied by a constant. Examples: square conduit, use square or rectangular beam; round conduit, use round or elliptical beam.

On the scale of m, not cm, Swengel had already addressed the question of area-averaging around 1947. He proposed waveguides or transducer arrays to approach 100% sampling (Swengel, 1950, 1956). However, in view of the high accuracy obtained with four-path and eight-path quadrature systems in large pipes and conduits in the 1970s and 1980s, it appears that adequate accuracy may be achieved in many large penstocks without sampling 100% of the profile. Refer to Fig. 4-14, which represents a converging duct having parallel sidewalls (typical for many hydroelectric turbine penstocks). The main objective is to sample the flow such that the computed total "flux" closely approximates $Q = \int_S V \cdot n dA$ where S is a surface perpendicular to all the streamlines and n is the normal unit vector. The approach suggested by this figure is to divide the converging volume into four (or more) segments of roughly equal volume bounded by streamlines. V_i is measured along a path P_i likely to yield V_{ci} at the center of each volume segment. $Q_1, Q_2, \ldots$ are calculated for each segment, and summed to yield Q. Due to physical constraints it may be impractical to install transducers relative to a mathematically-derived curved surface S. It is easier to install them if they are aligned vertically. As an example, in 1987 the four-path temporary pipe frame arrangement of Fig. 4-14 measured *converging* flow in a ~5-m × 7-m penstock at turbine generator unit 3 at Ontario Hydro's Chenaux Plant on the Ottawa River. Comparison of the ultrasonic readings and multipoint independent reference readings indicated that accuracy of about 1% might be obtainable with a calibrated ultrasonic four-path system. But eight to twelve paths were recommended for permanent installations in hydroelectric penstock ducts of similar geometry to reduce uncertainties due to flow profile down to approximately the 1% level.

The *upper limit* on conduit ID comes not from boundary or cut-off conditions but from attenuation. In the farfield, diffraction (beam spread) accounts for a 6-dB loss for every doubling of distance. Added to this geometric loss is the exponential loss due to scattering and absorption. Lowering frequency lowers these latter losses but increases the diffraction loss. In view of the long distances successfully interrogated in rivers, >1 km, it is not surprising that diagonal path measurements have been made in pipes of diameter up to 10 m. For clean water in pipes of 1 or 2 m, a

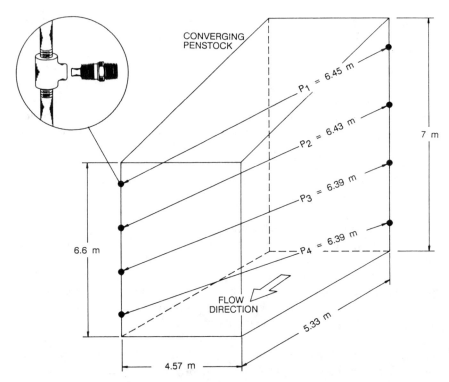

FIG. 4-14.   Converging penstock, a concrete conduit 4.57 m wide, with height reducing from 7 to 6.6 m in the space available for an ultrasonic measurement, required a different distribution of transducers than in round pipes or in constant area penstocks. A four-channel transit time flowmeter interrogated four pairs of standard 1-MHz wetted transducers that were installed in frame made from one-inch pipe and standard fittings in a 1987 feasibility test conducted at Ontario Hydro's Chenaux Plant. Paths P_1, P_2, P_3 and P_4 were used to measure the flow velocity near the centerlines of each of four volume segments into which the converging section was divided. Despite the intrusive pipe frame and transducers, and non-ideal transducer locations, agreement of about 1% was obtained between the ultrasonically-determined volumetric flowrate Q and that computed by an independent method. The trial also showed that for a complex section, care must be exercised in the selection of the number of transducers as well as their placement.

frequency on the order of one MHz is suitable. In a clamp-on application on a 4.7-m diameter pipe, 500-kHz pulses were used. Q determined in these measurements was found to agree within essentially 1% of an independent determination based on a 64-point flow survey of the cross section (Matson et al., 1987). This application is illustrated in Figs. 4-15 and 4-16.

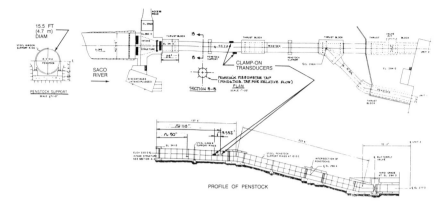

FIG. 4-15. Intake and penstock of Hiram Unit 1 showing locations of current meters, clamp-on transducers and penstock piezometer tap used in acceptance tests. The transducer spacing $S = 2.91$ m (9.55 ft). The ultrasonic instrument was a Model 6001A with preamplifiers, operated in the transit time mode. Illustration courtesy Central Maine Power Company.

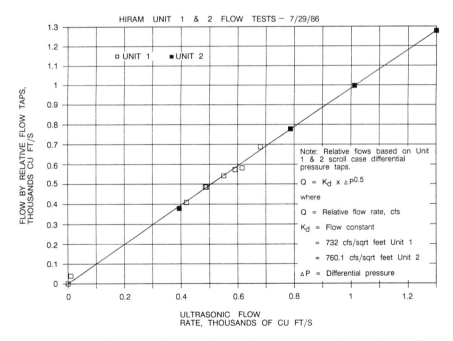

FIG. 4-16. Linearity test of clamp-on ultrasonic flowmeter system in comparison to flow in cu ft/s (cfs) derived from differential pressure taps. This particular ΔP method is considered a code-accepted reference with respect to proportional flow. The ultrasonic clamp-on data includes the standard smooth-wall correction for flow profile, i.e., meter factor $K = 1/(1.119 - 0.011 \log Re)$, where Re = Reynolds number. Source of graph: C.P. Clark, 1986. See Matson et al., 1987.

267

Multipath "Profile-Immune" Flowmeters

If the conduit in which flow needs to be measured were round, >100 diameters long, smooth-walled and straight, with no significant thermal gradients or flow disturbers, then one might expect the flow profile to be parabolic at $Re < 2000$ (laminar flow) and in accordance with the power law for turbulent flow. In practice one often encounters departures from these ideal hydrodynamic boundary conditions. Elbows, valves and branches generate flow disturbances such as swirl, cross-flow or other asymmetries which can propagate tens of diameters. Most real pipes are rough. The relative roughness determines in part how quickly the disturbances will be damped.

Multipath ultrasonic flowmeters appear to be the best solution so far, for accurately measuring flow ultrasonically in large conduits ($\gtrsim$ 1-m diameter) in which the flow profile is unknown and not necessarily axisymmetric. If swirl and cross-flow are absent, four paths usually suffice. Otherwise two crossed sets of four-path interrogations are recommended. [It may be shown that crossed paths, if properly located, average out the nonaxial components (Forster, 1987, priv. comm.) O'Hair and Nolan, 1984, use crossed *interlaced* paths to reduce the response to crossflow. Suzuki et al., 1972, used a clamp-on *vee* path to cancel crossflow response.]

Positions and weights for four-path Gaussian quadrature flowmeters are listed in Table 4-3. To show the effectiveness of this approach, we present test results for water and oil applications. First, results from two hydroelectric installations are described from Fain's 1986 TVA report. Second, we present Erikson's 1987 account of four-path ultrasonic flowmeters installed in 1976 on the Trans Alaskan Pipeline. (See also, Erickson and Graber, 1983.) Fain's remarks on *hydroelectric installations* follow.

During construction of TVA's Raccoon Mountain Pumped-Storage Plant, transducers for a four-path acoustic flowmeter were installed in a 10-ft [~3-m] diameter conduit connecting the spiral case and the spherical valve (March, 1981). The TVA Engineering Laboratory later made flow measurements using the acoustic flowmeter simultaneously with WK [Winter-

Table 4-3.   Positions and Weights for Gaussian Quadrature.
After Malone and Whirlow, 1971

	Three Path		Four Path	
X_k	± 0.7745	0.0000	± 0.8611	± 0.3399
W_k	0.5555	0.8888	0.3478	0.6521

Kennedy] taps and with a volumetric method during steady-state, one unit operation. The volumetric method consisted of timing upper reservoir level changes. Discharge was about 6,000 cfs [170 m³/s]. In the generating mode, flow rates measured using the WK taps averaged about 1.9% lower than those using the acoustic flowmeter. But flow rates using the volumetric method averaged about 1.7% higher than those using the acoustic flow-meter.

Outside TVA, the Electric Power Research Institute (EPRI) sponsored field tests in 1983 at the Kootenay Canal hydroelectric plant to provide data for direct comparison of acoustic flow measurement systems with other commonly used systems. Two independent systems using acoustic methods were tested in the same 22-ft [6.7-m] diameter penstock with two systems using time-pressure methods, three systems using dye-injection methods, and one system each using salt velocity and current meter methods. The WK taps were also monitored. Discharges were 4,000 to 7,000 cfs [113 to 198 m³/s]. The collected data were intended to help provide a basis for the acoustic method's acceptance by the utility industry for its consideration in test codes. The tests were essentially conducted in accordance with power test codes ASME PTC-18 and IEC Publication 41. The final report (EPRI, 1986a) concluded that the acoustic method results were at least as accurate as any of the code-accepted methods tested.

Based on the Kootenay tests, EPRI, 1986a, concluded that for utilities with hydroelectric resources, an acoustic system could be an excellent investment, regardless of whether the instrumentation is permanently attached to each penstock or moved from site to site in a monitoring program. Once the technique gains acceptance by the various performance code committees, it can be formally used to measure the guaranteed performance of newly installed hydroelectric turbines. Until then, according to this EPRI report, this method can be utilized to optimize the operation of hydroelectric facilities.

In another study (EPRI, 1986b) investigators found that Westinghouse and ORE (Ferranti Accusonic) four-path systems compared well in accuracy and repeatability with conventional methods at a power plant at Grand Coulee Dam. Acoustic flow measurement systems, that study concluded, offer utilities an inexpensive, real-time method for optimizing hydroelectric plant efficiency.

In the UK, the CEGB conducted four-path laboratory and site tests on a 500 MW turbine generator (Sochaczewski et al., 1981; Clay, 1985). Calibration results confirmed the accuracy of 1% or better, even under partly developed flow profiles, and supported the extension of the method to measuring changes in turbine-generator operating parameters and measuring thermal efficiency continuously. Temperature of the water was measured from the sound speed to an accuracy of about 0.1°C.

Trans Alaskan Pipeline Installation

It is now over a decade since the Trans Alaskan Pipeline (TAPS) went into operation. The flow sensor in the leak detection system is an ultrasonic flowmeter developed by the Oceanic Division of Westinghouse Electric Corporation (Erickson, priv. comm., 1987). This application followed extensive testing on crude oil at a Texas refinery in 1973 and at a Louisiana pipeline in 1974 and 1975. The end product was an improved LEFM which, according to Erickson, 1987, is certified accurate to 0.5% of flow and 0.1% repeatability. The ultrasonic technique itself had been developed over the preceding seventeen years, and is well established. Previous systems, however, were almost exclusively devoted to *water* flow measurement.

The installation of the special meter sections took place in the fall of 1976 (Chaney and Johnston, 1976a,b). When the oil did begin to flow on 30 June 1977, all 23 flowmeters were on-line and operational. As the line filled, performance testing began. After a brief shakedown period, the system performance was verified and the flowmeter was accepted for the leak-detection system. Since then several million barrels of crude oil have passed through the line, and all of the original standards have been maintained, according to Erikson. In particular, the 3σ repeatability = 0.1% (Graber, 1982).

The 48-inch (1.2-m) diameter of the pipeline made the cost of a turbine system prohibitive due to the manifolding and pig bypasses required. This made the ultrasonic technique especially attractive.

The flowmeter used on TAPS employs four acoustic paths in one plane in special welded-in meter sections. (See Chapter 2, Fig. 2-4.) There are 23 such sections welded along the pipeline between Prudhoe Bay and Valdez. The ultrasonic transducers are mounted in titanium holders which provide a pressure seal, enabling the transducers to be removed and replaced while the line is in operation (Fig. 4-17). The installation also includes ball valves which make it possible to remove the holders from an operating line through the use of a special extraction tool. A special Alyeska-designed system is mounted on the meter section to remove wax from transducer cavities impacted by the pig.

The electronics are installed in the control room of each of the twelve pump stations on TAPS. They are as far as 1000 ft (305 m) away from the meter section. Each electronics cabinet (except at Pump Station 1) contains two flowmeters, one for the suction side of the line and one for the discharge side. Flow rate and totalized flow along with various status indications are sent to the supervisory-control system where the leak detection algorithms are executed.

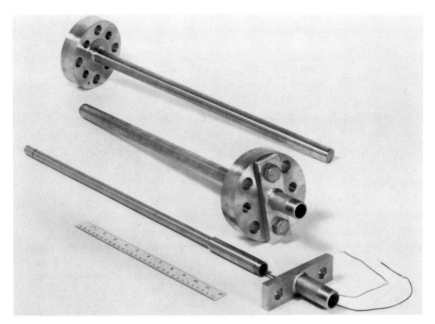

FIG. 4-17.   Titanium transducer housing similar to those used in Trans Alaska pipeline. Titanium forms an all-metal pressure boundary, with a gasket seated under the flange. Transducer module, 500 kHz, is removable from the housing while the housing is still installed in the pipeline. Housing can be withdrawn using an insertion mechanism and an isolation valve.

This four-path system demonstrated that ultrasonic metering of crude oil is practical and can provide advantages of accuracy and nonintrusiveness compared to more conventional metering systems, particularly in large pipeline applications.

[It should be noted that despite the high accuracy obtained in many four-path applications, O'Hair and Nolan, 1984, found that if the profile is very distorted, errors of a few percent may be encountered. "Very distorted" profiles occur in spiral flow produced by a double elbow such as that found in piping systems where they are used to absorb differential thermal expansion. See Fig. 4-2. It may also be noted that the better known four-path arrangements (Gaussian, Chebychev, etc.) are not the only arrangements possible. Alternative multipath arrangements include midradius chords (Baker and Thompson, 1975; Lynnworth, 1978) and generalized multipath interrogation geometries. One generalized proposal is due to Johnson et al., 1977a. Another, due to Pedersen, 1981, is illustrated in Fig. 4-18.]

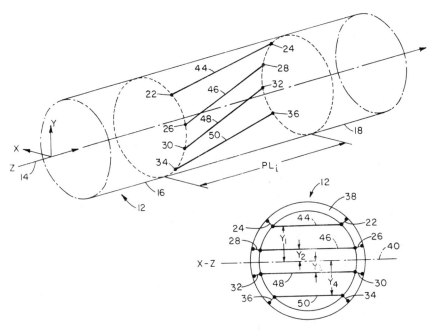

FIG. 4-18. Oblique and end view of a pipe in which flow is measured by a generalized interrogation method that uses four paths between four pairs of transducers located in planes parallel to the xz plane. All the paths do not lie in one plane. After Pedersen, 1981.

Nuclear-Powered Steam Generator in Electric Utilities

Measurement of liquid flow is one of the fundamental needs in all steam power plants and it is a particularly critical parameter when the steam power is generated by a nuclear reactor. For the power industry, according to Erickson, 1987, priv. comm., Westinghouse offers two LEFM (leading edge flowmeter) time-of-flight ultrasonic systems for in-plant use. The hot strap-on portable system can be used for plant diagnostics and a four-chordal-path system provides a feedwater flow measurement. Both systems have been tested by independent third party hydraulic laboratories.

The portable LEFM system is rated at 450°F (232°C), is flexible and allows plant operators to quickly make flow measurements on pipes throughout the plant by strapping a pair of acoustic transducers to the outside wall of the pipe. This system provides flow measurement accuracies of about 2% of flow and repeatability of 0.5% for plant diagnosis and trending analysis.

The four-path LEFM system (Table 4-3) uses wetted transducers inserted into a pipe section, can measure as many as four pipe sections simultaneously and is certified accurate to within 0.5% of actual flow. Additionally, this unit includes a personal computer which performs the nuclear plant calorimetric calculation.

In electrical power production by nuclear power plants, the limiting factor is the licensed thermal rating of the reactor core. To ensure that the licensed thermal rating of the plant is not exceeded, calorimetric tests establish the optimal operating conditions for the plant as close as possible, but just below, the 100% power level.

The calorimetric test is essentially a mass and energy balance performed on the secondary side of the power plant to determine thermodynamically the value of the reactor thermal power. To achieve optimum plant efficiency and maximize plant saleable power the reactor is operated at the 100% power level as determined by the calorimetric calculations. Because the feedwater flow measurement is the most significant value used in this calculation, the reliability and accuracy of this measurement becomes of great importance in maximizing the electrical output of the plant.

The standard feedwater measurement system used in plant feedwater system loops has been the venturi meter. This is a pressure drop (ΔP) device. Unfortunately venturi type meters are prone to a phenomenon known as fouling. This is a metallic plating usually of either copper and/or magnetite in the internal venturi throat area. This constriction of the throat causes the meter to show a greater ΔP and a corresponding erroneously high flow rate indication. This phenomenon has been well documented on numerous nuclear plants since ~1977. The erroneous flow reading effectively derates the plant's output when operating at the apparent 100% power rating.

When the LEFM calorimetric system is installed on plants experiencing this problem, the LEFM system provides a true flow measurement which is lower than the fouled venturi reading, allowing the reactor to power up to the designed 100% power level. This results in increased saleable power and higher plant efficiency. Unlike the venturi meters the LEFM is not prone to fouling.

In addition to providing a more accurate flow measurement and continuous plant calorimetric calculation, the LEFM also measures the *feedwater temperature* ultrasonically. This feature replaces the RTD (resistance temperature detector) measurement and provides a drift-free reading, which unlike the localized RTD is an integrated average across the pipe volume, accurate to within 1°F (0.6°C) or better. Data in Sochaczewski et al., 1981, indicate temperature accuracy of 0.1°C. [According to data in McDade et al., 1959, at $T = 400$°F (204°C) the sound speed $c = 1320$ m/s and $dc/dT = 1.75$ m/s per °C. See Section 3.8.]

In 1987, eleven nuclear plants had the Westinghouse LEFM system installed. It has been found that a small increase in power production can pay for the entire LEFM system in less than a year.

Another method of using ultrasonics to measure feedwater flow is to apply the transducers in the clamp-on mode, using the existing venturi as a reference only when it is freshly scoured and thereby known to be in calibration. [Venturis do *not* foul in *all* feedwater applications. Alternate reference methods include non-radioactive chemical tracers such as lithium.] The idea is to periodically calibrate a one-, two-, or four-path clamp-on against an acceptable reference meter, and then use the ultrasonic meter to monitor flow continuously until the next calibration with a re-cleaned venturi or other reference, perhaps every 12 or 18 months. [Crossed vertical and horizontal transit time paths are used in a four-path diametral clamp-on arrangement developed by Controlotron (Apex-4N system) and Combustion Engineering (C. French, priv. comm., 1988). The multipath arrangement of Fig. 4-19 similarly provides four transit-time paths and additionally provides options for a reflection mode (Section 4.3) and a tag cross-correlation mode (Sections 4.5.2 and 4.5.8).]

A fast-response (10-ms) four-buffer-rod ultrasonic flowmeter developed in Japan for pressurized water reactors is reported by Ikenaga et al., 1983.

Among the multipath clamp–on designs that originated over fifteen years ago, the 1973 design of Thomson–CSF may be mentioned. In this design one of the paths was a diametral path along which the sound speed was measured.

High-Temperature Clamp-On Measures Flow of Quench Oil

Consider the problem of measuring flow at 260°C, with the following constraints: carbon steel pipe diameter = 30 inches (~0.75 m); pipe location relatively inaccessible due to high overhead pipe rack, insulation and high temperature; liquid is quench oil, which is very viscous when cold but of low viscosity at 260°C; unknown acoustical properties of the liquid; hazardous area. The need to measure this flow at this particular chemical plant is related to the equalization of pump loads (Matson et al., 1987).

————————————————————————————————→

FIG. 4-19. Vertical and horizontal (or ±45°) crossed diametral paths provide for four-path transit time clamp-on interrogations. This tends to eliminate errors due to crossflow and reduce errors due to flow profile asymmetries. This multipath clamp-on ultrasonic system, as well as single-path clamp-on systems, can be calibrated periodically against an independent reference method. Cross sectional views of this Panametrics multimode design show details of transducer arrangements that provide, in addition to the transit-time mode, options for reflection and tag cross-correlation interrogations. The sound speed in the fluid can be determined from transit time data obtained across diametral or tilted-diameter paths. Legend: 1-Pipe riser clamp. 2-Guide channel. 3-Transducer assembly. 4-Oblique transmission paths. 5-Diametral paths. 6-Oblique reflection paths.

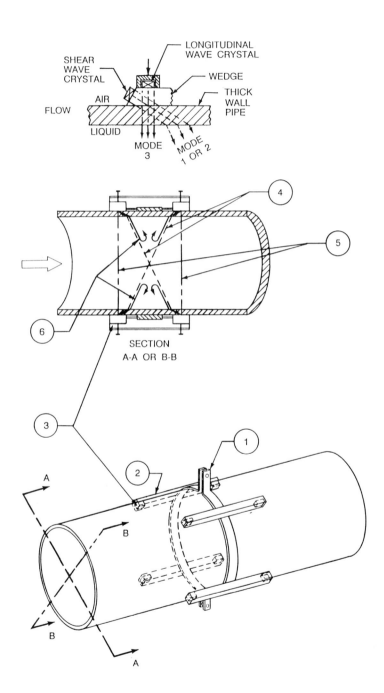

SECTION
A-A OR B-B

To accommodate the hazardous area requirement, the crystal was potted in a protective enclosure (Lynnworth, 1988b, c). The high temperature coaxial cable exiting from the transducer was coupled to a standard RG62 cable in an explosionproof junction box mounted on the transducer. The transducer elements were isolated from the high temperature by using graphite buffer rods similar to those in Chapter 3, Fig. 3-12. The buffer rods were installed when the pipe was at 260°C and they were pressure-coupled to the pipe using Teflon tape. Pressure was applied with screws in two welded-on yokes spaced at S = pipe OD.

This equipment was put into service in November 1986. Signals were detected immediately at the S = OD spacing, but the observed low sound speed, below 400 m/s, implied that a closer spacing might have been more appropriate. However, it was not necessary to move the transducers. A system check one year later indicated that no deterioration of the couplant or transducers had occurred despite continuous use at 260°C.

Each assembly in Fig. 3-12 contains a shear mode piezoelectric crystal bonded to the relatively cool end of a buffer rod. The buffer material is selected to meet acoustical, thermal and mechanical specifications. The surrounding pipe sleeving protects and electrically shields the transducer and buffer.

Among the transducer design objectives were the seemingly divergent goals of keeping the crystal reasonably cool while not removing too much heat from the pipe, and without resorting to a buffer rod of impractical length. The illustrated design accomplishes this to some degree by breaking the SS sleeve into pipe-coupled segments, by attaching the optional heat exchanger cooling coil to the sleeve segment furthest from the hot pipe, and by choosing a buffer material and coupling medium which restrict heat transfer out of the hot pipe. The design also provides for pressure coupling to the hot pipe, yet accommodates differential thermal expansion between buffer and sleeve. Another aspect of the design also accommodates changing sound speed in the pipe wall and in the water as temperature changes.

During their development, the buffered transducers of Fig. 3-12 were tested on a hot plate. Their contact faces were C-clamped together with a single layer of ordinary Teflon tape between them as the couplant. The intended test was to thermally cycle the wedge-shaped ends of the transducers ten times to 260°C, while monitoring pulse-echo and through-transmission signals. Inadvertently, on the tenth cycle, the temperature reached 360°C (680°F). However, no deterioration was observed at this temperature nor upon returning to room temperature.

To put this high-temperature clamp-on result in perspective, note that Karplus, 1977, previously developed clamp-on transducers and a gold foil pressure coupling technique (Section 3.3) which was used for at least five years at approximately 600°C to measure the flow of liquid sodium in a stainless steel pipe. See also, Drescher–Krasicka et al., 1985.

Cryogenic Application—Liquid Nitrogen

Information on the use of ultrasound to measure sound speed and flow velocity in *liquid cryogens* is rather limited. For laboratory studies the papers by Younglove, 1965, Straty and Younglove, 1973, and Lacovat, 1986, are helpful, as is the chapter by Heiserman, 1981. The results of flow calibration tests using liquid nitrogen, $Q = 0.13$ to 13 liter/s, reported by Brennan and Takano, 1982, indicated that most of the meters tested were applicable to cryogenic temperature.

Sondericker, 1984, determined the He mass flow rate from his V and c data at 6, 10 and ~300 K, using an NBS correction formula in the cryogenic region. In this work, commercially-available (Massa) 40-kHz wetted transducers were used. About 1% agreement with a hot wire mass flowmeter at room temperature and about 2% agreement with a calorimeter were reported.

Wastewater Application—Plant Effluent

In the early 1980s, Chevron U.S.A., Richmond, CA, was in the final stages of a 3-year project to design and build a revised refinery wastewater treatment system when it became evident that three large flowmeters would have to be added. The flowmeters would have to monitor the flow of two 30-inch (~750-mm) refinery effluent lines from an aerated bioreactor pond into a nearby bay. Data on the monitored flow rates was to be submitted regularly to a government regulatory agency. One flowmeter would also have to monitor the flow through a 42-inch (1.05-m) line from the refinery's adjoining chemical plant into the bioreactor pond.

Special difficulties soon became evident when the selection process was initiated. In addition to the diameters being ≥30 inches, the flowcells had to be fabricated from concrete, delivery was required within 60 days, the flow rate ranges were unusually broad, required accuracy was ± 1%, and the meter for the 42-inch line had to withstand wide pH fluctuations and potential chemical fouling.

All requirements were met by installing three Nusonics Model 8000 wetted-transducer flowmeters. Compared to competitive venturis or magnetic flowmeters, capital cost savings were over $25,000 and continuing operating cost savings are being achieved in the form of lower power costs (only about 10 watts per flowmeter), and low maintenance. When this application was reported (Owen et al., 1984) all three meters were operating satisfactorily and did not require maintenance in the three years since they were installed. From this 1984 report, it appears that the flow-meters met the plant's need for an instrument which can satisfy state and federal government accuracy requirements at relatively low cost. The selection was made after evaluation of several types of instruments. Most difficult of the challenges was how to measure the extremes of high and low flow rates with the same flowmeter in large-diameter concrete pipes.

The flowmeters were supplied in kit form, with the transducers pre-mounted in saddle plates for on-site installation onto flowtubes provided by the refinery.

For the 42-inch line, the flowmeter is calibrated at 0 to 30 million gallons per day (gpd), but flow is normally in the 1.5 to 3 million gpd range (velocity only 0.25 to 0.5 ft/s, or <0.1 to <0.2 m/s). During the heavy rainfall months, the maximum flow-rate measuring capability of 30 million gpd is needed. Flow measurement is also required in two 30-inch lines which control the level in the refinery's aerated bioreactor. Range of these flowmeters is 0 to 20 million gpd, with an average flow of 10^7 gpd (5×10^6 gpd in each line).

Another group of applications for which Nusonics' kit was appropriate is described by Duder and Powers, 1983. In Kennecott's Salt Lake City ore processing operations, flow of water had to be measured in pipes from 30 to 54 inch, with 1% accuracy, no excess pressure drop, for V from 1 to 10 ft/s (0.3 to 3 m/s). The kit solution met the technical requirements at substantially lower cost than alternative (nonultrasonic) methods.

4.2.1.2 Open Channels: Rivers and Streams

The main purpose in including this section in a book on process control, is to deliver a clear message that the path lengths that are amenable to interrogation by ultrasound can be as long as 1.2 km (Laenen, 1984). Other points of interest extracted from Laenen's review include:

- The first acoustic velocity meter to begin operation in a large nat-ural channel was at The Dalles, Oregon (Fig. 4-20.) This installa-tion, in 1968, initially used a path of 552 m. The path was later reduced to about 400 m to increase the signal.

- In a survey conducted in 1981, it was found that worldwide at least 88 acoustic velocity systems were used to measure streams >10 m in width.

- The problems encountered include: multipath interference; ray bending due to salinity or thermal gradients; attenuation due to scatterers such as sediment or entrained air in the stream and bi-

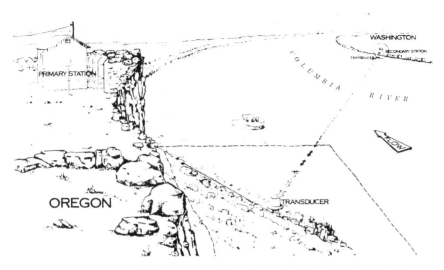

FIG. 4-20.   Artist's sketch of structures and transducer locations at The Dalles, Oregon. After Laenen, 1984.

ological build-up on the transducers; changes in streamflow orientation.

- At a frequency of 100 kHz, it is estimated that propagation over a 1-km path can tolerate a sediment concentration of at least 350 mg/liter. At 1 MHz, that same concentration of sediment would appear to limit the path to slightly over 50 m.

- Resolution is typically ± 0.05 m/s.

- Accuracy of the computed discharge appears to be $\pm 3\%$.

Large paths and natural environment magnify problems that occur on a smaller scale in industrial piping systems. "Magnified" problems such as highly attenuated, highly variable signals and multipath signals motivated Laenen to suggest correlation analysis in his 1984 paper, as a promising means to solve such problems. [This suggestion contributed to the decision at the author's laboratory in the mid-1980s, to embark on a flow-meter development program keyed to correlation detection. The main objectives of that 1985 program were not large-path open channel flow measurement. As later demonstrated in four 7-m paths in the Ottawa River in 1987 (Ontario Hydro Chenaux Plant, Fig. 4-14), however, long paths are not precluded even though the initial goals emphasized paths <1 m. The correlation suggestion by Laenen served as a starting point for developing a transit-time meter which could accurately measure both single-phase and two-phase flows (Jacobson et al., 1987b). Previously, the choice among commercially-available ultrasonic approaches for two–phase fluids was limited almost exclusively to Doppler.]

To measure with *only two transducers* both the *liquid level* and *flow velocity* of a liquid flowing with a free surface in a channel, the arrangement of Fig. 4-21 due to Paulsen and Birker, 1982, may be considered. Beam spreading at the cone angle α accommodates changes in liquid level. If the sound speed c is not known, their method could be modified to incorporate an additional transmission over a known path. Their design may be contrasted with an earlier *multi*transducer design of Suzuki et al., 1972.

4.2.1.3 Millisecond Response

Suppose a step change occurs in transit time. If the upstream and down-stream interrogation pulses were launched *simultaneously* (as proposed, for example, by Ono, 1948, but seldom done in practice) what is the minimum prf that can determine V within 1 ms of the step? If one neglects the time spent calculating V from the transit times, and if the upstream transit time $t_U < 1$ ms, the simultaneous prf answer is $1/(1 \text{ ms} - t_U)$. For a water filled pipe of 10-cm diameter interrogated with wetted transducers along a 45° path, $t_U \approx 100$ μs and the "simultaneous" prf ≈ 1.1 kHz for 1-ms response. If the pipe diameter increases to 1 m, where $t \approx 950$ μs,

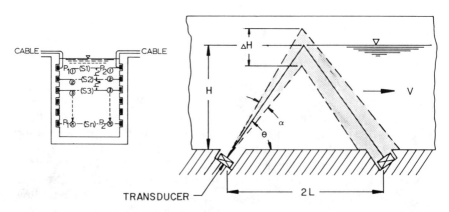

FIG. 4-21. Measurement of flow velocity V and liquid level H in an open channel, using only two transducers in contrast to multitransducer measurement shown in inset. Two-transducer method due to Paulsen and Birker, 1982. Multitransducer method due to Suzuki et al., 1972, © 1972 Peregrinus.

to maintain a 1-ms response time the "simultaneous" prf must increase to 20 kHz.

If the upstream and downstream pulses are launched *alternately*, as is usually the case, then a 1-kHz prf can indicate that *some* change has occurred within 1 ms of the occurrence of the step, but the determination of V takes two complete interrogations after the step. So in this case, an alternating prf above 2 kHz is needed for a 1-ms response time. Here, we take "response time" to mean the time required for V to settle within the one-shot resolution limit, e.g., ± 10 ns for a 100-MHz clock. If the upstream travel time exceeds 1 ms (as it would for a water path >1.5 m) then even with cw bidirectional interrogation a 1-ms response time as defined here is not achievable.

In cases where flow is *presumed* to be steady, a high sampling rate can yield data to check that assumption. One needs to be careful in relating fluctuations along the acoustic path to those parallel to the conduit axis. As an extreme example, oblique interrogation of a liquid immediately after a flow has been stopped by closing a valve is likely to indicate nonzero flow, due to eddies.

Ultrasonic flowmeters designed for measuring flow dynamics with frequency components up to the order of 100 Hz (response times on the order of ten ms or tens of ms) are discussed by Pedersen and Lynnworth, 1973, Herr and Vann, 1977, Carrington, 1976, Carrington and Martin, 1970, Carrington and McCloy, 1978, Lynnworth et al., 1985 and Buess et al., 1986.

Flow dynamics are of interest whenever flow is unsteady. Pulsating, oscillating, accelerating/decelerating and transient flows may occur during the launching of large manned space vehicles (Pogo phenomena); in testing quick-acting valves; in torpedo launch studies; in measuring pulsed flows such as those used to reposition orbiting communication satellites (Jacobson et al., 1987a); in natural gas pipelines due to reciprocating compressors, pressure regulators, process or control variations, or flow-induced disturbances (McKee, 1987; Sparks and Durke, 1987); and in blood and respiratory studies. Ultrasonics can yield faster response than most other flowmeters; it responds *linearly* to flow and hence yields the correct time-averaged flow, not the rms value as generated by nonlinear sensors that respond to the square of flow velocity; is the only nonpenetrating fast-response method applicable to optically-opaque closed conduits (e.g., clamp-on onto steel pipe); and it is applicable to liquids, gases and multiphase media. For these reasons, ultrasonics will probably emerge as the method of choice for many *dynamic* flow applications. An example of an ultrasonic flowmeter with fast-response digital outputs related to the difference in transit times Δt and separately, to the sum of the transit times

Σt (or to just one of the transit times, if $V \ll c$), is given in Fig. 4-22. An external computer is required in order to compute flow from these outputs. Sometimes, if the *dynamic* flow includes intervals of zero flow, these intervals can be used to compute V even by an internal computer that requires several hundred ms per V calculation (Jacobson et al., 1987a).

In the laboratory, allowing a water column to drop vertically unimpeded for a few tenths of a second provides a simple demonstration of flow dynamics (Fig. 4-23). At NASA Langley, a mass, spring and bellows arrangement generated oscillatory flows to test dynamic flowmeters (Herr and Vann, 1977). [Their simulator is illustrated in Lynnworth, 1979, page 493]. In Fig. 4-24, a flowmeter is switched rapidly from one pipe in which $V = 0$ to another in which $V \neq 0$, to simulate a step change in V, with c's equal or unequal in the two pipes.

4.2.1.4   Multiplexing

Multiplexing is motivated by the need to reduce the cost, size or weight associated with a flow measurement along a particular path. These advantages are gained in exchange for slower measurements and the risk of losing data from all the multiplexed (muxed) paths in the event of instrument failure.

Different Pipes

In the 1980s Controlotron introduced a four-channel muxed clamp-on flowmeter (Chapter 2, Fig. 2-18). This enabled users to measure flows in four pipes of different size.

As a spin-off from a muxed ultrasonic thickness gage developed for a communications satellite study (Jacobson et al., 1987a), a 6-channel muxer (Fig. 4-25) became available in 1986 for use with certain existing single channel flowmeters. However, this muxer was limited to applications for which the received signals were sufficiently similar on all six

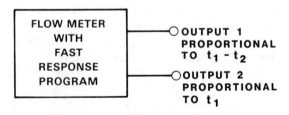

FIG. 4-22. Fast-response program generates transit time outputs with response times in the millisecond range.

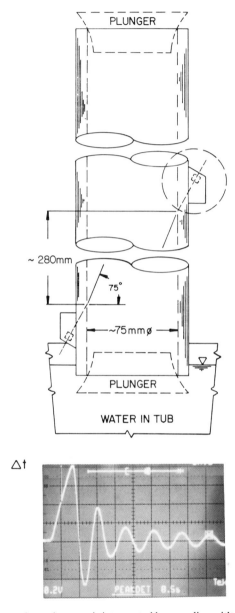

FIG. 4-23.   Water column drop test is interrupted by resealing with plunger, to demonstrate the dynamic response of an ultrasonic flowmeter, transit-time type, operated at a prf = 3000 Hz. Storage scope record shows nearly free fall during first half second, until flow is arrested by plunger. Note rapid deceleration, reverse flow, then oscillations at about 1.4 Hz. Vertical axis, Δt, is proportional to flow velocity.

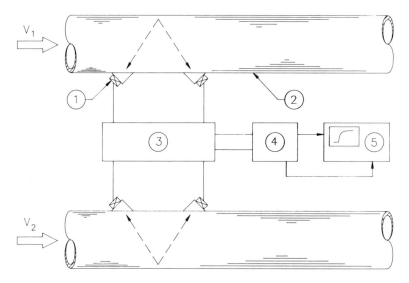

FIG. 4-24.   Electronic switch allows flowmeter's response to a step change in flow to be determined, without any acceleration of the fluid in the pipes. 1–Transducer. 2– Pipe. 3–Double-pole single-throw switch. 4–Flowmeter. 5–Oscilloscope.

pipes or paths, so that, in the electronic flowmeter, one group of windows, arming levels and filters allowed all the signals to be detected reliably. This limitation proved to be no hindrance in cases like a pair of matched 10-inch (~250-mm) diameter spoolpieces or a set of six offset style flowcells constructed of nominally identical parts (Figs. 4-26 and 4-27).

A correlation detection method was developed by Jacobson et al., 1987b, that removed the restriction requiring similarity of signals. The objective in the four-channel Model 6468 was to achieve a cost-per-point competitive with nonultrasonic meters in wide use, while offering technical advantages such as clamp-on, tolerance for single or multiphase

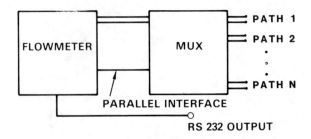

FIG. 4-25.   Multiplexer used by Jacobson et al., 1987a.

FIG. 4-26.   Examples of midradius flowcells that use single or multiple paths. Spoolpieces shown here use triple traverse interrogation path between ball-valved ports. Also shown are diagonally-opposed welded yokes which accommodate clamp-on transducers for tilted diameter interrogation. Application of midradius flowcells: high-accuracy measurements of water flow. The flanged pipes are 10 inch schedule 40 × 1.05 m long.

PIPE

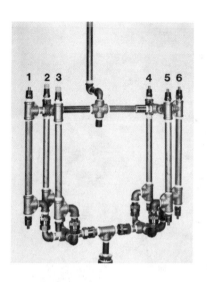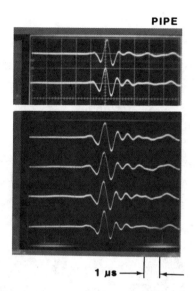

1 µs

FIG. 4-27. Examples of similar signals obtained using four wetted and two hybrid transducers installed in six offset flowcells for Model 6001A application. Same concept applies to clamp-on. Hybrids are in pipes 2 and 3.

fluids, ease of use, analog and digital outputs, etc. With an optional keypad installed, this instrument, while not battery-powered, found use in flow survey applications where pipes of different sizes and materials are encountered.

Different Paths in One Pipe

The Gaussian quadrature four-path flowmeters such as those discussed in Section 4.2.1.1, or open-channel applications where even more paths are required, provide an opportunity to reduce cost by muxing. See Fig. 4-28.

In some cases it is possible to connect transducer pairs in parallel and obtain with one instrument the average flow over such paths *without* muxing. Using midradius chords for illustrative purposes, Fig. 4-29 shows a parallel pair of single traverses obtained by transmitting the sound wave between pairs of transducers.

4.2.2 Gases

Although ultrasonic contrapropagation flowmeters at the present time are used in water or other liquids far more frequently than in air or other gases, the first measurements of sound speed in which *motion* of the

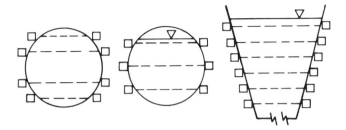

FIG. 4-28. Multiplexed multipaths can be single paths in N different pipes or multiple paths in single conduits as shown here.

medium was recognized to be of importance occurred in air (Derham, 1708; reviewed by Lindsay, 1966 and 1972; remarked upon by Rayleigh, 1877, p. 2).

An early contrapropagation design for an acoustic or ultrasonic air speed indicator is due to Wolff, 1942. This invention, filed in 1939, included upstream and downstream interrogation but did not recognize how these two measurements alone could be used to eliminate the influence of temperature. That important step awaited the subsequent independent work of Swengel (1950-1955) reported by Hess et al., 1950, Ono (patent filed 1948), and Kritz (1955-1958). In the 1940s, Swengel and Kritz found practical ways of rapidly alternating the directions of interrogation of pulses, or ways of using the difference in frequencies from upstream and downstream oscillating loops, respectively, and obtaining from such measurements, V independent of c. (See also, historical reviews cited in Section 4.1, and Chapter 10.)

Table 4-4 summarizes the contrapropagation applications in gases that are treated in this section.

**PARALLELED PAIR
OF SINGLE
TRAVERSES**

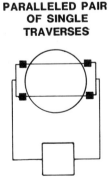

FIG. 4-29. Paralleled pair of transducers interrogate two midradius paths.

Table 4-4.   Summary of Contrapropagation Applications in Gases

Application	Reason(s) for Choosing Ultrasonics
Anemometer	Can resolve flow in x, y and z directions. Rangeability. Bidirectional. Can also yield temperature. Fast response.
Circulation of a vortex	No disturbance of flow region under test. Fast response. No need for special calibration.
Breathing dynamics	Fast response. Accurate. Bidirectional. Wide Range. Small flowcell volume. No obstruction to flow.
Auto engine inlet air	Accurate. Low cost (somewhat dependent on quantity and chips available).
Natural gas	Low installed cost compared to orifice station. Linear. Accurate. Little or no maintenance. Can be retrofitted on buried pipelines with access from top only.
Flare gas	No previous method could continue to operate satisfactorily without frequent maintenance. Adequate accuracy. Yields flow velocity, molecular weight (to locate source of leaks) and mass flowrate (for energy balance, steam control). Retrofittable by hot tapping despite proximity of other pipes or structures. Depending in part on manufacturer and model, flow range is ~1 cm/s up to at least 70 m/s.
Flue gas	Nonintrusive; transducers can be recessed and can be protected from direct exposure to high temperature by pumping cool air around them.

4.2.2.1   Natural Gas

A 1987 paper by Scelzo and Munk described field tests of an

ultrasonic flowmeter which had been developed for the measurement of flow in natural gas pipelines. This meter used the transit-time principle in the same manner as do similar meters for liquid flow measurement in completely filled closed conduits. Transducers may be installed, using a wetted configuration, by either hot tap or spoolpiece. The meter has been tested in both single-path and multipath configurations.

Data collected over a three-year period [1983 to 1986] confirms that a 50:1 turndown ratio and ±2% of reading accuracy are attainable for the single-path configuration, installed using hot taps, when the meter is located at a point in the pipeline where flow profile is fully developed.

In 1987, Scelzo and Munk concluded that further testing was needed to specify the performance of the four-path configuration. Improved accuracy, however (better than ±1%), and immunity to flow profile disturbances were expected, based on earlier results obtained by ORE and Westinghouse with their four-path meters in liquids and based on independent four-path results in natural gas reported by O'Hair and Nolan, 1984.

Background. In the eleven-year period preceding publication of the above-cited paper by Scelzo and Munk, Columbia Gas System Service Corporation (principally W.D. Munk) and Panametrics (principally N.E. Pedersen and J.E. Bradshaw) collaborated on the development of a transit-time ultrasonic flowmeter for natural gas pipelines. Early results were reported by Pedersen et al., 1977. Returning to Scelzo and Munk:

Early development led to a prototype meter which was tested extensively at a major Columbia Gas Transmission Measurement Station. In 1982, the cooperation of the natural gas transmission industry at large was solicited in order to perform a long-term field test of the meter in its wetted-transducer, single-path configuration.

As a result of this solicitation, nine meters were installed in the field, in various pipeline sizes at locations in the United States and Canada. Additionally, testing was performed at flow research laboratories in two European countries. The conclusions drawn in this [summary] are based on the results of the industry tests and other tests of the flowmeter using various hardware and installation schemes.

The transit-time flowmeter, Fig. 4-30, is capable of measuring and indicating flow in either direction. Once the velocity of flow is determined, the built-in flow computer of the meter uses AGA supercompressibility equations to compute the volumetric flow rate in standard cubic feet per hour or per day. Additional inputs required for the supercompressibility calculation are mole percent CO_2 and N_2, and specific gravity. Absolute pressure and temperature must be provided by appropriate transmitters (4 to 20 mA current loops).

In the single path configuration, the meter consists of one pair of ultrasonic transducers which are inserted into the pipeline, through isolation valves, by means of gas tight insertion mechanisms. All of the meters tested in the test program used two 2-inch taps and 100-kHz ultrasonic transducers. A later development, however, led to a smaller transducer (200 kHz) which may be installed via [a] 1-inch tap.

These smaller transducers [could be used in pipelines from 6-inch to 24-inch OD if the pressure is greater than 50 psi (~3 bar) and if the electronics was of the Model 7000 type.]

For pipes smaller than thirty inches, the conventional "across the pipe" installation geometry (Fig. 4-30, inset) need not be used. Instead, a "single bounce" of the ultrasonic pulse, off the opposite inside pipe wall, may be used (Fig. 4-31.) The single-bounce installation method has the advantage of doubling the ultrasonic path and thus doubling the transit time difference, which improves the time measurement resolution and therefore meter accuracy. Additionally, both taps are made at the top of the pipe rather than

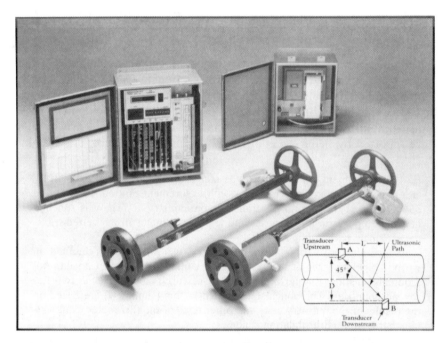

FIG. 4-30. Flowmeter Model 7000, printer, 45° transducers, and gas-driven insertion mechanisms used in 1983-1986 field test program to determine performance in a variety of natural gas pipeline installations. Inset: geometry of transducers installed perpendicular to pipe wall. In this work the flow velocity along the path was calculated from the equation $V = [(L^2 + D^2)/2L][1/t_1 - 1/t_2]$.

one on each side. The single-bounce method, in conjunction with the smaller transducers, greatly simplifies and reduces the cost of installation.

The secondary equipment consists of an electronics console, housed in a NEMA-4 enclosure, which makes the measurement and calculates standard volumetric flow. Indicated parameters available at the electronics console include: standard volumetric flow rate per hour or per day, velocity of flow, transit times, time difference, pipeline pressure and temperature.

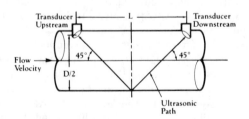

FIG. 4-31. "Single bounce" ultrasonic flowmeter geometry.

To verify that the spacing of the transducers in the pipeline is correct, the measured sound speed is compared with that calculated from the pressure and composition of the gas, the latter determined chromatographically.

Goals and Conclusions.   The goals of the test program were to

verify results obtained on the prototype meter, demonstrate the reliability and durability of the meter under actual operating conditions, demonstrate long-term repeatability of the meter, establish a "meter factor" for each test installation and determine if the meter factor is a function of pipe diameter or site configuration. Orifice or full bore turbine flowmeters were used for the "standard" measurement.

The flowmeter has proven to operate reliably in the long term. At one location, the meter was in continuous operation for over two years without maintenance. Other flowmeters in the program also demonstrated reliable operation from the time of their installation to the end of their test period.

The meter factors [determined at] different test program sites are not identical but they are consistent and repeatable. Typically, the scatter in the comparison data is ±1.5% or less, which is within the general measurement accuracy of the orifice measurement used at each site.
[Other conclusions are:]

- Using a constant meter factor and an upstream straight pipe run of at least 25 diameters, the Model 7000 has an absolute accuracy of ±2% of reading in pipelines from 6 to 30 inches in diameter.

- Repeatability is 0.25% of reading over ranges of 60:1.

- The flowmeter may be applied to pipelines 6 inches or more in diameter.

- The meter may be installed and removed from service without interrupting gas flow.

- Using 45° taps, the meter may be installed so that there is no transducer protrusion into the pipeline. With this installation configuration, the pipeline may be pigged without removal of the transducers.

- The meter has a turndown ratio in excess of 50:1. Actual flow rates from 1 ft/s to over 60 ft/s [~0.3 to 20 m/s] have been measured in various pipe sizes, from 6 to 30 inches.

- The meter causes virtually no pressure loss in the pipe.

- The meter measures flow in either direction and indicates the [flow] direction.

- The meter is portable. Only two people are needed to install, calibrate and remove it. After the site is prepared, installation may be completed in less than one working day.

Two other studies of natural gas flowmetering may be cited, in which a principal objective was to achieve profile immunity. In the four-path approach of O'Hair and Nolan, 1984, sponsored by British Gas, paths are crossed as in Fig. 4-32 to minimize the effects of jetting or swirling flows. See also, Nolan et al., 1986.

At Nederlandse Gasunie, Broekgaarden and Lammerse, 1986, investigated the errors with undisturbed profiles and also the errors due to swirl as a function of single, double and quadruple reflections between but a single pair of top-of-pipe transducers. (Fig. 4-33). Broekgaarden and Lammerse point out that if the desired long straight run is unavailable, measurements in the clockwise and counterclockwise (cw and ccw) senses could be used to measure swirl and hence eliminate its effect on the determination of the axial component of flow. [Hoyle, 1984, previously concluded on the basis of theoretical considerations, confirmed by his experiments on disturbed air flow profiles in a 10-inch pipe, that the triple midradius path (Fig. 4-33) was the path least subject to profile errors, if one is limited to two transducers. However, the converse of Broekgaarden's point is that: if swirl exists, then unless cw *and* ccw interrogations are used, off-diameter paths will yield erroneous results for axial flow. (See also, Baker and Thompson, 1975; Lynnworth, 1978.)] Referring to Fig. 4-33, note that all the illustrated reflection paths can be interrogated with the *same* transducers using the *same* ports. This possibility was discovered for the midradius case (Fig. 4-34) by J.E. Bradshaw and explained in Lynnworth et al., 1986, p. 584, and discovered for the quadruple reflection by Broekgaarden (Broekgaarden and Lammerse, 1986). Measurement of swirl in a pipe by using cw and ccw interrogation along the triple midradius path is discussed in Durgin and Roberti, 1987 and in Lynnworth, 1988c.

A discussion of 40-kHz ultrasonic determinations of *natural gas heat*

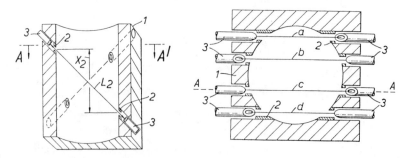

FIG. 4-32.   Four-path system due to O'Hair and Nolan, 1984, used to measure the flow velocity of pressurized natural gas.

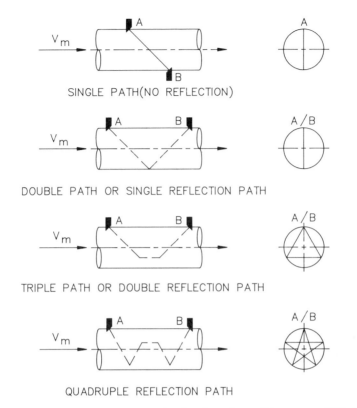

SINGLE PATH(NO REFLECTION)

DOUBLE PATH OR SINGLE REFLECTION PATH

TRIPLE PATH OR DOUBLE REFLECTION PATH

QUADRUPLE REFLECTION PATH

FIG. 4-33.   Paths used in Gasunie tests. After Broekgaarden and Lammerse, 1986. © 1986 AGA.

content, computed from the measured sound speed, is due to Watson and White, 1982. Assuming the heat content to vary linearly with molecular weight, for methane-ethane blends, heat content was determined with an accuracy of 0.4%. Ambiguity due to noncombustible gases (N_2, CO_2) is mentioned as one of the limitations of the linearity assumption. (Compare with Babikov, 1960, pp. 162-165.)

In 1987, Jaescke and Hinze reported that the resonant vane error in *density* measurements of natural gases in the range 20 to 60 kg/m^3 could be reduced from 0.6% down to about 0.1% if the data were corrected by a term depending only on the sound speed c in the gas. They proposed that c be computed using a polynomial having ten coefficients. Instead, suppose a "flowmeter" were in the same pipeline as the errant resonant vane densitometer. If that flowmeter senses both V and c, then the c can correct the density reading, and then the correct density ρ, multiplied by V, would lead to the *mass flow rate M* of natural gases.

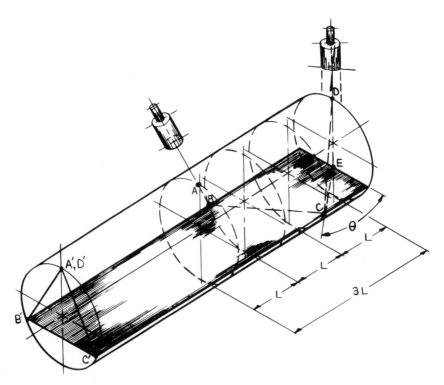

FIG. 4-34. Triple-traverse midradius path generated by nonintrusive straight transducer or 45° slightly-intrusive transducer. After Lynnworth et al., 1986. © 1986 IEEE.

4.2.2.2 Flare Gas

In the petroleum industry, protection against overpressure of vessels, piping, etc., is provided by safety valves at the various processing stations. The discharges from these stations are collected in headers and are directed from there into a main large discharge conduit. Typically, single headers collect discharges from a small number of safety valves, for example ten to twenty. A plurality of these single headers can be collected into a large collection header for an entire manufacturing unit. The gases are burned at an ignited flare or burner pit and from there are vented safely to the atmosphere (Figs. 4-35 and 4-36).

At any particular time, many of the safety valves may be leaking. Usually, however, the leakage rate from any one valve is very small and of no great concern. At times, however, the valves may leak excessively, for example from operation at a pressure too close to the safety valve setting, from mechanical damage during an overpressure incident, or from

FIG. 4-35.   Flare gases burning at a chemical plant, off-shore oil platform, and a refinery.
Photographs courtesy Airoil-Flaregas Ltd.

deterioration due to corrosion, erosion, fouling, or some other cause.
Leakage from a safety valve can be costly due to the loss of a valuable
product, degradation of the product being manufactured, or by creating
problems from operation of the flare system. Furthermore, unless monitored, the leakage might not be corrected for an intolerably long period
of time.

There have been many attempts to detect leakage in the collection head-

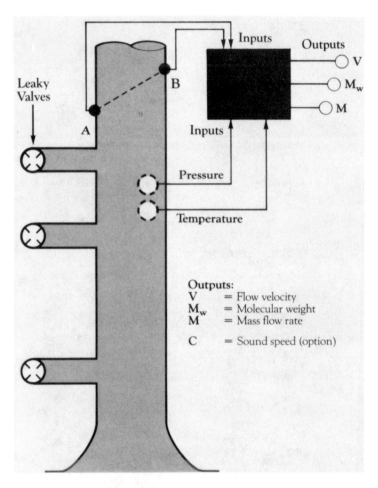

FIG. 4-36.   Typical installation of an ultrasonic flare gas flowmeter. A and B are ultrasonic transducers. Flows into flare system are from various sources. An electronic instrument (such as Panametrics Model 7100) alternates the direction of interrogation 50 to 100 times per second, so that, in accordance with the Sampling Theorem, transit times can be determined accurately despite rapid fluctuations (up to 25 Hz) in sound speed and flow velocity. Upper limit on prf is imposed by transducer ringing or reverberations in the gas path. After Smalling et al., 1984.

ers. These previous methods have generally not proved successful, according to Smalling et al., 1984, and Potter and Stadig, 1987. The malevolent atmosphere in the flare headers is one of the principal problems. Because the flare stack receives material from many sources, the process conditions are hostile to instrumentation. Commonly, there can be fouling, corrosion, precipitation of high molecular weight polymers, and over the passage of time, various combinations of these conditions. Thus, for example, turbine meters have been employed but tend to be inadequate due to clogging of their moving parts.

Also, most ordinary engineering materials fail because of corrosion. Further, the impact resulting from sudden pressure surges, temperature transients, steam, long distances from sensors to electronics, vibration, etc., all combine, as noted above, to present an unfavorable environment which tends to render inoperable prior (i.e., nonultrasonic) measurement systems. Furthermore, flow detection methods for finding leaks using differential pressure instruments are ruled out by the safety system back-pressure considerations which forbid pressure increases in the flare headers. In addition, the flow conditions within the header include both positive and negative directions of flow. The instrument systems noted above, in their simplest forms, do not distinguish flow direction nor do they compensate for negative flow. These instruments generally are unable to measure low flows accurately.

A further consideration in connection with measuring flare stack flow is the location of and limited access to existing pipes and the inability to take a particular pipe out of service merely to install the flowmeter. This means that it is important to be able to retrofit instruments such as flow-meters to an existing facility using, for example, an on-line hot tap procedure. Furthermore, the hot tap must be positioned accurately so that subsequent fluid interrogation occurs along a predetermined path, so that the sampled portion of the flow profile bears a calculable and/or reproducible relationship with the area-averaged flow velocity.

In addition to the hostile environment presented by the flare stack headers, there is a further difficulty, namely, the gas flow characteristics within the header can change rapidly.

Reviewing the above complexity, it may be said that the flare gas application differs fundamentally from the natural gas application of Section 4.2.2.1 in that the flare gas case involves both unsteady V and unsteady c. According to the Sampling Theorem, interrogation directions must be reversed at least two times faster than the highest frequency of significance in the V or c fluctuations, yet not so high that transducer ringing becomes a source of noise. (See Section 4.2.1.3; Ovchinnikov, 1987; Slepian, 1974.) In flare gas applications that use Panametrics equipment, the direction of interrogation is alternated 50 to 100 times per second. From the transit times, V is calculated as for liquids. It turns out that for flare gases tested so far, after taking gas temperature into account, c is nearly uniquely related to the molecular weight M_w. That is to say, the specific heat ratio γ is *not* independent of the molecular weight M_w. Thus, while classical theory predicts an uncertainty in the $c - M_w$ correlation of $\pm 25\%$

due to γ ranging from 1 to 1.67, in practice the uncertainty encountered to date, for M_w from about 6 to 60, is only about 2 to 5%. If the N_2 concentration is more than about 10%, however, corrections are required in the empirical $c - M_w$ correlation in order to obtain accurate results for M_w and M because the $M_w - \gamma$ relation for nitrogen differs from that for flare gases.

Having determined flare gas M_w from c and temperature, if one also measures pressure, then the gas density ρ is readily calculated. Then it is a short step to the flare gas mass flow rate M. Knowledge of M_w is also useful in finding the source of the leaking gas which is flaring.

Another practical design problem in flare gas work has been the need to use special paths (Figs. 4-37 and 4-38) either to avoid high attenuation in the long tilted-diameter path across large pipes, or to avoid obstructions caused by adjacent structures such as pipes in a pipe rack, or to utilize ports that exist or that can only be hot-tapped perpendicular to the pipe axis. Assuming Nikuradse profiles, meter factors have been calculated for such special paths. Graphs of meter factor K versus several geometric parameters are given in Lynnworth and Lynnworth, 1985; one example appears in Fig. 4-3.

For smooth-wall steady flow conditions, the estimated error in the calculated meter factors is $\pm 2\%$ $\pm$ the effect due to the disturbance of Nikuradse profiles by the transducers.

For temperature extremes, buffers with or without angled reflectors are of interest. Thermal gradients may still cause problems, however, due to material condensation, or sound speed gradients in the buffer or in the flowing medium.

Discussions of the *electronic arming and timing techniques* used during the first five years of the abovementioned flare gas flowmeter applications appear in Section 3.5.3.3, in Lynnworth et al., 1985, and in Wallace et al., 1985. In 1988 it was demonstrated at the Colorado Engineering Experiment Station by Jacobson and McGrath (priv. comm.) that by using a Barker-coded transmission and correlation detection (Jacobson et al., 1987b) air flow velocities in pipes could be measured with a repeatability of about $\pm 1\%$ of reading up to 83 m/s. (See also, Mylvaganam, 1989.)

By mid-1987 equipment of the type shown in Figs. 4-36 to 4-38 had been installed in over twenty pipelines in the largest US refinery (Baytown, Texas) and in a number of other refineries and chemical plants in the US, Australia, Canada, Europe and Japan. The annual savings in hydrocarbon products or steam typically equal or exceed twice the installed cost of the ultrasonic flare gas flowmeter. In one case reported in 1986 detection of a major H_2 leak translated into a payback period of about one day (Lynnworth and Ulte, 1986). In another case a minor meth-

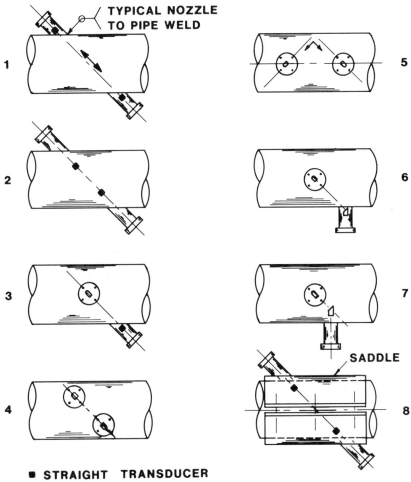

- **■ STRAIGHT TRANSDUCER**

- **▭ RIGHT ANGLE (90 DEG) TRANSDUCER**

- **◁ OBLIQUE (45 DEG) TRANSDUCER**

FIG. 4-37.   Special paths used to overcome access constraints or to create short path in large pipe in which the fluid has a high attenuation. See Smalling et al., 1984.

FIG. 4-38.   Example of hot-tapped installation of flare gas flowmeter transducers in a 20-inch pipe despite access constraints imposed by nearby pipes.

ane leak from a 1-inch pipe into a 24-inch pipe was discovered because it influenced the molecular weight in the larger pipe, even though the flow velocity in that pipe due to the small leak was undetectable.

Referring again to the installations in Exxon's Baytown refinery, the very first pair of Ti-housed transducers installed in Flare Stack 4 were removed temporarily for visual inspection after six months of service. Only a very thin deposit was observed. They were re-installed promptly, and continued to operate *without any maintenance*. By the end of 1987 over five years of continuous service had been logged in an environment which is generally considered transducer-unfriendly. Regarding the reliability of these flare gas flowmeters (transducers and electronics), in nine flare lines in the Baytown refinery where flare instrumentation is considered a high priority, the service factor exceeds 99%. (*Service factor* is defined as the number of days in the month the meter was working divided by the total number of days in the month.) In the most recent year for which service factor data are available, counting all twenty-two ultrasonic flare gas flowmeters, the overall service factor is 86%. The reasons for this low service factor are not known, but may be related to waiting for replacement parts or a relatively low priority (Willson, 1987, priv. comm.)

The following specific example is based on the 1987 report by Potter and Stadig.

Problem. The Exxon Chemical Americas (ECA) plastics plant in Mont Belvieu, Texas, produces approximately 600 million pounds per year of polyethylene, subsequently releasing hydrocarbons and inert gases into a flare system. When the plant was built in 1982, a system was installed to monitor the total flow of the process vent gases, emergency releases, and the inert gases used to purge lines or blanket vessels in order to help minimize the loss of valuable materials.

To help locate sources of increased gas flow occurring in the flare line, spot samples were frequently taken and analyzed by gas chromatography. Not only were sample collection and analysis costly, but the elapsed time before the source of a leak was detected contributed to high gas losses. At times, spot flare samples alone were insufficient in determining the source of a leak.

Compounding the problem, the thermistor type flow metering system originally installed has difficulty detecting very low flow velocity and could not handle the severe service. The sensing probes and associated lines were subject to plugging by entrained solids during upsets and sometimes even during normal flow conditions. At high blowdown rates, the probes would bend, forcing them out of calibration. On two occasions, the probes were physically destroyed.

To assure reliable, accurate readings, it was necessary to remove the thermistor probes for repair/replacement and recalibration at least every two to three weeks. Since the probes were located in the flare line approximately [8 m] above ground, replacement took two instrument men eight hours: four [hours] in a basket and another four to calibrate the unit. In practice, there were long intervals where there was no information on the identity, source, or magnitude of leak losses to the flare.

Since flow rates within the flare line varied and the rate could not always be measured reliably, excess steam was injected into the flare tip to maintain pollution control regulation compliance by assuring smokeless operation. The excess steam represented a potentially significant [waste of energy].

Solution. To overcome the shortcomings of leak detection on the flare gas line, in July 1985, ECA replaced the original monitoring system with an ultrasonic flare gas flowmeter [of the type shown in Fig. 4-36].

To withstand the corrosive flare line environment and the severe mechanical stress, the piezoelectric transducers (resonating near 100 kHz) were housed in a [titanium body joined to a 12.7-mm diameter tube, a construction similar to that represented in Chapter 3, Fig. 3-4. The probes were acoustically isolated from the pipe and precisely positioned by the hot tapping procedure described by Smalling et al., 1986. The ultrasonic flowmeter has a turndown ratio that accommodates flows from as little as 6 mm/s up to 15 m/s.]

Results. The total installation cost about $25,000 and was accomplished without flare shut-down by using hot tapping methods. Strict safety procedures were [followed] during installation since the location is a Class I, Division 2 area.

Just two days after installation, a savings of about $14,000 was realized when the system alerted ECA to the inadvertent loss of a costly solvent through a leaking valve at the rate of $600/hour. If the loss had not been discovered and stopped when it was, it would have continued for another 24 hours until detected by routine tank gaging of the solvent supply tank.

Accuracy of the "hot tap" installed instrument has been consistently within 5%, determined by injecting known volumes of nitrogen on several occasions. Changes in the average molecular weight of the flare gas are detectable to about the same level.

[Using the mass flowrate signal, ECA was able to] automatically control and optimize the flow of steam to the flare tip to ensure smokeless operation. Long term data on mass flow, molecular weight and clock time are being collected so that, in conjunction with analyses of flare gas samples, a more comprehensive interpretation of the plant operations can be made. The data will also be used to help close the overall plant material balance.

[In December 1986] during a routine plant turnaround, the probes were found to be clean with no further attention being required. Because the calibration of ultrasonic probes can be readily checked [electronically in the field, the plant realizes savings of over $5600 per year compared to the labor costs to remove the former probes for monthly cleaning and calibration.]

North Sea Platform, Full Flare Flow Velocities Up to 80 m/s

The material in this section is based on information supplied by Mylvaganam (1987, priv. comm.; 1989) of the Chr. Michelson Inst., Fantoft, Norway.

A flare on the North Sea platform, STATFJORD B, with 36-inch diameter pipe, has a minimum flow rate of 500 to 1000 m^3_N/hour, corresponding to velocities in the range of 0.3 to 0.5 m/s. The same flare with full flare has a flow velocity of about 80 m/s. The low flare is activated for 95% at the time, whereas the high flare occurs only 5% of the time. As a result, only 10% of the gas is flared in normal conditions. Almost 90% of the gas is flared during high flare and this happens in 5% of the production time. Therefore a need exists for higher rangeability and faster response than in the previously-described on-shore flare gas applications. The concept of ray rescue angle for the orientation of the ultrasonic transducer needed for single beam interrogation was introduced by Mylvaganam to overcome the beam drift at high velocity flows. He found beam drift significant in large pipelines especially at high flow velocity (Fig. 4-39). To overcome problems associated with noise at high velocities, he used a chirp pulse compression technique. To preserve accuracy of the meter at low velocities near zero flow, an adaptive signal processing combination of chirp and cw was used to interrogate the flow. Overall system performance was determined from wind tunnel tests. Results include, for a 36-inch pipe: 3% uncertainty at 95% confidence level, for $V = 1$ to 70 m/s. (Mylvaganam, 1987a,b; 1989).

The principle of the chirp signal and the pulse compression technique can be explained as follows. The chirp signal is a rectangular pulse of duration $\tau = t_2 - t_1$. The instantaneous frequency of the rectangular pulse can be selected as a suitable function of time t. (In order to avoid attenuation at high frequencies and noise at low frequencies, and also because of transducer bandwidth limitations, frequencies were chosen to lie within the 50- to 100-kHz octave.) Assume a linear frequency function, with instantaneous frequency $f = f_1$ at time $t = t_1$ and $f = f_2$ at $t = t_2$. The difference $B = f_2 - f_1$ is the bandwidth of the chirp signal. At the receiver the chirp signal is pulse-compressed through a process of correlation, thus delivering a sinc-function output with peak value $A(B\tau)^{1/2}$ where A is the constant amplitude of the frequency-modulated signal in the rectangular pulse of duration τ and bandwidth B, and $B\tau$ is commonly known as the time-bandwidth product. The pulse compression technique has an inherent processing gain of $(B\tau)^{1/2}$. The correlation peak is effectively of duration $2/B$. [The pulse compression technique is described in detail in Brookner, 1977, or Dixon, 1984.]

The transit-time difference at low velocities, particularly at near-zero velocities, will be low. Because of the low noise level at such low velocities a conventional tone burst of suitable frequency can be used, aided by chirp to avoid phase ambiguity. Thus, Mylvaganam's transit time measurement system used chirp interrogation alone for high velocities, ≥ 10 m/s, and a combined chirp-cw interrogation for low velocities, < 10 m/s.

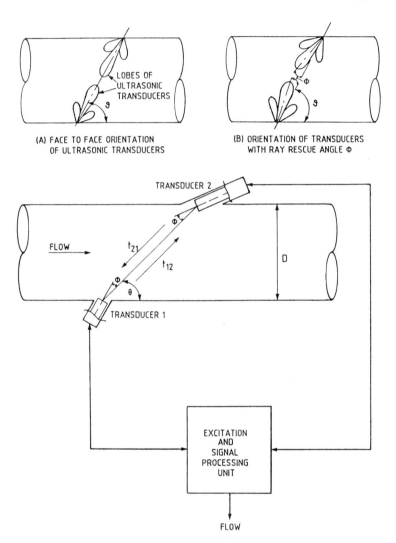

FIG. 4-39. Method used by Mylvaganam, 1989, to measure flare gas flow velocity up to
at least 70 m/s.

The resolution in 36-inch pipe for the meter is 0.01 m/s. Finally, the
repeatability of the meter was found to be 1% of volume flow for velocities
from 0.3 to 70 m/s in 36-inch diameter pipes. The restricted velocity range
for repeatability is a result of the windtunnel used for repeatability studies
and the reference flowmeter and is not due to the ultrasonic meter. The
ratio of the maximum velocity tested, ~100 m/s, to the resolution in 36-

inch pipe, 0.01 m/s, is obviously much larger than for any other flowmeter that might be considered for this application.

4.2.2.3   Air

Proven transit-time applications in air include anemometry, measurements of vortices in wind tunnels, breathing dynamics, and flue gas flow. Automotive inlet air applications are pending.

In an ultrasonic *anemometer*-thermometer reported by Larsen et al., 1979, the vertical component of wind was measured during the International Turbulence Comparison Experiment held in Australia in 1976. The transducers were commercial microphones operated in cw phase locked modes at ~19 and ~39 kHz. An example of strip chart data recorded over a 2-minute interval, compared with that from a different sonic anemometer from the USSR, is given in Fig. 4-40, where the full scale wind velocity is 6 m/s. This article states that many such instruments are in use for meteorological applications. References on earlier three-component sonic anemometers are cited.

De Cicco et al., 1982, built a 250-kHz anemometer and obtained flow readings up to 6 m/s in wind tunnel tests. One of their purposes was to demonstrate that a relatively low-cost ultrasonic anemometer could provide reasonably accurate results. The reported accuracy was 3% full scale from -25 to $+75°C$, and short term stability was 1.5% of reading at 17.5 and 300 cm/s. A three-axis version was reported the following year by Aprilesi et al., 1983 [Fig. 4-41(a)]. In the December 1988 issue of *Sensors*, p. 40, Applied Technologies illustrate their three-axis "sonic wind system" that measures wind velocity V and temperature T. Ranges are ± 20 m/s for V_{horiz}, ± 5 m/s for $V_{vertical}$ and -20 to $+50°C$ for T.

In 1986, Novex introduced a relatively-low-cost air flowmeter for ducts. This equipment included an insertable flowcell containing ultrasonic transducers and was designed for installation through a port in a rectangular air duct [Toal, 1987, priv. comm.; Fig. 4-41(b).]

Starting in the mid-1980s, KVH Industries conducted R&D to produce a solid state acoustic instrument to measure windspeed and wind direction that would be accurate (± 1 knot up to 60 knots and ± 2.5 knots from 60 to 100 knots) and be able to measure wind angle within 2°, while alleviating serious at-sea maintenance problems. According to Kits van Heyningen, 1987, potential applications for this type of anemometer include marine aviation, meteorological and industrial areas. Other potential applications include: high-accuracy sensor as part of a wind shear warning system for airports and for meteorological stations (both on ocean buoys and on land-based buoys) where maintenance and/or environmental considerations make mechanical devices impractical.

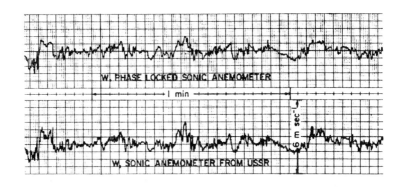

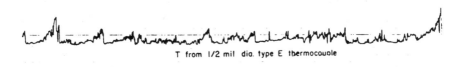

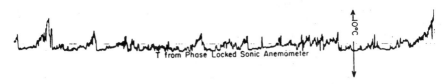

FIG. 4-40. Top: Comparison of data from US and Soviet sonic anemometers used in 1976 Turbulence Comparison Experiment. Note: full scale for each graph is 6 m/s. Bottom: Comparison of temperature obtained from the sound speed or thermometric part of the phase locked loop ultrasonic device, with a 12.5 μm type E thermocouple mounted in the center of the transducer array. Note excellent agreement in detail. After Larsen et al., 1979. © 1979 American Meteorological Society.

To measure the circulation Γ of *vortices in a wind tunnel,* Schmidt, 1975, and Engler, 1982, have used broadband transducers arranged in parallel paths, Fig. 4-42. (Compare with other differential path measurements, e.g., Section 3.6.1.4.) Γ has been measured behind models in a 3-m × 3-m wind tunnel at *V* up to 30 m/s. The following advantages were reported: (1) no disturbance of the flow region under test by the measuring probes; (2) short measuring times (e.g., ~0.15 s to obtain a mean value from 100 samples; (3) no need for special calibration. Schmidt's measurements also lead one to recognize a limit of the contrapropagation method, as depicted in Fig. 4-43.

Biomedical applications for *fast-response* contrapropagation flowmeters include air and medical gases (Kou et al., 1984) and *breathing dy-*

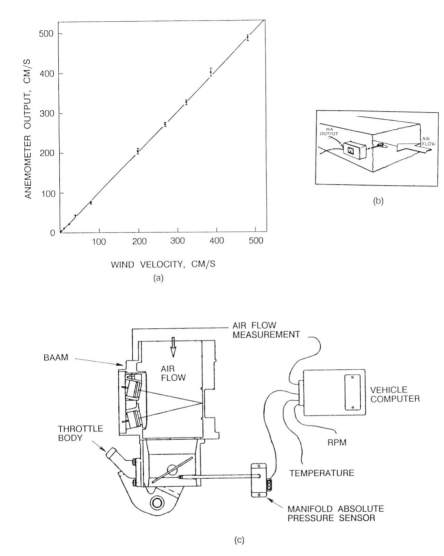

FIG. 4-41.   Examples of ultrasonic flowmeters for air that were designed to be manufactured at low cost and that were demonstrated at least in experimental versions by 1987. (a) Wind tunnel calibration curve for one channel of a three-component μP-controlled ultrasonic (250-kHz) anemometer. After Aprilesi et al., 1983. © 1983 IEEE. (b) Insertable flowcell installed through a 50- × 125-mm cutout in duct measures air flow up to about 20 m/s. Courtesy Novex. (c) Air flowmeter designed as part of an automobile engine air mass flowmeter. After Gutterman, 1987, priv. comm.

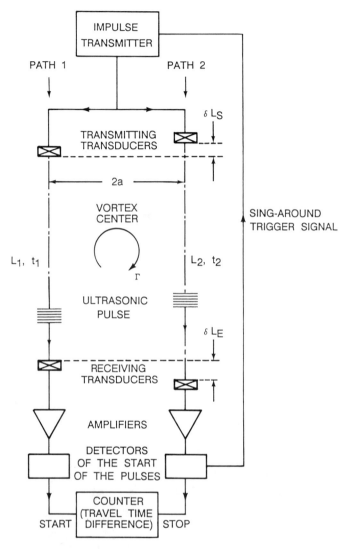

FIG. 4-42. Diagram of the measuring arrangement with two paths used to measure the circulation Γ of a vortex. After Schmidt, 1975.

namics of animals and human subjects (Brusasco et al., 1986; Buess et al., 1986).

Questions about the *breathing dynamics* of astronaut respiration following prolonged exposure to hypogravity led to a set of flowmeter requirements that included: ~240 Hz sampling rate, ±0.02 liter/s accuracy,

500:1 bidirectional range (± 0.02 to ± 10 liter/s), sample volume ≤ 50 cm³, and small pressure drop. In a laboratory feasibility study reported in 1985, Lynnworth et al. used an industrial ultrasonic flowmeter prototype and metal-housed 100-kHz transducers similar to those used in a flare gas application (Smalling et al., 1984; Lynnworth et al., 1984). The pulses were alternately transmitted upstream and downstream at a 300-Hz rate, thereby interrogating inhaled or exhaled air flowing through a 2- × 2-cm square duct or a 1.27- × 2.54-cm rectangular duct. The flowmeter electronics in the prototype form then available was limited in response time by the time to compute V to about 30 ms, or about 30 Hz. By making software changes and by generating an output proportional to Δt, the sampling rate was increased to 300 Hz, as seen in the stepped response to rapid inhaling and exhaling (Fig. 4-44). Apart from hypogravity applications (Farhi, 1976 and 1982) the "fine structure" in the breathing cycle of anesthetized patients, or in athletes, racehorses, etc., is thought by some investigators to contain important information relating to subtle chemical changes in the bloodstream.

In a subsequent study of breathing dynamics, Buess et al., 1986, showed that with better-damped transducers, even faster response is achievable. Using ultrasonic pulses simultaneously transmitted upstream and downstream at a 650-Hz rate, their flowmeter's characteristics included: response time <2 ms; noise <9 ml/s; bidirectional flow range from 0 to 9 liter/s; linear frequency response up to 70 Hz. (In gases at high pressure,

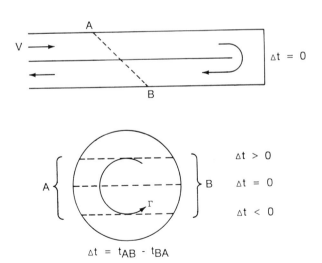

FIG. 4-43.   Two examples of contrapropagation paths where $\Delta t = 0$ despite flow being nonzero.

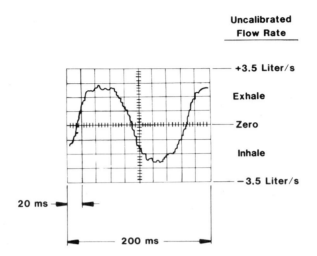

FIG. 4-44. Illustration of dynamic flowmeter response. Note steps in DAC (digital-to-analog) output occur at a rate above 300 Hz in this demonstration recorded on April 4, 1983. After Lynnworth et al., 1985. © 1985 IEEE.

or in liquids, if attenuation is low enough so that ~1-MHz well-damped transducers may be used, then sampling rates up to several kHz become practical. See Section 4.2.1.3.)

The following remarks on *flue gas* measurements are taken from Traina, 1985, 1986 of United Sciences Inc. (USI). He used recessed electrostatic transducers to measure flue gas temperature and flow rate. To protect the transducers from the hot gas, ambient air was blown around the transducers and into the flue. Path length in the first reported installation was about 5.5 m. Traina's system included electronic self-calibration for zero and span, automatically exercised once per day. Contrapropagating transit times were resolved to a few μs by a boxcar integrator, which he found to be reliable despite the high ambient noise due to flow in the flue.

The following account of an *automobile engine air inlet flow* application is due to Gutterman, 1987, priv. comm. To meet the emission and fuel economy demands of automobiles, many auto manufacturers have elected to use fuel injection (FI) systems to power their engines. Air flowrate information is combined with preprogrammed air fuel calibrations stored in the on-board computer which determines the exact fuel rate required for any driving condition. The computer then outputs an electrical pulse to fuel injectors, thus producing the desired fuel flow.

Some FI systems obtain the air flowrate by reading several engine sensors (e.g., RPM, throttle position, engine manifold pressure, etc.), and calculating the mass flowrate. In an air mass flowrate FI system, the computer obtains the air flowrate information directly from an air flow sensor. Since the sensor is the primary means for determining air flow, it is essential that it maintain accuracy throughout the life of the vehicle and that it operate in severe vibration, temperature, and pulsating flow environments.

One such meter is the Bidirectional Acoustic Air Meter (BAAM) by Rochester Products Division of General Motors (Gutterman, 1985a,b,c, 1987). It alternately fires pulse streams of 40-kHz ultrasonic waves against and with the air flowing into the engine. The small phase difference between these two systems is measured and processed along with pressure information to yield an output to the computer. In tests, departures from linearity are less than about 2% for air mass flowrates from 4 to 100 g/s.

A diagram representing this application is given in Fig. 4-41(c). It is anticipated that this type of ultrasonic flowmeter may reach production quantities, $\geq 10^6$/year, when its cost and performance are judged superior to existing nonultrasonic sensors.

4.3 Reflection Applications of Frequency-Domain (Doppler) and Time-Domain (Speckle-Tracking) Techniques

In contrast to the applications covered in the previous section, 4.2, and which generally rely on the fluid being single-phase and/or "sonically transmissive," the applications in this section rely on the fluid being polyphase (BHRA, 1987) and/or "sonically reflective." Traditionally such applications probably would be considered Doppler type, but in 1985 a second possibility became available.

The second possibility, speckle tracking, is mathematically related to the Doppler method. But instead of measuring the change in the frequency of the scattered waves, it uses multiple pulses to track the change in range to scatterers in the fluid. It is analogous to a moving target indicator commonly used at airport air traffic control centers. Recent activity in time-domain techniques, according to Kristofferson, 1988, is more a matter of *renewed* interest (stimulated in part by advances in cardiac color–flow mapping) than interest in a totally new approach to extracting information from the back-scattered Doppler signal.

4.3.1 Doppler

Doppler flowmeters usually are based on the principle that the frequency of ultrasonic waves reflected from scatterers within the moving medium is shifted in proportion to the velocity of the scatterers. It is usually presumed, and sometimes verified, that the scatterers travel at the local fluid velocity, but this condition is not always satisfied (Figs. 4-4 and 4-5).

The earliest available documents on applications of ultrasonic Doppler principles and methods include the patents of Chilowsky and Langevin, 1923, and Chilowsky, 1932. These patents relate, respectively, to measuring submarine motion from an observation point and measuring the speed of one's vessel with respect to the sea bottom or with respect to the sea itself. It may not have been anticipated at that time, that if the location or absolute velocity of the vessel were known precisely (e.g., from satellite navigational aids) then Doppler shifts from the sea can be interpreted in terms of sea currents.

While not often cited in reviews of medical or industrial Doppler flow-metering, sea navigation Doppler work continued in parallel with its land-based counterpart more or less simultaneously from Satomura's 1957 classic paper on using the Doppler method to study cardiac functions, in which he suggested using Doppler to measure the flow of blood. An abbreviated history of navigation Doppler work in the US is outlined as follows. According to Winston (priv. comm., 1988) Raytheon built a Doppler system in the late 1950s. Between about 1958 and 1962, a small company named Janus investigated Doppler. The work continued when Janus was acquired by General Applied Science Laboratories (GASL), until GASL was acquired by Marquardt. Some of the Doppler-oriented individuals transferred to Sperry and EDO Western. Ametek Straza subsequently entered the Doppler navigational field too. Winston recalls that the early navigational Dopplers were limited by the lack of a high power transducer material until PZT-type materials became available in the 1960s. Winston also reports that clamp-on Doppler experiments were conducted in EDO Western's laboratory around 1970, the objective being the measurement of flow of corrosive materials in pipes. A number of other organizations and individuals investigated Doppler for navigational, oceanographic, recreational and other applications since 1960. Recent papers on Doppler in the sea, for example, are listed in *The Journal of the Acoustical Society of America*, Suppl. 2, Vol. 83, Summer 1988, p. 73.

EDO Western's 1970 flowmeters for pipes are mentioned in Chapter 2. These used wetted transducers. Clamp-on Dopplers gained acceptance in the latter 1970s. One manufacturer (Polysonics) reported worldwide sales of over ten thousand such instruments between 1976 and 1987. Polysonics appears to have been the first to introduce a clamp-on Doppler in the US, in mid-1976, for industrial applications (G. Dorflinger, W. Smith, priv. comm., 1988). A summary of Polysonics' first decade of industrial experience appears in their *1986 Guidebook to Doppler Flow Measurement in Liquids*. This *Guidebook* stresses the importance of (a) using crystals that are sufficiently large; (b) transmitting sufficient rf energy; and (c) digital filtering, including tracking of Doppler signals, to discriminate against unwanted scattered frequencies while retaining the correct flow-velocity-dependent Doppler frequencies. Doppler equipment manufactured by Polysonics and others are shown in Figs. 4-45 to 4-51.

Knowledge of ultrasonic Doppler flowmeters has benefited from the numerous studies published over the past thirty-odd years, largely motivated by biomedical applications. Most of the interrogation geometries used in research and industry between 1970 and 1987 are represented in Chapter 2, Fig. 2-8.

As examples of the biomedical Doppler literature see Satomura, 1957; Barber et al., 1974a,b; Herrick, 1959, 1977; Baker and Daigle, 1977; Atkinson and Woodcock, 1982; Azimi and Kak, 1985; Magnin, 1987; Christensen, 1988. In the decade 1976-1986 many industrial users viewed the Doppler flowmeter as complementary to the contrapropagating type, the Doppler being considered appropriate for fluids which were too scattering for the transmission techniques available at that time. But just as Dopplers made inroads into single-phase liquid applications, so too did contrapropagation meters enter the multiphase arena in 1987. Besides "competition" from transmission methods in two-phase media, even within the reflection mode alternatives like range-gated speckle tracking (Embree and O'Brien, 1985; Bonnefous and Pesqué, 1986; Trahey et al., 1987a,b) arose in which frequency shift of the reflected waves was not determined, but rather (one could say equivalently) the change in transit time to a moving group of targets is determined, analogous to a moving target indicator used to control air traffic (Mason, 1964; Brookner, 1977; Kolano and Rowlands, 1977).

A Doppler approach from a new direction, perpendicular to the flow axis, was reported by Newhouse et al., 1985, 1987, and Censor and Newhouse, 1986. Using a focused beam, advantages were found in cases where classical oblique interrogation was not possible. Flow information was obtained from the spectra associated with the beam's off-axis components. See Chapter 2, Fig. 2-8(h).

Other Doppler transducer arrangements for duct flow are given in Chapter 2, Fig. 2-8. For liquids, interrogation frequencies usually range from ~0.5 to ~10 MHz. For industrial applications frequencies below 1 MHz, and typically near 0.6 MHz, are often used.

Commercially available clamp-on Doppler flowmeters often use a two-transducer wedge arrangement as represented in Fig. 2-8(a). Readings are dependent on the distribution of scatterers within the volume defined by the intersection of the transducer beams. Sometimes range gating is employed, to respond to a number of discrete zones within the volume of

FIG. 4-45. Polysonics Doppler flowmeter, serial number 0011, photographed after nearly nine years of measuring the flow of wastewater on an 8-inch pipe with an average flow of 1100 gpm. Installation date: 1976. Illustration courtesy *Polysonics/M&C News*, p. 45 (April 1987). Industrial and municipal liquids are reported to have been measured with this company's Doppler flowmeters in over 7000 installations as of 1987.

FIG. 4-46.   Ultrasonic Doppler flowmeter Model UFT-603 has adjustable flow ranges of 0.1 to 20 ft/s to monitor line sizes of ¼-inch and up. Internal calibrator reference and battery check feature are standard. For reliable scattering of ultrasound this meter requires 25 ppm of 30 μm or larger suspended solids and/or bubbles. Indicator light blinks to confirm acceptable application. Illustration courtesy Dynasonics.

intersection, from which a profile-weighted measurement is computed. (The possibility of responding to *one* zone in which the average flow is close to the area average in the pipe is investigated by Pfau, 1970.) In some cases, one transducer operates as both transmitter and receiver.

If two transducers are used they may be located on opposite sides of the duct, Fig. 2-8(c), or orthogonally as in Figs. 2-8 (b) and (d). Raptis and Karplus, 1978, 1981, used shear wave buffer rods to separate the transducers from a high temperature pipe (Fig. 4-47) and to launch a zigzag wave in the pipe to create the effect of an array [Fig. 3-18(e)].

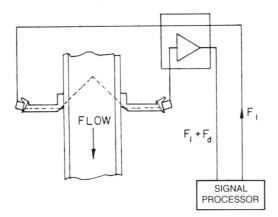

FIG. 4-47.   Doppler flowmeter with buffered transducers for high temperature applications, developed at Argonne National Laboratory by Karplus et al., 1981.

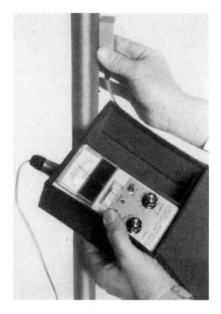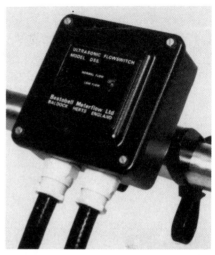

FIG. 4-48.   Portable Doppler flowmeter. Courtesy Bestobell.
FIG. 4-49.   Doppler flow switch strapped to a pipe. Courtesy Bestobell.

Apparently not yet in as widespread *industrial* use as Doppler clamp-ons, Doppler catheters saw considerable *biomedical* use starting in the 1970s. The catheter represented in Fig. 2-8(f) and in Fig. 2-13, due to Martin and Watkins, 1980, measures the product of flow velocity in the pulmonary artery times the cross sectional area in a manner that is essentially insensitive to catheter tilt. Some catheter features might eventually appear in downhole tools for measuring flow and other measurands (Section 9.10).

While not a "catheter," the intrusive two-transducer probe of Fig. 2-8(g) represents an insert Doppler flowmeter such as EDO Western's that was used in the early 1970s (Fig. 2-11). K-Flow (Fig. 2-12) and others introduced their insert Dopplers in the early 1980s.

An arrangement analogous to that in Fig. 2-8(a) was used by Jartti and Luukkala, 1977, to measure, *without contact*, the speed of paper sheet in *air*. Interrogation frequencies were between 100 and 200 kHz. Previously, Jakus and Coe, 1975, used a "passive Doppler" method in *air*, correlating the approaching and departing spectra of an automobile's

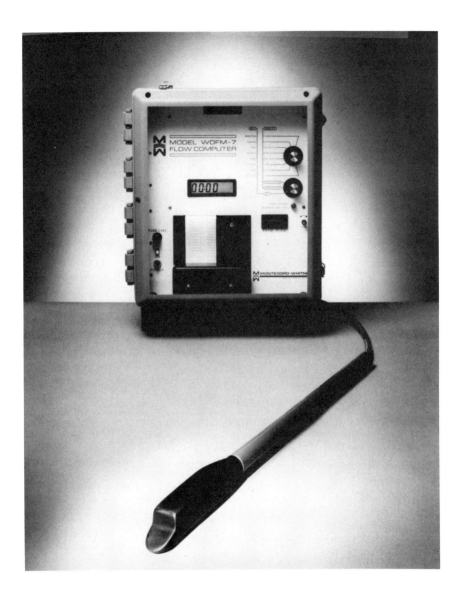

FIG. 4-50. Bidirectional flow Doppler velocimeter is combined with a piezoresistive pressure transducer in Soniflow™ open channel flowmeter from Montedoro-Whitney. Minimum depth for velocity measurement is 38 mm. Velocity range is up to 6 m/s. (Ouwerkerk, 1987, priv. comm.) Illustration courtesy Montedoro-Whitney.

FIG. 4-51.   Example of a μP-based Doppler flowmeter available in 1982. Courtesy Leeds
& Northrup.

noise, interpreting the extracted Doppler shift in terms of the vehicle's
speed.

Among the limitations that may influence Doppler performance in particular situations the following have been cited by several investigators:
spatial resolution (including profile skewing); spectral broadening due to
transit time effects of scatterers; velocity gradients; finite beam width;
nonaxial flow components such as turbulence; distribution of scatterers;
$|\Delta f|$ is not direction-sensitive; nonuniformity of interrogating beams; non-
uniform weighting due to attenuation; path ambiguity if sound speed gra-

dients are severe. The most important of these are discussed by Baker and Daigle, 1977, Atkinson and Woodcock, 1982, and Azimi and Kak, 1985.

On an almost monthly schedule, however, the Doppler literature documents theoretical and experimental developments, especially in the biomedical area. (Examples of mid-1988 papers include Kristofferson, 1988, Ashrafzadeh et al., 1988, and Tortoli et al., 1988.) Some of the listed limitations on Doppler therefore appear to be avoidable at least in certain cases. No doubt analogous advances will be reported in their industrial counterparts (e.g., Waller, 1984).

Practical guidelines from Waller's 1984 paper, and from an earlier paper by Yost, 1982, follow. Naturally some "exceptions to the rule" can be found in special or nonstandard situations. According to Waller, for clamp-on Doppler flowmeters of the types commercially available in 1984 to operate accurately, the following conditions needed to be met:

1.   The measured flow must be a liquid.
2.   To obtain volumetric flow, a full pipe is required.
3.   The pipe material must allow penetration of the ultrasonic signal. Various instruments have different capabilities. Some penetrate pipe liners, others do not.
4.   The flow velocity must be above a specified minimum, which varies from manufacturer to manufacturer and may range from about 0.2 to 0.7 m/s.
5.   Sonic discontinuities for the reflection and reradiation of the ultrasonic beam are required. Some meters require suspended solids or air bubbles. Others obtain their reflections from turbulent interface swirls or eddies.
6.   For accurate measurement, the flow profile must be well developed. This requires sufficient upstream and downstream piping runs.
7.   If the liquid contains suspended solids, the slurry velocity must be high enough so that the solids travel at the same velocity as the liquids.
8.   Most Doppler meters require turbulent flow, i.e., Reynolds number over 4000.

Waller cited several problems related to installation, which he explained as follows (see also, Fig. 4-52):

Doppler flowmeters are generally used to measure flow rates of liquids containing suspended solids, e.g., raw sewage, sludge, paper stock and mining slurries. When the liquid contains particles, the meter measures the velocity of the particles since they act as the reflectors for the ultrasonic beam. Some liquids travel so slowly that the particles float to the top or settle at the bottom of the pipe, effectively decreasing the inside diameter of the pipe. If the meter is to infer volumetric flow from the velocity measurement, this effective change in pipe sizes is a source of error which will probably result in a high flow indication. [Pipe area errors degrade the accuracy of *any* meter, Doppler or otherwise, in which volumetric flowrate Q is to be computed from the measured flow velocity V.]

If particles are not uniformly distributed throughout the pipe and accumulate in groups that are traveling at a rate considerably different from the average, a Doppler flowmeter may "lock-on" to the heavy concentration, again causing an error. [Similarly, as Yost, 1982, points out, one must be careful to stay away from velocity-increasing devices like orifice plates, partially-closed valves, venturis, etc.] On the other hand, there are a number of flow [disturbers] other than elbows that provide the signals needed for measurement. Some of the devices to look for are: piping tees, thermowells, flanged connections or other objects that disturb the flow.

After [selecting] the device that will provide the source of signals, one should next find the optimum distance from the disturbance yielding a signal representative of the average velocity of the flowing liquid. One would not want to place the transducer too near a turbine meter, for example, because there the reading primarily responds to the velocity of the liquid coming off the turbine blades. Rather, move a few diameters away where the dis-

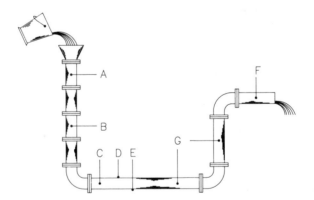

FIG. 4-52.   Right and wrong places to mount Doppler flowmeter transducers. Wrong places A through F: A–Pipe may not be full. B–Down-flow. C–Too close to elbow; profile may not be developed. D–Air may agglomerate. E–Solids may settle. F–Pipe may not be full. Right place: G–Pipe is full, profile is developed. After Waller, 1984. © 1984 ISA.

turbance created by the meter will have diffused into the flowing stream and will be traveling at about the same rate as the stream.

Waller advised that one should try to avoid flow measurement in vertical downward flows. Aside from the fact that the pipe may not be full, air bubbles may be introduced. Too many bubbles may hamper penetration of the ultrasonic beam and the bubbles will tend to rise against the flow. (Vertical upward flows can be troublesome too.) According to Waller:

> Some meters will operate with lined pipe, others will not. No clamp-on meter will operate properly is the liner is not solidly attached to the pipe wall, because of the air gap. Similarly, loose scale may block the beam.

Waller, 1984, also discussed some problems related to the process stream. According to Waller:

> If there are no sonic discontinuities, there are no reflectors, and most Doppler flowmeters will not function. Some bacterial slurries and chemical slurries may have many suspended solids but cannot be measured with all Doppler flowmeters. Gross changes in the process may also cause the meter to cease proper operation. For example, a meter of the type requiring suspended solids had been operating satisfactorily for a long time, but then stopped working. The problem was traced to a particle filter that had been placed upstream of the sensor, effectively eliminating the reflectors from the stream. Process changes in the other direction are also possible. The meter responds to bubbles up to a point, but if the liquid starts to foam there may be too many air bubbles to allow the ultrasonic beam to adequately penetrate the flow, causing erroneous measurement. Similarly a sludge may become too dense to allow proper penetration.
>
> Extremely high acoustic noise levels in the measurement area may modulate the ultrasonic waveform, causing errors.
>
> The nature of the materials introduced into the flow can be detrimental. Consider the case of the salad dressing that was being measured accurately and consistently [until] it was decided to add onions to the dressing. The onions absorbed the acoustic signals transmitted into the stream. Corn starch is another acoustic absorber and a correspondingly poor reflector.
>
> River water requires a meter capable of measuring both clean and dirty liquids. One [could] expect full-scale accuracies of ±2.0% to ±5.0% and repeatabilities of ±0.1% to ±1.0% with most Doppler meters [available around 1984]. Linearities usually range between ±0.5% and ±2.0% in applications for which the meter is recommended. [Specifications in 1988 were about the same but the range of applications and ease of use showed improvements.]

Generally speaking, if one finds that the location of the transducer and the reading produced does not provide what is believed to be the correct velocity, three suggestions are (Yost, 1982; Waller, 1984):

1. Move the transducer upstream and downstream of the initial site. If readings are fairly constant from location to location, the initial site was acceptable. If one gets different readings at the various transducer locations, move to a different section of the pipe.
2. Consider other reasons for velocity readings being incorrect. Could the pipe ID information be wrong? If the evidence shows that the calculations are correct, have the flow rate changed to see if that yields a proportionate change in reading. Alternatively, find another location for the transducer which provides a more accurate reading.
3. When testing an application where there are control valves in the line, be sure to partially close the valve as part of the test. One wants to be sure that the indicated flow velocity does not increase when the volumetric flow is actually decreasing. This would occur if signals were obtained mainly from the higher-velocity portion of the liquid as it goes through the partially-closed valve.

By 1988, well over ten thousand Doppler flowmeters were estimated to be in industrial use, in industries that included: paper, minerals, food, power, textile and municipal facilities.

Just as transit-time transducers are available for wetted *or* clamp-on service, so too are Doppler transducers. The advantages of *wetted* Doppler transducers include:

• Eliminates wall interferences.

• For large pipe, allows sensor to interrogate region where local velocity is good approximation to area-average velocity, i.e., depth of insertion about $\frac{1}{8}$ of pipe inside diameter, for developed turbulent flow profile (Miller, 1983, p. 5-10; Ginesi, 1987).

• For dense thick sludge and mud, allows sensor to get into flow and sense a representative velocity. This is to be contrasted with clamp-on Dopplers, for which response would be chiefly from the low-velocity boundary layer if the beam suffers from high attenuation.

As an example of wetted Doppler instrumentation, K-Flow's DM700 series of velocity sensors consist of a pair of completely encapsulated ultrasonic transducers and a temperature sensor. Several different packaging configurations are available for different applications:

a. DM701—Foil probe, for liquids below approximately 200 centipoise.
b. CM710—Angled probe, for highly-viscous contaminated liquids or slurries.

c. DM730—Flush probe, for relatively clean liquids where the pipe must not contain any obstruction.
d. DM740—Small-line probe, for liquids in pipes from 1.5-to 8-inch diameter.

For further general discussions consult: Faddick et al., 1979; Addie et al., 1980; Waller, 1984, or a recent in-depth text or review such as Atkinson and Woodcock, 1982, Azimi and Kak, 1985, Magnin, 1987, or Censor et al., 1988.

To the extent that *crack growth* falls within the scope of a flow measurement, it is appropriate to note in this Doppler section the NDT work of Leon and Scheibel, 1987, on a *Doppler technique for monitoring LP (low pressure) turbine blades.* As Leon and Scheibel explain:

> The technique is aimed at locating those blades with a propagating crack. It does this by detecting the resonant events which occur when some natural frequency of the blade or blade group shifts into coincidence with an integral order of running speed as a result of the growing crack. The monitor would not only detect the resonant event, but indicate the blade or blade group involved, as well as its vibrational mode. Only two, non-interfering, stationary sensors are required per monitored blade row.
>
> Initial testing on the L-1 stage of a 0.6-scale Westinghouse test turbine [was reported in 1987] to be encouraging. Work is proceeding on the development of the computer hardware and software for a prototype on-line system for long term installation in an operational power turbine.

At the time Leon and Scheibel's paper was presented, there was no practical on-line monitoring technique for detecting resonating blades in an operating turbine. As they pointed out:

> The Doppler technique, which is intended primarily for use in the crack-prone latter stages of an LP turbine, relies on two stationary, non-interfering sensors downstream of the monitored stage to detect the sound radiated by the vibrating blades. This signal is characteristically Doppler shifted, and of such low level that it requires considerable enhancement to extract it from the high background noise. After the enhancement process, the resonating blade is pinpointed, its vibration amplitude is indicated, and the vibration mode is determined.
>
> The importance of being able to detect resonating blades or blade groups is two-fold. First, design-problem resonant blade vibrations leading to eventual crack indication can be uncovered. Second and possibly more important, as a crack propagates it can cause resonant events to occur, each the result of some natural frequency of the blade group being shifted into coincidence with an integer multiple of running speed. These resonant events may last only days or even hours as the resulting high vibration causes even

more rapid crack growth, quickly pushing the blade group natural frequency below the integral-order driving frequency.

An on-line acoustic blade monitor would keep track of these resonant events. One such event on a given blade or blade group would be cause for concern; two or more might be cause for alarm. Replacement blades could be ordered in the interim making replacement a rapid and far less costly procedure. Additionally, the blade being replaced is almost sure to be cracked, instilling confidence in the monitor's capability for spotting blades in need of replacement.

Though the turbine manufacturers and others are pursuing new design approaches and new materials aimed at reducing blade cracking, it appears that this problem will continue to plague the utility industry for some time, making the development of an on-line crack propagation monitor of significant importance.

4.3.2   Speckle Tracking

The speckle-tracking method of measuring flow arose in studies of blood flow by Embree and O'Brien, 1985, Bonnefous and Pesqué, 1986, Bonnefous, 1988, and Trahey et al., 1986, 1987. In this method, pairs of interrogation pulses are transmitted into the scattering medium, and the echoes are correlated. As long as the interrogating pair are not spaced at too great a time interval, most of the scatterers contribute to both echoes which are to be correlated.

In an industrial implementation of this method, Jacobson et al., 1987b, programmed a flowmeter instrument (Model 6068) to transmit pairs spaced at intervals such as 1, 2, 5 or 10 ms. Delays were selectable manually (keyboard) or automatically. In this instrument the transmitted pulses consisted of one or more cycles or bits, a preferred pulse being an 11-bit Barker code. After a somewhat greater interval (e.g., ~100 ms) a second closely-spaced pair were transmitted. By differential analysis the common mode (nonrandom) echoes from stationary reflectors are largely eliminated.

Speckle tracking shares with Doppler an applicability to scattering fluids, and a fundamental response to V/c, not V. Compared to cw Doppler, speckle tracking offers advantages of range gating for profile measurement, and rangeability down to extremely low flow velocity merely by increasing the delay between the interrogating pair of pulses. Compared to the contrapropagation transit time clamp-on method, speckle tracking can measure flow in small pipe and/or low flow in cases where the contrapropagation Δt would be less than, perhaps much less than, 100 ns, such that accuracy of 1% of reading would be difficult to obtain.

Speckle tracking would appear to be more expensive to implement than cw Doppler. However, in the aforementioned industrial instrument, the speckle reflection mode was available either as a software alternative to the transmission mode, or as an optional add-on in a multimode version of that instrument (Section 4.5.8).

Aside from measuring the average flow velocity of scatterers, speckle tracking may find applications in mapping flow profile. From the measured profile, the meter factor K can be computed to provide a dynamic correction to transmission measurements of flow velocity along a single path. (See table in Section 4.5.8.)

Since the hardware necessary to implement speckle tracking can be essentially the same as used for transmission and tag cross-correlation, speckle tracking lends itself more easily than most Doppler methods to being incorporated in one instrument with multimode interrogation and flow analysis programs. If sound speed c can be obtained from a transmission measurement, then c can be eliminated from the V/c speckle result. Alternatively, if V is determined from a tag cross-correlation measurement (independent of c) then that result, combined with the V/c speckle result, can yield c.

Based on speckle-tracking laboratory experiments in which flow was measured in air/water mixtures containing void fractions up to 25%, as well as in non-aerated tap water, it was concluded that speckle tracking was likely to find industrial applications resembling Doppler applications, i.e., in fluids known to contain scatterers. Specific examples include slurries and liquids with entrained gas bubbles (BHRA, 1987). A further class of applications includes cases where the fluid may be nonscattering part of the time but nontransmissive or scattering at other times.

4.4 Vortex Shedding

Vortex shedding ultrasonic velocimeters that became available during the 1970s were based on two principles:

1. A series of vortices are formed periodically in the wake of a bluff body when there is sufficient relative velocity V between the body and the surrounding fluid.
2. The vortices modulate the transmission of an ultrasonic beam which interacts with them at a frequency proportional to V.

The following explanation of vortex shedding flowmeters in general, due to Miller, 1983, is instructive as an introduction to the subsequent remarks on *ultrasonic* vortex shedding flowmeters. As Miller explains:

Vortex shedding is a common flow phenomenon in which rotating zones of fluid are shed downstream of a barrier placed in a moving stream. This *roll-up* of fluid, first from one side of the barrier and then the other, is responsible for the waving of flags and the singing (aeolian tones) of telephone wires. The subject of vortex shedding is treated in most fluid dynamics texts and in numerous technical papers. To date, however, no complete theoretical explanation of the fundamental nature of vortex formation and the resulting alternate shedding pattern has been advanced.

Figure 4-53 shows the basic vortex-shedding principle. As the flow is split into two streams, the instability of the shear layer causes the fluid to roll up into a well-defined vortex, the time for complete vortex formation depending on vortex-element (barrier) geometry. After it is formed, the vortex sheds, and a second vortex begins to form on the opposite side of the element. If the vortex shedding is stable, the time for complete formation of the second vortex is the same as that for the first vortex, with the formation time [inversely] proportional to velocity. The pattern is then repeated alternately from side to side, resulting in the familiar downstream *von Kármán* (1912) *vortex street*. The vortex shedding results in pressure and velocity changes around and downstream of the vortex element. By placing pressure, thermal, or ultrasonic detectors in a location where the [effect is strong], the vortex-shedding frequency may be measured.

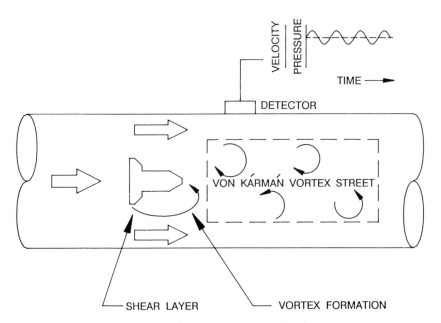

FIG. 4-53. Vortex shedding from a bluff body. After Miller, 1983. Modulation of an interrogating sound beam by the wake is utilized in J-Tec's flowmeters. © 1983 McGraw-Hill.

Strouhal, 1878, made the first experimental observation of the shedding phenomenon. He showed that the shedding frequency of a vibrating wire in the wind was related to the wind velocity and wire diameter. The Strouhal number is still widely used as the basic vortex-shedding correlation; it is computed as $S = fd/V$ where S is the Strouhal number, d the barrier width, or strut diameter, and V the free-stream velocity.

Experimental evidence indicates that the Strouhal number is essentially constant over wide velocity ranges and is independent of fluid density. Data obtained with industrial meters on liquids (Inkley et al., 1980), steam, natural gas, and air (White et al., 1974) substantiates this density independence. The pipeline velocity profile does, however, influence the constancy of the Strouhal number.

Returning now to the ultrasonic vortex shedding flowmeter due to Joy and Colton, 1972, referring to Fig. 4-54 and recognizing that the Strouhal number S is approximately equal to 0.2 in the linear range of normal flowmeter operation, it becomes evident that the vortex shedding frequency f is approximately equal to $0.2V/d$. Since the Strouhal number becomes linear for Reynolds numbers, based upon the width of the vortex generator or strut, of about 300, then linear operation can be designed to occur at any desired velocity by choosing the correct strut size. Strut sizes from 0.2 to 25 mm have been used with ultrasonic vortex meters. The choice of the strut size then is determined by the minimum desired velocity and by the characteristics of the vortex detection process.

As explained by Joy (priv. comm., 1988) it is in this detection process where the ultrasonic technique has the unique advantage that the sonic properties used are modified by the velocity of the vortex, which is *proportional* to the liquid velocity. Since this velocity is independent of the vortex strut size, the choice of the strut size can be made independent of the vortex detection process. Most of the "nonultrasonic" vortex meters detect the passage of vortices by measuring the pressure difference between opposite sides of the vortex strut or forces produced on the strut. These pressures or forces are proportional to the size of the strut and the *square* of the velocity so additional compromises must be made to accommodate the available pressure detection device having the desired sensitivity for the lowest velocities and a dynamic range equal to the square of the velocity or flow range.

Joy points out that typical ultrasonic vortex flowmeters now use vortex struts with widths of from 2 to 5 mm. In gaseous devices, the velocities are generally from 0.3 to 60 m/s, which produces vortices of 50 to 6000 per second. Resolution (as with all digital or quasi-digital devices) is one pulse so the resolution of the flowrate depends on the counting period. The time response should be the time for the formation of a single vortex

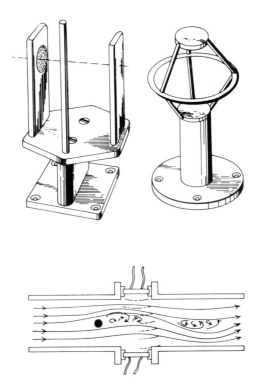

FIG. 4-54.   Examples of vortex shedding ultrasonic flowmeter configurations. Top: Diagrams from Joy and Colton's 1972 patent, in which an ultrasonic method and apparatus are
described for determining the relative velocity between an object and a fluid stream. A
vortex strut attached to the object is located in the fluid stream so as to generate Kármán
vortices at a frequency proportional to the relative velocity between the object and the
stream. In one embodiment, a transmitting transducer is located on one side of the strut's
wake and a receiving transducer is located on the other side of the wake. The transmitting
transducer's signal is modulated by the Kármán vortices and received by the receiving
transducer. An electronic detecting system is connected to the signal-receiving transducer
for detecting the modulations created by the Kármán vortices. An alternate embodiment of
the invention combines the transmitting transducer and the receiving transducer in a single
transducer structure. The combined transducer structure is mounted in the strut. A burst
of pulses is transmitted and reflected by a Kármán vortex back to the combined transducer
structure where the reflected signal is detected. Bottom: Bourns "Ventilation Monitor" is
used for patients connected to a mechanical ventilator or anesthesia circuit. The tube-strut
combination in the Bourns LS-75 sensor is designed to generate one beat, or vortex, each
time one milliliter of gas passes the strut.

or in the example above, would be from 20 ms to 0.16 ms. Response tests
were performed for an airspeed sensor using this concept, and showed
measured response times of from 0.5 to 10 ms.

   Besides the use of this technique in industrial process control, several
unusual applications include the measurement of engine cylinder blow-

by where the fast response enables the observation of each cylinder's leakage; airspeed sensor for dirigibles and helicopters as well as for fixed wing aircraft up to speeds of Mach 0.8; use as anemometers. With suitable designs, Joy continues, it can provide directional information as well as speed and such devices are being used on the U.S. Army's M1 Battle Tanks for artillery corrections.

Although the variety of successful applications reported with this method is less numerous than for the transit time and Doppler methods, in terms of quantity of ultrasonic instruments in use, ultrasonic vortex shedders appear to easily outnumber all other industrial ultrasonic meters combined, at present.

The following information on practical aspects of ultrasonic vortex meters is due to Joy, 1984. Illustrations and background references on this equipment may be found in his paper, from which this material is extracted, essentially as he wrote it.

Two forms of modulation are possible in this technique. One of these is an amplitude modulation of the received signal caused by coherent scattering of the beam as vectorial addition of the vortex rotational velocity and the sonic ray propagation velocity results in a deflection of the beam. The second method is a phase or time delay modulation produced by the effects of the transverse velocities between the vortices. Both types are used although they require a somewhat different form of vortex generation.

By correct sizing and placement of the ultrasonic transducers, the modulation of the energy can be controlled, such that in gases, 100% amplitude modulation can be obtained at the desired maximum flow. The positioning of the transducers is not critical.

The modulation level varies approximately with the flowrate. In liquids the modulation depth is reduced due [in part] to the increased velocity of propagation.

The choice of the ultrasonic frequency is influenced by a number of factors. In general, as the frequency is increased, the acoustic noise from valves or similar sources decreases. So, consistent with other factors, including absorption, it is desirable to use the highest practical frequency.

As with any imaging technique, objects or disturbances much smaller than the wavelength of the energy are essentially transparent. Therefore the lower limit of the ultrasonic frequency is related to the vortex size which in turn is related to the size of the vortex generating strut. The upper limit is basically determined by the desire to "not see" minor fluid turbulence and the absorption effects.

As examples, for vortex strut widths of ~6 mm operating in air, the optimum ultrasonic frequency is on the order of 60 kHz. For liquids, the frequency will be higher due to the higher propagation velocities.

One two-inch (~50-mm) gas meter has a strut of 0.092 inches (~2.3 mm). This unit has a range of 5 to 200 ACFM (actual cubic feet per minute). [1 ACFM = 472 cm^3/s, so 200 ACFM $\approx$ 0.1 m^3/s.]

While the vortex is basically measured at one point in the flow, the frequency is influenced by the velocity over the full length of the vortex strut, so that some averaging of velocity profiles is made. However, if the strut length to width ratio becomes too large, then a coherent sheet vortex is no longer formed and the frequency measured becomes more localized. In this event, the smaller strut size is accompanied by a lower pressure loss due to strut blockage, and may well be an acceptable trade-off.

An examination of the applications where ultrasonic vortex flowmeters are being used will tend to highlight where this type meter is most useful. Up to 1984, the greatest application of the gaseous vortex flowmeter in the industrial market had been in pipeline monitoring, in air compressor measuring and in low pressure natural gas energy audit applications. By 1988, automotive applications were the most numerous. A more revealing insight can be gained by some of the more unusual applications. Until 1984, by quantity the greatest number of this type meter had been in the medical field for pulmonary function or breath measurement applications. The attractive features leading to this include the low pressure drop associated with the small blockage of the meter. As an example, the unit being used for this function has a tube diameter of 12.7 mm with a vortex strut width of 2.3 mm. At full flowrate of 250 liters per minute, this has a pressure drop of less than 3 inches (~75 mm) of water. The low back pressure permits the patient to breathe more normally for either long term forced breathing or permits a more accurate analysis of the pulmonary functions. The tube size and vortex strut were selected to produce one vortex pulse per milliliter of flow, permitting easy readout and totalization. By 1984 more than 25,000 of these units had been manufactured.

A second application illustrating this type meter's characteristics is an on-board intake air measurement for automobile engines. The principal requirements for this application include:

- Fast response—ability to accurately measure the air flow during each engine valve opening as well as be able to follow the transient flow of kicking the throttle open (typically 10 CFM per millisecond).

- High data rate—in order to obtain an accurate measure of the air in each cylinder requires that a large number of pulses be generated.

- Long term accuracy be maintained to meet the 50,000 mile federal requirement.

- Low production cost.

- Low pressure loss.

- Unit-to-unit interchangeability.

[Mitsubishi was reported in 1986 to have been using an ultrasonic vortex shedding airflow meter for several years. The vortex in this case is created by a cone-shaped piece in the air passage (Yamaguchi, 1986).]

While these applications are not in the normal industrial environment, they show that reliability in a very severe environment, or in life support systems, can be obtained.

In any ultrasonic usage in gases, problems may be encountered due to molecular absorption. This is not a problem for most pure or simple gases but does pose limitations for some of the polyatomic gases. Gases such as carbon dioxide or freons exhibit this effect. The most common operational problem involves upstream valve noise. The use of bends or silencers may be necessary if excessive noise is encountered. In liquids, cavitation behind the vortex strut must be avoided both because of the noise produced in the meter output and erosion of the strut.

Vortex meters have not been widely used in waste water applications where fibers can wrap around the vortex strut. However, contamination of the strut from particle build-up is generally not a problem as the scouring action of the vortices tends to keep the strut clean. The units also should *not* be used in two-phase flow or where large amounts of entrained bubbles are encountered. Each bubble in the acoustic beam causes additional acoustic loss and if the bubbles are large enough, the meter will count them as if they were vortices.

Vortex meters have the desirable characteristic that, like orifice plates, calibration can be verified by the physical measurement of size. The frequency signal can be linearized to 1% of reading over a 40:1 dynamic range at stated conditions.

Abrasion of the strut width translates directly into a change in the meter factor K. In some of the meters, however, the strut is removable and can be checked or replaced.

Figures 4-55 to 4-57 illustrate some of the hardware available from J-Tec in 1987. (See also, Joy's remarks in Chapter 10 concerning automotive applications and quantities estimated at the end of 1987.)

Referring to the passive/active comparison in Chapter 2, Table 2-3, the question arises: Is there a *passive* acoustic way to sense the vortex shedding frequency? (If the shedding frequency is in the audible range, below 20 kHz, passive detection of the fundamental does not qualify the device to be called an *ultrasonic* flowmeter, strictly speaking.)

In the Yokogawa vortex shedding flowmeter, the shedding frequency is sensed by a piezoelectric crystal embedded inside the vortex shedder but outside the pipeline, Fig. 4-58. Models are available for process temperatures up to 400°C. Accuracy is reported to be ±0.8% of reading for liquids and ±1.5% of reading for gas or steam, provided the pipe size and Re requirements are satisfied.

As with most other vortex meters (Miller, 1983) accurate operation is normally restricted to the wide but finite range of Re in which the Strouhal number is essentially constant. For the Yokogawa vortex flowmeter, the

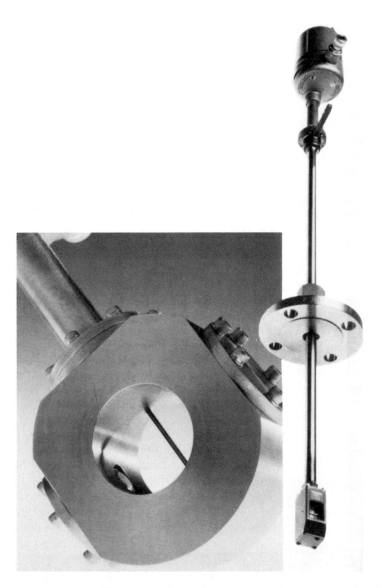

FIG. 4-55. When a fluid flows past an obstruction or strut, vortices are created in a fixed, predictable pattern. The number of vortices shed downstream per unit time is proportional to the flow rate. One can determine the flow velocity by counting the vortex shedding frequency. In J-Tec's meters an ultrasonic beam is transmitted through the vortex pattern downstream of a tiny strut. As the vortices travel through the beam, they modulate its carrier wave. The modulated signal is processed electronically to give an accurate reading of flow rate. In contrast to some nonultrasonic vortex flowmeters which can only detect vortices created by large, restrictive struts, J-Tec's designs use strut diameters of only a few mm. Illustration courtesy J-Tec.

FIG. 4-56.   The J-Tec high-pressure insertion meter is designed to measure gas flows up
to an ANSI 600 lb rating. This meter measures velocity flow rates of 0.2 to 150 ft/s (~0.06
to ~45 m/s) depending on line pressure. The hot tap model is designed for use where line
pressure cannot be shut down. A crank configuration allows the customer to measure the
varying flow across the pipeline, establish the average and thereby compensate for any
variations. A hydraulic version is also available using gas line pressure for insertion and
retraction. Courtesy J-Tec.

FIG. 4-57.   Utilizing vortex sensing, the VA220C sensor provides true airspeed data for
avionics applications. Having no moving parts, the VA220C measures airspeed over a wide
dynamic range and is essentially unaffected by temperature, density, altitude or other environmental parameters. Its small size and light weight make it ideal for use on remotely
piloted vehicles, drones, or airships. It also detects low airspeed on helicopters. Applications
include remotely piloted vehicles, drones, airships and fixed-wing or rotary-wing aircraft.
Courtesy J-Tec.

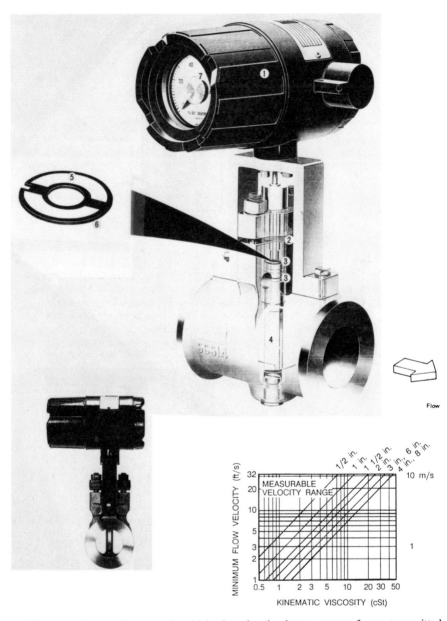

FIG. 4-58. Vortex flowmeter in which piezoelectric element senses forces transmitted through a strut. Top: 1–Transmitter. 2–Gasket. 3–Sensing element. 4–Vortex shedder. 5–Electrode. 6–Piezoelectric element. 7–Output signal indicator (option). Bottom left: strut detail. Bottom right: graph of minimum flow velocity versus kinematic viscosity. Courtesy Yokogawa.

minimum Re is 20,000 (40,000 for 6-inch and 8-inch meters), in order to obtain the rated accuracy. Disregarding nonlinearity, flow measurement is possible for $5000 < Re < 7 \times 10^6$. Charts are available from the manufacturer to guide users on minimum and maximum Re and flowrates as a function of the medium, its conditions and pipe size.

The basic 1972 patent by Joy and Colton covered the measurement of flow by generating a vortex by a strut, directing a "sonic signal" towards the vortices formed in the wake of the strut so as to generate a modulated sonic signal, and detecting the modulating frequency induced on the received signal. In the "wake" of that now-expired patent, it will be interesting to see whether any new vortex shedding inventions enter the field and achieve comparable success. J-Tec's patents on post-1972 improvements include: Joy et al., 1980, Mahany and Johnson, 1982, Thorne and Johnson, 1982, and Joy, 1984.

One design under investigation at the author's laboratory attempts to combine a removable or welded-in vortex shedding strut with clamp-on transducers. This is another form of hybrid which tries to retain the accuracy advantage usually associated with a wetted sensor with the convenience of noninvasive and removable transducers.

Another possibility is to use a vortex shedding strut which also acts as a torsional wave densitometer. In this case the shedding frequency would yield V, the torsional sound speed would yield ρ, and the combination would lead to mass flow rate M proportional to the ρV product.

4.5   Applications of Other Ultrasonic Flowmeter Methods

This section deals with the remaining methods listed in Chapter 2, Fig. 2.1. Coverage generally will be brief for two reasons. (1) Liquid level is treated as a separate topic as part of Chapter 7, Section 7.5; (2) The "other" methods are not yet in wide use for industrial process control applications.

Section 4.5.2 on tag cross-correlation contains previously unpublished details on one particular type of high-temperature (316°C) application. These details, due to Little, 1987, priv. comm., may suggest solutions to problems even when correlation is *not* utilized in the detection or processing of the signals. Extracts from the works of others in this field will give the reader an idea of the potential of the method.

4.5.1 Liquid Level in Weirs, Flumes and Ducts

In many preferred weirs or flumes, the volumetric flow rate Q is typically proportional to $H^{2.5}$, where H = liquid level. The most common *ultrasonic* method for determining H in such cases is to mount one or two transducers above the water surface, and measure time-of-flight down to the water and back. Since a change in air temperature of 6K at 300K results in a 1% change in sound speed in air, temperature compensation is needed. It is usually derived from a thermocouple or by a reflector in the air path.

In ducts which are not full, or open channels, H is needed in order to convert flow velocity V to Q. Examples of H measurements in conjunction with V measurements by a contrapropagating method are shown in Figs. 4-59 to 4-61.

Equipment similar to that shown in these figures to measure H is available from several manufacturers. It should also be pointed out that, depending on access (from top, side or bottom), the number of transducers (1 or >1) and the ultrasonic wave type initiated (longitudinal, shear, extensional, torsional, flexural, Rayleigh or Lamb) there are many more options than those shown in Figs. 4-59 to 4-61. See, for example, Chapter 2, Section 2.4.3, or Chapter 7, Section 7.5. Section 4.6.1 deals specifically with converting a flow velocity measurement in a partly-full pipe to a volumetric flow rate.

A submerged, upward-looking ultrasonic sensor used with Advanced Instrumentation's open channel flowmeters can transmit through at least several feet (>1 m) of wastewater under conditions typical of sewage collection systems. In such applications, the sensor is normally mounted at the bottom of the flume.

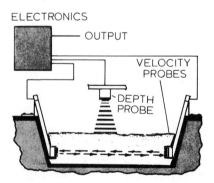

FIG. 4-59. Example of liquid level measurement in conjunction with contrapropagating transmission measurement, from which volume flow rate is obtained. Source: Badger Meter, UF 300 brochure, 1977.

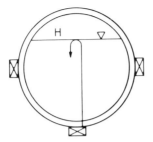

FIG. 4-60.   Clamp-on measurements of *V* and *H* for a pipe which is at least half full.

In industrial outfalls, however, conditions sometimes arise which degrade or prevent ultrasonic echo ranging. These conditions are:

- Severe aeration

- Suspended dendritic precipitates (metal snowflakes) resulting from heavy metals removal

- Solids loading (suspended and settlable) above 0.5%

- Sustained sedimentation and coating of the flume.

According to Gates, priv. comm., 1987, and Advanced Instrumentation's literature:

Such conditions are usually the result of turbulent flow, and are often accompanied by foaming. To provide dependable flow monitoring under such conditions, the submerged sensor may be mounted in a stilling pocket. This pocket is substantially smaller than a stilling well and is molded directly to

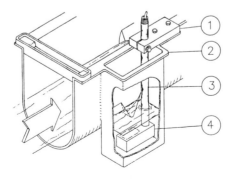

FIG. 4-61.   Stilling pocket for measuring liquid level in flumes. 1–Clamp. 2–Stilling pocket. 3–Pipe nipple. 4–Submerged sensor. Illustration courtesy Advanced Instrumentation.

the flume. Pocket and flume are connected by a hole in the flume at the level of the step. By positioning this hole above the bottom of the flume, passage of sediments into the pocket is minimized. Also, the pocket has adequate space below the sensor to accumulate solids which do enter. In contrast with bubble tubes, floats or capacitance probes used in stilling wells, the submerged ultrasonic sensor is not severely affected by buildups of oil or grease on the sides of pocket, so maintenance is minimized. See Fig. 4-61.

4.5.2   Cross-Correlation of Tags, Transmissions and Reflections

Ultrasonic cross-correlation *tag* flowmeters measure the time for a characteristic tag to be carried downstream a known distance. Tags include those that are inherent in the fluid, due to turbulent eddies, or those that may be induced (as by focused ultrasound) or enhanced by the proximity of an upstream elbow, obstacle, discontinuity, heat source or sink. Density fluctuations in a two-phase fluid also constitute tags. The major effect of turbulence is to randomly modulate the phase of the interrogating beams, whereas density fluctuations primarily cause amplitude modulation. (See Beck and Plaskowski, 1987).

In typical ultrasonic cross-correlation tag flowmeters two or more acoustic beams interact with naturally-occurring tags such as turbulence or density fluctuations in the fluid. The scale of turbulence sometimes restricts the two parallel transmission paths from being chosen closer than $D/2$ while turbulent mixing decorrelates the sought modulations if the paths are separated by more than about $2D$, where D = duct diameter. Even for close paths, swirl can decorrelate the signals. Tornberg et al., 1983, were able to use a 40-mm spacing on a 250-mm diameter pipe, i.e., a spacing of less than $\frac{1}{6}$ of the diameter.

To put the following cross-correlation tag applications in perspective, it may be observed that, prior to 1987, no single correlation approach had evolved as a "general solution" to a wide class of turbulent or two-phase problems. Rather, solutions emerged in response to particular problems presumably significant to the investigator(s) concerned. This situation might have been due in part to the high cost of building a correlation flowmeter prior to ~1985, and to the limited commercial availability of such equipment. [Equipment manufactured by Kent (UK) is described by Keech, 1982.]

As an example of a special problem for which tag cross-correlation of diametrically-transmitted waves might be the best of the acoustic techniques known in 1988, consider the accurate measurement of flow velocity of *turbulent air* in a duct at ordinary temperature and pressure, *by clamp-on*. If the duct were made of a material so attenuating that acoustic short circuit is absent, yet thin enough so transmission through it is possible, then tag cross-correlation might be a practical way to achieve the sought clamp-on flow measurement. To the extent that a real duct is simulated by an open-ended styrofoam coffee cup, the oscillogram in Chapter 3, Fig. 3-40 demonstrates that transmission across a diametral air path is possible using 100-kHz transducers gel-coupled to the outside of the "conduit," without short-circuit interference.

Compared to Doppler or speckle tracking, tag cross-correlation has the advantage that flow velocity can be determined without knowledge of the sound speed.

Ultimately, it may turn out that although cross-correlation first attracted attention in flowmetering as a means of tracking tags, it major use may be elsewhere. For example, between 1985 and 1987 cross-correlation became important as a means of detecting flow-interrogating signals even when "tag jitter" is absent (Embree and O'Brien, 1985; Bonnefous and Pesqué, 1986; Trahey et al., 1987 and 1988; Jacobson et al., 1987b). The transmitted signal may be a tone burst; it may be modulated or coded by random noise, by a computer-generated pseudo-random noise code, or by a swept frequency chirp. Received signals typically traverse the medium alternately in opposite directions (standard contrapropagation method). Alternatively, and equivalent in some respects to a Doppler, echoes (speckle) from scattering centers in the fluid may be tracked by range gating. The possibilities of combining cross-correlation detection principles with a variety of acoustic/hydrodynamic interaction principles underlie the multimode method of Section 4.5.8. Referring to Section 4.5.4 on Noise methods, note that noise can be used as a signal whose transit time can yield information on sound speed c and flow velocity V. The noise may be present in a system due to an undesired leak or generated intentionally. Detection at two locations A and B, digitization and cross-correlation yield the travel time from A to B, or B to A, from which V and c are determined. In some cases, c yields the temperature T of the medium averaged over the path AB (Chapter 5).

In early 1987, two prototype clamp-on cross-correlation flowmeters were tested at Oregon State University, in a project sponsored by the Bonneville Power Administration. One of these meters used the cross-correlation *tag* principle; the other used cross-correlation to detect coded tone bursts combined with *time-of-flight* (upstream-downstream) principles. Calibration data are given in Fig. 4-62 (Trimmer, 1987; Trimmer et al., 1988). These meters appear to be the first two ultrasonic correlation types designed to be manufacturable at a price comparable to contemporary clamp-on contrapropagation meters.

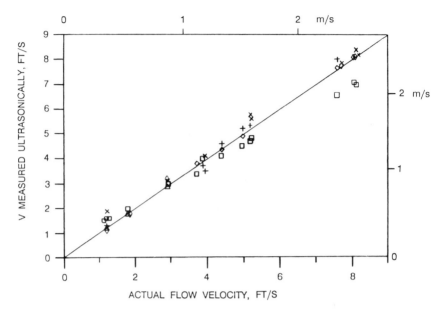

FIG. 4-62. Clamp-on calibration data for several ultrasonic instruments including a prototype in which cross-correlation is used to sense the flow velocity of tags in the fluid, and,
in another instrument, is used to aid in the detection of contrapropagating pulses whose
travel times are sought as a basis for calculating flow velocity. Pipe: 14-inch (356-mm) OD
carbon steel. Legend: □-BPA Doppler. +-USBR Doppler. ◇-TK UFP-1000 transit-time.
△-Panametrics Model 6068 prototype transit-time, correlation detection. ×-USGS tag cross-
correlation prototype instrument. After Trimmer, 1987, and Trimmer et al., 1988.

Let us now review the pre-1985 work on cross-correlation *tag* ultrasonic
flowmetering. In some of Flemons' 1977 work, the transducers were coupled to diametrically-opposed points on the steel pipe using a silicone
rubber spacer and silicone oil couplant. The rubber spacer permitted flat-
faced transducers to be used. Signals were averaged for about 10 s because, similar to the Doppler effect, the statistical nature of the acoustic
flow interaction imposes trade-offs between accuracy and response time.

Flemons, 1977, Runde and Joffre, 1979, Raptis and Sheen, 1981, Tornberg et al., 1983, Keech, 1982, Jacobson et al., 1985, and others found
special cases where the cross-correlation flowmeter may be better suited
than those Doppler or contrapropagating types where signal detection is
unaided by correlation. Flemons, Runde and Joffre studied nuclear reactor applications where *V* is considerably higher and water flow is therefore more turbulent than in the usual industrial situation. They also studied
cases where a second phase is present (bubbles or slurry), distributed
randomly and where both phases have the same velocity profile.

The travel time of inherent tags can be determined using transmission or reflection (scattering) techniques based on propagation in the fluid. [If density (ρ) fluctuations occur, then a pair of axially-displaced density-sensing devices (such as density-sensing waveguides) would yield V if the ρ data can be cross-correlated.] Ultrasonic detection of fluid-borne tags was initially practiced by comparing transmissions over two axially-displaced diametral paths. Later on, paths other than the diameter were investigated, including multiple pairs of paths. See Figs. 4-63 to 4-67.

According to unpublished remarks by Flemons, 1985, priv. comm., clamp-on tag correlation possesses several inherent advantages over the clamp-on transit-time (contrapropagation) method. For example, the axial distance parameter in the flow calculation can be simply the distance separating the upstream and downstream transducer sets, instead of a more complex function involving the beam angles, the velocity of ultrasonic propagation in the coupler, the pipe wall and the fluid, all of which vary with temperature.

In the mid-1980s only one correlation flowmeter was offered commercially (Kent Process Control; see Keech, 1982). This instrument reportedly could meter a wide range of corrosive and hostile fluids in two-phase and multiphase conditions. The flowmeter consists of two main elements: a wetted transducer sensing head with associated electronics and a correlation signal processor. The transducer and associated electronics can be adapted in several ways to optimize the system for a given application. If rapid response times or improved accuracy are required, additional sensing channels can be added. At low flow rates, the addition of extra channels extends with operating range of the instrument. But the response time of the instrument reduces as the number of channels increases. The resulting cross-correlation signal-to-noise can be doubled, tripled, etc., in a given integration time by combining the input data in different ways depending upon the application. See Fig. 4-64 and Coulthard, 1981. Despite the paucity of commercially-available flowmeters based on correlation in the mid-1980s, a continuing interest in this technology is indicated by Beck and Plaskowski's 1987 book and papers such as those by Braun, 1988, and Sidney et al., 1988, scheduled for presentation at IMEKO XI (International Measurement Confederation), Houston, Texas in October 1988.

In Finland, the correlation tag principle has been successfully applied by Karras and co-workers to the measurement of pulp suspensions using longitudinal wave, normal incidence transducers attached to the outside of pipes. Karras' work has occasionally utilized obliquely incident shear wave clamp-on transducers.

Karras' group, like one at the Argonne National Laboratory, also cor-

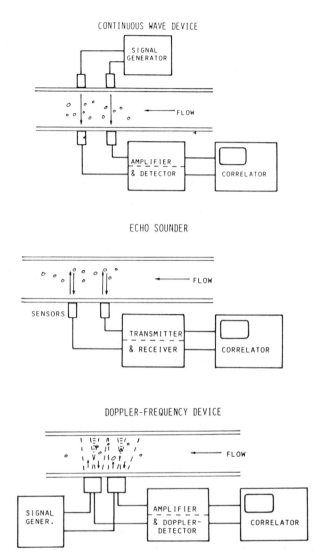

FIG. 4-63.   Top: Two-channel cw system using clamp-on transducers. Middle: Echo sounder and back wall measurement using clamp-on transducers. Bottom: Measurement setup for the Doppler correlator clamp-on flowmeter. After Tornberg et al., 1983. © 1983 IEEE.

related Doppler returns. Some details will be given below, extracted from Tornberg et al., 1983. See also, Fig. 4-63.

Flemons states that "leakage" of ultrasonic signals through the pipe wall from transmitting to receiving transducers has proved troublesome, especially on small diameter pipes. To some extent this limitation can be

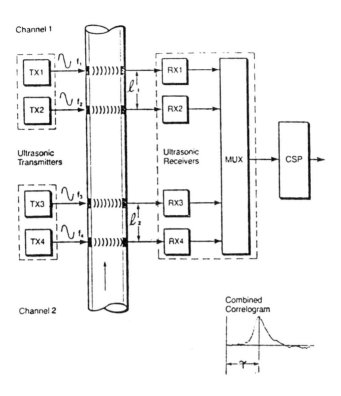

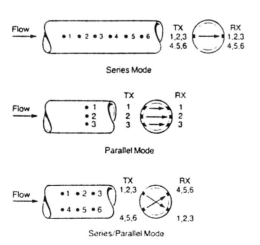

FIG. 4-64. Multichannel concept and transducer orientations usable with Kent correlation flowmeter. Courtesy ABB Kent Inc.

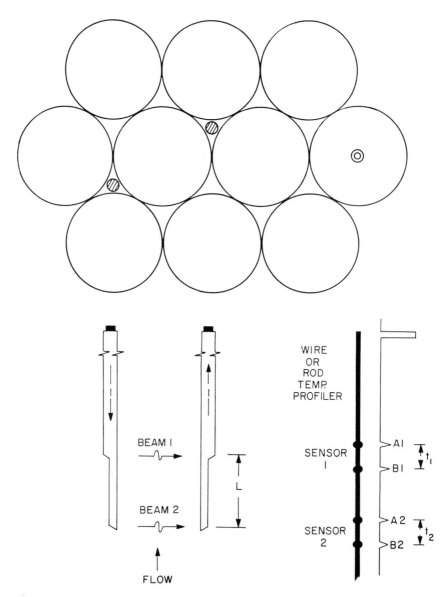

FIG. 4-65.   Correlation time between transmissions beamed over two parallel paths a distance L apart provides a measure of flow. Multi-bevel probes are shown located in triangular cusps formed by intersection of close-packed round ducts. Analogously, a single wire or rod may contain temperature-sensing zones spaced L apart. Here, it is proposed to correlate the transit times t_1 and t_2. (After Lynnworth et al., 1974b). © 1974 ISA.

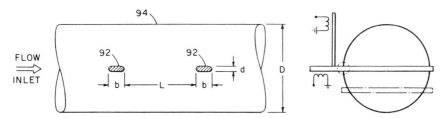

FIG. 4-66.   Correlation of densitometer readings at two axially-displaced locations could serve as basis for mass flowmeter, according to proposal by Lynnworth, 1980. Cross sections labeled 92 represent torsional mode densitometer waveguides. Transducers are shown with coils outside the pipe.

overcome by quadrature demodulation (Jacobson et al., 1985). In cases where the transducers for large pipes are no more expensive than for small, the ultrasonic system becomes attractive for large-pipe applications where conventional nonultrasonic systems can be relatively expensive. Process temperature changes up to 250°C have been experienced with no need for adjustment nor change in calibration. Canadian experience prior to 1986 was limited to a portable system.

One potential application suggested by Flemons (and others) is to permanently attach ultrasonic transducers to high-accuracy calibrated pressure-difference flow elements. If the two systems were calibrated together and compared frequently thereafter, any failure of either system would

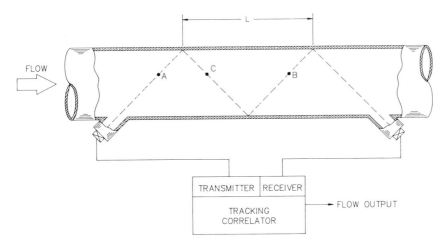

FIG. 4-67.   Zigzag interrogation path proposed by Jacobson (1985, priv. comm.) as a means of reducing the number of transducers needed in a clamp-on tag cross-correlation flowmeter. (Jacobson et al., 1987b).

be immediately apparent. Because the two systems operate on entirely different principles, a significant increase in reliability could be obtained with a small increase in the cost of the system.

At the 1985 stage of development, operation at $Re < 2 \times 10^4$ was uncertain even near elbows. Accuracy of 2% was reported by Flemons for high Reynolds numbers, $Re = 10^5$ to 5×10^7.

Accuracy suffers on small pipes due to the difficulty of accurately defining the spacing parameter. The laminar flow regime had not been explored. With two-phase fluids, using amplitude demodulation, laminar flow measurements might be possible, according to Flemons, 1985.

In the work reported by Tornberg et al., 1983 and 1984, and by Tornberg, 1986, a flow measurement system using cross-correlation was studied and constructed for use in the pulp and paper industry. (See also, Karras et al., 1982, 1985 and 1988.) The primary signals were produced by the modulation action of suspended particles on the intensity of two ultrasonic beams. Both continuous and pulsed intensities were used. The signals were received either as attenuated beams penetrating the tube or as echoes scattered from the backwall. The intensity of the Doppler-shifted frequency of the scattered ultrasound was also used as a primary signal in order to avoid the ringing of the transmitted frequency within the crystal and pipe wall. The radial flow profile, the nature of the correlation method and the frequency filtering of the primary signals made the correlation function asymetric and caused some measurement errors. The system, however, has been used successfully for the measurement of pulp suspension flows.

In tests using cw, Tornberg et al. found that the majority of correlating information was within a frequency range of 10 to 100 Hz when the transducer spacing was 320 mm and the flow speed was 1 to 5 m/s. They concluded that the correlation is obtained from suspended inhomogeneities about 5- to 10-cm wide, while the single fibers of flocks are not resolved for correlation. These results showed that the size of the correlating inhomogeneity in the average flow stayed nearly constant and the turbulence turned and rotated them without quickly breaking them down.

In practice the transducer spacing was 40 mm even for large tubes (250-mm diameter). This yields a shorter measuring time. When the attenuation pattern of the back-wall echoes can be interpreted in terms of particle density or consistency of the suspension the instrument effectively measured mass flow rate, or production rate, of the paper-making plant (Karras, 1984).

Jacobson (1985, priv. comm.; 1987b) proposed a correlation flowmeter that in the tag mode would need but a single zigzag path interrogated by one pair of wetted or clamp-on transducers (Fig. 4-67). Consider a disturbance (turbulence, particle, bubbles, etc.) at point A which modulates the ultrasonic beam in phase and/or amplitude. Some time later the disturbance will have travelled to point B and modulates the beam in a similar manner. The modulation is extracted from the ultrasonic signal and a tracking correlator is used to track the time delay at which the modulation is most similar. The flow velocity will then be given by $V = L/T_o$. A tracking correlator samples two signals at time t_1 and t_2, respectively, and using feedback the tracking correlator maintains a delay $T = t_2 - t_1$, at which time the correlation is a maximum. In the illustrated zigzag path case the two signals are derived from the same transducer pair.

The disturbance will also pass through point C and the disturbances at points A and C should correlate at a different time delay. This correlation should have a small effect because the ultrasonic beams are in opposite directions and Doppler/phase shifts will be different, and also the delay varies across the diameter of the pipe, tending to smear out any correlation.

A *one*-transducer correlation flowmeter geometry might be practical using a reflection technique similar to that illustrated in Fig. 4-68.

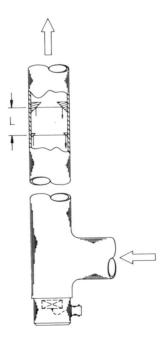

FIG. 4-68.   Proposed one-transducer cross-correlation tag flowmeter in which annular reflectors are spaced a distance L apart.

The following unpublished description of ultrasonic transducers as used with a correlation flowmeter up to 316°C (600°F) is due to Little, 1987, priv. comm.)

Since 1979, Combustion Engineering (C-E) has been measuring flow in pressurized water reactor piping by ultrasonic methods. Coolant flow rates on the order of 500,000 gpm are measured at temperatures up to 600°F and pressures up to 2200 psi. The system performs three main functions: Transmitting and receiving ultrasound across axially-displaced diametral paths in the pipes; electronically processing the received signals; and performing cross-correlation between two received signals to determine the transit time of the fluid between the two diametral paths. This data is used to accurately calculate coolant flow rates as part of the reactor plant acceptance testing. Typical measurement uncertainty of coolant flow taken in the field has been ±2.5% (2σ) due to the undeveloped velocity profile in the 50-inch OD (1.25-m) pipes. An independent calibration laboratory test showed that the flow measurement uncertainty for symmetric pipe profiles was less than ±1.0% at the 2σ level and the flow measurement error for diametric average velocity determined ultrasonically was less than ±0.5%.

A transducer array was designed to be mounted along the 50-inch OD pipe in an area which has specially ground flat surfaces 180° apart. These surfaces were machined on a pipe section prior to plant construction. The arrays consist of an aluminum bar holding six transducers and three 8-pole magnets. Magnets of 150-pounds force were chosen to still have sufficient holding power even at an elevated temperature of 600°F. Avoidance of transducer misalignment due to "walking" during pipe vibrations was ensured by application of high temperature silicone rubber adhesive around the edges of the magnets and transducers at the pipe surfaces.

The transducers are comprised of spring loaded Teflon blocks holding lithium niobate crystals which make direct contact with the flat surface of the pipe. An ultrasonic couplant is needed between the crystals and the pipe surface that will function at temperatures above 500°F. C-E uses a couplant which lasts long enough to complete the flow measurements. This installation is considered temporary since flow measurements are taken within a one-month period.

The output of lithium niobate crystals tends to fall off as pipe temperatures increase. A high-voltage pulser (2000 V) has been developed which allows for a good signal-to-noise ratio at temperatures above 500°F.

Data are obtained by sequentially exciting four transducers from upstream to downstream and peak-detecting the first half-cycle of each 1-MHz burst received from transducers on the opposite side of the pipe. The data from two received signals are processed to determine the transit time of the coolant. Primary system flow rates can be calculated and used to validate reactor design assumptions.

Further improvements to the transducers were in process in 1987 such as mechanically damping the crystal ringing to avoid interference on adjacent transducers in the array. The interference appears as a 1-MHz burst superimposing itself in the same time frame as the half-cycle being measured. Another improvement has been in the mounting of transducers on curved surfaces to eliminate the need for expensive, specially-ground flat surfaces on the pipe. The use of curved "shoes" made of material of the proper acoustic impedance to match the crystals to the pipe is desired in order to increase the transfer of acoustic energy.

Use of Correlation Detection in a Transit Time Flowmeter

Laenen, 1984, suggested correlation detection as a way of overcoming attenuation and jitter problems that have frustrated users of transit time flowmeters when applied to the long, inhomogeneous and variable-index paths encountered in rivers and streams. Correlation detection is well known in radar as a means of improving the likelihood of reliably detecting weak signals in the presence of noise, especially if the signal is fluctuating. The use of correlation detection in ultrasonics has been found effective (23-dB improvement in signal-to-noise ratio, Gorfu and Hayward, 1986 and Hayward and Gorfu, 1988) in an NDT project where it was necessary to minimize peak power levels to achieve intrinsic safety. Nuspl et al., 1986, used correlation detection in their acoustic thermometry work at frequencies on the order of 0.5 to 3 kHz, discussed in Section 5.1.1.5. Mylvaganam, 1987a, b, correlated chirp transmissions; see Section 4.2.2.2.

At Panametrics work began on an ultrasonic flowmeter using correlation detection in the mid-1980s. Some of the ideas in this project were borrowed from earlier investigators, among them Pedersen, who demonstrated in the early 1970s some of the advantages of using a pseudorandom code to modulate the carrier in a transit-time ultrasonic flowmeter (Pedersen and Lynnworth, 1972). The correlation flowmeter which emerged from this instrument development project at Panametrics was termed the Model 6068 (Jacobson et al., 1987b).

A Model 6068 prototype was tested during a comparative study of ultrasonic clamp-on flowmeters at Oregon State University in 1987 (Trimmer, 1987; Trimmer et al., 1988). Included in this study were three different clamp-on ultrasonic flowmeter technologies—Doppler, transit-time and tag cross-correlation. The meters were tested on water on a long straight run of pipe and also close to elbows and bends. On the straight run the transit-time flowmeters (Tokyo Keiki UFP-1000 and Panametrics 6068) proved more accurate than the Doppler meters or the tag cross-correlation meter.

A graph of the meter readings versus actual flow is shown in Fig. 4-62. The Doppler meters and the tag cross-correlation meter rely on the reflection or modulation of the ultrasonic beam by a second phase or turbulence. They showed an improvement close to elbows and bends in the pipe. The correlation detection used by the 6068 transit-time flowmeter proved fairly immune to the turbulence at the bends and accuracies of about 5% were obtained within one or two pipe diameters of an elbow. The errors are primarily due to uncertainties in the flow profile.

Doppler and tag cross-correlation flowmeters have a distinct advantage where the fluid contains a high percentage of bubbles or particles which strongly attenuate the ultrasonic signal through the fluid. The "acoustic short circuit" signal transmitted through the pipe wall tends to swamp the liquid-borne signal, causing large errors in a transit-time meter. The Doppler meters, which measure a frequency shift, and the tag cross-correlation meters, which utilize the fluctuations of the ultrasonic signals, are relatively immune to acoustic short circuit. [The tag cross-correlation meter may exhibit a "phase inversion," while the unpredictable penetration of the ultrasonic beam may cause errors in simple Doppler meters. There are, however, electronic means to compensate for these acoustic short circuit and uncertain penetration errors such as, respectively, quadrature demodulation (Jacobson et al., 1985) or range-gated Doppler (Brandestini, 1978).]

The 6068 consists essentially of a programmable transmitter, receiver and digitizer, and a fast processor. The transducers may be installed to obtain range-gated returns reflected from the particles or bubbles in the fluid. The returns may then be digitized and processed to determine the flow velocity either using a FFT (Fast Fourier Transform) to determine the Doppler frequency shift (Azimi and Kak, 1985) or a "time domain" method (Embree and O'Brien, 1985). The FFT method may be superior for high flow velocities/high turbulence but the time domain method in principle is superior for moderate or low (including laminar) flows and viscous fluids.

Lastly, if two transducer pairs are multiplexed the 6068 provides the essential hardware for a tag cross-correlation meter. If the transducer transmitter and receiver pairs are mounted in parallel paths, upstream and downstream, the multiplexer allows the 6068 to rapidly interrogate each pair in turn. The received signals are digitized and the processor extracts phase or amplitude information. The phase or amplitude modulation from the upstream and downstream pairs are correlated to determine the velocity of "tags" as described earlier. Therefore, if different transducer configurations are provided, the 6068 may be programmed to automatically select that measuring mode likely to yield the highest accuracy, based on flow and signal conditions.

4.5.3   Phase Fluctuations Due to Turbulence

This category overlaps correlation, but also is distinct in that correlation methods are not required in order to sense the phase fluctuations. To date, applications have been of an R & D nature. Examples include the work of Schmidt and Tilman, 1970, and Dinkelacker and Stiewitt, 1971, whose experimental arrangements are included in the author's 1979 review of ultrasonic flowmeters.

Bruner, 1985, found that by making contrapropagating measurements orthogonal to the flow, scalar (pressure, temperature) contributions to the phase fluctuations would be cancelled out. The resulting rms phase jitter was found to be proportional to flow velocity, over the ~3:1 flow range investigated. It appears that this method might provide a useful approach for low-cost flowmetering on a *relative* basis. But considerable calibration data might have to be amassed before flow velocity could be measured on an *absolute* basis to high accuracy. (Compare measurements of bearing noise, however, where a database is used to relate noise to incipient failure—Chapter 2, Section 2.6.5.) It may be of interest to compare Bruner's phase jitter versus flow rate correlation with that observed by Green et al., 1979. Referring to the flow of solids conveyed pneumatically, Green et al. regard the flow as a mixture of steady flow and a superimposed smaller irregular flow. Regarding the irregular flow as "flow noise," it was found experimentally that there was a direct correlation between the flow noise and the mass flow rate. In the *non*acoustic device of Green et al., the flow of particles past an electrode generates changes in capacitance, convertible into an electrical signal. Hamade, 1982, p. 135, notes that the amplitude of ultrasonic signals received after transmission across a 30-cm path of turbulent-flowing coal/air mixtures exhibited fluctuations that increased with increasing velocity. In Hamade's work, the frequency was 459 kHz and the velocity ranged up to 60 m/s. The relation between flow velocity and amplitude fluctuations was nonlinear. Atkinson, 1975, found that the rate at which the amplitude of the echo backscattered from blood passed through its mean level is related to the flow velocity of the scatterers. The possibility of relating pressure fluctuations to flow is considered by Dawber and Sinclair, 1977, as well as by other investigators, some of whose works are cited in the next section.

4.5.4   Noise Due to Flow, Leaks or Impacts

Raindrops, running water, wind, whistling, materials sliding down a chute, musical instruments, etc. all provide us with flow indications by the amplitude and/or frequency distribution of noise. The relationships, however, are generally highly nonlinear and are difficult to calibrate reliably. (Examples of flow- and pressure-related laboratory calibrations of noise appear in Gopal et al., 1978. See also, Blake, 1986.) Tanisawa and Hirose, 1988, reported experiments where air bubbles were introduced into the fluid and an obstacle was set in the flow to enhance the generation of sound. As a result, they found a significant improvement in linearity, stability, and reproducibility of sound pressure as a function of V. This suggested the possibility of *measuring* V, whereas in the past, noise has served mainly as the basis for flow switches (Fig. 4-69), for detecting and locating leaks, and for detecting the transition from laminar to turbulent boundary layers during reentry. Data on noise versus V are available well above Mach 1, in contrast to the other "subsonic" ultrasonic methods reviewed here. [Boundary layer acoustic monitors available from Kaman Sciences in the mid-1970s are shown in the author's 1979 review, p. 465 (Sachs et al., 1977). Hellbaum and Garner, 1988, discussing sensors for hypersonic flight, mention the continuing need to improve the understanding of laminar to turbulent transition dynamics. Pressure fluctuations in the boundary layer, a result of Tollmien-Schlichting waves, include frequencies from a few hundred Hz up into the MHz range.] Noise flow switches and ultrasonic leak detectors have each been manufactured by several companies. (See Chapter 2, Section 2.1.3.)

Mann and Crosby, 1977, used a piezoelectric transducer in a particulate-laden gas stream to generate an output voltage related to the rate of particle collision and hence mass flowrate. Yeack-Scranton, 1987, in her studies of microscopic surface irregularities in air bearings, found that certain piezoelectrically-detected plate mode transients were due to slider-to-disk contacts. In noncontact tests at rotational speeds of 3600 rpm (120π/s), she reported a microturbulence signal that appears to be due to air fluctuations induced by slider to disk roughness variations. Laville, 1987, proposed a rainfall measurement device using underwater acoustic signals. By means of spectrum analysis, it was expected to yield an accuracy of 20% when applied to large bodies of water. Earlier, Nystuen, 1986, reported that the relationship between spectral level and rainfall is quantifiable, and that rain's noise spectrum can be distinguished from other noise sources. Nystuen used a broadband hydrophone to record underwater spectra due to rainfall and compared these data with independent references underlying measurements obtained with a distrome-

FIG. 4-69.   Noise flow switch is primarily used to detect a major change in flow rate, or flow stoppage in a solid particle or liquid flow stream. Transducers clamp on or may be of the insert probe type. The transducer is an ultrasonic microphone that picks up ultrasonic noise generated by solids in the flowing medium. Filters in the unit block out noise in the audio range to provide immunity from normal acoustic signals in the environment to permit amplification of the signals within the ultrasonic spectrum. Integration of this spectrum provides an analog signal that varies according to the average flow velocity. A high or low flow alarm point can be set to indicate (locally and/or remotely) when the flow rate deviates from the defined limit. The transducer should be mounted in a location of high flow stream turbulence or noise. The preferred location is on the outside radius of an elbow. Other possible locations are on a valve, nozzle, orifice or flange. It is important to keep the transducer out of areas with high ambient noise such as near pumps or compressors. Source: Smith Meter Div./Geosource, Bull. 8.7.0, Jan. 1983.

ter. (A *distrometer* is an acoustic instrument which indicates drop size from the pressure pulse due to the drop striking its surface.)

The flow switch from Smith Meter shown in Fig. 4-69 is used to detect a major flow rate change (e.g., flow stoppage) in a liquid or solid particle flow stream. The flow monitor is available with either a nonintrusive clamp-on transducer or an insertion probe.

The nonintrusive transducer clamps onto the outside wall of the conduit. It detects flow of solids that are heavy enough to create noise when sliding along the wall of a pipe, hopper, or inclined trough. It is typically used to detect flow stoppage of plastic pellets, limestone or cement slurries, etc. It is also a convenient way to detect passage of a pipeline displacer or "pig."

The passive insertion probe is typically used to sense major changes in concentration of small particles (e.g., due to a flow stoppage or filter failure) in a duct or air blown dust, fly ash, abrasives, coal dust, etc. By monitoring the exhaust air from a "Bag House" the breakage of a "bag" can be promptly detected, according to the manufacturer.

The manufacturer of the "ultrasonic stethoscope" shown in Fig. 4-70 lists a dozen or more applications for their battery-powered probe, Model 500C:

- steam traps
- motors
- bearings
- valves
- pumps
- gears, gearboxes

- line blockage
- compressors
- hydraulic systems
- heat exchangers (on-line)
- underground leakage
- leaks behind walls

Some of these are discussed by Bandes, 1987.

In cases where a sudden leak occurs in a gas pipeline, expansion waves propagate upstream and downstream. If these waves are received by two spaced-apart transducers, the location of the source of the leak can be determined (Rocha and Finch, 1986; Chapter 3, Fig. 3-62). Expansion waves that produce a well defined transient can be timed by ordinary techniques.

Less obvious, perhaps, is the idea that if a leak produces not expansion waves but merely noise other than a pure tone, the digitized signatures of that tone as received at different distances from the source can be cross-correlated so that the source location can be found (Kupperman et

FIG. 4-70. Valve leak testing by passive method. If a valve is seated properly, there will be no flow and therefore, no signal. If there should be a break in a seat, the liquid or gas flowing through will create a detectable turbulence. Source: UE Systems.

al., 1987). As a corollary, if the travel time of the noise can be measured as it propagates in upstream and downstream directions, one would expect that a passive contrapropagation flowmeter is obtainable.

Detecting Wear of Plasma Cutting Torch Nozzles

Since its invention in 1955, according to Braeuel et al., 1987,

> . . . the plasma arc cutting (PAC) process has gained wide acceptance as a cost-effective means of cutting steel, aluminum, and stainless steel.
>   Like other cutting processes, PAC works by locally applying heat and momentum on the workpiece to melt it and expel the molten material. A cutting gas (such as nitrogen) flowing out of a nozzle at high velocity provides the momentum while an electric arc, burning through the same nozzle between a tungsten electrode and the workpiece, provides the heat.
>   The orifice of the nozzle must sustain the high-velocity flow of the cutting gas, which is heated to a plasma of more than 30,000°C. Thus the nozzle is a very critical component of any PAC system because it tends to wear rapidly under such extreme operating conditions, and the resulting change in the geometry of the orifice affects the hydrodynamic behavior of the plasma jet, which consequently degrades the quality of the cut.
>   The detection of [typical nozzle defects] is currently [1987] performed visually, either through direct inspection of the tip or indirectly when a deterioration of cut quality or unreliable arc initiation is observed. No sensing technique that could replace human intervention exists or has been reported. This is a problem, particularly with large automated PAC systems, where a faulty tip can cause substantial waste of time and material due to bad cuts. The idea of using an *acoustic sensing method* to detect faulty tips was based on the hypothesis that since those defects affect the gas flow sufficiently to deteriorate the process, this must be associated with a change in the jetlike sound of this flow. Steady laminar flows are silent and it is the instabilities of unsteady flows (turbulence, vortices) that generate sound waves, the power of which is less than 1% of the flow energy. Sound power increases with the eighth power of the jet velocity and is generated mainly in the mixing zone, an area at the jet exit four to ten orifice diameters long. The source of energy for turbulence is shear arising from the nonzero viscosity of the fluid. Within the randomness of turbulent flows there are often quasi-deterministic structures, commonly referred to as "coherent structures," dominantly axisymmetric in circular jets. These structures are more efficient noise generators than totally random turbulences.
>   [Based in large measure upon such considerations, Braeuel et al. developed an] off-line acoustic technique for the automatic remote detection of nozzle wear in plasma cutting torches. [A] laboratory prototype able to sort out the degraded tips from a batch with 100% accuracy was built. The instrument "listens for" the strong resonant tone that unworn tips were

found to generate under certain flow conditions. The amplitude and spectral structure of this tone are extremely sensitive to any changes in the orifice geometry, and thus nozzle wear can be monitored from a very early stage. An experimental study revealed conclusively that this sound is produced by the periodic shedding of ring vortices at the exit of the sharp-edged cylindrical orifice. A linear relationship between peak frequency and the bulk average velocity through the orifice for three diameters of tips was observed.

4.5.5 Thermal: Heated Paths, Heat-Pulsed Fluids

Hot wire anemometers typically indicate V or mass flow rate based on a measurement of the current required to maintain the wire at a fixed average temperature T (e.g., fixed resistance). One can consider measuring either the average T, or the T distribution, in an electrically heated waveguide based on the T dependence of sound speed. The T distribution is obtained by reflecting a small portion of the interrogating sound pulse at each of a number of equidistant points along the waveguide, and measuring the time intervals between the echoes. In principle, one could compute the V profile from the T profile (Lynnworth, 1979). Although ultrasonic equipment has been available for over ten years for automatically scanning a multizone probe to determine the temperature profile, it has not yet been demonstrated as a flowmeter. Potential applications exist in ducts and wind tunnels. (See Chapter 5, Section 5.2.1.)

Since about 1985, perhaps motivated in part by growing interest in SAW sensors for various measurands (including temperature), several investigators explored the combination of SAW temperature measurement with a heated substrate or heated path, to achieve a flow velocity or mass flow sensor. Early steps in this direction include those of Ahmad, 1985, 1987, Brace et al., 1987, and Joshi, 1988. Applications presumably would mimic some that were previously addressed by nonacoustic hot wire anemometers and heat transfer devices.

Heat pulse methods may be categorized in this section or in the correlation section. According to a 1987 announcement from Molytek, electronic (*non*acoustic) generation and detection of a 0.5°C heat pulse yields 0.1% precision for $Q = 0.05$ to 100 ml/minute. The possibility of generating a tag in an industrial fluid by means of a focused pulse of ultrasound was suggested in the tag cross-correlation section. Thermoacoustic pulsing and detection may be viewed as analogous to hyperthermia treatment of some tumors in conjunction with c-based ultrasonic thermometry (Davis and Lele, 1985).

4.5.6 Mechanical: Ultrasonic Sensing of Position or Motion of an Indicating Element

The vortex shedding ultrasonic flowmeter employs a strut that interacts with flow. Ultrasound detects that interaction. Similarly, the float in a rotameter rises in a tapered tube as a function of flow, and its height could be sensed ultrasonically. An impeller rotated by flow can periodically modulate an ultrasonic beam, or reflect it. A commercial form of this latter combination is available from Baird Controls. Claimed advantages include lack of drag by ultrasonic sensing of impeller rotation, and the probe does not collect metallic or rust particles, according to the manufacturer. Other mechanical/ultrasonic interactions might include: springy target, the deflection of which increases with flow; sensing of pressure drop ΔP by ultrasound (position of a diaphragm; sensing of differential column heights); etc.

4.5.7 Beam Drift

At flow velocities up to several tenths of the Mach number M_S, the downstream transport of sound by the flowing stream has on some occasions been used to measure the flow velocity. The earliest accurate use of this method is due to Dahlke and Welkowitz, 1960, Figs. 4-71 and 4-72. After a two-decade hiatus, the method reappeared in the dissertation of Hamade, 1982, and subsequently, in an instrument developed by Available Energy for testing in a Detroit Edison plant (Leffert and Weisman, 1988).

According to Leffert, 1987, Available Energy developed a two-phase,

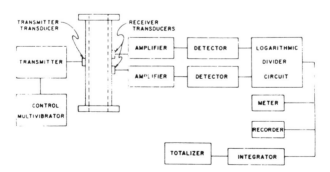

FIG. 4-71. Block diagram of the beam deflection flowmeter system. After Dahlke and Welkowitz, 1960. © 1960 ISA.

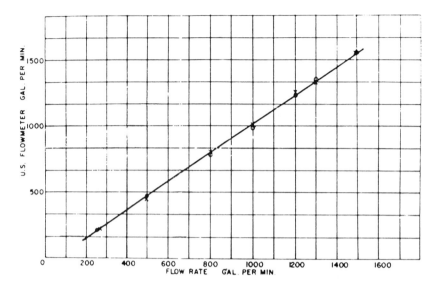

FIG. 4-72. Measured versus actual flow rate for two test runs of the beam deflection
flowmeter. After Dahlke and Welkowitz, 1960. © 1960 ISA.

solids-in-gas ULTRAFLO℠ ultrasonic mass flowmeter. This on-line nonintrusive instrument measures the velocity from the *downstream drift* of
the ultrasound sent across a pipe diameter and it measures the solids
loading from the sound attenuation due to the suspended particles (Hamade, 1982). The instrument was reported to be in operation on a large
(50 ton/hr) coal pulverizing mill at the Detroit Edison Monroe Power Plant.
Transducers, mounted on all eight burner lines (16-inch OD) are multiplexed to one mainframe which, under control of a fully-dedicated mi-
crocomputer, services all eight lines and presents values of the output
parameters on a color graphics terminal and strip-chart recorder. After a
sweep of measurements on all eight lines, the instrument also automatically sums and outputs the total mass flow rates of air and coal for comparison to the power plant control room values of air and coal feed rates
to the pulverizing mill. Good agreement on total mass flow rate is reported
between the ULTRAFLO℠ instrument and mill feed rates of both air and
coal over the entire allowable range. It is also reported that, confirming
a problem long suspected, the instrument showed that the coal flow is
not balanced into the eight burner lines and furthermore, flow instabilities
exist in both space (burner line cross section) and time that probably
degrade combustion performance.

4.5.8   Multimode

The main advantages and limitations of four of the foregoing flowmetering modes are summarized in Table 4-5. From this table it may be concluded that no mode is best for all fluids, flows and conduit conditions.

Table 4-5.   Comparison of Four Ultrasonic Measuring Modes, All of Which Can be Contained in One Instrument. A Chart Representing the Multimode Algorithm Appears in Fig. 4-73

Mode	Advantage	Limitation(s)
Transmission	Yields highest accuracy. Yields V and, independently, c. Sometimes yields composition- and density-related information, e.g., mass flow rate if density and c are related by a known function.	Requires fluid to be sonically transmissive; errors increase at very low and very high velocity. Need to know path geometry. If path is short, the upstream-downstream time difference may be too small to measure accurately.
Reflection (speckle tracking)	Tracks motion of scatterers even at very low velocity and in highly attenuating media; can obtain profile along acoustic path by range gating even if duct geometry is ill-defined; can estimate the flow velocity even when transmission all the way across the path is impossible due to high void fraction. Usable even when path is short.	Not appropriate for pure single-phase non-turbulent (non-scattering) fluid. Wetted transducers respond to Mach No. V/c, not to V alone. (Clamp-on transducers, however, generally compensate for c by using a wedge of known sound speed.)
Tag cross-correlation	Tracks motion of acoustic or thermal discontinuities or eddies, typically created near hydrodynamic disturbances. Cross-correlates modulation due to the flow of scatterers such as tiny bubbles or particulates. Can be range-gated and can be used in ducts or passageways of small or ill-defined geometry. Measures V independent of c. May provide practical advantages in profiling flow in a direction normal to the flow.	Not appropriate for laminar flow unless multiphase conditions prevail. Despite several successful applications having been reported, no general solution has been announced yet. Confusion likely if different tags move at different velocities.

Table 4-5. (*continued*)

Mode	Advantage	Limitation(s)
Noise spectrum	No upper limit on velocity; can indicate presence or absence of leaks, and can locate their source; can indicate transition from laminar to turbulent flow in boundary layer; useful for relatively low-cost flow switches; may yield a velocity estimate if enough other data is available.	Accurate calculations of flow from noise magnitude or specrum alone are generally not possible because of complexity of noise generated, by dependence on geometry of noise source, and interference from other (non-flow) noise sources such as nearby machinery, pumps, etc.

It has been recognized by a number of investigators that an ultrasonic flowmeter that can interact with flow in more than one way has a better chance of dealing with surprises such as occurrence of a second phase, cavitation, wide flow excursions, applications with different materials, changing geometry as encountered by a downhole flow tool (Chapter 9, Section 9.10), etc. In a design due to Jacobson et al., 1987b, the "best" mode is automatically selected according to acceptance criteria such as standard deviation, acceleration and others. The chart guiding this autoselection is represented in Fig. 4-73.

Just as "two heads are better than one" (Section 3.9) two modes may be combined to yield results unattainable by either mode alone. For example, a range-gated reflection mode might yield profile data and a dynamically-corrected meter factor K that subsequently could be used to improve the estimate of flow obtained by a contrapropagation measurement across the diameter.

4.6 Flow-Velocity-Related Parameters

In this section we include volumetric flow rate, mass flow rate and heat flow measurements based on flow velocity, and a boat speed application which is mathematically similar to a flow velocity measurement in a closed or open conduit.

4.6.1 Volumetric Flow Rate in Partly-Full Pipe

Referring to Fig. 4-21 in Section 4.2.1.2, or Figs. 4-59 and 4-60 in Section 4.5.1, this application would appear to require one to compute Q from the product of the average flow velocity $\mathbb{V}$ times the area A of the cross

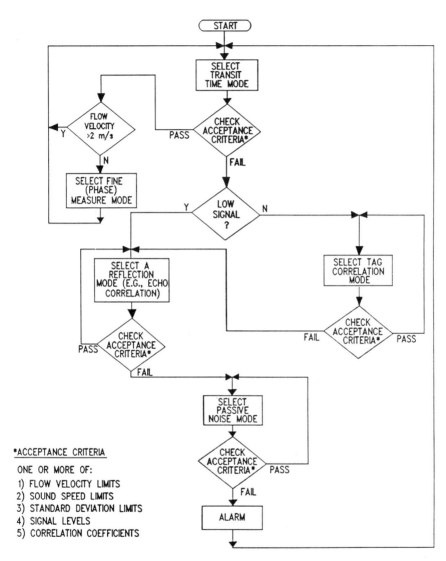

FIG. 4-73.   Diagram shows how one flowmeter might select the best measuring mode from its four-choice main menu. The four choices correspond to modes identified in Table 4-5 as: Transmission, Reflection, Tag cross-correlation, and Noise spectrum. For very low velocities that generate $\Delta t < 100$ ns, a phase measurement of a 64-cycle or 128-cycle tone burst could provide sub-nanosecond resolution of Δt, with 0.1 ns calculated as a practical limit for one design that was analyzed in 1988.

section of the flowing liquid. Depending on the accuracy required and the uncertainty in pipe conditions, it may be possible to determine Q from the liquid level alone (Nagel, 1987). On the other extreme, a series of V's measured at levels H_1, H_2, . . ., H_n together with a measurement of level would appear to provide the highest accuracy. See Fig. 4-74 or Malone and Whirlow, 1971, Suzuki et al., 1972, or Lowell, 1977. As a practical matter, measurement of V along with one or two special chords, together with an H measurement, often suffice.

The following remarks are extracted from a paper on open-channel flow measurement by Kyser and Heyden, 1984. See also Fig. 4-75, due to Badger Meter.

> In open channels or partially filled pipes the presence of a free surface in addition to the attendant problems of stagnation, backwatering, and flow reversals established a need for development of an instrument capable of handling these constraints.
>
> The normal gravity-driven measurement techniques such as weirs and flumes have difficulties with such phenomena. [Before 1984] needs were expressed in Milwaukee and Green Bay, Wisconsin to measure flow for billing purposes in sewer collection systems where the hydraulic conditions made the use of the more traditional gravity-driven devices difficult. The first contrapropagating partially-filled metering system was installed in 1973 and since then viable devices have become commercially available.
>
> Traditional techniques such as weirs and flumes, which convert flow energy into differential head/pressure relations, are practical unless flooding and submergence are present. Partly because they are unidirectional they may not suffice for the more critical applications involving process control and custody transfer of water and sewage.

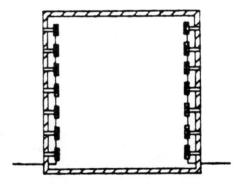

FIG. 4-74.   Area-averaging ultrasonic flowmeter was designed by Swengel, 1956, for *full* large conduits or penstocks, but the multi-transducer arrangement shown here could equally well be applied to *partly*-full ducts.

The form of contrapropagating flowmeter under discussion is shown in Fig. 4-75. The unit is designed to obtain fluid velocity measurement along with a depth measurement. This information is then related by the general continuity equation.

The area of flow is determined from the depth of the liquid and the shape of the channel. Depth is normally measured with an acoustic depth sensor. The velocity of flow is obtained by the contrapropagating meter provided the liquid depth is sufficient to submerge the sensors. The velocity sensors are normally mounted at 25% of maximum depth, but this limits the lowest flow that can be measured ultrasonically. To minimize this problem, the meter can utilize other information at low depths. For example, Manning's equation [Sellin, 1970] or a small version of a flume or weir can augment the measurement at low depths. Stagnation and backwatering usually fill a channel rapidly so the employment of the gravity device at low levels is practical.

As flow increases the depth of flow also increases and the instrument is designed to automatically determine the proper mode of measurement. For levels which fill the pipe completely, the unit operates as a normal full pipe flowmeter. [Based on results] from an extensive test of a configuration at Colorado State University in late 1970, in which a range of flows over 200:1 were measured, the error was less than ±3%.

This unit can be installed in existing flow conduits of many shapes, usually with a minimal amount of construction effort. The effort requires mounting the sensors and then routing the cables to the electronic unit. The user must

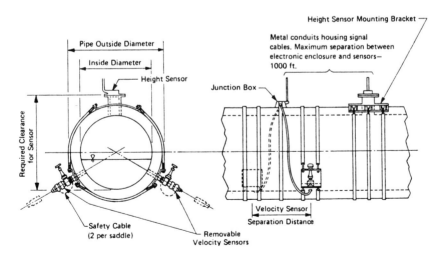

FIG. 4-75.   Measurement of flow in partly-full pipe using contrapropagation to measure V in one plane, and another ultrasonic time of flight from the top transducer to measure liquid level. Courtesy Badger Meter.

be prepared to provide the manufacturer with accurate site measurement data.

Open-channel or partially-filled pipe flowmeters require additional straight runs prior to the site to assure stable developed profiles. The presence of a free surface removes a strong frictional force for a section of flow perimeter, [slowing the] development of the predictable uniform profile. Under a joint EPA/Badger Meter Project (Anderson et al., 1976, Contract EPA-600/2-76-243) research was conducted to calculate the correction factors for single-beam velocity measurements in circular and [noncircular] conduits. The resulting data show that the required correction factors are orderly, uniform and highly predictable over the cross-section areas of meter application. For circular pipes a value of 0.96 serves for depths from 30% to 90%. Small variations are noted below and above this range. Further, with constant head and changing velocities, the correction factors are essentially constant. To achieve accuracies of ± 2 to $\pm 4\%$ the correction factors must be programmed into the flowmeter.

4.6.2   Mass Flow Rate of Liquids

In duct flow through a conduit of area A, the mass flow rate M may be obtained from ultrasonic measurements of density and flow velocity, or from acoustic impedance and Mach number, as follows:

$$M = \rho \mathbb{V} A \qquad (4\text{-}1)$$

or

$$M = (\rho c)(\mathbb{V}/c)A, \qquad (4\text{-}2)$$

where ρ = fluid density, c = sound speed, $\mathbb{V}$ = area-averaged flow velocity, ρc = characteristic impedance, and $\mathbb{V}/c$ is taken as the Mach number. For convenience of presentation, ultrasonic approaches to M determinations may be categorized according to the above pair of equations. This categorization recognizes, for example, that Doppler methods respond fundamentally to $\mathbb{V}/c$, not V.

Ultrasonic methods of measuring ρ are discussed briefly in Chapter 2, Section 3, and in Chapter 6. These methods include various sensor geometries, or, for well-defined liquids, use of c as a measure of ρ (Section 6.1.1). In principle, M could be determined by combining a torsional ρ sensor and a square-duct axial-path area-averaging flowcell for measuring $\mathbb{V}$.

Ultrasonic methods of measuring ρc include measuring the electrical impedance of a transducer loaded by the liquid (Kritz, 1955a,b; 1959, 1961) or reflectometer or impedometer methods (Lynnworth et al., 1982). But, if the fluid is not homogeneous, or if deposits form on the face(s) of the reflectometers, errors in M will result due to the locally sensed ρc not being representative of average properties in the fluid.)

The mass flow rate into or out of a container may also be derived from volumetric (liquid level or interface locations) and density profile data.

Returning to the first mass flow rate equation, $M = \rho \mathbb{V} A$, it will be understood that a *partly*-ultrasonic mass flowmeter could be constructed by using ultrasonics to measure either ρ or $\mathbb{V}$, and a *non*ultrasonic device to measure the other term. Examples might be: ρ by torsional sensor, $\mathbb{V}$ by turbine; ρ by dielectric-constant-based instrument (e.g., Clausius-Mosotti law applied to jet fuels) and $\mathbb{V}$ by a contrapropagation method (Lynnworth, 1979, p. 511); ρ from temperature, and $\mathbb{V}$ by any ultrasonic flow velocimeter.

4.6.3   Heatmeters for Hot Water

In principle, determining the heat carried by hot water is an obvious combination of flow velocity and temperature measurements. But in practice, finding the best technical and economical solution is not so obvious. In the Westinghouse flowmeter (Erickson, 1987, priv. comm.), sound speed in the water is used to measure an average temperature across the same path used to measure flow velocity in large pipes. Controlotron's clamp-on Thermal Energy Meter Model 960E was reported (ca. 1985) to measure the flow of thermal energy for pipe diameter $D > 10$ cm.

For small pipes, ~1-inch diameter, an insertable ultrasonic heat meter made by Siemens (Mágori, 1985) combines a "lambda locked loop" flow velocimeter with cold-leg and hot-leg temperatures measured electrically (RTD). This solution is used on thousands of European applications to monitor heat delivered from central water heating systems.

The temperature difference across the customer's installation and the used volume represent the actual thermal energy consumed.

A commercial heat meter containing an ultrasonic flow sensor with lambda locked loop and interdigital transducers, produced by Siemens, covers a flow range of 50 to 1500 m^3/h at operating temperatures from 20 to 90°C. The resolution is obtained by interrogating along an appropriate acoustic path by using triple reflection of the ultrasonic waves at the sidewalls (Fig. 4-76). By an $n + 1$-fold acoustic path, influence of crossflow can be minimized because its influence on the ultrasound before the reflection is compensated for by the inverse influence after the reflection. The sensor's performance is nearly independent of the flow profile because the acoustic field fills the whole cross-section of the pipe. The accuracy of the heat meter is better than $\pm 2\%$ of the actual value in a 30:1 range from 20 to 90°C. As an illustration of the low power con-

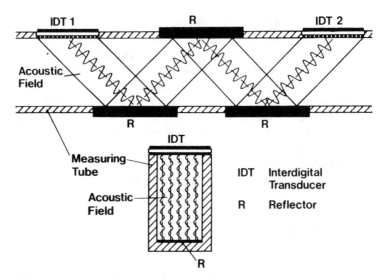

FIG. 4-76.   Longitudinal section and cross section of the ultrasonic flow sensor used in a Siemens heat meter. After Mágori, 1985. Illustration courtesy Siemens.

sumption possible with ultrasonic flowmeters, the Siemens heatmeter will operate with one built-in lithium battery for six years.

4.6.4 Boat Speed

Just as Chilowsky and Langevin's 1916 Doppler methods for measuring boat speed are related to subsequent Doppler flowmeters, so too can one draw parallels between contrapropagation measurements in pipes and the speed of a boat through water. Furthermore, the ultrasonic contrapropagation measurement of boat speed, and separately, depth, is analogous to measurement of V and H in a partly-full pipe.

The following remarks on the need for boat speed and depth measurements, and an example of transducer locations for these purposes, are substantially as presented in literature from Brookes & Gatehouse, 1986.

The *needs* for accurate speed and log are mainly safety, performance and fishing efficiency. Accurate measurements of distance travelled through water is essential for basic navigation and dead reckoning. Comparing one's dead-reckoned position with the position indicated by Loran yields set and drift, and, more importantly, yields an independent standard to accept or reject the Loran-indicated position. Without accurate and repeatable data on speed through the water, it is almost impossible to assess actual performance versus "promised" performance. An accurate speedometer can show the boat's response to small changes in RPM, trim tabs and load distribution virtually instantaneously.

Accurate and repeatable speed data, when compared to RPM, reveals the actual effect of prop changes and provides a basis for evaluating changes in engine performance and fuel consumption.

A common misconception, according to Brookes & Gatehouse, is that a speedometer is unnecessary when a boat is equipped with Loran. Unfortunately, Loran speed is misleading for performance monitoring. The response time is too slow and more importantly, speed over the ground is unrelated to boat performance through the water.

Accurate and repeatable speed affords the sportfisherman much better control of trolling speed, within ¼ knot of the desired speed. In principle, this should correlate with improved fishing.

Regarding *depth information*, knowledge of the depth of water is necessary for the safety of the crew and boat. Shallow and deep alarms can warn of sudden changes in depth.

Regarding *fast response*, Brookes & Gatehouse state that while many

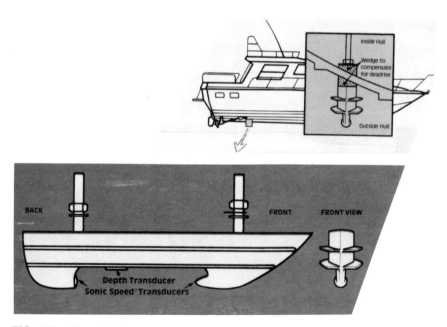

FIG. 4-77.   Example of transducer locations for a boat speed and depth system having the following characteristics: Boat speed up to 35 knots, calibration ±4%, depth to 200 m, prf = 2 to 20 Hz. Overall length of illustrated contrapropagation transducer housing = 488 mm. Source: Brookes & Gatehouse, 1986.

depth sounders work adequately at slower speeds, steady readings at higher speeds are even more critical with respect to safety. The deflecting wings on the Brookes & Gatehouse HS911 and HS911-XTL direct the hull turbulence away from the transducer face to improve the accuracy of depth measurements at all speeds.

To control one's efficiency in the process of fishing, whether inching into the shallows, trying to return to a known "hot spot," or trolling along a deep shelf or canyon, accurate depth is important.

Examples of transducer locations designed to meet the foregoing needs are illustrated in Fig. 4-77.

5. Temperature Applications

Ultrasonic thermometers are based on the temperature dependence of sound velocity. The "sensor" may be *the medium itself* whose temperature is to be measured or a *foreign body* immersed in the medium. Techniques have been developed that, when applied to inert gases, have provided temperature measurements from cryogenic levels as low as 2 K up to the order of 10^4 K. This range should not be surprising in view of Mayer, 1873, who over one hundred years ago tabulated his calculations of c in air from absolute zero to 2000°C in support of his suggestion to utilize the temperature-dependence of c in an acoustic pyrometer. Today we recognize that extrapolating c's dependence on the square root of gas temperature below the liquefaction temperature (-196°C for N_2) is invalid. Mayer's 1873 pyrometry suggestion, however, is indeed valid, and his analysis of and experiments on heated resonators remind us that some fairly precise acoustic measurements were possible a century ago despite the number of sensors and scientific instruments being rather limited.

Over "short" paths the *response time* for acoustic measurements of temperature can be less than 1 millisecond, a feature of significance to one studying shock-heated or combusting gases. Note that in air at 20°C, $c = 343$ m/s. Therefore the transit time is 1 ms for a path of 343 mm ($\sim$1 ft). At higher temperature, say 2000°C, an air path of $\sim$1 m can be traversed in 1 ms.

By using combusted gases as their own sensor, temperatures up to ~2000 K have been measured with audible (0.5 to 3 kHz) tone bursts over paths of tens of meters. At cryogenic temperatures, 2 to 20 K, and over the short paths associated with a 400 kHz cw laboratory interferometer, absolute accuracy in He gas on the order of ten millidegrees K (10 mK) is suggested by data reported over twenty-five years ago by NBS researchers Cataland et al., 1962. Sometimes an acoustic measurement of gas temperature occurs as a coincidence of another acoustic measurement. For example, Gregory et al., 1985, while developing a Helmholz resonator technique to measure piglet volumes (Chapter 7, Section 7.6), found that monitoring the resonance frequency of the empty cavity proved to be a convenient way to determine its temperature more accurately than with liquid-in-glass thermometers.

By using a hot steel body as its own sensor, internal steel temperatures were measured in laboratory specimens from 300 to 1500 K by the early 1970s. Momentary-contact dry-pressure-coupling techniques were used in order to avoid the temperature limits on transducers and couplants.

As examples of *foreign sensors*, thin-wire *nonresonant* sensors have been used in the 300 to 3000 K range in nuclear and industrial applications where conditions preclude the use of thermocouples, resistance devices or optical pyrometers. At 2500 K, temperature sensitivity of 0.1% was obtained in Re sensors 5-cm long by timing five round trips with an electronic instrument that resolved the time interval between selected echoes to 0.1 μs. Magnetostrictive transducers have been operated at rotational speeds over 1000 rpm and in noisy environments. Temperature profiling of up to sixteen regions using only a single guided path has been reported by using Ti sensors designed for use from room temperature to at least 500 K. In five-zone doped or thoriated-W profilers, results have been obtained up to over 3000 K. Here, sensors are scanned according to their distance from the transducer.

Resonant sensors based on bulk- or guided-wave devices have been available commercially with millidegree sensitivity for use at intermediate temperatures. Experimental resonant sensors have been studied for high-temperature applications. Besides temperature determinations, if one measures simultaneously the amplitude as well as the transit time of the received signal, energy absorption processes can be studied as a function of temperature. In SAW devices designed to measure etching (Joshi, 1987) or a variety of other parameters, it is convenient to obtain a temperature-compensation signal from an ultrasonic reference path which becomes part of a resonant temperature sensor. Resonant SAW temperature sensors are now available commercially (ETA Industries; Hauden, 1987).

5.1 Medium as Its Own Sensor

Solids, liquids, gases and plasmas can be used as their own temperature sensor. Theory is simplest for an *ideal gas* at low pressure, where sound speed obeys a square root dependence upon temperature T (Newton, 1687; Mayer, 1873; Cataland et al., 1962.) In many *liquids*, c decreases linearly as T increases. (Antimony, some aqueous solutions and water are among the few exceptions. See Section 3.8.) In many polycrystalline *solids*, particularly those having the best potential as temperature sensors, c is nearly a linear function of T up to their recrystallization temperatures. Linearity, while convenient, is no longer an important requirement of the sensor, since electronic linearization is readily available, given adequate calibration data. A *repeatable c* versus T characteristic is more important.

While most pre-1985 work in ultrasonic thermometry used active methods, passive methods are possible too. For example, flow, vibration or noise, if sufficiently intense, ought to cause a resonant sensor (solid resonator or fluid-filled cavity) to generate ultrasound at a frequency that depends on the sensor temperature. (See fluidic thermometer, an edge-tone resonator, in Chapter 2, Fig. 2-28.) In some materials, acoustic emissions (AE) are generated when phase transformation temperatures are reached. In studies of the AE spectra for non-premixed turbulent gas combustion, Seshan, 1986, found that the spectra exhibited a consistent shape over the Reynolds number range 8200 to 82000. Seshan sought a predictable frequency domain in which AE correlated with combustion efficiency. Seshan concluded that the ratio of the average acoustic power output in the most responsive band to the acoustic power output at the loudest frequency increased "significantly and unmistakably as combustion efficiency approaches 100%." The most responsive band covers the range from a transition frequency, $f_k/2$, up to $1.5f_k$, where f_k = Kolmogorov frequency. The Kolmogorov frequency f_k depends on temperature, as well as on the size of the smallest eddy.

5.1.1 Gases and Plasmas

In this section five applications are treated, to indicate the scope and limitations of temperature measurements in gases and plasmas.

5.1.1.1 Cryogenic Standard

Why consider ultrasonics for cryogenic temperature measurements? The answer, in part, as expressed by Cataland et al., 1962, is:

the conventional gas thermometer has limitations in this temperature range which have not been circumvented. The existence of these limitations is apparent when one considers the spread of both isotherm and gas thermometer data for determined temperatures in the liquid helium range.

Important experimental quantities associated with conventional thermometric methods, such as volume determinations, gas absorption, dead spaces, and known masses of gas, are avoided in the velocity of sound measurement. Also, the required accuracy of pressure determination is greatly reduced, and the need for a reference temperature is eliminated. In the limiting case (as the pressure approaches zero) the sonic temperature determination depends upon the experimental measurement of the sound velocity, the correct value of the gas constant R, the mass of the helium atom, and the ratio of specific heats at zero pressure. In practice, the measurements can be conducted at a pressure sufficiently low that pressure and virial coefficients are second-order quantities. For helium gas the relating equation can be expressed in a pressure expansion,

$$c^2 = \left(\frac{C_p}{C_v}\right)_{p=0} \left(\frac{R}{M}\right) T \, [1 \, + \, \alpha p \, + \, \beta p^2 \, + \, \cdots], \qquad (5\text{-}1)$$

where c is the velocity of sound in helium gas; $(C_p/C_v)_{p=0}$ is 5/3; R, 8.315 $\times$ 10^7 (erg/deg mole); M, 4.0024 g/mole; p, the pressure; α and β, appropriate constants into which the virial coefficients enter; and T, the temperature to be determined.

The object of the experiment reported by Cataland et al., 1962, is to measure the velocity of sound in helium gas

> at constant temperature and sufficiently low pressures so that a fitted curve can be extrapolated to obtain the velocity (in an ideal gas) at zero pressure. Then the above equation can be used to determine temperatures independent of the virial coefficients. If the virial coefficients are sufficiently well-known the experiments can be conducted at higher pressures.

By using an acoustical interferometer, Cataland et al. reported reproducibilities of 2 mK and agreement with earlier nonacoustic temperatures to within ~10 mK. (The accuracy of the earlier methods was not uncontested.)

5.1.1.2 Atmospheric Sonic Thermometer

Why use sound speed in the air itself to measure temperatures in the atmosphere? Reasons include fast response and thermal equilibrium more readily achievable than with foreign sensors, particularly at high altitudes. Problems include acoustic isolation of transducers; disturbance of the measurand by the transducers or their supports; flow noise due to ascent, descent or wind; increasingly difficult transmission in rarefied atmosphere as altitude increases; ultimately, fundamental interpretive problems when wavelength is comparable to the mean free path.

For details on laboratory experiments or balloon tests the following references may be of interest: Barrett and Suomi, 1949; Larson et al., 1964, in which laboratory experiments were conducted using a 50-mm air path in a bell jar using 23-kHz Massa transducers, to simulated altitudes of 60 km; Larsen et al., 1979.

5.1.1.3   Plasma Transport Properties

This laboratory application illustrates the use of pulse techniques to $\sim 10^4$ K, and the simultaneous measurement of transport properties at the temperatures determined from the sound speed.

To determine the transport properties of high-temperature gases ultrasonically, using the method proposed by Carnevale and later executed by Carnevale et al., 1967b, one basically measures two acoustic properties of the gas: sound velocity and sound absorption. Sound velocity can be determined by measuring the transit time required for the sound wave to traverse a known path length. Sound absorption can be determined by measuring amplitude as a function of path length.

To accomplish these two measurements at temperatures above the melting point of the ultrasonic probes, the probes are inserted into the gas only momentarily (~ 0.1 s). The transit time and amplitude of an ultrasonic pulse which traverses the gap between the probes are measured simultaneously. This procedure is repeated for different gaps. The sound velocity is calculable from the slope of a graph of distance versus time, and the gas temperature is then obtained from the theoretical temperature-sound velocity relation. Absorption is determined by the rate of change of amplitude with respect to gap length.

A simplified block diagram of the equipment is shown in Fig. 5-1. The circuits to the left of the gas path generate a high-voltage pulsed oscillation

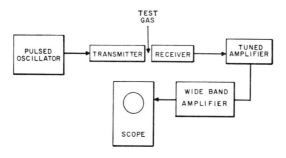

FIG. 5-1.   Block diagram of equipment used by Carnevale et al., 1967b.

which is converted to an ultrasonic pulse by a piezoelectric transducer cut to resonate at ~1 to 3 MHz. Typically, the sound is transmitted through a 50- to 150-mm long, 12.7-mm diameter fused-silica buffer rod, across the gas path, and then through a similar rod on the receiver side.

Differential-path ultrasonic measurements in plasmas were first conducted in inductively heated argon. Here, flow velocity was low, typically less than 1% of the sound velocity. The argon plasma exhausted vertically out of a fused-silica tube. Probes were manually swept through the plasma. Later, pneumatically actuated pistons drove the acoustic probes into and out of the test gas.

Although testing at the highest temperatures requires momentary contact, some testing of argon plasmas at T~8000 K was accomplished using water-cooled aluminum and fused-silica probes in continuous contact with the gas.

At lower temperatures, in muffle-tube experiments, steel and fused-silica buffer rods continuously contacted gases heated to 1300 K. Alumina could be used to ~2000 K, and other refractory materials such as oxides, carbides and nitrides appear promising for use to 2500 to 3000 K, provided outgassing and attenuation in the probes would not be excessive at these temperatures. In muffle-tube experiments, continuous contact permits the probe tips to come to thermal equilibrium with the test gas, thereby eliminating the thermal boundary layer in the gas.

Ultrasonic Probe Design Considerations.   The ideal ultrasonic buffer rod for plasma diagnostics must satisfy a number of requirements, some of which are divergent. Most of the data reported by Carnevale et al., 1967a,b, were obtained using rods of fused silica.

Experimentally, it is desirable to obtain about a 10-dB change in amplitude over a region of substantially uniform temperature. In 7000 K argon, measurements at $f = 3.5$ MHz would yield a 10-dB attenuation over a differential path of 2.5 mm. As the ultrasonic frequency is increased, shorter differential paths provide 10-dB attenuation. The shorter the path is, the smaller the temperature difference is over that path. Thus, an important experimental objective is to study the gas at the highest possible frequency. In this way, the measured attenuation and the derived transport properties can be more accurately associated with a particular temperature T. Use of higher frequencies also permits smaller-diameter probes to be used without introducing beam spreading. A disadvantage of higher frequencies, however, is that the acoustic power output is less at higher frequencies (limited by mechanical failure of the thinner crystal transducer). Also, the shorter-wavelength pulse is more attenuated by boundary layers and turbulence.

Experimentally, a compromise frequency is chosen that provides reasonable attenuation, ~10 dB, over a differential path length in plasma-heated gases of 2 to 3 mm. To probe ultrasonically ac- and dc-heated plasmas of $5000 < T < 15000$ K and $p = 1$ atm, ultrasonic frequencies in the range ~1 to 3 MHz were used. When other high-temperature sources are employed, e.g., muffle tubes or shock tubes, such that T is quite uniform, and pressure may depart considerably from 1 atm ($0.1 < p < 10$ atm, for example), the total path length x may be as large as 100 mm and frequencies from ~0.5 to 3 MHz become useful.

Carnevale et al. concluded that, at elevated temperatures, the differential-path ultrasonic technique could measure velocity and absorption in argon and helium, from which viscosity was calculated. These results agreed within a few percent with independent viscosity determinations up to 1300 K, thereby establishing the validity of this ultrasonic technique. At still higher temperatures (8000 K), the viscosity results calculated from the ultrasonic measurements are in reasonable agreement with accepted values. See Fig. 5-2.

The technique was later extended to mixtures of monatomic gases to determine diffusion coefficients, to polyatomic gases to determine relaxation times and rotational collision numbers, and to plasmas at temperatures approaching 20000 K (Carnevale et al., 1967b).

5.1.1.4 Shock-Heated Gas, <100 μs Response Time

Similar to the foregoing application, this laboratory measurement has the added difficulties of noise due to the bursting of the diaphragm, fixed paths and short measuring times behind the shock reflected from an end wall. Carnevale et al., 1967c, summarized the work as follows:

Temperatures of the gas behind reflected shocks have been measured simultaneously by a line-reversal method and by an ultrasonic method. Line-reversal temperatures were determined for shocks in neon doped with chromium carbonyl concentrations of 0.11% and 0.33% for a temperature range from 4400 to 6800 K. The ultrasonic method was used to measure temperatures (from 4500 to 7700 K) of pure neon as well as the mixtures of neon plus chromium carbonyl. The average relative deviation of the ultrasonic measurement from the simultaneous line-reversal measurement was 2.1% for the smaller concentration of chromium carbonyl and 2.7% for the larger concentration. The best least-squares fit of the ultrasonic temperatures measured in the reflected shock region versus measured incident shock speed is within 1% of the temperature predicted from the shock jump relations for pure neon. Similarly, the best fits of the ultrasonic temperatures versus Mach number and the line-reversal temperature measurements versus Mach number are also within 1% over the entire temperature range. The trans-

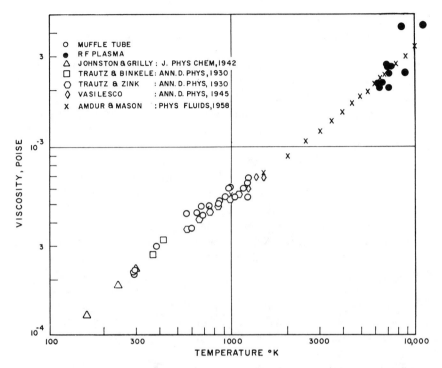

FIG. 5-2.   Viscosity versus temperature in argon. After Carnevale et al., 1967a, b. Ultrasonic data: open and closed circles. Sources for other data are given in Carnevale et al., 1967b.

lational modes and the electronic states corresponding to the CrI lines measured, were in local thermodynamic equilibrium with each other.

The principles and characteristics of conventional shock tube operation can be found in the cited 1967c paper. The shock tube used in these studies, as described by Carnevale et al., 1967c,

consists of a 0.91-m long stainless steel driver with a 7.62-cm inside diameter and a 5.08-mm wall thickness which is connected to a driven section consisting of a 6.1-m long stainless steel tube of 5.08-cm square cross section and 0.635-cm wall thickness. The two chambers are separated by an aluminum diaphragm (0.081 cm thick) which is cross scribed to insure uniform bursting.

In this work the region behind the reflected shock has been used as a stationary high-temperature gas reservoir for the ultrasonic and line-reversal temperature measurements that were recorded 1.90 cm from the end wall.

Incident shock speeds were measured between two thin film platinum heat transfer gages located 201.6 and 140.6 cm from the end wall. Signals from the gages were amplified and recorded on a rasterized Tektronix type 545 oscilloscope. Shock speed measurements were obtained primarily to compare temperatures calculated from the ultrasonic method with those obtained from the shock speed. The temperature of an ideal gas can be determined from its sound speed by the well-known relationship,

$$c = (\gamma RT/M)^{1/2}, \tag{5-2}$$

where c is the sound speed, T is the temperature (in K), M is the average molecular weight, γ is the ratio of specific heats of the test gas, and R is the gas constant.

In a real gas one must take into account the frequency dependence of the sound speed. This is due to the finite relaxation times required for the internal degrees of freedom to adjust to the small fluctuations of the translational degrees of freedom caused by the sound wave. The equilibrium between the various degrees of freedom is not established at the higher frequencies and a dispersion of sound speed results.

In general, if a particular process occurs with the relaxation time τ, it produces a dispersion for sound waves in the neighborhood of a relaxation frequency. For sound frequencies which are far from any of the relaxation frequencies, the sound speed is given by the usual formulas for a nonrelaxing gas, except that those processes which are too slow to follow the sound wave are neglected in calculating the gas properties.

In the case of pure monatomic gases for temperatures up to about 8000 K, $\gamma R/M$ is constant and known. The addition of small concentrations of $Cr(CO)_6$ to the inert test gas causes a decrease in $\gamma R/M$. There is also an increase in the average molecular weight M over the pure neon molecular weight. The average molecular weight is independent of the degree of dissociation and depends only on the initial mole fraction (x) of $Cr(CO)_6$:

$$M = (1 - x)M_{Ne} + xM_{Cr(CO)_6}. \tag{5-3}$$

The change in M causes the sound speed in the neon, chromium carbonyl mixture to be about 1.3% lower than the pure neon sound speed in the most serious case ($x = 0.33 \times 10^{-2}$).

The specific-heat ratio (γ) is lowered by less than 1% by the presence of the $Cr(CO)_6$ in the concentration range of the present experiment. The decrease in γ is due to the presence of the polyatomic species (vibrational and rotational specific heats). Between 800 and 5000 K the major polyatomic molecule is CO. As the CO begins to dissociate at high temperatures, the γ increases once again to the monatomic gas value 1.67. Since dissociation and vibrational relaxation are so slow that they do not fully [respond to] the 1-MHz sound wave, the specific-heat ratio deviates even less from the monatomic value than one would expect from equilibrium thermodynamics.

Summarizing, the change in M may be calculated exactly from the initial composition of the test gas. The decrease in γ depends more on the p-T conditions behind the shock but causes less than 1% change in γ. For the work reported here, the decrease in $\gamma R/M$ due to $Cr(CO)_6$ concentrations of 0.11% and 0.33% diluted in neon is 0.78% and 2.34%, respectively. These corrections [were] included in the temperature determinations.

The ratio of the temperature behind the reflected shock to room temperature may be calculated from Eq. (5-2) by noting that the transit time (t) across the fixed shock tube width is inversely proportional to the sound speed. The reflected shock temperature is then

$$T_5/T_1 = (t_1/t_5)^2(\gamma_1/\gamma_5)(M_5/M_1), \qquad (5\text{-}4)$$

where the subscript 5 indicates reflected shock equilibrium conditions and the subscript 1 indicates reference conditions (room temperature in neon).

A block diagram of the ultrasonic apparatus is shown in Fig. 5-3. During a temperature determination, the experimental sequence is as follows. An upstream heat transfer gage is used to simultaneously trigger an rf pulse ($\sim$1-MHz) generator and an oscilloscope. The pulse wave train is converted

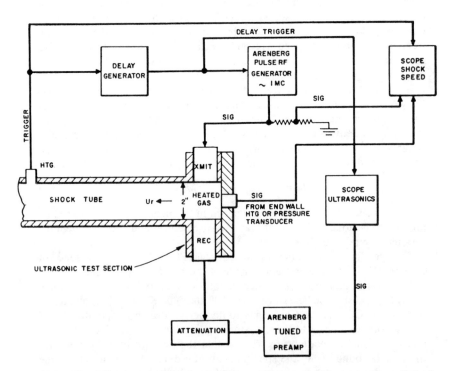

FIG. 5-3.   Block diagram of ultrasonic apparatus used to measure the temperature of shock-heated gases. After Carnevale et al., 1967c. © 1967 AIP.

to an ultrasonic pulse by a lead zirconate titanate transducer. The sound is transmitted across the shock tube and is received by an identical lead zirconate titanate transducer which converts the sound wave back into an electrical signal. The signal is then amplified and displayed on an oscilloscope. Sound velocity and hence temperature are determined from a measurement of the transit time required for the sound wave to traverse the 5.08-cm gas path across the shock tube. Prior to the shock measurement, the transit time t_1 of the pulse is determined by calibration with a gas whose sound velocity is well known at room temperature (e.g., neon or helium).

The details of the ultrasonic probe are shown in Fig. 5-4. The experimental problems associated with the use of ultrasonic probes in shock tubes include isolation of the transducers from the shock tube walls, damping of the shock front impulse to the ultrasonic probes, and overcoming disturbance in the test gas itself. Acoustic power is coupled into the shock tube walls from the shock front and also from the breaking of the diaphragm. In addition, the sound wave travels faster with much less attenuation around the walls of the shock tube test section than it does across the shock tube gas path, thereby effectively "short-circuiting" the sound wave received through the gas path. Therefore, it is important to mechanically isolate the active parts of the ultrasonic transducers. Isolation is accomplished by mounting the

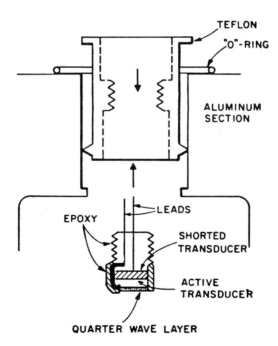

FIG. 5-4.   Transducer used for temperature measurements in a shock tube, designed by C.A. Carey and reported in Carnevale et al., 1967c.

probe in a spool which is made of an acoustically attenuating material such as Teflon. Mechanical contact is also minimized since the spool contacts the test section on only two small rings at either end. The vacuum seal is made by a cover plate (not shown) at some distance from the spool. Isolation is improved by mounting the active part on an epoxy rod which is connected to the Teflon by a screw thread.

When the transducer is used in the reflected region of the shock tube both the incident and reflected shock fronts pass over the transducers in which case impulses from the shock fronts cause the transducers to ring. This ringing must be damped so that an acoustic signal may be transmitted during the test time in the reflected shock region. This is accomplished with minimum loss of acoustic signal by mounting a second crystal, which is short circuited with conducting paint, behind the active transducer. The acoustic energy trapped in the short circuited crystal is then converted into Ohmic losses in the short circuit. The transducers used in the present experiment ring for 50 to 100 μs, depending on the Mach number and pressure.

Finally, there is the problem of disturbances in the test gas itself during the test time of the reflected shock region. Since the pressure variations in the sound wave are of the order of 10^{-6} atm, small disturbances [in] the shock tube flow which are negligible for most purposes cause noise problems in the ultrasonic system.

In order to obtain a usable signal-to-noise ratio, the use of narrow band amplifiers, high-power oscillators, and impedance matching between the transducer and the test gas are necessary. Impedance matching is accomplished with a quarter wave layer. The matching layer is composed of an epoxy resin loaded with silica microballoons to obtain an acoustic impedance (ρc) which is as close as possible to the geometric mean between the transducer and the gas. This resin is then cast on the surface of the transducer and cut on a lathe to the thickness which gives maximum signal. The ultrasonic system is generally calibrated with helium due to the expense involved in using large amounts of neon. The true transit time t_1 in neon is determined by multiplying the true transit time in helium by the sound speed ratio of helium to neon.

Ultrasonic temperature determinations which have been obtained in neon over the temperature range from 4500 to 7700 K are shown in Fig. 5-5 as a function of M_s^2. The solid line is the theoretical reflected shock temperature for a pure neon firing,

$$T_5 = (T_1/4)(M_s^2 + 1)(3M_s^2 - 1)/M_s^2, \qquad (5\text{-}5)$$

where T_5 is the reflected shock temperature, T_1 is room temperature (297 K), and M_s is the Mach number.

The experimental ultrasonic temperature determinations are indicated by the circles. The average temperature deviations were within 1.8% of the calculated temperatures for pure neon. The least-squares fit deviated less than 1% from the line represented by Eq. (5-5) over the entire temperature range. See Fig. 5-5.

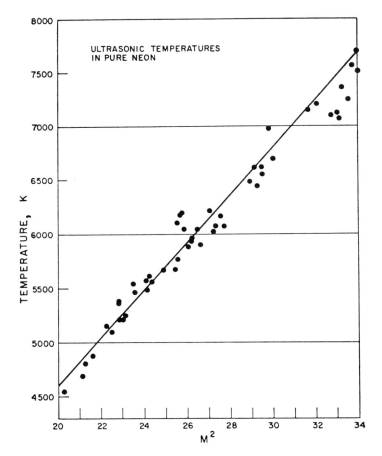

FIG. 5-5. Temperatures in reflected shock region in pure neon obtained by ultrasonics from 4500 to 7700 K as a function of the square of the Mach number. The solid line is the theoretical reflected shock temperature T_5 according to Eq. (5-5). After Carnevale et al., 1967c. © 1967 AIP.

The results of the simultaneous ultrasonic and line reversal temperature determinations for $Cr(CO)_6$ concentrations of 0.11% are shown in Fig. 5-6. The solid line represents the theoretical temperature for pure neon.

Although it appears that gas temperatures behind reflected shocks in pure neon in this shock tube can be determined within 2% by measurement of shock speed, the direct temperature measurement of shock-heated gases as reported here would seem to be more desirable than a calculated temperature. A direct measurement insures that nothing unusual occurred during the test shot which might have adversely affected the gas temperature, a condition which cannot be accounted for with a calculated temperature.

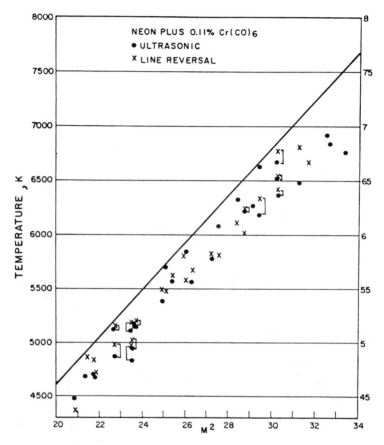

FIG. 5-6.   Simultaneous ultrasonic and line-reversal temperature determinations in neon
+ 0.11% Cr(CO)₆. After Carnevale et al., 1967c. © 1967 AIP.

The two methods can be used interchangeably, depending on the test gas
composition. For pure neon the ultrasonic method was used since neon
lines were not excited during the equilibrium period at the temperatures
utilized (≤8000 K). For Cr(CO)₆ concentrations greater than about 0.1%
both methods have been used to measure gas temperatures. For concentrations less than about 0.1% the ultrasonic method is generally preferred
to line reversal due to a decreasing background lamp absorption within the
shocked gas as the concentration is lowered, which accounts for some uncertainty in the measured emissivity value. The line-reversal method offers
the advantage that the measured temperatures are independent of gas composition whereas the ultrasonic measurement cannot be obtained without
prior knowledge of the gas composition.

Since both the ultrasonic method, which is a measure of the kinetic or translation temperature, and the line-reversal method, which is a measure of the excitation temperature, are in agreement, the results are consistent with the existence of local thermodynamic equilibrium in the reflected region of shock heated gases for the operating conditions used in these experiments. [Regarding *total* gas temperature, see Coats et al., 1989.]

5.1.1.5   Large Power Station Furnace—Acoustic Path ~10 to 20 Meters

Techniques for measuring hot gas temperature T based on the propagation velocity of acoustic waves between one or more pairs of transmitters and receivers have been developed by Green and Woodham, 1983, and Green, 1985, of the CEGB (Central Electricity Generating Board) and their colleagues or licensee in the UK, and in the US by Nuspl et al., 1986, 1987. See also, Whitten and Kitchen, 1988. This noninvasive method permits continuous real time measurement of gas temperatures in hostile environments existing in large electrical utility boiler furnaces. Despite the high-background-noise environment associated with the combustion process and soot blowers, turbulence and sources of sensor contamination, T measurements were made above 1000°C over paths in the ~10 to 20-m range. (For some boilers, the path lengths encountered may be 35 m. Equipment developed in the CEGB work was specified in 1988 for paths of 1 to 10 m. Nuspl's system, installed permanently in some twenty power plants and industrial boilers by the end of 1988, has operated on paths up to 27 m. For the equipment in use in 1988, the limit on path length was imposed by background acoustic noise.)

To appreciate the need for acoustics, and its potential not only in the utility boiler applications reviewed in this section but also in a variety of other high temperature gas situations, consider the nature of this generic problem and the limited choice of T sensors. None of the available solutions is totally satisfactory.

The following material in this section is largely based on material supplied by Nuspl (1987, priv. comm.) and by the CEGB and their UK licensee (Combustion Developments Ltd., 1988).

According to Nuspl et al., 1986, 1987:

Flue gas temperatures are important operating parameters during boiler start-up and after the boiler goes on line. During start-up, furnace exit gas temperatures (FEGTs) must be continuously monitored to avoid overheating superheater tubes until steam flow is established. At higher operating loads, gas temperatures can be monitored at different locations in the furnace, and between heat absorbing tube banks to provide indications of surface cleanliness.

Poor quality fuels have a tendency toward increased slagging and fouling characteristics which degrades overall boiler performance. Furnace wall slagging increases the FEGT, which if near or at the ash fusion temperature causes superheater slagging. Fouling can occur within tube banks throughout the convection pass that may penalize boiler efficiency through increased flue gas temperature leaving the boiler.

It would be clearly advantageous to continuously measure gas temperatures at certain locations in a boiler as a way to monitor heat transfer surface cleanliness. A decrease in flue gas temperature differential across a tube bank would indicate decreased heat transfer in the boiler.

High temperatures coupled with an erosive environment make it impractical to measure gas temperatures on a continuous basis with conventional thermocouples anywhere but at the economizer outlet.

Water-cooled high velocity thermocouple (HVT) probes are used to measure temperatures above 1,000°F (538°C). They have been successfully used for many years, but become impractical when boiler widths exceed 40 to 50 ft (12 to 27 m) since the maximum working length of a probe is 24 ft (7.3 m). Often there is insufficient access to ports or external clearance to permit a complete traverse.

To utilize the HVT to determine average temperatures requires that many point-by-point temperature measurements be taken in a very short time span within a measurement plane. Individual measurements are then averaged to obtain a final temperature.

The water-cooled probes are difficult to maneuver and support, and the test procedure is quite labor intensive. Extreme care must be taken to protect thermocouple junctions, extension wire splices, ceramic shields and insulators from mechanical damage. This temperature measurement method is suitable only for periodic test measurements.

Optical pyrometers have been used for several years as a way to make spot readings, primarily in combustion zones. For one to be reliably used, it must be responsive to two or more optical wavelengths. To use it, one requires preliminary knowledge of thermal gradients through the gas medium to the point where the temperature measurement is to be made. Optical pyrometers have not been widely used as continuous measurement devices in boilers.

According to Nuspl et al., optical pyrometers do not work below 1,600°F (871°C) and would not be useful for monitoring FEGTs during boiler start-up.

As mentioned earlier, measurement of the temperature of hot gases goes back to the nineteenth century. Besides Mayer's work, the possibility of measuring high temperatures by acoustics had been investigated experimentally in *small*-scale (short path) situations by Suits, 1935, Livengood et al., 1954, Carnevale and co-workers, 1961 to ~1970, Apfel,

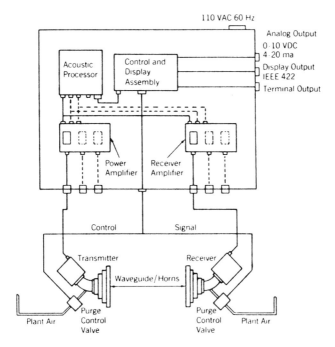

FIG. 5-7.   System control box detail. After Nuspl et al., 1986.

1962, Longstreet, 1972, and others. Green and Woodham, 1983, measured temperature distributions up to 1500°C in a full-size boiler of maximum dimension 25 m. Nuspl et al. were probably the first in the US to conduct extensive acoustic T determinations over long paths including independent HVT probe measurements in a number of boilers. See Fig. 5-7.

To obtain T from a measurement of sound speed c requires, even for an ideal gas, that the gas molecular weight M and specific heat ratio γ be known. Nuspl et al. list γ as a function of T, for the following conditions (Table 5-1):

Specific heat values for air are based on a 1.0% moisture content by dry weight (lb moisture/lb dry air). Flue gas specific heats are based on 12.0% carbon dioxide, 6.0% oxygen, and 88.0% nitrogen by dry volume basis with a 5.0% moisture content by weight (lb moisture/lb dry gas).

In the work reported by Nuspl et al., 1986, the transducer was a standard electromagnetic type, isolated thermally from the hot gas by horns made in part of steel electrical conduit (compare Apfel, 1962). The horn's

Table 5-1. Specific Heat Ratio in Air and in Flue Gas, After Nuspl et al., 1986

Temperature		Specific Heat Ratio	
°F	°C	Air	Flue Gas
70	20	1.40	1.37
500	260	1.38	1.35
1000	538	1.34	1.31
1500	816	1.32	1.29
2000	1093	1.31	1.28
2500	1371	1.30	1.27
3000	1649	1.29	1.26

delay was determined by sweeping frequency through resonance. As Nuspl et al. explain:

> For this application, the input consisting of the desired signal plus noise is cross-correlated with a stored sample of the transmitted signal. Because a stored sample of the transmitted signal is free from noise, the cross-correlation function will produce a greater output auto-correlation function for any given input signal-to-noise ratio.
>
> It was ultimately found that the auto-correlation function was useful in defining a data qualifier that was used to determine the "goodness" of received data. This term is the amplitude of cross-correlation divided by the square root of the auto-correlation function at zero lag. It has a value between 0.0 and 1.0 with values approaching 1.0 representing higher qualifiers and therefore better data.
>
> The cross- and auto-correlation signal processing approaches coupled with a variable frequency acoustic signal transmission proved to be a significant advancement for processing acoustic signals in noisy environments. Computer simulations showed it was possible to operate with background noise slightly above the received signal, or a signal-to-noise ratio [less than one]. Sootblower noise was still a problem to be addressed, but not insurmountable because measurements could be made by monitoring sootblower noise and then sampling between blowing cycles.

Field tests of the acoustic pyrometer system, and including, for comparison, HVT probes, were conducted on two pulverized coal-fired boilers, as reported in 1987 by Nuspl et al. One boiler, a 250 MWe unit, was 42 ft (12.8 m) wide and the other, a 650 MWe unit, was 69 ft (21 m) wide. All measurements were made in the upper furnace region. Following Nuspl's explanation:

> In the smaller boiler, operating at low load, the acoustic pyrometer system measured an average gas temperature of 1,500°F (816°C) for each of two test runs compared to HVT data at 1,594°F. The comparison was not as

good as expected assuming a ±25 to ±50°F uncertainty in HVT data and a 1% uncertainty in acoustic pyrometer data.

Eighteen traverse points at 20-ft (6-m) intervals were used on the smaller boiler. A review of the test method revealed that HVT measurements should have been taken just inside of the boiler sidewalls to compensate for the steep temperature gradient that existed there. This gradient had been reported to be as much as 1000°C/m within about 0.5 m of a wall and 100°C/m beyond this distance up to 1 m.

Figure 5-8 represents a load drop during which time the wall blowers were operated [and] then followed by an increase in load. The vertical distance within the "hysteresis" loop represents the reduction of furnace exit gas temperature as the result of clean furnace walls at a given load.

Figure 5-9 shows a comparison of instantaneous temperatures with a running average. The dots are snapshots of gas temperature along a specific acoustic path. Note the large scatter in individual measurements indicating that FEGT is quite dynamic. HVT tests showed extremes of as much as 200°F above and below a mean value at any given probe position. Figure 5-10 represents boiler shutdown data.

On the basis of such results and calculations, Nuspl et al. concluded that:

Field tests in utility and industrial type boilers have shown that acoustic pyrometry is a practical approach for on-line continuous temperature measurement in hostile environments. This technique gives an average line-of-

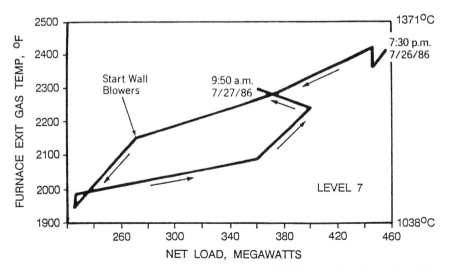

FIG. 5-8. Furnace exit temperature versus net load (MW) for a load swing in July 1986: 7:30 pm to 9:50 am. After Nuspl et al., 1987.

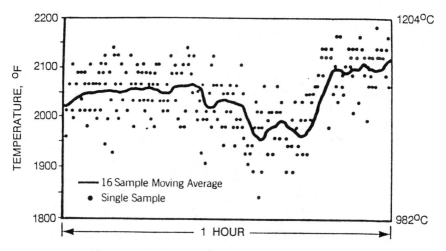

FIG. 5-9. Instantaneous temperatures versus moving average. After Nuspl et al., 1987.

sight measurement between an acoustic transmitter and receiver [neglecting path bending—see Green, 1985].

- A nonintrusive system has been developed to withstand severe operating environments both inside and outside a boiler. In many cases, it can be installed while the boiler is in service. Its outputs are compatible with a variety of recording and control devices.

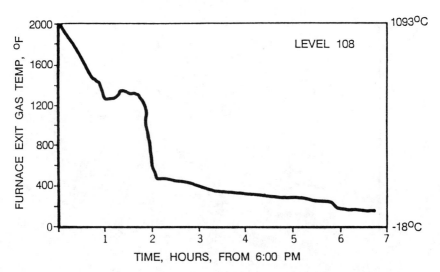

FIG. 5-10. Typical furnace exit gas temperature during boiler shutdown. After Nuspl et al., 1987.

- Test data indicate that acoustic pyrometry can be applied to boilers and perhaps to other hostile environments as wide as 80 to 90 ft (24 to 27 m) using electromagnetic transducers.

Technical details on how the post-1985 CEGB system operates were not supplied in time for inclusion here. By way of summary, however, Fig. 5-11 shows CEGB-licensee CODEL's typical arrangement of eight transducers. Each transmitting transducer, which may be a spark discharge, fires about every 20 s (Wyber, 1975). In the multiple transducer case, the time interval to make a complete sequence of transmitter/receiver pairs is less than one minute. A total of ten to fifteen complete sequences is necessary to obtain sufficient data for an accurate distribution update, although the mean temperature across the plane of measurement can be updated every cycle. Temperature profile is computed by representing the temperature distribution by a two-dimensional Fourier series whose coefficients are determined from the series of simultaneous equations using multipath data. Figures 5-11c,d show temperature distributions for a 500-Mw front-wall-fired boiler, without and with path bend correction, respectively. Pre-1988 furnace gas boiler trials in the UK included the power stations at West Burton, Fiddlers Ferry, Cottam, Padiham and Willington. Field trials investigated factors such as effect of firing pattern; pulverized coal size distribution and slagging effects on the thermal performance; effect of installing low NO_x burners, and efficiency comparisons of two supposedly identical boilers. The profiles identify cold spots and hot spots. Applications may be extended in the future to other cases, e.g., cement kiln secondary air temperature for process control. As a result of these investigations and other tests, the following specifications were listed by CODEL, 1988:

Measuring range:	Selectable, 0 to 3000 K
Accuracy	
Mean temp.:	2% of reading
Profile temp.:	4% of reading
Repeatability:	0.5% of reading
Response	
Single path:	Selectable, 20 to 200 s
Multiple path:	Typically 15 minutes for complete profile update
Path length:	1 to 10 m
Outputs	
Single path:	4 to 20 mA
Multiple path:	Video display and printer port
Power:	110 V AC/240 V AC

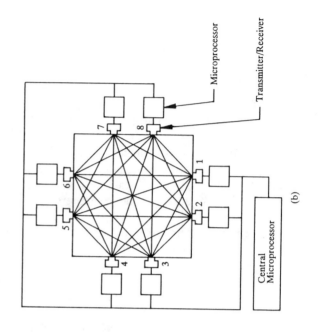

Microprocessor

Transmitter/Receiver

7

8

6

5

1

2

4

3

Central
Microprocessor

(b)

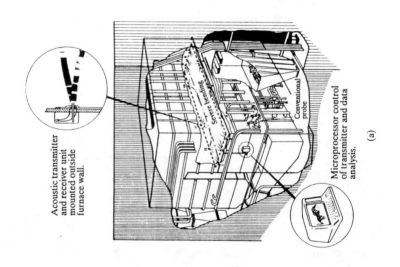

Acoustic transmitter
and receiver unit
mounted outside
furnace wall.

Severe bending

Conventional
probe

Microprocessor control
of transmitter and data
analysis.

(a)

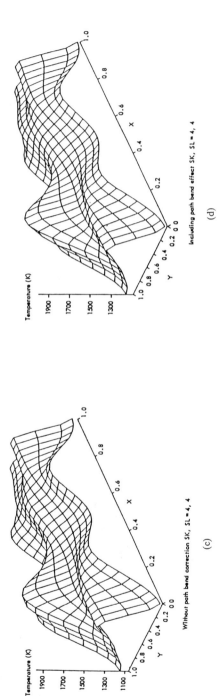

FIG. 5-11. CEGB/CODEL gas paths and measured high temperature profiles. (a) Sonic pyrometry at furnace exit plane. (b) Typical CODEL system arrangement, courtesy Combustion Engineering Developments, Ltd. (CODEL), 1988. (c) 500-MW front-wall-fired boiler, without path bend correction. (d) Effect of path bend correction. (a), (c) and (d) after Green, 1985, © 1985 AIP.

391

Operating temp.:   0 to 50°C
Construction: Housings fully sealed

For this equipment in the intended application, Green, 1985, listed the following potential sources of error (Table 5-2):

Table 5-2. Summary of Measurement Errors, after Green, 1985. © 1985 AIP

	Cause	Typical Error
Gas composition	Variation in mean molecular weight and changes with temperature of the specific heat ratio.	For given fuel s.d. on $M \simeq$ 0.4%. Specific heat ratio variation with temperature linearized in range 300–1800 K, s.d. $\simeq$ 0.6%
Path curvature	Fermat principle. Dependent on angle of path to temperature gradient.	Worst case may give 2% error. Most paths very much less. Can make correction
Gas velocity	Transit time for pulse traveling downstream reduced, upstream increased.	Measurements in both directions used to eliminate error. Can be used to determine gas velocity
Particles in gas	Reduction in speed of sound related to mass concentration and specific heat of particles.	1% by weight gives 0.2% change in effective R
Timing error	Variations in pulse rise	Estimated s.d. 0.3%
Path length error	Measurement errors and thermal distortion	Estimated s.d. 0.2%
Input and output of sound pulse	Variation of gas temperature in cooled mounting tubes	Design and furnace configuration dependent. Typical s.d. 0.1%
Speed of sound pulse	Increases with increasing pulse pressure ratio	10 kNm^{-2} pulse travels at Mach No. of 1.04. Typical timing reduction for furnace measurements $\simeq$ 40 μs
Temp. variation in time	200 K oscillations at approx. 1 Hz may occur in a furnace	Expectation of error s.d. 2 K
Imperfect gas	Second and higher viral coefficients affect speed of sound as pressure increases	Negligible at temperatures and pressures under consideration

5.1.2 Liquids

Compared to gases, liquids generally are much easier to get ultrasound into and across. This is because impedance mismatch and attenuation are much less severe in liquids than in gases. Practical difficulties stem from: lack of wetting (a problem with liquid metals); corrosive action; container curvature; bubbles and other foreign obstacles; geometric, phase and composition uncertainties; access restrictions; inhomogeneties; wall erosion; deposits on walls; unmelted charge; stirrers and other mechanical structures inside the vessel; and other environmental considerations. Typical or potential high temperature applications would be: measuring the temperature of water or sodium coolant for nuclear reactors; liquid steel temperature in a ladle, tundish or basic oxygen furnace; molten glass temperature in a pool; and nickel-base alloys in a tilting ceramic crucible in an inert environment.

When acoustic access to a known (pure) liquid is obtainable, the technique can be straightforward. Sodium temperature was measured non-invasively by the author at ~600 K in a 250-mm diameter × 10-mm wall stainless steel pipe at ANL (Argonne National Laboratory) by using 3-MHz rf bursts. There was no penetration of the pipe (Chapter 2, Fig. 2-34). The reason for trying to make such measurements is that in fast breeder reactors, sodium's temperature and flow provide a measure of reactor power output.

A typical temperature coefficient for *hydrocarbon fuels* is ~4 m/s K. Their sound velocity may typically range from ~800 to 1600 m/s. Thus, the velocity changes by a few tenths of 1% per K. Therefore, if the pulse time interval is measured to ±1 ns, the temperature of liquids across paths of ~1 mm can be resolved to ±1 K and paths of ~1 m, to ±1 mK.

Piche et al., 1987, studied propagation near the melting point of *semicrystalline polymers*. For a number of *liquid metals*, the isobaric temperature coefficient of velocity is in the range −0.2 to −0.6 m/s K. For many organic liquids, the magnitude of this coefficient is about ten times larger, typically ranging from −2 to −5 m/s K.

Coefficients in several *liquified gases* (CH_4, Ar, N_2, O_2) have still larger magnitudes, about −8 to −9 m/s K.

A specially prepared Ti buffer rod was found by Mansfield, 1981, 1982, to be wettable and survivable and therefore useful in inspecting molten Al. Mansfield's Ti probe design *could* be used for measuring T in molten Al, but in fact its use in molten Al is for detecting inclusions and inhomogeneties, not T (see Section 7.8). [Compare with wetting preparations of Smith et al., 1974. Note too that a "noncontact" transducer such as an emat (Chapter 3, Table 3.7) does *not* need to be wetted by the liquid in which sound speed is to be measured.]

Ultrasonics is a particularly convenient way to measure liquid temperature when ultrasound is being transmitted through the liquid for other purposes, such as measuring flow or liquid level. (See, for example, remarks by Erickson in Section 4.2.1.1.)

5.1.3 *Solids*

Bulk solids may be defined as solids in which the compressional wave sound velocity does not depend on lateral boundaries. This implies cross-sectional dimensions greater than 5λ, where λ = wavelength. Bulk solids are usually interrogated using longitudinal and transverse shear waves, i.e., bulk waves, rather than extensional and torsional guided waves. The bulk waves in the specimen or workpiece being processed may, however, be launched and detected by guided waves that are coupled to it.

Ultrasonics is one of the few methods capable of measuring the average *internal* temperature of a body. Temperature determinations on steel in the mill are now largely confined to surfaces. These measurements give no information about the internal temperature, which is the important factor in hot working. Further, excessive soaking or reheating times may be avoided if the internal temperature can be determined, thereby increasing production and economizing on fuel. An example of the effect of temperature on the pulse-echo ultrasonic transit time is shown in Fig. 5-12. This result was obtained by using the momentary-contact high-pressure dry coupling method. [The idea of using sound speed to indicate the internal temperature of solids undoubtedly occurred to a number of investigators who measured the elastic properties of solids at temperature extremes. The author's suggestion to measure steel temperature ultrasonically was presented in 1967 (Lynnworth and Carnevale, 1969).]

Although the momentary-contact pressure-coupling technique has been used successfully in some pilot plant (Droney and Pfeiffer, 1980) and full-scale steel mill operations at high temperatures (Krupski, 1979; Jeskey, 1987, priv. comm.—see Section 7.4) there is always motivation to avoid direct physical contact with red-hot steel. See, for example, Boyd and Sperline, 1988 or Boyd et al., 1989. Wadley et al., 1986, have used laser transmitters and electromagnetic (emat) receivers in their ultrasonic studies of internal temperature distribution in hot stainless steel 304, up to 750°C.

As Wadley et al., 1986, point out:

> The processing of materials from the most basic, such as the casting of steel, to the most advanced, such as the epitaxial growth of optical or elec-

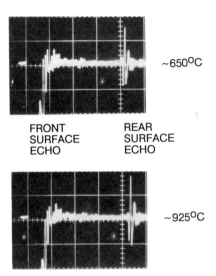

~650°C

FRONT   REAR
SURFACE   SURFACE
ECHO   ECHO

~925°C

FIG. 5-12.   Temperature measurement in a 100-mm thick steel billet, using 1-MHz longitudinal pulse, momentarily pressure-coupled to billet surface. Oscilloscope sweep speed, 10 μs/cm. After Fowler and Lynnworth, 1970.

tronic materials, is undergoing a radical change. Recent scientific advances have greatly improved the understanding of many of the phenomena involved in processing. This, in turn, has resulted in the emergence of sophisticated process models capable of predicting the effect of process variables upon, and the microstructure properties of, processed material. These process models are now becoming the heart of automated control systems for materials processing. These systems promise both productivity and quality enhancements for existing processes and the more reliable processing of many advanced materials that hitherto were considered too unstable for commercial development.

As expressed by Wadley et al., 1986, it may be anticipated that future automated control of materials processing

will consist of three interlinking systems. *Sensors* will continuously measure, in-process, critical microstructure and process variables without [perturbing] the process. Sensor data will then be utilized by *predictive models* to evaluate the current state of processing and, with the aid of expert systems and other artificial intelligence techniques, a new, optimum set of process conditions will be determined. Finally, *control* systems will stably implement these conditions.

While major strides have been made in both predictive process modeling and process control, the development of materials-processing sensors has lagged behind and is the limiting factor for many potential applications of automated process control. The required sensors for materials processing are quite unlike those of other industries.

Implementation of advanced materials processing strategies are presently hampered by a lack of adequate process-control sensors. For the processing of basic materials, such as steel and aluminum, an internal temperature-distribution sensor is needed. For these metals, the velocity of ultrasound is well known to be a strong function of temperature.

Wadley et al. considered determining the velocity-temperature relation for a particular steel grade so that an ultrasonic-velocity distribution (tomogram) for a hot steel body could be converted to an internal temperature distribution. To overcome the usual need in conventional tomography for thousands of path-integral measurements along intersecting ray paths, Wadley et al. incorporated *a priori* heat-flow information into the reconstruction algorithms.

As a result, for uniformly cooling bodies with the simple geometries usually encountered (e.g., circular, square or rectangular cross sections), the number of measurements can be reduced from many thousands to as few as three to five. Initial experiments on austenitic steel samples of cylindrical and square cross section by using non-contact ultrasonic techniques have shown that this approach yields acceptable temperature profiles at temperatures of up to 750°C.

(For comparison with an analogous problem in the biomedical area, refer to Johnson et al., 1977; Davis and Lele, 1985; Gibby, 1980; and Galkowski, 1983. For a more general presentation of principles and topics in tomographic imaging see Kak and Slaney, 1988.)

Regarding temperature reconstructions in a square geometry, Wadley et al., 1986, proceeded as follows.

Time-of-flight measurements were performed on a 304 austenitic stainless-steel block nominally 152 mm square and 37.5 cm high. The block was mounted on a translation stage so that it could be moved past the sensing coil of a stationary emat (Fig. 5-13, top). A horizontal laser beam was aligned with the emat coil and brought to focus at the surface of the block. The line joining the focal spot with the center of the emat coil remained fixed in space as the block was translated so that measurements could be made on a series of parallel flight paths.

If independent TOF measurements for propagation along different ray paths are made, an internal temperature image may be reconstructed using tomographic algorithms. However, it turns out to be better to use a least squares inversion procedure that facilitates incorporation of *a priori* information for the reconstruction. The information is a thermal model which exactly predicts the internal temperature distribution when the initial and

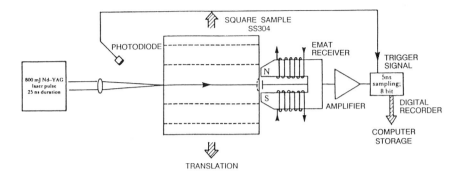

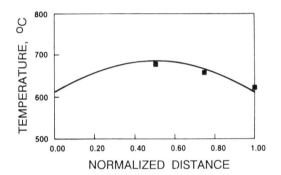

FIG. 5-13.   Top: Schematic diagram of the apparatus used by Wadley et al., 1986, to measure TOF (time of flight) values for square cross section experiments. Bottom: Internal
temperature distribution reconstructed from noncontact ultrasonic measurements (curve)
compared with embedded thermocouple measurements (squares) for AISI 304 stainless steel.
The normalized distance corresponds to a line through the center of a 150- × 150-mm square
billet. After Wadley, 1986 and Wadley et al., 1986.

boundary conditions are known. In practice, the TOF measurements are
used to determine boundary conditions by comparison against predicted
TOF values based upon successive interactions of the temperature model.

These investigators found that reconstructed temperature profiles for
the square geometry, Fig. 5-13 (bottom), show good agreement with data
from embedded thermocouples.

In contrast to the foregoing non-contact and momentary-contact MHz
methods, one can also consider lower frequencies and in some cases more
permanent coupling means. By welding, bonding, screwing or otherwise
pressure-coupling acoustically-slender flexible waveguides to the hot
workpiece, it may be possible to interrogate red hot bodies at frequencies

on the order of 100 kHz and measure surface and internal temperature less expensively than with laser and emat methods. Bonding and mechanical coupling methods are not restricted to electrically conductive solids, nor do they require optical access, thereby avoiding some of the limits of emats and lasers, respectively.

Consider, for example, the extensional-mode acoustically slender waveguide hardware and techniques developed for applications including some at temperature extremes (Figs. 2-67, 6-17, 8-14, 9-23, and 9-42, right). It occurred to the author in 1988, in response to a US Department of Energy solicitation on measuring the surface and internal temperature in hot workpieces, that waveguides even as large in diameter d as 10 mm, if operated at 100 kHz ($\lambda \approx 50$ mm in steel at room temperature, and about 35 mm at 1000°C) would still be relatively free of dispersion. [Near room temperature, for $d/\lambda = 0.2$ and $\sigma = 0.3$, according to Eq. (3-6), $c_e = 0.991 \, (E/\rho)^{1/2}$. Near 1000°C, where approximately, $d/\lambda = 0.29$ and $\sigma = 0.35$, $c_e = 0.975 \, (E/\rho)^{1/2}$, still close to the limiting value for a thin rod (Section 3.1.1).] Accordingly, at low frequency, acoustically slender waveguides might provide, in some cases, an easier way to convey ultrasound to and from a hot object. It is planned to begin investigating the high-temperature practicality and limitations of this extensional-longitudinal-extensional method at Panametrics and at a steel company in 1989.

5.2   Foreign Sensors

In contrast to using the medium itself as its own T sensor, in this section the sensor is added to the system in order to measure T. Such *foreign* ultrasonic sensors may be classified as *resonant* or *nonresonant*. The former would be expected to be more accurate, of higher resolution thermally and spatially, and more immune to noise. The latter tend to be more versatile, and lend themselves to profiling, to higher temperatures, to averaging the temperature over paths that are long and not necessarily straight, and are usually easier to manufacture for new situations, at least in small quantities.

5.2.1   Nonresonant Thin Wire Sensors

The thin-wire sensor in principle is traceable to Frederick's notched bar (1947, 1948) and more directly to Bell's notched wire (1957). Following conversations in 1963 between the author and J.F.W. Bell, E.A. Thorne and others associated with the Dragon Project at the A.E.R.E., Winfrith, Panametrics began a series of programs aimed at measuring temperature automatically in the core of nuclear rocket engines and nuclear reactors. Some effort also went into sensing liquid sodium using bare wire sensors, measuring furnace profiles, etc.

The thin-wire probe (Chapter 2, Figs. 2-29 and 2-30) handles much like a thermocouple and often borrows much from thermocouple technology and materials experience.

In principle, the ultrasonic thin-wire sensor offers the following *inherent advantages* over thermocouples, for high temperature applications:

- Only one material needs to be selected—can be nonmetallic

- Sensor is non-electrical in nature

- No electrical insulator needed in sensor or lead-in regions

- Average T obtained over the length of the sensor

- Faster response usually possible, because of lower mass

- Simpler construction for T profiling, because of using only one wire

- Immune to gradient errors along lead-in

- Feedthrough can be all-metal, i.e., no glass-to-metal seals required

In practice, in comparison with thermocouples, the following *disadvantages* are experienced:

- In high-temperature applications a sheath is usually required, and the ultrasonic sensor tends to stick to the sheath

- Electronics is more complicated, and if not mass-produced, tends to be relatively expensive

- T is not as localized

- Possibility exists for picking up mechanical noise

- Calibration data not as complete as for thermocouples, with respect to drift as a function of time or environment

- Technology is not familiar to instrumentation engineers, so more time and attention are required to make ultrasonics work properly

- Ultrasonic wire T sensors have received serious attention only for a limited number of specialized applications, so that their general utility has not yet been demonstrated

In a paper by Jacobson et al., 1987a, it was pointed out that a mass-produced ultrasonic *flowmeter* had been modified for use in timing echoes from waveguide probes. In 1988, the flowmeter due to Jacobson et al.,

1987b and referred to in Chapter 4, Section 4.5.8 was similarly programmed for use with wire waveguides. The availability of a general-purpose intervalometer version of an otherwise-standard instrument (e.g., Model 6468 "flowmeter") tends to offset the high-cost disadvantage listed above.

Hellbaum and Garner, 1988, point out a potential advantage of acoustic fibers over fiberoptic devices with respect to measuring surface temperatures during hypersonic tests, namely, ability to withstand much higher temperatures.

While there are a number of documented success stories (e.g., Table 5-3 and corresponding Figs. 5-14 to 5-16) a general-purpose T sensor of the nonresonant type is not yet available commercially. It appears that, at least for the next few years, or until ultrasonic instrumentation and sensor design is taught more widely, those who wish to take maximum advantage of this technology will have to do some of it themselves and get involved in designing and testing. Potentially significant ultrasonic thermometry applications may in fact be overlooked in some market surveys because they occur only in combination with *other* waveguide sensor applications, e.g., liquid level (H) or density (ρ) sensors where T-sensing is required for compensating the H or ρ readings.

As an example of an H application where T was measured ultrasonically, consider the determination of H in water from 20 to 250°C by Miller et al., 1980. In another application, the objective is aircraft fuel-tank gag-

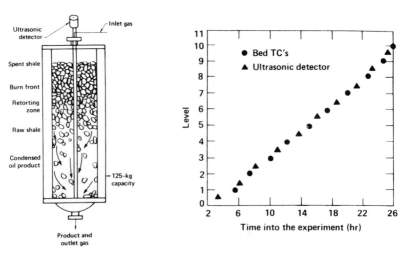

FIG. 5-14.   Left: Details of ultrasonic detector installation in oil shale retorts for tests during small 125-kg retort run S-22. Right: Arrival times of the retorting temperature (400°C) as measured by bed thermocouples and ultrasonic thermometer. After Ronchetto, 1981.

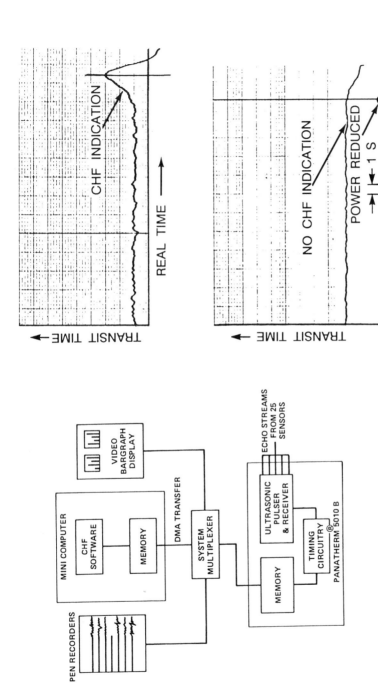

FIG. 5-15. Critical heat flux (CHF) experiment. Diagram illustrates signal processing equipment and examples of pen recorder responses, showing zones with and without CHF indications. After Barber et al., 1979.

401

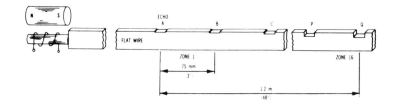

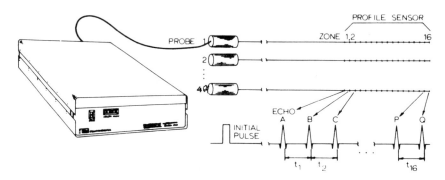

FIG. 5-16.   Details of ultrasonic equipment used in CHF experiment. Top: sixteen-zone Ti ribbon waveguide, blackened and notched. Bottom: Panatherm® 5010C, a scanning inter-valometer developed ca. 1978 to measure time intervals in up to sixteen zones per probe, and in up to forty probes, at scan rates up to about 5000 zones/s.

ing, that is, determining the mass of fuel remaining. As will be discussed in Chapter 6, this requires a measurement of the ρ-profile as well as H in the fuel tanks. If these measurements utilize a ρ-sensing waveguide in which the sound speed is not independent of temperature, then the tem-perature needs to be measured in the waveguide. What better way to do this than by ultrasonics? Thus, as demonstrated in Miller et al., 1980, the torsional wave travel times respond to ρ, H and T. The extensional travel times respond to T only, in the same waveguide, and so provide a con-venient means of T compensation.

5.2.1.1   Single Zone

Nonresonant sensors may be categorized as single zone or multizone. *Single*-zone sensors are used where one temperature suffices (local, or averaged over the length of an extended zone), or where attenuation or spurious echoes would preclude the multizone design. Since c can indicate T right up to the melting point of the sensor, single-zone sensors could be used to temperatures over 3000°C, e.g., W, Re and some other refractory materials. To generate echoes of equal magnitude from the beginning and end of a one-zone sensor, in the absence of attenuation, the magnitude of the sound pressure reflection coefficient is $|R| = 0.618$.

The transducer material used with slender waveguides is sometimes piezoelectric but most often magnetostrictive. The most common magnetostrictors are nickel and iron/cobalt alloys such as Remendur or Permendur. Some other common alloys are sufficiently magnetostrictive to be used as transducers, and may be preferred when their other attributes are sought such as oxidation resistance, corrosion resistance or expansion coefficient. Examples include some stainless steels, Kanthal and Kovar. To produce microstrains $>10^3$ the rare earth alloys may be considered (Clark, 1980; Clark et al., 1984). Sources include Edge Technologies (US) and Feredyn (Sweden).

5.2.1.2 Multizone

Multizone sensors are usually made by periodically notching a wire or ribbon waveguide. The magnitude of the reflection coefficient R at the notch must be kept small, preferably ≤ 0.1, or else reverberations in early zones interfere with echoes in later zones. By making zone lengths unequal, the interference problem can be reduced.

Besides notching, multiple zones can be created by reflective mounting points, kinks or sharp bends, added mass, sudden change in cross-sectional area, or use of a fiberacoustic bundle having elements of unequal lengths. Up to 32 zones have been sensed on one notched waveguide in a laboratory environment, and 16 zones in critical heat flux experiments. But 8 to 10 zones is considered a practical maximum for equal zone lengths in an easy-to-make stainless steel waveguide, and 5 was initially the limit for short (~10-mm) zones used at Sandia to measure T up to nearly 3000°C (Carlson et al., 1977).

Although the idea of using a filed notch, small drilled hole, kink, bend, added mass or other local discontinuity as an echo source may be intuitively obvious, the rigorous analysis of such discontinuities is yet to be completed. Field, 1986, points out the difficulty of such analysis in his report that presents the most recent thorough discussion of the many details involved in building waveguide thermometers. It is the attention to such details that led to the following results at Sandia:

- Ultrasonic thermometry has been developed as a high-temperature profiling diagnostic for use in the LMFBR (Liquid Metal Fast Breeder Reactor) Debris Coolability Program at Sandia National Laboratories.

Table 5-3. Ultrasonic Profiling Applications Reported Since 1977, Compared in Terms of Transducer, Lead-In, Sensor, Joints, Seals and Supports, and Electronics. Adapted from Lynnworth, 1982c.

Author and Application	Reason for Selecting Ultrasonic Thermometer (UT); Conclusions Based on Test Results	Transducer	Lead-In
Tasman et al., 1977–1982. Centerline temperature profile in nuclear fuel.	Intense radiation precludes opt. pyro. Tc's have short life above 2000°C and decalibrate under irradiation. Fuel structural changes yield imprecise T data. Re bulb gas thermometer too bulky. Noise thermometer not yet successful over 2000°C. UT's OK for 400 h at 2500°C. Accuracy limit due to unavoidable mechanical contacts.	Remendur, 1.4 mm ϕ × 12 mm long. Alnico-V magnet at end, 8 mm ϕ × 22 mm long. Coil, 180 turns of 0.18 mm ϕ Pt wire insulated with ceramic E ($\sim Al_2O_3$). Ceramic bobbin.	2% Thoriated W, 1 mm ϕ
Carlson et al., 1977. In-core molten UO_2 fuel pool experiment. Steep temp. gradient, $\sim$30°C/mm. Debrisbed experiments; steel melting by fissionheated UO_2.	High temp., lack of visual access precluded conventional diagnostics. UT provided temp. data to about 1000°C higher than was achieved with tc's. Also, axial temp. profile data was obtained.	Remendur, 0.7 mm ϕ, + Alnico magnet, 5 mm ϕ × 25 mm long. Coils, 2 mm long, with 100–150 turns of #44 Cu, 15 to 30 μH. Ferrite disks at coil ends. Delrin bobbin 2 mm long × 1.1 mm ϕ.	2% thoriated W, 0.5 mm ϕ Estimated temp. error due to heat condition <10°C with ThO_2 sheath.
Barber et al., 1979. Detect critical heat flux (CHF) in experiments that simulate a nuclear reactor fuel assembly, in a tube bundle heated electrically to generate heat nonuniformly in the axial direction. The tube walls are intentionally tapered so that an axial temp. gradient is produced.	Max power output of a nuclear reactor core is limited by CHF. In experiments it is necessary to detect CHF quickly enough to protect the bundle from overheating and also locate the point of CHF. System rapidly measured significant change in temp. System scanned $\sim$4000 zones/s, resolved 2°C. No failures despite 500 CHF events in 5 separate	Remendur, 1.6 mm ϕ × 22 mm long. Parallel magnet. Coil, commercially-available type, 12.7 mm long. Al housing. Teflon insulators. Not affected by strong stray fields. Floating ground.	SS304, 1.6 mm ϕ + Ti ribbon between brazed joint and first notch.

Sensor	Joints, Seals and Supports	Electronics
Sol-gel prepared, straight-drawn 2% thoriated W. Three zones: 50 × 0.85 ϕ; 30 × 0.72 ϕ; 50 × 0.60 ϕ. Two zones: 30 × 0.80 ϕ; 50 × 0.65 ϕ. Thoria-coated. Above 1600°C, sensitivity ~0.9 ns/°C-cm. Cal. stabilized by grain-boundary pinning by ThO_2.	Remendur/thoriated W joint brazed with Au-8% Ni.	Modified Panatherm® 5010A with ~25 ns resolution; programmable scanner for up to 10 zones on 1 or 2 probes, digital readout on teletype, 1.2 s per zone. Subsequently, data acquisition system controlled by desk-top computer, for up to 4 zones on each of 4 probes.
2% thoriated W, 0.5 mm ϕ. Circumferential notches cut by 0.4 mm diamond wheel define 5 zones each 10 mm long, yielding echoes about 4 μs apart. Notch $|R|$ = 0.2 to 0.3.	Percussion-arc butt weld between Remendur and W. Thoria sheath OK for up to 1 hr at 2800 to 2900°C. Sheath ID = 0.7 mm ϕ for 0.5 mm ϕ sensor. Rotated probe a few turns/s to avoid contact echoes.	Pulser + preampl.; bandpass video ampl. and limiter; six gate control servoes; multiplex system; and timer (commercially available), with BCD output to computer. Up to 10 zones in 16.7 s.
Ti ribbon, 3.2 × 0.25 mm, Ni-plated, black-Cr coated. Sensor lengths 114 mm. Length for 16 zones = 1.83 m, the full length over which CHF could occur. $|R|$ < 0.1 for most notches. Each of 25 tubes in bundle contained a UT.	Rubber stopper supports probe. W weight keeps ribbon straight, inside tube of 4.6 mm ϕ. Insulating soft fibrous washer gaskets on some notches center Ti in tube. All joints silver brazed in air.	Panatherm 5010B + minicomputer. Computer programmed to recognize CHF. Pen recorder and bargraph displays supplement transit time display and readout to computer.

Table 5-3. *(continued)*

Author and Application	Reason for Selecting Ultrasonic Thermometer (UT); Conclusions Based on Test Results	Transducer	Lead-In
	simulations of a nuclear reactor fuel assembly, over a 4-yr period.		
Kneidel, 1982. Same as Barber et al., 1979, except: shorter thermal response time, make probes easier to build, more durable and avoid spurious echoes due to sensor nonuniformities, to achieve higher S/N.	Same as Barber et al. Concluded that 16 zone technique is OK for detecting rapid temp. changes but not simultaneously useful for accurate temp. measurement.	Same as Barber et al., 1979	Lead-in flattened and tapered to match ribbon by means of a stamping die. Ti portion of lead-wire length selected to be 1.67 × zone length to minimize interference reverberations.
Miller et al., 1980. Water level sensor over range of 750 mm, from 20 to 250°C and pressure up to 15.2 MPa. Initially, no temp. gradients; 1 zone. Eventually, multizone system to compensate for temp. distribution.	Need unambiguous indication of inadequate core cooling; need device which could be verified even when its reading is constant for long times. Conclude extensional waves can provide temp. compensation data in same sensor in which torsional velocity indicates water level.	Remendur, 1.6 mm ϕ. Two separated coils, 1 for Joule, 1 for Wiedemann effect. See Dress, 1983, regarding transducer improvements.	SS304, 1.6 mm ϕ × ~300 mm long. Lead-out rod length ~1.5 m, above transducer. See Rogers and Miller, 1982 for steel wool transducer damping and membrane feedthrough details.
Ronchetto, 1981. Oil shale retort: need to monitor progress of retorting temp. in order to control process, increase efficiency and improve oil yield.	Need to continuously monitor temp. profile within retort. Sheathed tc's have high failure rate *in situ*, where thermal, mechanical and chemical stresses are rigorous. Need 25 to 50°C accuracy at positions known to 1.5 to 3 m. Appears that UT can be a useful *in situ* instrument: inexpensive, rugged, accurate enough for the remote sensing application.	Remendur 27, 1.6 mm ϕ. Parallel biasing magnet.	SS304 (29 cm) or Kanthal A-1 (38 cm) long.

Sensor	Joints, Seals and Supports	Electronics
Ti cross section reduced to 2.54 × 0.127 mm. Coated with Zynolyte black spray paint. Thermal response shortened by factor of 2.5. 15 zones aligned with spacer grids.	Remendur tig-welded to SS lead-in. SS to Ti joint brazed with Ag/Cu alloy by resistance heating in an inert gas.	No differences from Barber et al., 1979, reported.
SS304 flattened wire, 1.6 × 3.2 mm. Data for only one zone reported, but multizone tests were conducted and are planned for inclusion in future probes for this application, for higher accuracy in water level determination when temp. gradients exist.	Silver braze all joints. Supported by lead-out rod above transducer; rod brazed to electrical connector pin. Probe is spring-tensioned for straightness inside perforated sheath. Also tied with 50 μm SS wire to lie near sheath axis. Sheath: 10 mm OD × 1 mm wall × 3 m long.	Panatherm 5010C with time-ratioed blanks for extensional and torsional modes. See Dress, 1983, for alternative electronics.
SS304 or Kanthal A-1. SS: notches at 75-mm intervals for 12 zones. A-1: notches at 150 mm intervals out to 1.9 m for 10 zones, installed in capped tube.	Remendur brazed to SS304 or Kanthal A-1 lead in. Sheath, 6-mm φ × 1.4 m-long SS304 tube, open ended or capped.	Pulser/Receiver + oscilloscope for 100-ns time resolution.

Table 5-3. *(continued)*

Author and Application	Reason for Selecting Ultrasonic Thermometer (UT); Conclusions Based on Test Results	Transducer	Lead-In
Gibby, 1980. Medical: Temperature measurements of tissues during hyperthermia electromagnetic treatment. (Compare with Johnson et al., 1977b, or Davis and Lele, 1985).	Accurate temperature measurements in the presence of strong rf or microwave fields. Probe must not affect or be affected by the heating field. It must therefore be made of a dielectric material. Desired probe diameter 1 mm or less. Desired accuracy is 0.1°C. Range from 25 to 50°C.	PZT-5A, air-backed.	Glass rod, 75 mm long.
Galkowski, 1983, 1987. Continuation of Gibby's work using 0.3- to 1.0-mm φ probes.	Accurate temperature measurements in the presence of strong rf or microwave fields. Probe must not affect or be affected by the heating field. It must therefore be made of a dielectric material. Desired probe diameter 1 mm or less. Desired accuracy is 0.1°C. Range from 25 to 50°C.	PZT-5A, air-backed.	Pulled silica tapered rod.
Galkowski, 1987, 1988, priv. comm. The measurement of the average internal temperature of an operating metal vapor laser.	The high temperatures (up to 1500°C) and strong electromagnetic fields within the tube prohibit the use of more traditional methods of thermometry. Minimal beam blockage, with small diameter probes, is also a useful feature. Desired accuracy is 10°C. The test system performed to the desired specifications. A multizone system (for temperature profiles) is yet to be constructed.	Ceramic, air backed.	Transducer mounted directly to an alumina rod.

Sensor	Joints, Seals and Supports	Electronics
Acrylic.	—	Laboratory type signal generator, digital synthesizer, commercial RF amplifier, custom phase-locked-loop circuitry and signal processing.
Acrylic.	—	Laboratory type signal generator, digital synthesizer, commercial RF amplifier, custom phase-locked-loop circuitry and signal processing. Plus improvements.
Al_2O_3 under consideration.	—	Pulse generator, 1-ns time-to-digital converter, custom switching and signal processing circuitry and a Digital Equipment Corp. LSI-11/23 microcomputer.

Table 5-3. (*continued*)

Author and Application	Reason for Selecting Ultrasonic Thermometer (UT); Conclusions Based on Test Results	Transducer	Lead-In
Field, 1984, 1986.	Similar to Carlson et al., 1977.	Analysis, construction details and results reported in detail.	
Jacobson et al., 1987b. Liquid level experiments conducted at known temperatures.	Convenient way to temperature-compensate the torsional mode that is used to sense density and liquid level.	Remendur, torsional mode	SS304, round, 1.6 mm diameter.
Kirkpatrick and Kuzniak, 1987. Liquid level, large-scale, tens of meters.	Same as Miller et al., 1980.	Remendur; then Ni tube Current-pulsed for torsion	SS304L
Findlay et al., 1988, priv. comm. Aircraft fuel gaging. Six-month flight test program in an MD87 test airplane.	Easy, accurate way to temp.-compensate the torsional mode that is used to sense density and liquid level.	Remendur, ext'l mode	SS304

- Temperatures approaching 3000°C with spatial resolution of 10 mm and indicated temperature gradients of 700°C/cm have been measured.

- Instruments have operated in molten sodium, molten steel, and molten UO_2 environments.

- Up to 14 measurement zones on a single probe in molten sodium have been used with 12-mm and 15-mm spatial resolution.

- Hermetically sealed units [have been] operated at elevated temperatures.

- The stability of the system varies from ± 1°C to ± 15°C depending on the sensor design constraints for a particular application.

- Doped tungsten sensors have been developed to permit operation of total measurement zone lengths of 40 cm at temperatures above 2500°C.

Sensor	Joints, Seals and Supports	Electronics
Doped tungsten.	—	Like Carlson et al., 1977
SS302, 0.75 × 3 mm.	Silver-brazed joint; seal: swaged Teflon ferrule.	Modified ultrasonic flowmeter (Panametrics Model 6001A)
SS304L, various shapes including flags.	—	Like Dress, 1983, except for Wiedemann pulse for transducer.
SS304, approx 2 × 6 mm.	Welded; swaged Teflon furrules + RTV/ epoxy.	Same as Jacobson et al., 1987a.

5.2.1.3   Phase Change or Melting Point Detector

Bell, 1957, and his colleague Thorne, 1963, reported, respectively, on their high-temperature measurements of solid-state phase changes and the detection of melting of filler metal deposited in cavities drilled into a waveguide at points remote from the transducer. Such measurements are interesting in that they respond strongly to a particular event, i.e., a discrete temperature rather than to a continuum of temperatures. Their colleague Hub, 1963, studied phase changes in iron up to the melting point and reported on significant changes in c and α on passing from one cubic lattice form to another as the melting point was approached (Fig. 2-62). Later, in 1968, the author electrically self-heated Mo, Re and W wires to their melting points while monitoring the sound velocity and thereby obtained estimates for their Young's moduli at 2610, 3180 and 3410°C, respectively. Aside from these laboratory applications and a few others, the ultrasonic detection of a phase change, particularly melting, has not yet found a niche in industry as a practical tool for thermometry.

In the biomedical field, however, there exists a need in hyperthermia treatment of deep cancer for accurate subcutaneous temperature measurements at one particular temperature, 43°C (Hirama et al., 1985; see also, Section 5.2.2.1). Companion et al., 1986, proposed that a modified formulation of beeswax be implanted as a bead or injected into tissue, where its melting could be detected remotely based on changes in its ultrasonic reflectivity. Initial tests in phantom tissue gel were reported as encouraging.

5.2.2 Resonant Sensors: High-Q Mechanical Resonators

A symphony orchestra contains many examples of resonant systems in which the resonant frequency is an *un*desirable function of environmental parameters, including temperature. It does not appear likely that musical instruments that require tuning in order to compensate for temperature will find applications as industrial thermometers. However, as pointed out by Benjaminson and Rowland, 1972, the application of the quartz resonator to temperature measurement was an outgrowth of fifty years of research aimed at eliminating the effect of temperature on the resonant frequency.

Of the three types of high-Q mechanical resonators to be discussed, the quartz thermometer is the most accurate, and at least until now appears to have enjoyed the widest range of commercial use in applications such as calorimetry, industrial process control and oceanography. Tuning fork resonators have been studied with respect to measuring high temperature, >1000°C. SAW resonators are of interest as thermometers because of their small size, low mass, low cost if mass produced, and their ease of incorporation into other circuits where accurate T compensation is required. Synthetic *fresnoite* ($Ba_2Si_2TiO_8$) was shown by Hou et al., 1988 to retain its piezoelectricity above 1000°C and accordingly is of interest as a thermometer.

5.2.2.1 Quartz Thermometers

As pointed out by Nakazawa et al., 1987, when temperature sensing was linked to frequency in the mid-1960s, this aspect of metrology was advanced considerably. Within a decade came the introduction of stress- and temperature-transient-compensated quartz resonators. Nakazawa et al. point out that

> for the unrotated Y-cut quartz resonator, as is well known, not only is the first-order frequency-temperature coefficient relatively large, but the second- and third-order frequency-temperature coefficients also are relatively

large. While this is indicative of a resonator in which the output frequency f will be a function of its temperature T, the relation between f and T will not be linear over a wide temperature range and as a result a special calibration is required when it is utilized. Moreover, the calibration is a function of time because of such factors as stress relaxation in the resonator electrodes and changes in the stresses occurring in the crystal mounting supports. These particular perturbations, and others, go under the collective name of *resonator aging*. Additionally, temperature transients applied to the resonator enclosure also lead to thermal gradients within the crystal and, as a result, produce large frequency excursions.

In view of the inherent limitations existing in non-rotated Y-cut quartz plates, a doubly rotated linear coefficient (LC) cut has been developed for thermometric sensing applications [Hammond et al., 1965; Benjaminson and Rowland, 1972]. The LC-cut resonator comprises a cut which is located at an orientation where the second- and third-order frequency-temperature coefficients are substantially equal to zero or at least negligible, while the first-order frequency-temperature coefficient is not zero. The LC-cut, therefore, provides a linear frequency versus temperature characteristic, approximately 1000 Hz per °C [or ~30 ppm/°C].

Hewlett-Packard (HP) introduced its first quartz thermometer in 1965. Within its T range, -80 to $+250°C$, according to its manufacturer, the HP 2804A quartz thermometer has advantages over other types of thermometers such as: the 2804A is an easy-to-use, high-resolution digital thermometer that approaches standards-level accuracy; it is more rugged and much easier to use than a standard platinum resistance thermometer and Mueller bridge, yet costs considerably less; there is no tedious bridge balancing or calculation of temperature from measured resistance. The 2804A is reported to outperform industrial platinum, thermistor, and thermocouple thermometers in terms of accuracy, resolution, repeatability, stability and probe interchangeability. The quartz temperature sensor is a precisely cut, polished, and gold plated quartz disk about the size of a contact lens. (See Chapter 2, Fig. 2-27). According to HP:

Each quartz sensor operates at approximately 28 MHz and has a frequency sensitivity of about 1000 Hz per °C. This high sensitivity allows measurement resolution of up to 0.0001°C. Quartz sensors are inherently stable and have exceptional long term stability. This results in repeatable measurements over long periods with minimum drift.

Each quartz sensor is mounted in a probe sheath and is connected to the 2804A by a 3.7-m cable. Since a frequency is transmitted through the cable, the accuracy of the temperature measurement is unaffected by noise or cable resistance effects. Remote measurement over a kilometer is possible with an external oscillator and line amplifiers.

Since each quartz sensor has a unique temperature-versus-frequency response, every quartz sensor is individually calibrated at approximately forty temperatures. This calibration data is fitted to a curve using a regression technique. The unique coefficients for this sensor are recorded into a PROM (programmable read only memory) which is mounted in a calibration module shipped with each probe.

The following applications of the HP quartz thermometer are taken from the 1972 review by Benjaminson and Rowland.

The quartz thermometer has found a wide range of application because of its rather unusual characteristics relative to other methods of thermometry. Among the most important of the characteristics are portability, stability over extended periods, temperature resolution, direct reading linear display, and rapidity of measurement. Because of these features, many instruments are used as working standards for the calibration and checking of ordinary mercury-in-glass thermometers, thermistors, thermocouple installations and temperature regulators and controllers.

Some other applications result primarily from the very high resolution of the quartz thermometer. The absolute accuracy of the instrument can never exceed or even equal that of the standard grade 25-ohm platinum resistance thermometer since temperature, in the range of interest, is defined in terms of the latter instrument. The quartz thermometer can, however, provide temperature resolution exceeding that of the PRT because of its mode of operation. Temperature is converted to frequency in the probe-oscillator section and a frequency may be measured to any desired resolution simply by extending the gate time of the counter. . . . Extended reading times are quite acceptable when the temperatures to be measured are steady state, or if the total temperature change is rather small during a fairly lengthy experiment. Examples of these two cases are the *in situ* efficiency testing of hydraulic machinery and precision calorimetry.

A pump converts shaft mechanical energy into flow work, kinetic energy, and potential energy of position. In a turbine the conversion occurs in the opposite sense. The design engineer's task is to make these conversions with the highest possible efficiency, while the operating engineer is concerned with keeping the efficiency high during the useful life of the equipment, and performing maintenance as required. The difficulty lies in evaluating performance without removing the equipment from service. It has been demonstrated that efficiency can be determined by measuring the pressure change and the temperature change between the inlet and outlet streams. The pressure measurement is simple but the temperature differences are quite small. Culver first applied the quartz thermometer to this problem and found that the combination of high resolution and direct measurement of temperature difference $(T_1 - T_2)$ made the instrument very attractive for field evaluation of hydraulic machinery. London and Nelson concluded that the quartz thermometer offered significant advantages in this application because of its high sensitivity and ease of use compared to thermocouple and resistance thermometer methods. This latter point is particularly important when measurements of the required precision are to be made under field, rather than laboratory, conditions.

The typical precision calorimetry experiment involves a temperature change of only one or two degrees over a total time of 30 minutes or so. The absolute temperature at which the measurements are made must be known with moderate accuracy; however, for the changes that occur during the experiment, the greatest possible precision and resolution are essential. The quartz thermometer with [selectable] gate time has proven satisfactory for calorimetry at [100 microdegree] resolution, one significant advantage being the integrating nature of the instrument that eliminates the necessity of frequent measurements during the main temperature rise period and greatly reduces the number of data points to be handled. Several methods have been used for collecting and reducing the experimental data, including digital printers or digital to analog converters driving strip chart recorders followed by off-line computer or calculator processing. An interesting system was described at the 1968 Calorimetry Conference in which the quartz thermometer was coupled directly to a digital computer. Temperature readings were triggered by the computer's time base and transmitted to the computer that constructed the temperature-time curve, performed the required calculations and reported the results to a terminal in the calorimetry laboratory within a few minutes after the end of an experimental run. The key feature here was the ability of the thermometer to take a temperature reading when commanded by an external signal.

Some temperature measurement problems require sensing at some distance from the parent instrument. A few examples are thermal profiling of rivers and harbors, monitoring of temperature levels in factory effluents and at locations downstream of the discharge point, and deep sea oceanographic studies. Because the quartz crystal thermometer converts temperature into frequency rather than current or voltage, the precision, accuracy, and resolution of its measurements are independent of the distance from the sensor to the thermometer proper, and no corrections for lead resistance are required. For distances up to 1700 m (5000 ft) the probe may be cable-connected to the instrument; at greater distances, encountered chiefly in the oceanographic field, a free floating instrument package containing a recording system is often a preferred solution.

The chief assets then of the quartz crystal method of thermometry are the stability of the sensor, extremely high temperature resolution, digital presentation of the data, compatibility with data recording and instrument control equipment, ease of use, freedom from lead length problems, and reasonable cost. Because of these it is often an attractive [alternative] to resistance thermometry, even in such demanding applications as precision calorimetry.

Despite the practical utility of the LC cut as used in the foregoing applications, Nakazawa et al., 1987, contend that this type of resonator is difficult to manufacture because the two orientation angles ϕ and θ in the polar coordinate system are such that the cut is not located near any x-ray planes of any reasonable strength. Furthermore, it is uncompensated for any in-plane stresses such as electrode stresses that lead to component aging.

Nakazawa et al., 1987, therefore set out to improve the piezoelectric resonators described above, i.e.,

> to provide an improvement in quartz resonators that are sensitive to temperature, to provide an improvement in quartz resonators that exhibit ultralinear frequency-temperature characteristics, and to provide a quartz resonator in which the frequency is compensated against the effects of stress as well as thermal transients and hysteresis effects.

These investigators found stress-compensated quartz resonators having ultralinear frequency-temperature responses (NLSC cuts).

> They consist of doubly rotated cuts in which the orientations, defined in terms of the polar angles ϕ_0 and θ_0 lie on a locus where the second-order temperature coefficients of frequency are zero in the range of $\phi_0 = 20° \pm 2°$ and $\theta_0 = 20° \pm 5°$. These angles of orientation are in close proximity to, or on the locus of, zero frequency coefficients of stress. In the vicinity of the NLSC cuts, a relatively strong $12\bar{3}1$ x-ray plane was found at the quartz crystal orientation of ϕ_0 substantially equal to 19.1° and θ_0 substantially equal to $+16.6°$. This x-ray plane simplifies the manufacture of NLSC cuts.
>
> Measured $f-T$ characteristics were found to agree well with theory. The first-order coefficient α had a magnitude, for different modes and T ranges, on the order of 13 to 39 ppm/°C. A thermal time constant of 11 s was also measured.

Biomedical applications use quartz thermometers too, but the crystal cuts, sensor mounts and excitation and detection methods are different. While such applications are generally beyond the scope of this book, a design due to Hirama et al., 1985, may suggest analogous applications in the industrial process control sector. The motivation in Hirama's work was to measure temperature at the locus of therapy during hyperthermia treatment for deep cancers. Since the effective temperature on cancer cells is reported to be 42.5°C, and since normal cells would be injured at 43.0°C, the accuracy of the T measurement is critical. The semi-invasive T-sensing system of Hirama et al., 1985, consists of a sensor unit, probe and mainframe. The sensor unit can be implanted for several weeks because of its batteryless operation and small size (2.5-mm diameter $\times$ 30-mm long). To measure the resonance frequency of a quartz crystal unit,

Hirama et al. used the transmission method, but transmission was achieved in a wireless manner. They used an electromagnetic coupling system using a coil on the signal generator side and an ultrasonic coupling system using an ultrasonic transducer on the detector side, as shown in Fig. 5-17. Experiments conducted in physiological saline solution and a phantom showed that a resolution of ±0.01°C can be achieved at a distance of 150 mm.

5.2.2.2 Tuning Forks

Tuning fork ultrasonic thermometers can be classified according to whether the electroacoustic transduction occurs within a few wavelengths of the resonator or occurs remotely.

A remote resonant gas-filled cavity had already been demonstrated as a high-temperature thermometer for use to 1000°C by Apfel as early as 1962. Apfel's work was reported shortly after a series of papers by Bell and colleagues appeared on the thin-wire *non*resonant thermometer. Apfel's remote resonator publication may have inspired Bell to begin a resonant line of pursuit which led to his invention of several tuning fork resonators. Examples of these appear in Table 5-4, from Bell's 1972 paper. This table also includes a more recent design capable of much higher Q, due to Tehon, 1987, and a torsional tine design due to Paros, 1983. Bell's 1972 designs, however, had already demonstrated calibration stability and probe-to-probe reproducibility of about ±2 K despite a Q of only ~50 and despite three-month exposures at ~1470 K.

Bell, 1972, reported that a number of applications were under investigation for the tuning fork sensors, such as T measurement of continuous cast brass using alumina sensors; measurement of T profile in jet engines in the 2300 K region; and use of Mo sensors in graphite-moderated reactors. It is interesting to observe that Tehon too initially started to build his tuning fork T sensors for breeder reactor thermometry, and later redirected his investigation towards jet engine T measurements.

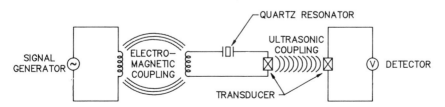

FIG. 5-17. Wireless transmission system using electromagnetic and ultrasonic coupling. After Hirama et al., 1985. © 1985 IEEE.

Table 5-4. Characteristics and Examples of Resonant Temperature Sensors. Numbers
            1–5 after Bell, 1972, number 6 after Tehon, 1987, and number 7 after Paros,
            1988

No.	Resonator	Q	Diameter ratio for Q = 50	Features
1	$\lambda/2$	$\pi/2r$	0.125	Can be supported at center, which is a node. Requires a high temperature junction.
2	$\lambda/4$	$\pi/4r$	0.087	Can be made an integral part of the line by chemical machining.
3	$\lambda/2$ $\lambda/4$	$\pi/2r^2$	0.421	Requires a minimum machining for given Q. Is integral with line.
4	$\lambda/4$ $\lambda/4$	$\pi/4r^2$	0.353	Can be supported on large diameter section.
5		Not readily analyzed		Flexure "tuning fork" mode. Identical calibration of batch. Shape inconvenient.
6		---	---	Undercut cylinder yields higher Q than most predecessors; manufacturable by shrink fitting.
7		---	---	Dual torsionally oscillating tines connected to mounting pad through a mechanical isolation system. Electrodes deposited on piezoelectric (quartz) resonator.

Tehon not only improved the Q substantially, by decoupling, but also
introduced to T probe design the use of *interior* surface wave (Rayleigh
wave) propagation as a means of permitting exterior mechanical support
without affecting the propagation of these interior surface waves. For this
isolation method to succeed the tube's wall thickness must be greater
than the depth of penetration, preferably several wavelengths thick. (See

also, Rosenberg and Coldren, 1977). Note, however, that Rayleigh waves can transfer from one side of a plate to the other as they propagate along a plate (Viktorov, 1967, p. 94; Bertoni, 1988, priv. comm.)

A miniaturized quartz tuning fork thermometer with *local* transduction is made in Switzerland (ETA Industries; see Hauden, 1987) and is available commercially under the name Thermopak. Its sensitivity, 34 ppm/°C, is about the same as HP's quartz thermometer. Its resonant frequency, ~262 kHz, is two orders of magnitude lower than in the 28-MHz HP sensor.

Resonant ultrasonic sensors apparently have not yet been incorporated into one multizone or multipoint waveguide to measure temperature profile. It might be possible to do so, however, using narrowband reflectance and/or transmission techniques analogous to those used in fiberoptics. For example, in-fiber Bragg-grating tandem temperature sensors have been formed in the core of a germanosilicate fiber, and the temperature is read out using a multiwavelength or scanned source (Meltz et al., 1988).

The characteristics and examples of a number of resonant sensors used for temperature measurement in the 1972 to 1988 period are listed in Table 5-4. These may be compared with resonators designed to measure parameters other than temperature, as well as temperature, shown in or referred to in a 1988 review by EerNisse et al.

5.2.2.3 Surface Acoustic Wave (SAW) Resonators

As reviewed in Hauden's 1987 paper, miniatured quartz SAW (surface acoustic wave) T sensors had been built with LST and JCL cuts by 1982. The SAW resonators exhibit faster thermal response than the bulk wave resonator used in the HP probe.

A SAW experimental gas flow rate sensor was investigated by Ahmad, 1985, using the principle of heat transfer as a function of mass flow rate. (This is analogous to hot wire anemometers.)

YZ-cut $LiNbO_3$ is used as a substrate for the SAW delay line. In addition, a thin film heater is fabricated adjacent to the SAW delay line. The substrate of the SAW delay line is heated at a constant power by this heater. The flow of gas carries away heat, thus lowering the temperature of the substrate and resulting in a corresponding change in the frequency of the SAW delay line oscillator. The sensitivity of the SAW oscillator frequency to flow rate at various pressures and temperatures of the substrate was determined in experiments reported by Ahmad, 1985, of Rexnord.

Examples of the experimental set-up and test data appear in Fig. 5-18. See also, Ahmad, 1987, Joshi, 1988, and Chapter 4, Section 4.5.5.

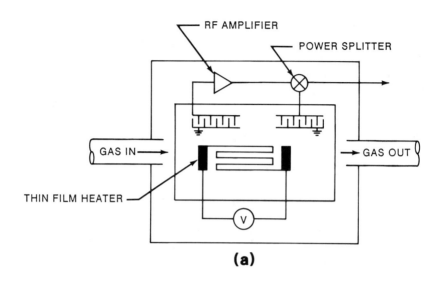

(a)

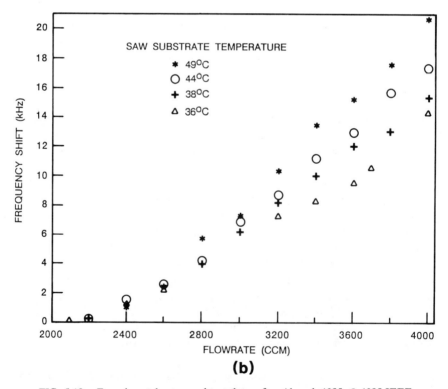

(b)

FIG. 5-18.   Experimental setup and test data, after Ahmad, 1985. © 1985 IEEE.

5.2.3 Impulse-Induced Resonance Sensors

If the isolation section is eliminated in some of the remote resonator designs of Section 5.2.2.2, one is left with a quarter- or half-wave resonator at the end of the lead-in line. If that resonator is excited with a broadband impulse, it will resonate, somewhat like a piano tuner's tuning fork when struck, or like a structure under resonance testing by the "bang" method (Kaufman, 1987).

Actually, the impulse induced resonator T sensor evolved as a limiting case of a single-zone nonresonant sensor. When Fowler was presented

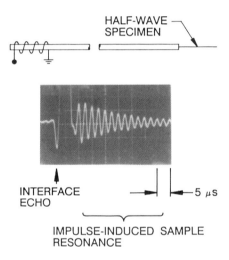

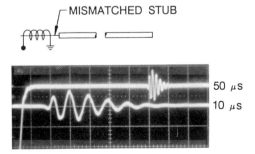

FIG. 5-19. Top: Impulse-induced half-wave resonance in a silicon-carbide whisker *specimen*, after Fowler, 1971. Bottom, resonant *transducer*, after Lynnworth, 1987.

in the late 1960s with the problem of measuring elastic moduli in very short, small-diameter specimens of SiC or other refractory fibers, he found that the hardware and techniques available at that time were not able to generate separated echoes. Turning this problem into a solution, Fowler, 1971, developed designs for sensing elastic properties or temperature in a remote sensor of acoustic impedance substantially different from that of the lead-in and excited into resonance with a broadband pulse.

In a related form, impulse-induced resonance can be used to convert a broadband electrical pulse to a quasi tone burst in a mismatched stub transducer. This transducer case can be viewed as the magnetostrictive analog of a narrowband piezoelectric transducer imperfectly matched to an adjacent buffer rod.

Examples of resonances induced in sensors or stub transducers are illustrated in Fig. 5-19.

6. Density Applications

As explained in Chapter 2, Section 2.3, the density (ρ) of solid, liquid or gaseous media can be measured or computed from measurements of ultrasonic propagation in the medium itself, in a foreign sensor immersed in the (fluid) medium, or in an adjacent member in acoustic contact with a boundary of the medium. Density can influence the sound speed c, the reflection coefficient R, the attenuation coefficient α, or the jitter in the phase or amplitude of the received signal after transmission across a two-phase medium. If ρ is to be sensed by a float or in a buoyant liquid system, the transit time can be related to the float's position, x, or to the interface location, H, from which ρ is derived. One rather unusual "float" method employs *acoustic levitation* to measure ρ remotely in liquids and low density solids (Trinh and Hsu, 1986).

6.1 Medium as Its Own Sensor

In some pipeline interface detectors the interface between two different hydrocarbon liquids is detected based on the difference in their sound speeds Δc, even when their ρ difference is negligibly small (Zacharias, 1972). However, in one given well-defined liquid, a correlation often exists between ρ and c, since both usually decrease as temperature T increases.

6.1.1 Sound Speed in Well-Defined Liquids

In a pipe, even though the medium can act as its own ρ sensor, a foreign probe, spoolpiece or cell often is required in order to measure c accurately. The controlled geometry provides a known path. (In principle the pipe ID could be determined by subtracting from the calipered OD the wall thickness measured ultrasonically at opposite points.) If the liquid is indeed well-defined, it would seem easier to obtain ρ from a T measurement. Thus it would appear that the motivation to pursue an *ultrasonic* measurement of ρ based on c in the medium itself would be lacking unless: (1) c is measured without any additional ports, for example, from the sum of the transit times in a contrapropagation flowmeter; (2) c is measured without any penetration of the pipe or container; (3) c correlates with ρ, even though T does *not* correlate with ρ, as is the case in some two-phase fluids.

Examples of ρ-c correlations are given in Figs. 6-1 to 6-3. As a numerical example, for water between 0 and 52°C, $\rho = 28.142 - 5.657 \times 10^{-2}c +$

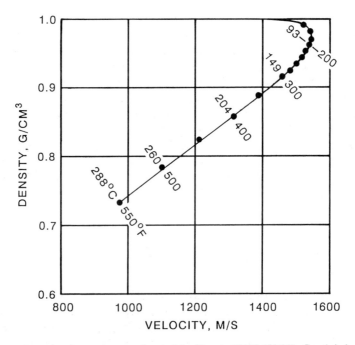

Fig. 6-1.   Water density versus sound speed for T up to 288°C (550°F). Graph is based on c data in McDade et al., 1959, and handbook data for ρ (e.g., Crane, 1980).

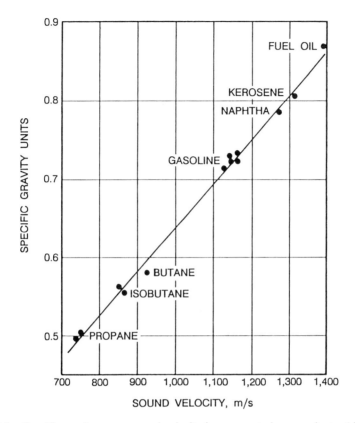

FIG. 6-2.   Specific gravity versus sound velocity in some petroleum products. After Zacharias and Ord, 1981. © 1981 Oil & Gas J.

$3.931 \times 10^{-5}c^2 - 9.105 \times 10^{-9}c^3$ g/cm^3 with c in m/s (Nguyen, 1985, priv. comm.). Between 93 and 288°C, the ρ-c relation for water may be approximated by the following polynomial, due to Nguyen (1988, priv. comm.): $\rho = 4.9954 \times 10^{-15}c^5 - 3.033 \times 10^{-11}c^4 + 7.354 \times 10^{-8}c^3 - 8.895 \times 10^{-5}c^2 + 5.4036 \times 10^{-2}c - 12.5367$ g/cm^3 with c in m/s. As is seen from Fig. 6-1, the ρ-c relation for water is double valued. This stems from water's peculiar c-T relation, which also is double valued. See Chapter 3, Fig. 3-80 and Table 3-14. An example of a ρ-c application in the brewing industry is provided by Fig. 6-4. A study aimed at determining the specific gravity of lead-acid battery electrolyte, due to Swoboda et al., 1983, reported advantages such as: continuous sample and display, instantaneous tracking of changes, and averaging.

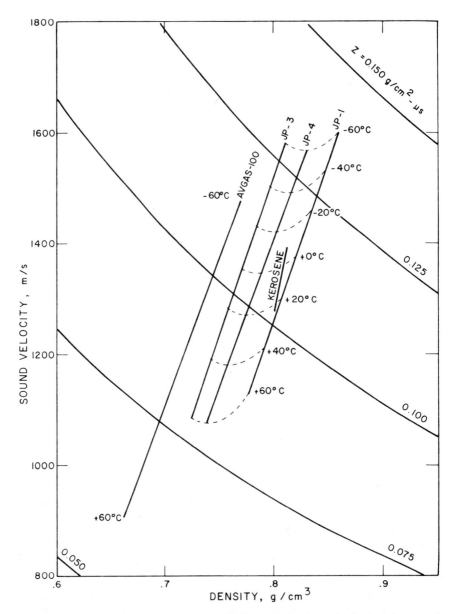

FIG. 6-3.   Graph for estimating sensitivity of indirect density determination based on sound speed and temperature data. After Lynnworth et al., 1973c.

FIG. 6-4.   Investigations on beer have shown a correlation between original gravity and sound velocity. Paar's SPR 4115 concentration analyzer measures sound velocity in main flow. From this, mPDS 4000 system calculates original gravity of beer. Site of installation is after the beer filter. System output can be used to control original gravity. Measuring range: 0 to 20% Plato. Temperature range: -2 to $+10°$C. Accuracy: $\pm 0.1\%$ Plato. Benefit: Continuous monitoring of beer quality, exact determination of filter first and last runnings, exact separation, reduced extract losses. Source: Paar.

Density Profile

In outdoor (including aircraft) fuel tanks in which the fuel is of uniform composition but not isothermal, the density varies spatially. The mass of fuel remaining in the tank, M_f, is then not immediately calculable from liquid level H alone. If the fuel is an unknown mixture, then c does not bear a known relation to ρ. However, to the extent that the fuel type can be identified from c and another easily-measured parameter such as temperature, the ρ profile may be deduced from the c profile. Figure 6-3 shows how ρ and c vary with temperature for several aircraft fuels. To determine M_f one could first measure c and T in a reference region to determine the fuel type (or fuel mixture). Knowing the fuel type implies that one also knows the T-dependence of ρ and c, and hence the ρ versus c relationship. In this case, a multireflector method (see Section 6.2.2.2) together with a T sensor yields both H and the c profile, and c at a known T. From this, it appears that the ρ profile and M_f could be derived, at least for simple tank geometries and settled fuel. In the particular case of fuel gaging in an airplane wing tank, where it is desirable to minimize the number of lead wires, it may be advantageous to use only one pair of wires, and use low-pass filtering for the temperature sensor signal and high-pass filtering for the ultrasonic transducer. If the liquid is unsettled because of in-flight maneuvers other than routine cruising, suitable corrections must be applied as functions of roll, pitch, yaw and accelerations, and H would probably be measured at more than one location in each fuel tank.

6.1.2   Attenuation in Sludge and in Drilling Mud

Suppose dense insoluble particulates of dimensions small compared to the wavelength λ and of density $\rho_2 > 2$ g/cm^3 and of a given size and shape distribution are added to an initially-pure liquid of $\rho_1 \lesssim 1$ g/cm^3. It seems reasonable to expect that the denser the mixture, the higher the attenuation, at least up to some moderate to high concentration. If this intuitively predicted relation is reproducible, it can serve as the basis of an ultrasonic densitometer. Limitations on such a correlation stem from particle size and shape distributions. For example, for a given mixture density, depending on the scattering regime (which depends on particle size compared to λ) it is not immediately obvious whether the attenuation coefficient α ought to increase or decrease as particle size decreases. In other words, α is not simply a single-valued function of ρ.

Another limitation stems from viscosity. If the carrier liquid's viscosity is strongly T-dependent, α can be expected to respond strongly to T, possibly more so than to ρ. However, despite these limitations, practical applications have been reported in sewage, sludge and drilling muds.

6.1.3   Consistency of Pulp Suspension in a Paper Mill

As stated by Härkönen et al., 1983:

> The fiber content of the cellulose-water suspension is an important industrial factor because it partly determines the thickness of the paper manufactured from the suspension. The amount of the material in the paper pulp suspension is measured in percentage and is called the consistency. Traditionally the consistency value is measured by mechanical meters using propellers, vanes and rotating cylinders. Optical consistency sensors have become [commercially available] recently and they have many advantages over the older mechanical systems.
>
> Ultrasonics is seldom studied as a means of characterizing the pulp suspension and no commercial instrument is available. However, if the ultrasonic consistency meter is combined with the ultrasonic flowmeter, the actual pulp flow rate can be obtained with a single instrument.
>
> The suspension particles attenuate ultrasonic intensity by absorption and scattering. A suspension of a certain consistency attenuates the ultrasonic beam [depending on] its particle size. Fibers have a tendency to agglomerate

into what are called flocks. Flocks are also of interest in the paper industry because they produce inhomogeneous paper causing so-called formation. Thus, paper pulp suspension is characterized both by consistency and flock size as well as by the content of the filling material. One can measure the direct beam attenuation as well as the sidewards scattering at different frequencies. The objective is to determine both consistency and flock size.

Among the experiments reported by Härkönen et al., 1983, consider one conducted in a flow test facility with a pipe of 250-mm diameter having wetted ultrasonic transducers at the inner surface of the pipe. Figure 6-5 shows the direct intensity measured at 4 MHz as a function of the consistency of pine pulp for flow velocities of 0.4, 0.6 and 0.8 m/s.

6.1.4 Coal Particulates in Air

The increase in attenuation of ultrasound at $f = 459$ kHz due to coal particulate matter in air was studied by Hamade, 1982. Hamade found that the greater the mass loading, the greater the attenuation. But the results also depended on particulate size, shape and other factors. This

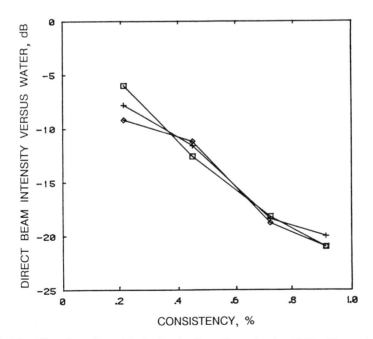

FIG. 6-5. Direct beam intensities in the pipe flow. Pipe pulp, f = 4 MHz. Flow velocities: ◇ 0.4 m/s. + 0.6 m/s. □ 0.8 m/s. After Härkönen et al., 1983. © 1983 IEEE.

makes it difficult to express ρ as a function of α. Empirical solutions to this problem nevertheless may be obtainable (Leffert and Weisman, 1988; see comments supplied by Leffert, 1987, in Chapter 4, Section 4.5.7).

6.2   Foreign Sensors

In this section we shall first consider applications of *cylindrical tube* ρ *sensors*, some of which have been commercially available for many years. (Examples were included in Chapter 2, Figs. 2-35 to 2-37.) Improvements in size, temperature-immunity or other characteristics still appear from time to time. Next, we shall consider the torsional mode ρ sensor which has been investigated over the past ten years in R&D programs and which may soon become a standard product. Lastly, we shall consider two experimental configurations based on floats or buoyancy, wherein the position of a sensor or interface yields ρ.

6.2.1   Cylindrical Tube Sensor

In the ITT-Barton Model 668 densitometer (Fig. 6-6) a vane lies in the diametral plane of a tube much like the "hyphen" within the letter θ. In this densitometer the resonant frequency f is converted to ρ from a quad-

FIG. 6-6.   Model 668 liquid densitometer from ITT-Barton.

ratic relationship (Eq. 2-4). Linearity is about 1 mg/cm^3, and repeatability about 0.3 mg/cm^3. The probe can be inserted directly into a pipeline. The manufacturer cautions, however, that this densitometer should not be used in pipes less than 4-inch diameter, since the minimum intrusion is about 56 mm. Liquid temperature range is -35 to $+95°C$, and viscosity, up to 20 cp.

If the fluid can be *bypassed* to an external U-tube, ρ can be measured with the greater precision of equipment such as Paar's Model mPDS4000 (Fig. 6-7). Precision of 10 μg/cm^3 and accuracy of 100 μg/cm^3 are attainable, depending on span. An example of a gas densitometer manufactured by Sarasota Automation is given in Fig. 6-8. ITT-Barton, Sarasota Automation and Solartron/Schlumberger supply densitometers for either liquid or gas applications. Spool orientation is critical in certain models. With respect to natural gas applications, Jaeschke and Hinze, 1987, reported that a generic correction derived from sound speed c could reduce ρ errors from ~0.6% down to ~0.1%.

If a resonant spool is too large for a potential application, a resonant *tuning fork* may be considered (EerNisse, 1988). Another alternative for

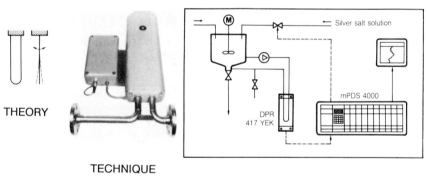

THEORY

TECHNIQUE

APPLICATION

FIG. 6-7.   (Left): The U-shaped tube is caused to oscillate in the Y-mode by an EV 512 excitation amplifier which is connected by a two-wire line to module TWR in the DPR 2000 electronic evaluation unit. The oscillation is associated with a moment of flexure which must be absorbed by a counter-weight. This is realized by a built-in seismic mass which makes additional mounting precautions unnecessary. The thermal inertia is small, and the oscillator is very insensitive to pressure. The pressure drop in the cell corresponds to the pressure drop of a tube of 500 mm length and 6.6 mm internal diameter. (Middle): The DPR 417 Y oscillator is built in a splash-proof, plastic-coated aluminum housing. (Right): Film manufacturing plant uses mPDS4000 system to measure concentration of silver in solution, to reduce expense and avoid waste. Measuring range: 1.0 to 1.4 g/cm^3. Temperature range: 20 to 40°C. Accuracy: ± 0.0001 g/cm^3. Pressure: 4 bars. Output from system is used to control rate of silver addition, to keep concentration at optimum level. Benefit: Silver concentration of high constancy, no waste, no need for spot checks because of continuous measurement. Source: Paar.

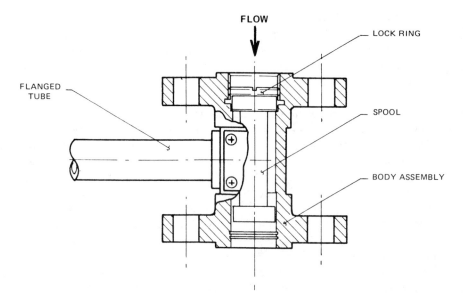

FIG. 6-8.   FD71 gas densitometer from Sarasota Automation.

applications in small pipes or in confined quarters is to use a torsional mode waveguide having cross sectional dimensions on the order of mm, not cm.

6.2.2   Torsional Mode Waveguide of Noncircular Cross Section

The torsional density sensor represented in Chapter 2, Fig. 2-39 originated in 1977 following a request to the author by A.E. Arave of EG&G-Idaho for an improved transducer to measure the void fraction in two-phase water/steam mixtures. Arave, 1970, had already developed novel leaky waveguide (impedometer) probes but their impedance-related response to ρ was expected to be somewhat blurred because of the influence of temperature and sound speed in the adjacent fluid, and also because of the difficulty of measuring attenuation accurately in the harsh conditions of this application. The idea of using the speed of torsional waves in a flattened waveguide to sense ρ probably was triggered in part by the author's earlier observation that the speed of extensional waves in a threaded steel rod of ~3-mm diameter was reduced when dipped into water (Section 3.1.13). This reduction is much more noticeable in a threaded rod than in a smooth rod. In a smooth rod the hydrodynamic mass loading effect for extensional waves is barely discernible. In any event, pulse-echo experiments with rectangular cross sections, conducted in February 1977 and reported later that year (Lynnworth, 1977) demonstrated rather easily that, for liquids of low viscosity, the fractional decrease in torsional velocity was approximately proportional to the density ρ of the surrounding liquid, inversely proportional to the waveguide density ρ_s, and dependent on a shape factor K. The following empirical expression was derived on the basis of data obtained with waveguides of rectangular cross section:

$$\Delta c/c \approx (\rho/2\rho_s)[1 - (1/K)]. \tag{6-1}$$

As a numerical example, if a stainless steel probe ($\rho_s = 7.8$) having a 3:1 aspect ratio (K $\approx$ 0.5) is dipped in water ($\rho = 1$) then $\Delta c/c \approx -0.06$.

It was evident from the experiments conducted in early 1977 that the increase in torsional transit time depended on the depth of immersion, and for a viscous liquid like cold glycerine, on the viscosity too. The small torsional echo generated in the waveguide at the liquid/air interface was noted but not utilized until some ten years later. (See, however, Smith and Junger, 1961, whose patent, filed in 1956, implies that the interface echo generated in a waveguide was well known over thirty years ago.)

With respect to the sensor's shape, it was recognized in a 1978 patent application (Lynnworth, 1980, filed Feb. 1978) and in studies at EG&G-Idaho by Arave et al., 1978, and Arave, 1979, that noncircular shapes other than rectangular, e.g., hydrodynamically contoured shapes, would be necessary in high-velocity flowing fluids, especially two-phase liquid/vapor fluids. The first rigorous optimization of the shape with respect to sensitivity awaited the contributions of Bau, 1986, and his graduate student at the University of Pennsylvania, J.O. Kim (Kim and Bau, 1986–1989). For example, they found that for a given aspect ratio, the diamond cross section is about three times more sensitive than the rectangular cross section: $S_\diamond \approx 3S_\square$.

The torsional waveguide ρ sensor was by no means a standard commercial probe in 1987, ten years after its invention. Nevertheless, it had already been configured in a variety of shapes to suit R&D experiments and to suit instrumentation requirements in *void fraction* tests up to 350°C (Arave, 1979; Arave et al., 1978); in *liquid level* tests of propellants at 1 g and at low g (Jacobson et al., 1987a) and in water up to 250°C and 15 MPa (Miller et al., 1980); and in two-phase *mass flowmeter* tests in a confined space (Lynnworth and Nguyen, 1986; illustration in Section 9.10). Experimental models were also built for a *fuel gaging* feasibility study wherein ρ profiling probes containing two and six zones were flight tested in 1987.

To determine whether a fluid at high temperature ($T \sim 500°C$) and high pressure ($P \sim 300$ bar, or ~ 30 MPa) is in its low or high density phase, a spiraled or straight torsional sensor such as that proposed in Fig. 6-9 may be considered, provided ρ changes sufficiently when the phase changes from low to high density (e.g., 10% change). In the experimental project for which this design was proposed, it was desirable that the density (or phase) sensor be small (maximum axial length <2 cm), minimally obstructing the viewing area in the limited-volume cylindrical cavity (to avoid compromising optical observations); that no electrical feedthroughs pass through the test fluid; that the sensor not prevent use of a small magnetic stirrer; and that the high T and P be sealed reliably. (Compare with float approach, Section 6.2.3.)

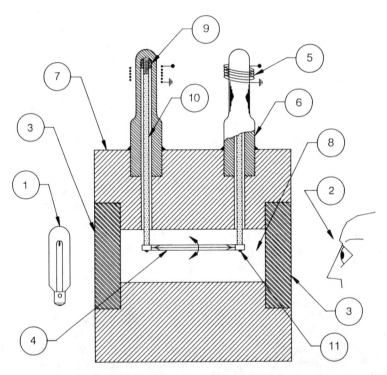

FIG. 6-9. Schematic of torsional ρ sensor proposed for a high temperature, high pressure experiment in which the phase of the test fluid is to be determined, based on the ρ difference for each phase. Legend: 1–Light source. 2–Observer. 3–Viewing port. 4–Torsional density sensor, 1.5 to 3 cm long, diamond cross section except near ends (spiral version not shown). 5–High temperature coil outside all-metal pressure boundary. 6–Metal housing (seal detail omitted). 7–Pressure vessel. 8–Fluid under test. 9–Threaded support welded to housing. 10–Magnetostrictive waveguide, one end threaded. 11–Welded or brazed joint.

6.2.2.1   Void Fraction Detector

Illustrations summarizing the sensing system, the planned sensor location in a reactor core inlet, sensor construction and test results for this application appear in Figs. 6-10 to 6-16. This application was the first extensive field test of the torsional waveguide densitometer. As explained by Arave et al., 1978, a void fraction detector or densitometer for measuring ρ in a loss-of-fluid test (LOFT) in a nuclear reactor core inlet application must be able to

function and survive under severe environmental conditions and certain spatial limitations. Some of the main requirements are presented in Table 6-1. Reactor operating conditions of 343°C and 15.5 MPa, a substantial radiation environment, and the standard LOFT chemistry combine to impose a very high-stress, corrosive environment for the subject instrument. Standard LOFT chemistry trace elements in the hot water include boron (100

Table 6-1.   Detector Design Requirements For the LOFT Core-Inlet Application. After Arave et al., 1978

Design Life			
Reactor	UDD	Temperature (°C)	Time (hr)
Nonoperating	Nonoperating	21	10,000
Nonoperating	Operating	343	2,500
Full power nuclear operation	Operating	343	500 to 1,000

Pressure Environment		
Reactor	Pressure (MPa)	Temperature (°C)
Normal operation	15.5	343
Design maximum (hot)	17.2	343
Design maximum (cold)	21.5	21

Nuclear Irradiation Environment	
Reactor	Neutron Exposure
Normal operation	7.0×10^3 n/cm^2 − s
Total exposure (maximum)	1.3×10^{20} n/cm^2 (acceptable)
	2.6×10^{20} n/cm^2 (design goal)
Reactor	Gamma Exposure
Normal operation	7×10^8 R/hr
Total exposure	3.5×10^{11} R (acceptable)
	7.0×10^{11} R (design goal)

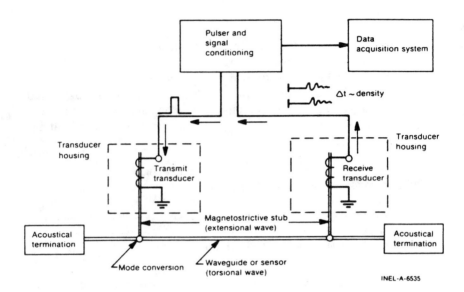

FIG. 6-10.   Ultrasonic density detector system components. After Arave et al., 1978. Illustration courtesy Idaho National Engineering Laboratory (INEL).

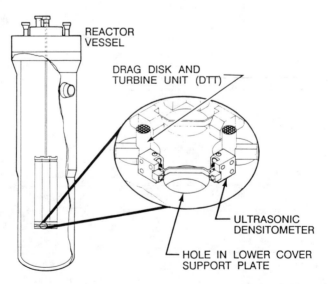

FIG. 6-11.   Planned ultrasonic density detector mounting location in LOFT core inlet. After Arave et al., 1978. Courtesy INEL.

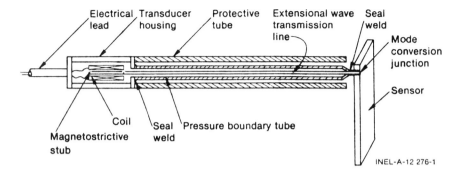

FIG. 6-12.   Method of maintaining the pressure boundary using SS304 as the only material in contact with the water. After Arave et al., 1978. Courtesy INEL.

to 3000 ppm), lithium (0.2 to 2.2 ppm) and others to form a highly corrosive environment. [See Table II in Arave et al., 1978.]

Fluid loadings and variations in temperature, pressure and flow velocity further increase the severity of the environmental conditions. Figure 6-11 shows the planned installation of each detector at a 7-cm diameter hole in the lower support plate. At this location the maximum steady state operational velocity of the coolant is calculated to be about 9.14 m/s with max-

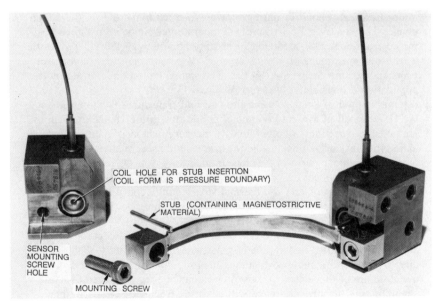

FIG. 6-13.   Ultrasonic density detector second prototype (side view). After Arave et al., 1978. Courtesy INEL.

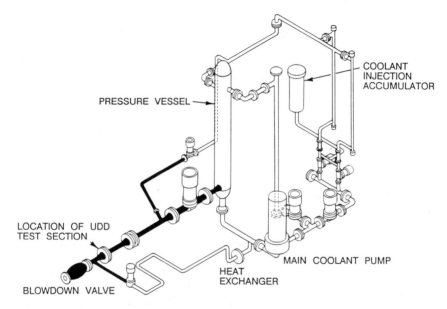

COOLANT
INJECTION
ACCUMULATOR

PRESSURE VESSEL

LOCATION OF UDD
TEST SECTION

BLOWDOWN VALVE

HEAT
EXCHANGER

MAIN COOLANT PUMP

FIG. 6-14.   Blowdown flow path and test section location. UDD is the ultrasonic density
detector. After Arave et al., 1978. Courtesy INEL.

imum transient velocities during blowdown of 68.6 m/s during the steam
phase. The steady state single-phase operational momentum flux is estimated from computer modeling to be approximately 29,130 kg/m·s^2, with
a transient maximum of 44,646 kg/m·s^2. Maximum survival temperature
transients for the design are 649°C/s. The spatial limitation imposed by the
application are visible in the inset of Fig. 6-11.

[One method of encapsulating the coil and transducer is shown in Fig.
6-12.] When all of the environmental conditions in the LOFT application
are taken into account, all other density measurement instruments are eliminated [because of] not being able to survive in the severe environment and/
or not being able to perform the measurement task. The detector is designed
with all stainless steel (304) interfaces, thus providing a workable solution
to the corrosion problem, the thermal stresses inherent in the application,
high survival temperature (649°C) and high pressure (15.5 MPa). A fully
assembled prototype unit is shown in Fig. 6-13.

Referring to the assembly in Fig. 6-13, it appears that, except for the stub
and the hard leads, the component most vulnerable to catastrophic failure
due to severe flow-induced loadings and/or vibrations is the sensor. To
avoid failure of the sensor because of resonance effects induced by vortex
shedding, the sensor is designed with a vortex frequency of approximately
778 Hz while the lowest natural frequency possible for the sensor illustrated
is approximately 2200 Hz. This sensor natural frequency is sufficiently

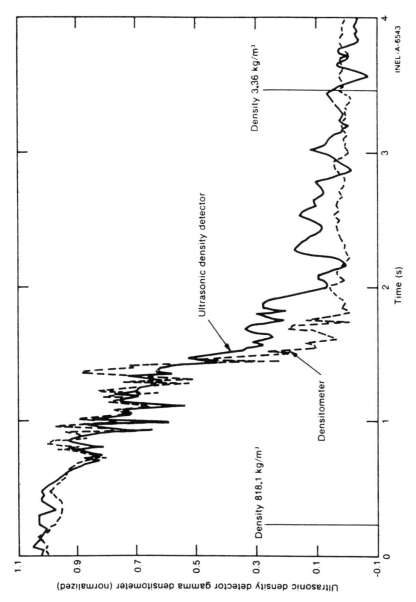

FIG. 6-15.   Example of ultrasonic density detector and gamma densitometer output versus
time for the 5F blowdown. After Arave et al., 1978. Courtesy INEL.

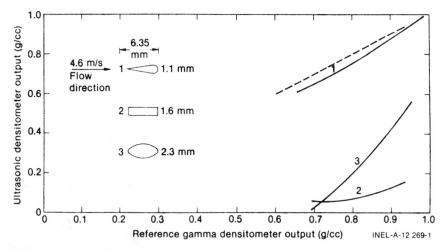

FIG. 6-16.   Sensor shape versus bubbly flow performance. After Arave, 1979. Courtesy INEL.

greater than the typical Strouhal vortex frequency to give high confidence that the sensor will withstand resonance effects due to fluid motion (R. Greif, priv. comm., 1977). To check the possibility of fatigue failure of the sensor due to vortex shedding and other forces induced during operation, a stress calculation was performed. A margin of safety for fatigue from fifteen to twenty was indicated by this calculation.

Test Results.   The main objectives of steam-water transient (blowdown) tests, according to Arave et al., 1978,

were to obtain steam-water transient data to provide a qualitative comparison between the ultrasonic density detector and a gamma densitometer, and to determine whether the instrument could survive the rigors of pressure and temperature and a water chemistry similar to the conditions in the [intended] application.

The blowdown test facility apparatus for simulating a pressurized water reactor transient environment is sketched in Fig. 6-14. It is equipped with a quick-opening blowdown valve capable of opening within 20 ms and operates with several selectable break orifice sizes.

The gamma densitometer was oriented to measure density along a chordal path across a diameter closely approximating the portion of the pipe traversed by the density sensor. However, the angled portions on each end of the sensor near the stubs constitute a significant deviation from a pipe diameter.

An example of one of the typical blowdown test results is given in Fig. 6-15. Differences between the torsional density sensor and the gamma densitometer were attributed by Arave et al. to differences in paths and to acoustic noise carried down the pipe and coupled into the transducer.

Despite these problems, temperature sensitivity (largely avoidable up to 350°C through use of Incoloy 902 or 903 instead of SS304, or through temperature compensation) and possible slight sensitivity to pressure, it was found that the

> performance of the ultrasonic system in the transient (blowdown) steam-water environment was quite satisfactory; besides operating in the blow-down environment the unit yielded data in qualitative agreement with the gamma densitometers used as reference instruments.
>
> [The investigators stated that the ultrasonic system] is capable of sur-viving a variety of harsh environmental conditions and variations, including thermal shock, two-phase blowdown conditions with corrosive water chem-istry, and pressure variations.
>
> The purpose of the tests reported was not to determine the accuracy of the ultrasonic system. However, based on the test data reported, the ul-trasonic system was expected to provide density measurements in the [LOFT] application to accuracies within 10% of range.

In a subsequent report (Arave, 1979), two-phase test results were pre-sented showing the advantage of a wedge-shaped hydrofoil with the angle of flow attack parallel to the sensor thin axis (Fig. 6-16). On the basis of these and other tests, it was concluded that the torsional sensor could measure the average ρ of the two-phase media surrounding the sensor in a 343°C pressurized water reactor environment.

6.2.2.2   Two-Phase Mass Flowmeter

The idea of combining the torsional ρ sensor with an ultrasonic V sensor can be realized in different configurations. An axially-oriented ρ sensor, for example, was combined with an offset-style axial path flow veloci-meter in Lynnworth, 1980.

In 1986, Lynnworth et al. investigated several configurations whereby ρ and V sensors could be combined in in-line cylindrically-enveloped packages of restricted diameter. In one of these designs (Chapter 9, Sec-tion 9.10), the stainless steel torsional sensor utilized a diamond-like cross section suggested by Bau and Kim, with an effective length of about 250 mm. In this design, V is measured at a longitudinal wave frequency of 1 MHz, and ρ is measured with 100-kHz torsional waves.

Figure 6-17 illustrates two experimental mass flowmeters wherein both V and ρ are measured near 100 kHz. In the V part (top flowcell), exten-sional waves in the axial waveguide are mode converted at normal in-cidence to longitudinal waves in the fluid. In the ρ part, extensional waves

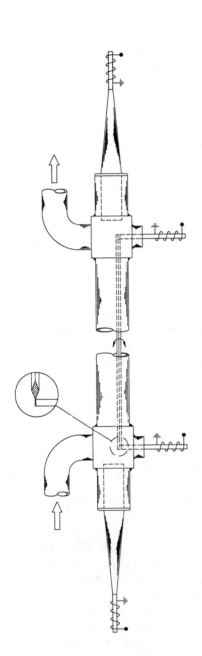

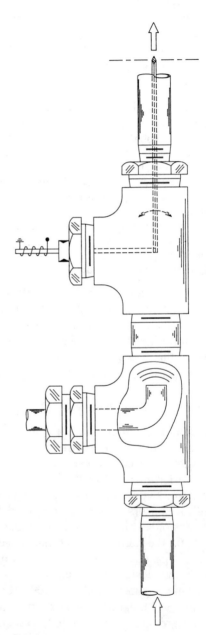

FIG. 6-17. Experimental mass flowmeters using ~100-kHz magnetostrictive transducers. Extensional waves are mode converted to torsion for the density measurement and are also mode converted to longitudinal waves for the flow velocity measurement. Top: offset style. Bottom: in-line style.

are mode converted to torsion at the orthogonal joint. The 100-kHz waves may be generated and detected magnetostrictively. Magnetostriction in Remendur is effective down to $-196°C$ (Lynnworth, 1982b) and probably much lower, and up to its Curie point, about 960°C. Therefore, a mass flowmeter using magnetostrictive transducers would appear to be usable over an unusually large temperature range. The lower part of Fig. 6-17 illustrates a proposed in-line version of a mass flowcell. It is one of several configurations being investigated in the author's laboratory in a NASA Lewis SBIR 1989 feasibility study on cryogenic instrumentation.

6.2.2.3 Fuel Gaging

According to a 1986 report (ARINC, 1986), there is an industry-wide, world-wide need for a significant improvement in the accuracy and reliability of fuel gaging in aircraft tanks. With the present (nonultrasonic) fuel gaging equipment, errors at the 3σ level as high as 7% can occur due to the combination of fuel density stratification and variability in fuels from different sources. Even errors in the 2 to 5% range have serious economic consequences because of the large amount of unnecessary fuel that must be loaded, to compensate for uncertainty in the density and mass of fuel remaining in the tanks.

As stated in this 1986 report, apart from the cost of departure delays due to discrepancies, there is a large hidden cost to the airlines caused by flying with more fuel in the tanks than is indicated on the gages. In practical terms, there is a tendency to avoid underfueling aircraft. Consider the London–New York route as an example. On a Boeing 747 an extra 1000 kg over New York will have required 1350 kg to be loaded at London, i.e., 26% of the excess fuel is wasted carrying the extra weight. On a Concorde the situation is even worse. Here, an extra 1000 kg at New York requires 1740 kg at London. In this case, 43% of the excess fuel is wasted. For the four major U.S. airlines, it has been estimated that the annual savings (in 1986 dollars) could be ~$20 million from a reduction in flight reserves of 1% of full scale.

An opportunity arose in 1987 to test prototype two-zone and six-zone torsional mode density profilers in airplane wing tanks (Fig. 6-18). The design that was available for these tests consisted of a stack of sensors that in side view resemble one or more ∃s. The horizontal parts are extensional mode segments. The vertical parts are torsion. In principle, a stack of ∃s ought to be able to sense the ρ profile provided the stack is fully immersed. When the fuel level is between two horizontal segments, however, the small echo generated at the liquid/vapor interface can interfere with the sought time interval measurements in vertical and horizontal segments.

FIG. 6-18. Six-zone torsional density profile probe installed in MD87 wing tank near fuselage. Illustration courtesy Douglas Aircraft.

For certain sensor proportions, it appears that one would be able to torsionally interrogate from above the interface in such a way as to separate the liquid level H from the local liquid density. (See, for example, Chapter 2, Section 2.4.3, or Chapter 3, Section 3.6.2.) On the other hand, a better solution may consist of orienting the ρ sensors in horizontal planes. In Fig. 6-19 the ρ sensors consist of torsional mode washer-like rings having a cross section whose shape and aspect ratio (diamond, b/d≈3) are reasonably responsive to ρ. As seen in the oscillogram inset in Fig. 6-19, $\Delta t = 3\mu s$ for the SS sensor in air and then in water.

6.2.3 Float-Type High Temperature Hydrometer

In studies of supercritical hydrocarbon fluids such as toluene at 400°C and 5 kpsi, it is difficult to discern the forming and disappearing of a discrete liquid/vapor or liquid/liquid interface. One approach to this level-sensing problem, due to Gall, priv. comm., 1979, utilized a float made of graphite as a liquid level indicator within an all-metal-sealed chamber, Fig. 6-20. An ultrasonic waveguide, at least partly-magnetostrictive, is attached to the float. The position of the waveguide inside the 3-mm OD × 0.5-mm wall SS tubing is determined from extensional wave transit times measured in the cool (ambient temperature) portion of the wave-guide. The measuring coils are outside the SS tube pressure boundary. The equipment shown in Fig. 6-20 was tested on water/oil mixtures at room temperature and was reported to have worked satisfactorily. In tests on hydrocarbons at elevated temperature and pressure, interfaces which *formed instantaneously* could be detected. But, disappearing interfaces or gradually-forming interfaces were not detectable within the time allotted to the experiments. This application demonstrates the conversion of ρ to sensor position, and also illustrates the magnetostrictive trans-duction across a metal-sealed pressure boundary, to sense the position of a component at high temperature and at high pressure. (Compare with Wiedemann linear displacement transducer, Section 7.9.2; which is also used as a float indicator of interfaces between media having different densities.) [If the density change accompanying a phase change is substantial, then a ρ sensor (in contrast to a float sensor) may be the most direct approach to monitoring the phase distribution. As an example, the transit time and amplitude of echoes returning from a torsional ρ sensor surrounded by H_2O depend on whether the H_2O is in the form of ice, water or vapor. Similar remarks apply to a vapor which may agglomerate into high-density droplets.]

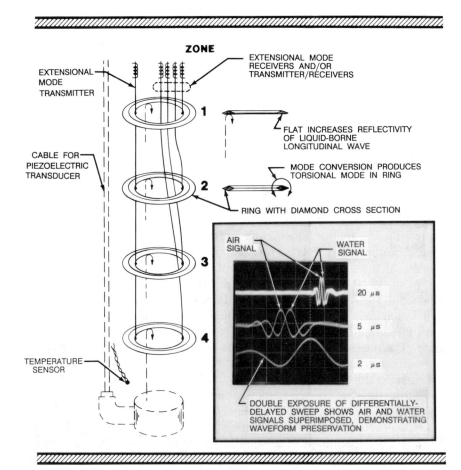

FIG. 6-19. Diamond rings engage density in middle of four zones, and, when immersed, also provide reference echoes for the longitudinal wave transducer which would sense liquid level from the bottom. Rings may be partly flat-faced to enhance their reflectivity, in exchange for some decrease in ρ sensitivity. Alternatively, straight vertical torsional waveguides (not shown) could sense H based on echo at interface. Inset: Oscillogram for a SS316 ring of diamond cross section, tested in air and then in water, shows $\Delta t = 3\mu s$. The bottom trace is a double exposure in which the air and water waveforms are superimposed to show the extent to which the nominally 100-kHz waveform is preserved upon immersion in a liquid of low viscosity. Ring nominal dimensions: 33.4 mm OD × 24.3 mm ID × 1.5 mm maximum thickness. Lead-in and lead-out waveguides were made of Remendur, 1.6-mm diameter × 300-mm long.

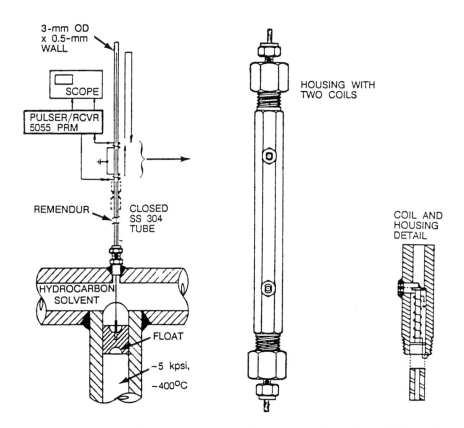

FIG. 6-20. Liquid level float. Remendur waveguide, attached to float, rises and falls within sealed SS304 tube. After Lynnworth, 1979. © 1979 IEEE.

6.2.4 Buoyancy-Type Densitometer for Immiscible Liquids

Intended for down-hole wellbore applications at high temperature and high pressure, a design due to Sweet, 1986, places a column of the test fluid above what he calls the calibration liquid, the latter being denser than and immiscible with the test fluid, and located in the equivalent of a manometer-like U-tube. Because the calibration liquid is immiscible with the test fluid, an interface between the two is established, and the vertical position of this interface, and changes in that position, are used to determine the density of the test fluid.

7.   Interface Measurement, Proximity Sensing and Gaging Applications

7.1   Types of Interfaces: [GLS]*[GLS]; Settled Versus Unsettled

In this section, we take as the type of problem to be solved, the detection and location of the extended interface between two media differing in one or more acoustic properties, such as sound speed c, attenuation coefficient α, density ρ, or terms derived therefrom. (We use the term "extended" to distinguish interfacial areas typically ten or more times the wavelength squared from "small" inhomogeneities like bubbles or particulates which are treated in Section 7.8.) The notation [GLS]*[GLS] is a shorthand representation or reminder of six interfacial combinations (Table 7-1). Specific examples include: air on water (GL); gas on gas (GG), e.g., heavy gas (SF_6) under air or light gas (H_2) over air; liquid adjacent liquid (LL), e.g., when one type of gasoline follows another down a pipeline; monitoring the solidification front during solidification of a steel ingot (SL); thickness gaging of paint, or of the cladding in a layered composite (SS).

Table 7-1.   Six Combinations of Interfaces Between Media Differing in One or More
             Acoustic Properties Such as c, α or ρ

	Gas	Liquid	Solid
Gas	GG	GL	GS
Liquid	GL	LL	SL
Solid	GS	SL	SS

The shape of liquid/gas interfaces is not always planar, especially if the liquid is agitated, frothing, unsettled due to low g as in orbital flights, or otherwise unsettled.

The term "gaging" includes measurement of a one-dimensional variable such as thickness, and also volumetric and mass quantity determinations such as fuel remaining in a tank (or ullage, the amount by which a container falls short of being full).

7.2   Approaches and Access Constraints— Transmission, Reflection, Resonance, Adiabatic Perturbation

In Chapter 2, Tables 2-4 and 2-5 categorized liquid level approaches and access constraints in several ways. To find the "best" approach to the problem, it may be helpful to compare proposed solutions that utilize access from top, side or bottom; clamp-on or wetted; one transducer or more than one; measurement of transit time or amplitude; etc.

In order to not repeat the general information already given in Chapter 2, we limit the present discussion to pointing out that in most industrial applications, the main *interface and gaging methods* have been *reflection*, *transmission*, and *resonance*.

Resonance methods include bulk wave (usually longitudinal), resonance thickness gaging, surface acoustic wave (SAW) [e.g., monitoring of etching (Joshi, 1987)] and level sensing by detecting the damping effect of a medium upon the otherwise-relatively-undamped vibrations of a probe. Resonance methods also include the use of a resonating cavity to measure the volume of a body therein (Deskins et al., 1985). In other words, the resonant frequency can be related to the dimensions (resonator thickness, for example) or to the volume of a body, and the energy decay rate, or Q, can be related to the presence or absence of an acoustic energy sink, and perhaps to its absorptive properties.

Adiabatic Perturbation

If a given volume of an ideal gas, Q_{vo}, at pressure P_o, temperature T_o, sound speed c_o and density ρ_o is compressed adiabatically to a new volume Q_v, with $\Delta Q = Q_v - Q_{vo}$, the following ratios hold, for the isentropic process:

$$T/T_o = (c/c_o)^2 = (\rho/\rho_o)^{\gamma-1}, \tag{7-1}$$

$$(Q_{vo}/Q_v)^\gamma = P/P_o = (\rho/\rho_o)^\gamma = (c/c_o)^{2\gamma/(\gamma-1)}, \tag{7-2}$$

where γ = specific heat ratio.

From these relations it is clear that

$$Q_{vo}/Q_v = c/c_o^{2/(\gamma-1)}. \tag{7-3}$$

It follows that

$$Q_{vo} = -\Delta Q/[1 + 1/(c/c_o)^{2/(\gamma-1)}]. \tag{7-4}$$

These equations suggest that if the gas is compressed adiabatically by a known decrement of magnitude ΔQ, or until c increases by a sought amount Δc, it ought to be possible to determine Q_{vo}.

The calculated results of adiabatically compressing an ideal gas of γ = 1.4 for several compression ratios are listed in Table 7-2. Example: Let $Q_{vo} = 1000$ cm^3, $\Delta Q = -1$ cm^3, and $c_o = 343$ m/s. The initial transit time t_o over a path length $x_o = 343$ mm will be 1 ms. Let the volume be decremented adiabatically by $|\Delta Q|$ but let the path remain fixed. The new transit time will be 1 ms/1.0002 = 999.8 µs, a change of $\Delta t = 200$ ns. If Δt can be measured to ±1 ns (a resolution commonly achieved in transit time flowmeters and thickness gages) then it would appear that c could be resolved to ±$\frac{1}{2}$%, and Q_{vo} to ±2.5%. In other words, the 0.1% compression by ΔQ raises the temperature by about 0.04% or about 0.12 K near room temperature, which raises the sound speed by 0.02%. Reversing the direction of calculation, in effect, yields Q_{vo}.

This proposed method would require calibration to check the validity of the ideal gas and other assumptions or approximations implicit in the

Table 7-2.   Calculated Isentropic Ratios for Ideal Gas Having $\gamma = 1.4$

ρ/ρ_o	Q_v/Q_{vo}	P/P_o	T/T_o	c/c_o
1.000	1.000	1.0000	1.0000	1.0000
1.001	.999	1.0014	1.0004	1.0002
1.01	.990	1.014	1.004	1.002
1.1	.909	1.143	1.039	1.019
2.0	.500	2.639	1.320	1.149

foregoing simplified analysis. Bidirectional interrogation would probably be necessary to eliminate flow effects, and the prf or bandwidth would have to allow the c determinations to be made quickly, before the adiabatic assumption is violated by the temperature returning to T_o due to heat transfer. (See also, Watanabe et al., 1988.)

7.3 Thickness of Solids Mainly by Pulse-Echo Techniques

This topic is typically treated in detail in NDT texts. See, for example, Krautkrämer and Krautkrämer, 1983. Many different forms of this problem occur in practice. If we choose to group these problems according to the boundary conditions, in particular, according to the nature of the adjacent media, we are led to the subcategories which follow. Note: we shall treat a boundary as "fluid" if that boundary is formed by a fluid or non-rigid couplant between a transducer, wedge or buffer rod and the part where thickness is to be measured.

7.3.1 Fluid-Bounded on Both Sides

This category may be subdivided into cases including: air on both sides; water on both sides; couplant on one side and gas or liquid on the other side.

7.3.1.1   Air-Bounded

Three noncontact acoustic possibilities of measuring sheet thickness in air are shown in Fig. 7-1. In (a) the solid sheet is opaque to ultrasound, so the method of obtaining x involves measuring air paths y_1 and y_2 and

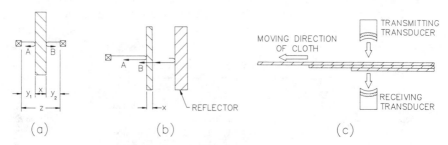

FIG. 7-1.   Solid sheet, air-bounded, noncontact. Configurations (a) and (b) include reflection; (c) relies on transmission only, based on an application due to Yano et al., 1987. © 1987 IEEE.

subtracting their sum from the transducer spacing z. Note that as x becomes a small difference between large numbers, accuracy suffers.

In (b) and (c), the solid is thin or porous enough to allow some transmission. In these cases, solutions for x may be derived from the *amplitudes* of the signals rather than their travel times. The amplitude-based solutions partly complement case (a) when x is so thin that method (a) becomes impractical. For example, if the thickness of plastic films <25 μm thick is to be determined ultrasonically, method (b) or (c) appears easier to execute than (a). Analysis for (b) is simplified if A and B represent the amplitudes of monochromatic tone bursts, incidence is normal, beam spread, mode conversion, and nonlinearities are neglected, and if $kx \ll Z_1/2Z_0$ where k = wavenumber in the sheet, $2\pi/\lambda$; Z_1 = sheet characteristic impedance and Z_0 = characteristic impedance of air. The echo ratio derived from the reflection and transmission coefficients is $A/B \approx (Z_1/2Z_0)(kx)$. In view of the approximations in the above derivation, it would be wise to calibrate any proposed installation.

Referring now to Fig. 7-1(c), Yano et al., 1987, used 500-kHz focused transducers (2-mm spot size) for measuring the boundaries of overlapping pieces of cloth to ±1 mm. For porous sheets, e.g., cloth or absorbent paper, the amplitude-based method does not yield x independent of porosity. However, for a given porous material, if the insertion loss is X dB/layer, then the received amplitude may be interpreted in terms of the number of stacked sheets. The dimensions of air-bounded steps may be determined by a comparison of echo profiles, or by deconvolution, even when the steps are small compared to wavelength (Kleinschmidt and Mágori, 1985). See Fig. 7-2.

7.3.1.2 Water-Bounded

Referring again to Fig. 7-1, the same geometries may be considered. But because of the relative case of transmitting high frequency ultrasound through water compared to air, frequencies can be increased by much more than the ratio of sound speeds (1500 m/s versus 343 m/s, or 4.4), and so higher resolution of x is generally attainable.

Because of the better impedance match between solids and water compared to air, it is practical in the water immersion case to detect front and rear surface echoes from the sheet. See Fig. 7-3.

Echo A will always be larger in magnitude than B, C . . . . B, C, . . . are of gradually decreasing amplitude. An accurate timing method used in some pulse-echo thickness gages times the interval between B and C. These echoes have like polarities and comparable magnitudes, while A and B are of opposite polarity and of very different magnitudes ($|A| \gg |B|$) if the sheet's characteristic impedance greatly exceeds that of water.

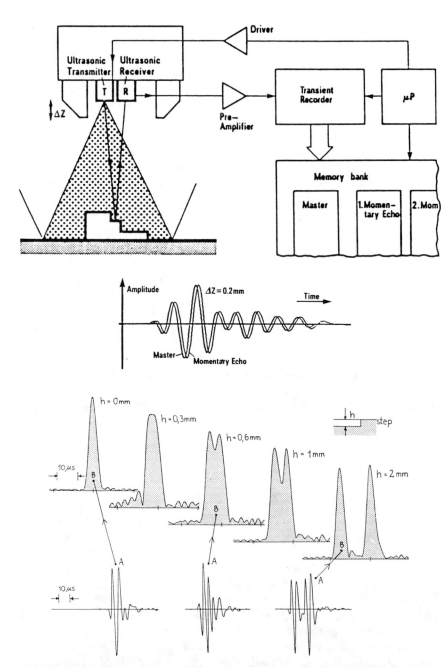

FIG. 7-2. Top: Measurement of a distance by comparison of echo profiles. Bottom: Distance resolution of direct (A) and deconvoluted (B) ultrasonic echoes obtained with L2QZ transducer on a step with variable height. In this method the System Function corresponding to the impulse response of the ultrasonic system is eliminated by mathematical operations. After Kleinschmidt and Mágori, 1985. © 1985 IEEE.

454

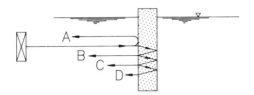

FIG. 7-3.   Solid sheet immersed in water.

Depending somewhat on frequency, the resolution would be ±3 μm in low-attenuation media, where $c \approx 6000$ m/s, assuming one uses an instrument having ±1 ns time resolution.

7.3.1.3   Couplant-Bounded on One Side, Air on the Other Side

This category ordinarily falls under "contact testing" in the NDT literature. One or two transducers are usually coupled to the part using a liquid, grease or gel, or a resilient coupling material like rubber, urethane or the like. (See also, Section 3.3; or Krautkrämer and Krautkrämer, 1983). In the pulse-echo mode, for example, using one transducer (Fig. 7-4) time interval may be measured from the initial pulse to the first echo, or for higher accuracy, between multiple back-surface echoes.

Buffer rods between the electroacoustic element and the couplant layer delay the echoes, enabling very thin parts to be measured. (See Chapter 2, Section 2.4.2 for numerical examples.) Buffer rods are also used to extend the temperature range of the method up (or down) to a temperature limit imposed by the couplant and the contact duration.

Two transducer elements, usually mounted within one housing on acoustically-isolated wedge segments, are often required when the part in question is severely corroded or pitted on the far side.

7.3.2   Fluid-Bounded on One Side Only

This category represents a coating or layer bonded to another solid, the substrate. The "fluid" is usually water, for immersed testing, or a couplant, for contact testing. The wave type used most often is longitudinal, at or near normal incidence. However, layered structures support other modes which may be preferred in special cases. As an example, consider the following application where the electroplated thickness to be measured was so thin, 2 to 5 μm, that Sezawa waves turned out to underlie the preferred test method.

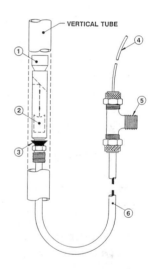

FIG. 7-4. Photographs illustrate how one measures thickness in a pulse-echo contact test using a couplant between transducer and various test objects. Diagram illustrates the measurement of the wall thickness of a small-diameter (10.3-mm ID) tube, from the inside. In one of the early applications for this design, the tube was SS347 or Inconel alloy 600, of 0.5- to 1.5-mm wall thickness. The transducer can be inserted up to 1.2 m into vertical straight tubes having inspection access from the bottom. The transducer is fitted for water delivery and water containment, fitted with coaxial cable, and interfaces mechanically with the customer's insertion mechanism/water delivery system. The wiper seal provides for water containment. The design uses a high frequency delay line transducer and a 90° reflecting mirror. Legend: 1–Centering device. 2–Focused transducer. 3–Seal. 4–R/G 178 B/U cable × 10-m long. 5–⅛-inch MNPT water inlet. 6–4.5-mm OD nylon tubing × 6-m long. Source: Panametrics NDT Div.

7.3.2.1 Sezawa Waves Measure 2- to 5-μm Electroplated Ag or Au

As explained by Tsukahara et al., 1986,

IC and LSI chips are usually mounted on lead frames. The lead frame is a thin substrate made of either copper or 42% Ni-Fe alloy (42-alloy). Typical thickness of the substrate is 225 μm. The substrate is electroplated with gold or silver on a surface. The electroplated layer is 2 to 5 μm in thickness. The production process of the lead frame consists of electroplating, drying and washing, and is continuously conducted in an automated production line. It is, therefore, desirable to measure the layer thickness immediately after the electroplating process and to feed back to the controller of the electroplating process, in order to keep the layer thickness precisely constant. The x-ray fluorescence method is currently employed for the layer thickness measurement of the lead frames. It requires, however, longer than 10 seconds for the measurement so that it is not suitably applied to the production line.

A pseudo-Sezawa wave method of layer thickness measurement was proposed in 1984, based on the discovery of dips in the frequency- and angular-dependence of reflection coefficients for layered surfaces. By 1986 the measurement could be executed in a short time with high resolution and without mechanical movement of the transducers.

Following Tsukuhara et al., 1986, one may explain the principle of the measurement by supposing that

an infinitely thick substrate occupies a lower half space, and a layer with thickness d covers the surface of the substrate. It is well known that a number of modes of surface waves exist in such a case, when the transverse sound velocity in the layer is less than that in the substrate. The fundamental mode is a Rayleigh wave, and the second mode is a Sezawa wave. The dispersion curves of these waves are shown in Fig. 7-5, top. In the figure, the phase velocity is normalized with respect to the sound velocity in water, and a dimensionless variable $kd/2\pi$ is used in place of the wave number k. The substrate is 42-alloy and the layer is gold. The Sezawa wave has a cut-off wave number, at which the phase velocity is equal to the transverse sound velocity in the substrate. The Sezawa wave is usually defined for the wave number greater than the cut-off wave number, as indicated by a solid line in Fig. 7-5. The dashed line in the figure is a pseudo-Sezawa wave, which is a kind of leaky surface wave. The pseudo-Sezawa wave, when excited by some means, would not propagate freely along the surface, but radiates its energy into the substrate and [its intensity] decreases exponentially.

Suppose, then, that the upper half space is filled with water and a plane wave is incident onto the layered surface with an incident angle θ. If the frequency of the wave satisfies the condition of excitation of the pseudo-

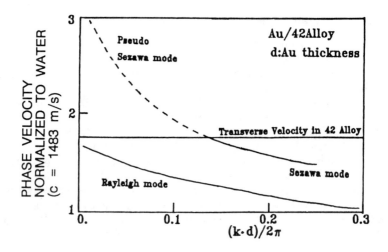

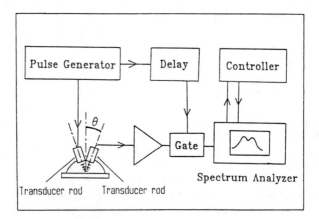

FIG. 7-5.   Top: Dispersion curves of surface waves excited on a layered half space, loaded with water. Bottom: Block diagram of instrument for layer thickness measurement by means of pseudo-Sezawa waves. After Tsukuhara et al., 1986. © 1986 IEEE.

Sezawa wave, then the majority of the incident energy [is transmitted] into the substrate, and therefore, the energy of the reflected wave [is reduced] drastically. [In experiments] the phenomenon is observed as a dip in the power spectrum of the reflected wave.

     This condition is practically realized by setting the incident angle at an appropriate value, such that the pseudo-Sezawa wave is excited with a wave number just below the cut-off wave number. This angle turns out to be 31° for a lead frame consisting of a 42-alloy substrate with a gold layer.

Once the incident angle and materials of the layer and substrate are determined, then [an] fd value, or a product of the dip frequency f and the layer thickness d, is uniquely determined, except for the temperature dependence. Therefore, the layer thickness d_o can be estimated by measuring the dip frequency f_o and calculating $d_o = (fd)/f_o$.

A block diagram of their instrument is given in Fig. 7-5, bottom. To understand some of the limitations of this method one needs to consider factors such as mechanical tolerances of the positions of the transducers, and, if the substrate is thin, the excitation of Lamb waves too, and their effect on accuracy.

After considering such factors, with respect to this application, Tsukuhara et al., 1986, were able to report on the performance of their instrument as follows:

- Measurable thickness range: 1-20 μm

- Stability: ±0.2%

- Accuracy: ±1%

- Measurement time: 1 s

7.3.2.2   Room Temperature

Manual Scan

To conduct this test the procedure is essentially like that in Section 7.3.1.3. In the present case, however, the substrate may have an impedance > or < that of the cladding, so measurement of round-trip transit time needs to take the polarity of the echo into account.

Automatic Scan

To avoid wear on the transducer, automatic scans are conducted in a large water-filled tank, if the part allows this. Bubbler probes and water-filled "wheel search units" are other ways of coupling in motion. Arrays permit scanning to be accomplished by moving the beam without physically moving the transducer or the part.

Extruded tubing can be scanned automatically for wall thickness eccentricity using fixed transducers as in Fig. 7-6.

For on-line eccentricity applications on cylindrical products such as pipe, tubing and cable, Panametrics offers a series of four-channel display units. These instruments will display and hold the last valid reading on

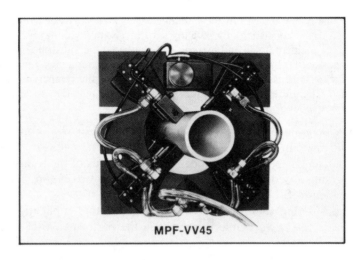

FIG. 7-6.   Top: Wall thickness eccentricity of extruded plastic tubing measured by four transducers. Bottom: Four-channel instrument Model 5215-4DU. Source: Panametrics.

each of four measurement channels, updating the information at a rate equal to the multiplexing rate of the gage. This provides a continuous visual monitor of product thickness in each of four quadrants. Display units can also calculate eccentricity measured between diametrically opposite positions on the product. The instrument may also be used to display readings taken at multiple locations on flat products such as sheet or plate.

Fixtures can be supplied with rollers to help stabilize products that are sufficiently hardened to permit roller contact with their surface. On softer products, no contact is required, provided that the product is relatively free from horizontal or vertical movement.

Both types of fixtures may be used either immersed in a cooling tank or free standing. If immersed, they will be supplied with water bubblers to keep transducers free of air bubbles, sediment, or other acoustic obstructions. If free standing, they are supplied with flow squirters to create a stream of water for coupling.

7.3.2.3 Low Temperature—Ice Accretion on Airplane Wings

As explained by Hansman and Kirby, 1985,

> aircraft icing remains one of the most severe aviation weather hazards. A system to measure aircraft ice accretion and accretion rate in real time could directly reduce this hazard. Real-time measurement of ice accretion rate can provide the pilot with a quantitative evaluation of icing severity. Therefore, the effectiveness of changes in flight path to minimize ice accretion can be determined. In addition, by measuring ice accretion on critical components such as wings, engine inlets, propellers, or rotor blades, an ice accretion measurement system can be used to automatically activate and optimally control ice protection systems. Although many schemes have been suggested for measuring aircraft ice accretion, there remains a need for the development of a practical system capable of performing real time, in situ measurement of ice accretion.
>
> [A study was initiated at MIT] to evaluate the feasibility and potential performance of an ice detection system using pulsed ultrasonic waves to measure ice thickness over a small transducer mounted flush with the aircraft surface. Since the technique of ultrasonic "pulse-echo" thickness measurement produces a real-time ice thickness signal, the ice accretion rate may be determined by electronically differentiating this thickness measurement with respect to time.

Figure 7-7 is a composite diagram showing how transducers have been flush-mounted in experimental plates and in airfoil surfaces such as an airplane wing, or on an ice probe such as that used in flight tests reported by Hansman and Kirby, 1986. As reported by these investigators:

> Following a number of experiments at MIT and in the NASA Lewis Icing Research Tunnel, a cylinder containing small flush-mounted ultrasonic transducers was installed on an airplane to determine sensor performance during natural icing flight tests. The results appear in Fig. 7-8. In contrast with the other protruding sensors used in that flight test, the ultrasonic system did not require special calibration, since earlier tests indicated an average ice sound velocity of 3.8 mm/μs, relatively independent of whether the ice accretion was of the wet or dry growth type.

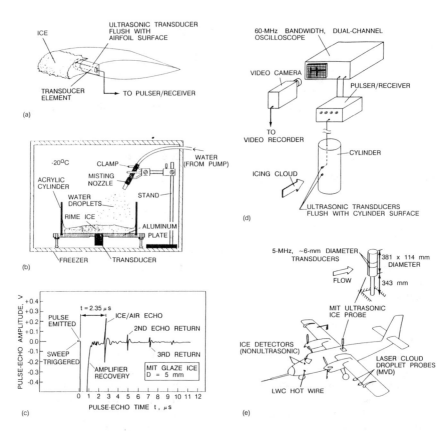

FIG. 7-7. (a) Pulse-echo ice accretion measurement concept includes flush-mounted transducer, and electronic functions such as a pulser/receiver, signal processor and a cockpit display. (b) Rime ice simulation method utilized at MIT. (c) Typical pulse-echo trace for ice. (d) Configuration of the experimental apparatus used for tests in NASA Lewis Ice Research Tunnel and for natural icing in flight. (e) Test cylinder installation for natural icing tests on NASA Lewis Twin Otter Icing Research Aircraft. The 11.4-cm cylinder instrumented with ultrasonic transducers was exposed to icing conditions at the end of an extension post that was vertically extended through the roof of the aircraft by an experiment carrier mounted in the aircraft. After Hansman and Kirby, 1985 and 1986. © 1985, 1986 AIAA.

Tests conducted on cylinders instrumented with ultrasonic transducers exposed to simulated and natural icing conditions led to the following conclusions:

1. Ultrasonic pulse-echo techniques can be used to provide a continuous measurement of ice thickness on an accreting body. Ice thickness was measured to within an accuracy of ± 0.5 mm (± 0.02 in.) using this technique.

2. The ultrasonic signals used for pulse-echo thickness measurement allow the

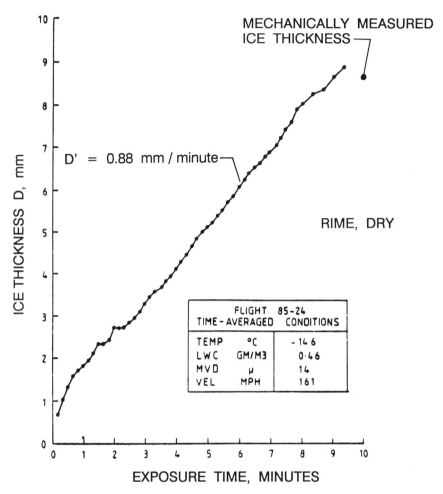

FIG. 7-8.   Flight test data of the ice thickness versus exposure time, showing measurement accuracy of ±0.5 mm confirmed by mechanical test at end of flight. Average accretion rate is 0.88 mm/minute. After Hansman and Kirby, 1985 and 1986. © 1985, 1986, AIAA.

following ice growth information to be determined:

a.   Ice thickness, from the pulse-echo transit time through the ice.

b.   Ice accretion rate, obtained by differentiating the measured thickness with respect to time.

c.   The presence of liquid water on the surface of the ice, from the time variation of the surface echoes.

d.   A measure of the surface roughness, from the width of the surface echoes received.

3.   For dry ice growth in natural icing conditions, fluctuations in cloud liquid water content result in corresponding changes in ice accretion rate measured with the ultrasonic pulse-echo technique. The time response of existing probe-type detectors is both dependent on icing severity and limited by the need to repeatedly deice the probe. The time resolution achievable using ultrasonic pulse-echo techniques is potentially superior since ice thickness can be directly measured many times a second. The ultrasonic system requires no cloud-dependent calibration.

Ultrasonic pulse-echo techniques therefore offer the proven potential for the development of an operational instrument for detecting and monitoring aircraft icing as well as providing previously unobtainable data on the fine temporal and spatial characteristics of ice growth in simulated and natural icing conditions. In addition, the unique capability of the ultrasonic pulse-echo technique to detect the presence of liquid water on the accreting ice surface allows the threshold between wet and dry ice growth to be experimentally determined; from this "critical impingement" data valuable information about the heat transfer at the ice surface may be inferred.

For additional background material on ultrasonic studies of ice solidification, the paper by Bailey and Dula, 1967, may be of interest. (See also, Section 7.5.2.4). An ice detector using flexural waves, proposed by Chamuel, 1984, was said to be more sensitive than if longitudinal waves were used; in one embodiment, both waves are used. Perkins, 1986, briefly discusses some of the in-flight ice-gaging alternatives. One of the alternatives happens to be another ultrasonic device, a resonant diaphragm vibrating between 20 and 50 kHz. Manufactured by Vibro-Meter, its resonance frequency is increased by a layer of ice, but decreased by water, oil or dirt. This resonant sensor design, Fig. 7-9, is included here for comparison purposes.

The sensor has two active elements: the detection element and a heating and cooling element, this latter being the key factor enabling warning to be given of impending ice accretion.

The cooling and heating element enables accurate temperature control of the detection element, and allows conditions to be created that permit advance detection and warning of ice formation on the airframe or engines. The heating capability is such that the ice formed on the sensor can be quickly melted and the detection process repeated for verification purposes.

The provision for local cooling allows one to determine the temperature margin between existing conditions and the onset of accretion; successive comparisons between the prevailing icing onset temperature and the ambient temperature allow predictions of the time remaining until ice formation will begin.

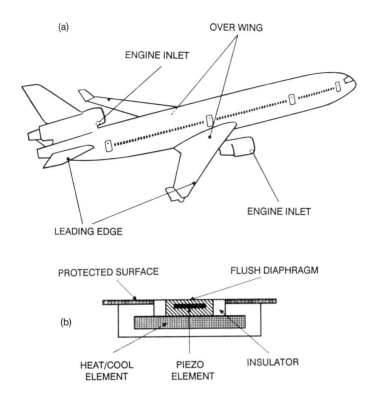

(a)

OVER WING

ENGINE INLET

LEADING EDGE

ENGINE INLET

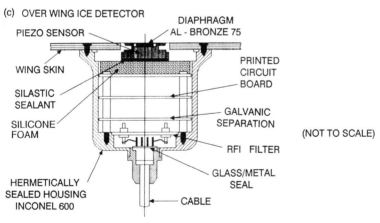

PROTECTED SURFACE

FLUSH DIAPHRAGM

(b)

HEAT/COOL
ELEMENT

PIEZO
ELEMENT

INSULATOR

(c) OVER WING ICE DETECTOR

DIAPHRAGM
AL - BRONZE 75

PIEZO SENSOR

WING SKIN

SILASTIC
SEALANT

SILICONE
FOAM

HERMETICALLY
SEALED HOUSING
INCONEL 600

PRINTED
CIRCUIT
BOARD

GALVANIC
SEPARATION

RFI FILTER

GLASS/METAL
SEAL

CABLE

(NOT TO SCALE)

FIG. 7-9.   Resonant diaphragm ice detector (advanced warning system, AWIDS), manufactured by Vibro-Meter. (a) Location of ice detectors (each about the size of a twenty-five
cent US coin) over wings, leading edges, and engine inlets. (b) Principle: the natural frequency of the diaphragm, which is forced into oscillation by the piezoelectric element, is
increased sharply by ice accretion (due to increased stiffness) but lowered by water or
contaminants (due to increased mass). (c) Diagram of an over wing ice detector, not to scale.
See Lustenberger, 1986, for details. Illustrations courtesy Vibro-Meter.

465

7.3.2.4   High Temperature

In this section three high temperature thickness gaging applications are considered: *ablation gaging* of heat shield material in the laboratory and during reentry; *erosion* measurement on hot steel pipe; and measuring the *wall thickness of red-hot steel tubes* in the production environments of the Watervliet Arsenal forge and a Timken steel piercing mill in Ohio.

Ablation Gaging

Whenever ultrasonic measurements are made at temperatures other than ambient, and especially at high temperature, the likelihood increases that sound speed gradients will be present. Sometimes T gradients are intentionally introduced, as in the work of Kalugin and Mikhailov, 1961, or Anderson et al., 1986. Kalugin and Mikhailov measured the transit time through a steel bar in which the axial distribution of T was measured by embedded thermocouples, Fig. 7-10. From this "tomographic" data they computed c versus T. In Anderson et al. the axial T gradient was introduced as a means of determining stress, Fig. 7-11. In ultrasonic ablation gaging, the sound beam travels from the transducer, which may be near 20°C initially, out to an ablating surface whose temperature may be thousands of degrees if the heatshield material is graphite, for example.

In Chapter 5, Section 5.1, where media serve as their own T sensors, the problem of extracting the T profile from measurements over a *multiplicity* of paths is addressed. The solution used there, however, is not applicable if only one path is available.

As an introduction to the subject of ablation, we may note that the interaction between the earth's atmosphere and a solid body traveling at hypersonic speeds, i.e., above Mach 5, has been studied for many years.

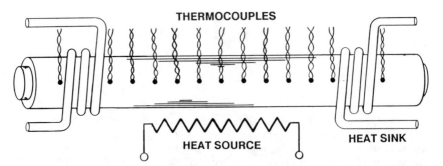

FIG. 7-10.   Schematic of a method of measuring elastic properties as a function of temperature, wherein a temperature gradient is intentionally introduced. High temperature gradient method due to Kalugin and Mikhailov, 1961.

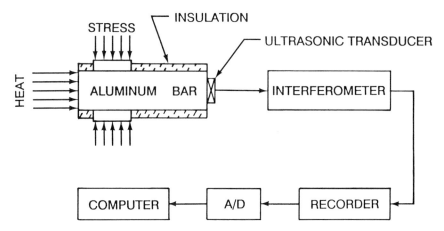

FIG. 7-11.   Experimental arrangement for the detection of subsurface stress in metals using ultrasound and time-varying temperature gradients. After Anderson et al., 1986. © 1986 IEEE.

Years ago, astrophysicists were concerned with this problem in connection with meteors (Lindeman and Dobson, 1923; Opik, 1933).

To help understand the phenomena associated with ablative heat-protection systems under development for reentry vehicles, as well as the burning rate of solid rocket propellants, surface recession data were needed. Numerous sensor technologies were explored during the search for this data (Legendre and Chase, 1974; McGunigle and Jennings, 1975; Elands et al., 1988).

Around 1960, ultrasonic experiments were begun at Avco, the purpose being to measure the thickness of reentry vehicle heat shields while they ablate during reentry. The interpretation of pulse-echo transit time data in terms of thickness requires knowledge of $c(x)$ [that is, $c(T)$ and $T(x)$, where x = coordinate along the axis of the sound beam]. Further, during ablation, one needs to determine which elastic interface first reflects the pulse. The optimization of an ultrasonic ablation gaging system ideally is based upon knowledge of $c(T)$, $\alpha(T)$, $T(x)$ and the noise to be encountered during those parts of the flight trajectory for which ablation data is sought. It is generally impossible to obtain sufficient data on c and α versus T up to the maximum T of interest. It is likewise impossible to obtain $T(x)$ and noise data over all the possible heating rates and other aerothermodynamic conditions associated with various trajectories. On the other hand, it is not always necessary to optimize a system in order to get it to work.

One type of laboratory experiment for calibrating ultrasonic round trip transit time t versus silhouette length s is shown in Fig. 7-12. In such experiments, for a so-called Type 1 resin-ceramic composite, it was found

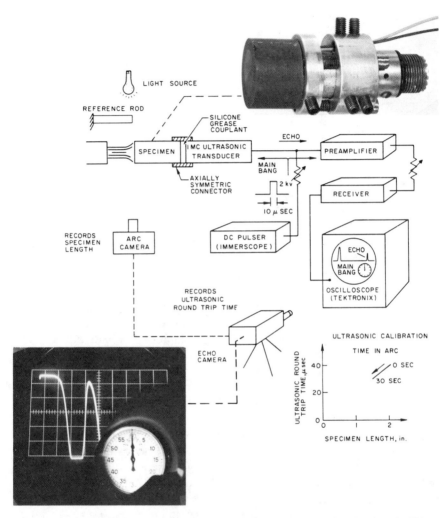

FIG. 7-12.   Experimental arrangement for ablation tests in one of Avco's arc facilities.
After Lynnworth, 1961.

that after a few seconds of heating in an arc jet, s exceeded $c_o t/2$ where
c_o = sound speed at room temperature. (See Fig. 7-13). For this heatshield
material, as for most solids, it was known that c decreased at high temperature. This meant that the echo was generated beneath the ablating
surface. The explanation for a subsurface elastic interface had to do with
a char layer formed during ablation, and of a thickness approximately
inversely proportional to the ablation rate. In some of the experiments

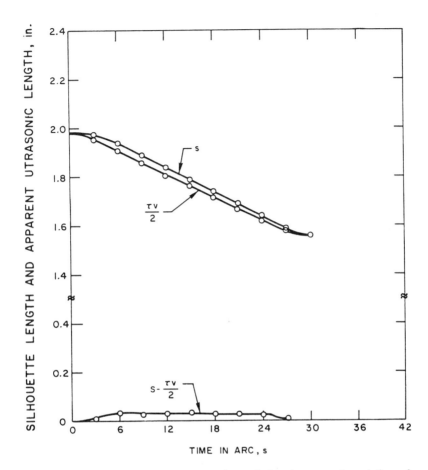

FIG. 7-13.   Silhouette length and apparent ultrasonic length versus elapsed time of arc exposure for a resin-ceramic heatshield material subjected to a stagnation enthalpy of 10^4 Btu/lb. After Lynnworth, 1961. © 1961 Academic Press.

the specimen was notched perpendicular to its axis and instrumented with a thermocouple in the plane of the notch, to confirm the source of the echo.

It may be remarked that, in the early 1960s, as some types of composite heatshield materials were improved with respect to thermal properties the attenuation became so high as to prevent ultrasonic pulse-echo thickness measurements from being conducted, given the limits of space, transducer and electronics technology available at that time.

For these high-α materials, a more practical solution seems to have been a step-type gamma ray ablation gage (Grund and Lynnworth, 1960, unpubl.; Prohaska and Lauletta, 1961; Chase and Galbiati, 1965; Legendre and Chase, 1974). But for other materials where α apparently was not so much of a problem, e.g., graphite (McGunigle and Jennings, 1975), or radar window material (Gieske et al., 1987) practical ultrasonic ablation sensors have been developed. As part of the database in support of these sensors, reference is made to c versus T measurements such as those of Gieske, 1978-1985, for graphite materials tested at Sandia up to 3370°C in some cases.

The ultrasonic ablation recession measurement system due to Mc-Gunigle and Jennings, 1975, of HTL-K West is comprised of a transducer and a signal processor. The transducer is bonded to the back surface of the nosetip or heat shield, and propagates a 0.2- to 1.5-MHz pulse through the material. The transducer then converts the echo returning from the ablating surface into an electrical signal. The PZT-5A transducer generates either shear or longitudinal mode, depending on the nosetip material. The remote signal processor, powered by 28 VDC, synchronously averages many echoes using "storage bins," and reconstructs the averaged echo waveshape at a low frequency suitable for telemetering to the ground on a channel of low bandwidth, 0.5 to 1 kHz.

The development, testing and flight experience reported by McGunigle and Jennings concentrated on nosetip stagnation point recession measurements. These authors point out that the aerodynamic noise occurring in rocket tests on the ground can be uncharacteristic of and more severe than noise occurring in reentry. These authors concluded that their ultrasonic ablation recession measurement system provided accurate recession data on a number of flight programs. While much optimization remained to be done, the concept was considered proven in flight by 1975. It was also used to provide unanticipated information such as crack detection and aerodynamic noise measurements.

The problem of *unsteady* regression rate, wherein the temperature profile is a function of that rate, is approached iteratively starting with an assumed profile, in a method due to Elands et al., 1988.

Erosion of Hot Steel Pipe

In programs extending over some ten years, ca. 1977-1987, Youngdahl and Ellingson of the Argonne National Laboratory developed a multiplexed waveguide system for measuring the thickness of high-pressure hot steel pipes subject to the erosion and corrosion found in coal conversion process equipment. This equipment operates up to ~540°C in liquefaction and gasification plants.

Some of the key elements in their systems may be summarized as follows:

Transducers. A dedicated (recoverable) transducer is bonded to the outer end of each waveguide with five-minute epoxy and clamped in place. The preferred transducer has a nominal diameter of ~10 mm and a frequency of 7.5 MHz, either resonant or moderately damped longitudinal wave (λ = 0.76 mm in steel at 20°C).

Waveguide Design. Waveguide design considerations include material, dimensions, method of attachment to pressure boundaries, and sidewall treatment. The acoustic waveguides in this system are solid metal rods utilized to thermally protect conventional PZT transducers while maintaining acoustic contact with the high-temperature component being monitored. The waveguides are typically mounted in groups, as shown in Fig. 7-14, at sites of anticipated erosive wear. The ultrasonic system described here interrogates the waveguides sequentially and can accommodate multiple groups of guides at sites each up to ~150 m from the pulser-receiver.

Austenitic stainless steel (SS) is favored for low thermal conductivity, and wrought material (ASTM grain size ~6) is used to avoid excessive ultrasonic attenuation encountered in large-grained SS. All waveguide material stocks are screened for acceptable acoustic transparency before fabrication of guides. Additionally, a test piece of each rod stock is used to ultrasonically determine the optimal ratio between the cross-sectional areas

FIG. 7-14. Series of five threaded waveguides on thermally insulated elbow fitting on 410°C slurry line at Exxon Coal Liquefaction Pilot Plant. Flexible conduit protects RG174 coaxial cables connecting transducers to relay enclosure. After Youngdahl and Ellingson, 1982. Illustration courtesy Argonne National Laboratory (ANL).

of the neck and shoulder (Fig. 7-15), because of differing radial microstructural variation between material lots. The typical neck diameter of 7 mm provides a suitable sampling area with small depression of local temperature.

Three sizes of waveguides have been developed and extensively tested. Medium-knurled, 130-mm long, 9.5-mm diameter Type 304 SS waveguides are most frequently employed (Fig. 7-15); these are attached by electric arc stud welding to the components being monitored. . . . . For attachment by pressure coupling rather than welding, waveguides of ⅜-11 threaded Type 304 SS rod up to 200-mm long are utilized (Fig. 7-16). These longer waveguides do not require the convection-cooling fins employed by the shorter guides but do require greater clearance from nearby components and are more susceptible to mechanical disturbances. Finally, waveguides of 0.6-m length, tapered in diameter from 16 to 13 mm over the length, have been developed and tested. These were equipped with support struts to preserve orientation and straightness in the environment of open flames through which they passed to the piping in a slurry preheater.

A 27-point *ultrasonic scanning system* [was] tested at a coal liquefaction pilot plant. Examples of *test results* appear in Fig. 7-17.

Other applications of the foregoing high-temperature technology may develop in erosion studies of furnace linings, or the wear of electrodes used in processing molten metals or molten glass.

Wall Thickness Gaged on Red-Hot Steel Tubes, 600 to 1100°C, by Momentary Contact

In using ultrasonics or any other technology for process control, it is clearly desirable to sense the measurand as early in the production cycle as possible, and thereby introduce corrective action with a minimum of waste. For materials made at high temperature, this means making the

FIG. 7-15. Knurled SS waveguide with Cu cooling fins, reference shoulder, and tip for stud welding. After Youngdahl and Ellingson, 1982. Courtesy ANL.

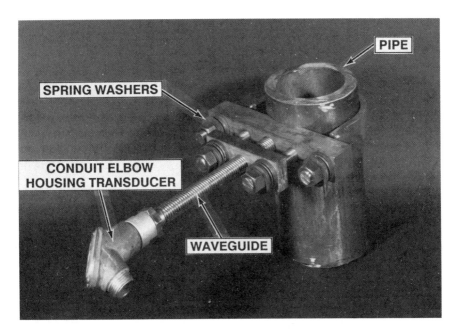

FIG. 7-16.   Threaded SS waveguide and clamp on 3-inch (89-mm OD) pipe. Cu or Au foil couplant is used at waveguide-pipe pressure-contact interface. After Youngdahl and Ellingson, 1982. Courtesy ANL.

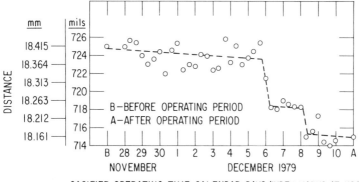

FIG. 7-17.   Erosive wear at selected site on 540°C Type 316 SS cyclone separator during operation in coal gasifier effluent system at Morgantown, WV Energy Technology Center. After Youngdahl and Ellingson, 1982. Courtesy ANL.

measurement at high temperature. In steelmaking, opportunities for high temperature testing and measurement have challenged ultrasonic investigators for many years, in fact, almost from the earliest days of ultrasonic NDT.

"Classic" red-hot problems in the steel industry include: measurement of the wall thickness of continuously cast steel, and of tubing made in piercing mills or on rotary forges; measurement of the extent of the "piping" defect (a shrinkage cavity) in ingots, billets and blooms; and measurement of temperature, microstructure or other internal conditions. Two of these applications are discussed below, namely, installations of ultrasonic transducers in a forge and on a piercing mill. In both cases, tube wall thickness is measured despite temperatures being as high as 600 to 1100°C.

Because of the difficulty of coupling to tubes, blooms, billets and similar large hot parts during their processing in a steel mill, considerable effort has gone into *noncontact* transduction by emats and lasers Wadley et al., 1986; Boyd et al., 1989. Ultimately such methods may prove successful, as the incentives are evident. In the meantime, however, some benefits apparently can be derived despite the expense and inconvenience of having to arrange for a pressure coupling momentary contact system to bring the transducer into brief contact with the part to be tested.

Installation on a Rotary Forge

In a report covering experiments and evaluations over the 1976-1979 period, Krupski, 1979, explains the need, method and results of installing an ultrasonic measurement system for gaging wall thickness of hot tubes "as forged." The following extract essentially follows Krupski's presentation.

> With the introduction of the rotary forge at the Watervliet Arsenal, a new line of inspection equipment was required. The need to design inspection equipment to measure hot tubes in the "as-forged" condition created gaging problems not often encountered in industry. [Forgings had to] be inspected immediately off the forging machine to monitor uniformity of wall thickness. Any tendency toward excessive wall thickness variation [had to be] detected immediately and corrected. The high production rate of forge operation could have resulted in the production of many rejectable tubes if inspection were delayed until the forging cooled. The inspection system required development of [an uncommon] technique for ultrasonics. Hydraulic pressure was used to bring transducers [into a brief, controlled contact (<10s)] with the hot forgings to establish coupling. Wall thickness measurements were taken simultaneously at four positions around the tube

circumference and a wall thickness variation at that cross-section was printed out.

The rotary forge is numerically controlled and is capable of forging a 105-mm M68 gun tube in approximately eleven minutes. At this rate eight-hours' production could exceed thirty cannon tubes. Standard (nonultrasonic) inspection equipment used on cannon forgings to measure diameters, straightness and wall thickness can not be employed until after the forging cools.

Wall thickness of rotary forged cannon tubes is critical. Because the forgings contain minimal excess material, wall thickness must be held uniform. Uniform wall thickness insures even material distribution and sufficient stock for final machining operations.

If a malfunction leads to the production of forgings with non-uniform wall thickness, this condition would not be picked up until the forging had cooled sufficiently to allow for the use of conventional inspection techniques. During the ten-hour period required for the ~1350-kg forging to cool to approximately 38°C, forty more tubes could be produced. These forty tubes could very possibly have the same unacceptable wall thickness characteristics and could be scrap. A definite need existed to develop a method to inspect forgings for wall thickness immediately off the forge machine in order to monitor the production process and take corrective measures if necessary.

System Description. The wall thickness variation measurement system supplied by Sonic Instruments consists of a traveling bridge riding on two tracks. The bridge supports the measurement transducers (Fig. 7-18) and moves over the tube along its length. The bridge is motorized and can be stopped at each measurement interval required.

The tube to be inspected is supported on three vee-block stands. The position of these supports minimizes the bending of the forging that may occur. Because the yield strength of these forgings at these high temperatures is substantially reduced, support positions are critical. These supports are repositioned for each different tube forging (Fig. 7-19).

The moving bridge carries four ultrasonic transducers with water-cooled delay lines, mounted 90° apart. Each is controlled by a long stroke hydraulic cylinder which holds the transducer against the forging with a force of approximately one ton. Mounting the transducers in this way allows for measurement of wall thickness variation at a particular cross-section. Transducers contact the forging surface at four equally spaced points. [In contrast to the Timken application described by Jeskey in the next subsection, wall thickness readings in the forge application were not accurate because of] the high temperature and corresponding decreased sound velocity. [To the extent that there exists] constant temperature around the forging at each cross-section inspected, an accurate wall thickness *variation* measurement, which was the objective, [is assured]. In the absence of a temperature measurement, the temperature gradient from one end of the tube to the other makes it difficult to get accurate wall thickness readings.

FIG. 7-18.   Hot forging in Watervliet inspection station. Four transducers on the traveling bridge are connected to readout electronics in a heat-insulated cabinet. If satisfactory signals are obtained, four wall-thickness readings are displayed and printed along with their measuring positions. A μP computes the variation between minimum and maximum readings and prints the wall-thickness variation. Hydraulic motors move the bridge to the next selected cross-section and the process is repeated. If one or more signals are absent the transducers are released and retracted from the forging. After Krupski, 1979. Department of the Army photograph.

Results. Tests were made at a number of locations on a representative tube, covering temperatures from ~600 to ~950°C. The variations in measured wall thickness were compared to values obtained the next day on the cold tube and were found to be within the ±0.8 mm of actual variation, as required for this application.

Installation on Timken Piercing Mill

The Timken Company manufactures seamless steel tubing for mechanical applications in the automotive, aircraft, railroad, anti-friction bearing, construction, mining and machine tool industries. The seamless steel tub-

FIG. 7-19.   Transducers contact tube. After Krupski, 1979. Department of the Army photograph.

ing is produced to hundreds of combinations of quality, steel type, finish and size.

As explained by G.V. Jeskey, priv. comm., 1987, seamless steel tubing is produced on a piercing mill at Timken's Canton, Ohio facility. The operation consists of a series of hot processing operations (steel temperatures ranging from ~800 to ~1300°C). Size specifications, however, are based on *ambient* temperatures. Measurement techniques that correlate with the final diameter and wall thickness specifications are therefore required for initial setup as well as for process control throughout the production run.

In the absence of a suitable wall thickness measurement at the high temperature of the process, the initial setup typically would be evaluated by processing one or more tubes, cutting samples off the tube ends, cooling the hot samples in a water bath, and then manually gaging the diameter and wall thickness using calipers and micrometers. These techniques are very slow, typically consuming fifteen to twenty minutes of production time per sample measured. Another limitation is that the sampling is re-stricted to a tube end.

Measurement checks during a production run are usually made the same way. This presents an additional problem. Because the measurements require so much time, only a small percentage of the tubes are sampled during a production run.

In a prior project reported by Jeskey et al., 1977, an ultrasonic technique was developed for real-time measurement of steel solidification (Section 7.4). That work included improvements over the earlier laboratory tech-niques of Carnevale et al., 1964 and Fowler and Lynnworth, 1970, for coupling ultrasonic energy to hot steel samples using the momentary con-tact technique, and provided a basis for direct thickness gaging of hot steel tubes.

Even though the feasibility of using the momentary contact technique was considered established, improvements were required to adapt it to the stringent accuracy requirements of tube thickness gaging. The primary objective was to develop the capability to measure the wall thickness of hot tubes with sufficient speed so as not to impede production capacity (measurement cycle time less than one minute) and to correlate the high temperature measurement to the wall thickness after cooling to ambient with an accuracy of 1%. For nominal wall thicknesses of 12.7 and 50 mm, the maximum allowable error was 0.13 and 0.5 mm, respectively. In order to achieve these goals, it was necessary to:

- Develop a transducer assembly to measure pulses reflected from the tube surfaces.

- Develop instrumentation capable of monitoring the appropriate ul-trasonic reflections and of making time delay measurements with accuracies of 10 ns.

- Design a mechanical system to control orientation of the ultrasonic transducer to maintain precise normality with production tubes under the relatively high contact pressures needed for coupling.

- Develop a hydraulic/mechanical system to apply and maintain con-tact pressures within 25 psi (~2 bar).

- Develop correlation algorithms for the wide variety of tubing prod-ucts.

- Calibrate and verify system capability.

Jeskey felt that his main technical improvement in this work was the development of the transducer assembly. It consists of a protective hous-ing, a piezoelectric element, and a water cooled delay line. The essential considerations were (1) maximizing the energy reflected from the tube inside diameter, (2) designing an echo-free buffer usable on hot steel in the pulse-echo mode of operation, and (3) preserving the integrity of the ultrasonic waveforms.

The hot tube gaging system consists of the following subsystems:

- An ultrasonic unit to gage the wall thickness of hot tubes.

- A laser unit to gage the OD of hot tubes.

- Optical pyrometers to measure the surface temperature of hot tubes.

- A programmable logic controller to automatically sequence all testing functions.

- A minicomputer to perform system calculations and to provide the operator interface.

The system (Fig. 7-20) was installed in a piercing mill in September 1985 and has replaced the manual method of tube gaging since that time. During this time, the gage has been applied to dozens of steel compositions ranging from 0.1 to 1.0% carbon, in numerous tube sizes ranging from 12.7- to 50-mm wall thickness and 100- to 300-mm OD and for tube temperatures from ~700 to ~1100°C. Statistical analysis has shown the system capable of measuring the hot tubing and calculating the dimensions at ambient temperatures within an accuracy of ±1% for the wall and ±0.1% for the OD.

Jeskey, priv. comm., 1987, concludes:

> The hot tube gaging system was originally justified strictly on reduced mill downtime as compared to the manual method of cutting and micrometering water quenched samples. While this was achieved, the benefits realized because of improved measurement accuracy and capability for more measurements per order were more far-reaching, as listed below:
>
> - Increased production capacity.
> - One or two fewer off-size trial tubes per setup.
> - Less resizing.
> - Increase in on-time deliveries.

7.3.3   Solid-Bounded on Both Sides

This category of thickness gaging of solids is exemplified by the problem of measuring the thickness of adhesive or "bond line" B between an accessible metal container A and an internal material C. In other words we have a solid ABC sandwich and want to measure the thickness of B.

FIG. 7-20. Timken's ultrasonic wall measurement system clamped on a hot tube at Piercing Mill Number Three. The photograph shows a close-up of both the top and the bottom inspection heads. The water cooled transducer assemblies are located in the center of the hydraulically coupled vee-block centering units. Source: Jeskey, 1987, priv. comm. Illustration courtesy Timken.

If B is at least one wavelength thick at the test frequency yet is thin compared to A, then in a pulse-echo test the reverberations in A straddle the echo from the BC interface. Further, if the AB mismatch is not too great and if the BC mismatch is not too small, and if attenuation and scattering in A and B are not troublesome, a simple pulse-echo test suffices.

Another example that may be included in this category is provided by an NDT contact transducer. Typically it includes a piezoelectric crystal B bonded between a thin ($\leqslant 0.1$ mm) wear plate A and an attenuating backing C. A, B and C have comparable acoustic impedances. If the bonding layers are too thick the crystal rings. Ringing can be observed in the time domain or, using a spectrum analyzer, in the frequency domain. Transducer assemblies also ring if a disbond exists, even if the bond thickness in the bonded areas is very thin. In other words, one can not always draw conclusions about thickness based merely on an observation of ring-ing.

We shall not pursue this category in depth. It is understandable, how-ever, that if all the above ifs are not satisfied, then other approaches may be required.

7.4 Wall Thickness of Solidifying Steel Ingot by Transmission Technique

An ultrasonic transmission technique due to Jeskey et al., 1977, was used to measure the rate of solidification of steel ingots, based on the depen-dence of through-transmission transit time upon the extent of ingot so-lidification. Using ultrasonic probes that were momentarily pressure-cou-pled to the hot surface of the solidifying ingot, laboratory experimental results were obtained for 20-cm square ingots (weighing approximately 225 kg) of induction-melted 4620 type steel. The ultrasonically determined solid thickness was verified in the range 2.5 to 6.4 cm by dumping partly solidified ingots and then calipering their actual wall thickness. Addi-tionally, the ultrasonically determined growth rate was found to compare closely with theoretical predictions and independent observations of other investigators. The thickness of the solidified ingot wall determined from the ultrasonic data appears accurate to within 5 mm, which is comparable to the ~4 to 7 mm theoretical width of the mushy zone bounded by the liquidus and solidus. It was concluded that the ultrasonic transmission technique is applicable to monitoring steel ingot solidification for solid wall thicknesses from about 2.5 cm up to at least 10 cm. Moreover, the abrupt change in slope of the curve of transit time versus solidification time can be used as the indicator of complete solidification.

Earlier ultrasonic investigations of steel solidification (Kurz and Lux, 1968) and at least one subsequent investigation (Parker et al., 1985) have been limited to pulse-echo or reflection approaches. In such studies one attempts to locate and follow the solidification front by measuring the round trip transit time for a pulse to propagate between the solid surface and the zone between liquid and solid phases. Reflection measurements have indeed been satisfactory in tracking the advance of relatively well-defined interfaces in solidifying, melting or ablating materials like ice, wood's metal and some heatshield materials. However, in the case of solidifying steel, ultrasound propagation problems associated with coup-ling, high absorption, grain scattering and the irregular, ill-defined liquid/solid interface, all tend to frustrate the reflection measurement. Although encouraging results had been obtained in steel using *reflection* methods, at least in some laboratory studies, the author suggested to Jeskey that measurement of transit time using an ultrasonic *transmission* approach might be a more effective way to monitor steel solidification.

Underlying this suggestion was the thought that since the sound speeds differ in liquid versus solid steel, the average sound speed across a solidifying ingot would be a weighted average of these, the weighting principally being a function of the liquid-to-solid ratio along the ultrasonic path. A further argument favoring the transmission approach, as opposed to the earlier reflection approaches, was based on the observations of several investigators that the reflection coefficient is small for longitudinal waves normally incident upon a solid/liquid steel interface. On the other hand an interface of low reflectance may be expected to exhibit high transmittance.

Suppose one wishes to resolve the interface with an uncertainty comparable to the width w of the mushy zone bounded by the liquidus and solidus isotherms. Reflection techniques suffer from destructive interference caused by scatterers distributed within the mushy transition zone. The interface or interphase reflection therefore would be smeared. The transmitted wave, however, may be less degraded on passing through the mushy zone. In fact, the solid/liquid transmission coefficient is augmented by the impedance-matching effect of the mushy zone. Other advantages of the through-transmission approach include: easier probe design, in that a longer, undamped buffer can be inserted between the transducer and the hot ingot; the receiver amplifier is isolated from the transmitter; lower frequency is permitted, even down to where the wavelength $\lambda \gg w$, provided beam spread is not excessive; lower frequency, in turn, lowers the attenuation coefficient, and may also avoid ambiguity in time (or electrical phase angle) measurements.

Disadvantages of the through-transmission technique include: the error in indicated solid wall thickness, expressed as a percent of reading, tends to be larger during the early stages of solidification; one path cannot distinguish symmetrical from asymmetrical solidification (one wall could be thicker than the opposite wall); access is required to opposite sides of the ingot.

Ultrasonic Transmission Approach

The theoretical basis for relating the thickness of the solid portion of a solidifying ingot to the transit time measured across the ingot was explained in Chapter 2, Section 2.4.4, in connection with Fig. 2-46. Using the notation of that section, it may be shown that the sensitivity of t to increases in x is expressed by $dt/dx = (c_y - c_x)/c_y c_x$.

Sensitivity of the measurement is thus proportional to the difference in sound speeds for the two phases. Suppose the model is adapted to approximate a solidifying steel ingot. The solid wall thickness is represented by $x/2$. Make two simplifying assumptions: take c_y as a constant for the molten steel phase and take c_x as another constant for the nonmolten steel. If the furnace is tapped with only a small degree of superheat, the first assumption is justified. Further, if the ingot surface temperature remains substantially constant during solidification, the second assumption is supported. Assume also that the mushy zone width w is negligibly small compared to $x/2$ and y. This third assumption appears reasonable except at the beginning and end points of solidification.

Experimental Method

The experiments consisted of a number of runs in which each ingot mold was filled with molten steel and then interrogated with ultrasound throughout solidification to determine the ultrasonic transit time versus solidification time relationship. Solidification experiments were repeated with the molds pivoted to facilitate rapid dumping of the liquid metal after allowing solidification for different time intervals. After dumping, the solid steel wall thickness was calipered at one of the transducer ports to provide an empirical calibration of extent of solidification with time.

Figure 7-21 shows three stages of the experimental procedure. Figure 7-21(a) shows the filled mold as solidification begins. Diametrically opposed transducer ports were located about halfway up, where the internal dimension of the slightly tapered mold was 20 cm. The ports were temporarily sealed with removable steel plugs. The plugs were held in place by the hydraulically-actuated transducer housings or probes, when the piston rod was in its partly extended position. In Fig. 7-21(b), solidification has proceeded far enough for the plugs to have been withdrawn without danger of a leak. The plugs, withdrawn by retracting the piston rod, fall down, out of the way. In Fig. 7-21(c), the ultrasonic transducer probes are advanced to be momentarily coupled to the exposed wall of solidifying steel, during which instant transit time across the ingot is displayed on an oscilloscope and recorded photographically. A second trace records the round trip time in one of the two matched probes, so that the buffer rod delays can be subtracted from the total travel time between piezoelectric elements. The difference equals the travel time t across the 20-cm path, i.e., $z = 20$ cm. The interrogating wave consists of a burst of cycles having center frequencies typically 0.5 or 1.0 MHz.

The momentary contact, pressure-coupled ultrasonic measurements could be repeated at frequent intervals during solidification but in these

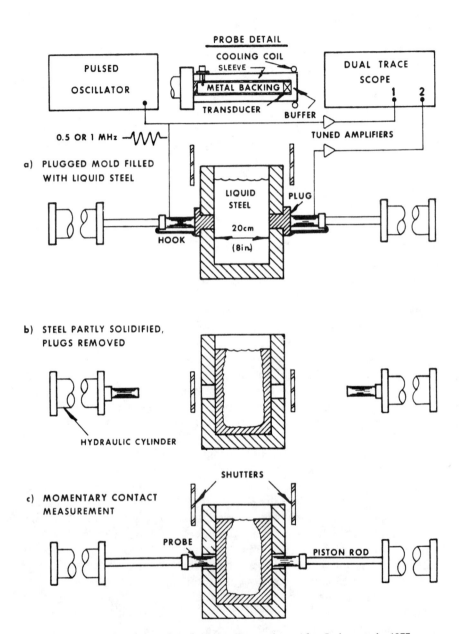

FIG. 7-21.   Three stages in experimental procedure. After Jeskey et al., 1977.

experiments, not before the plugs were removed. Therefore, ultrasonic data is absent for $T < 120$ s. Solidification time T is measured from the time the liquid level in the mold reaches the ports, which is about 15 s after the furnace is tapped.

One of the probes used in tests at 0.5 and 1 MHz is shown at the top of Fig. 7-21. Probe contact time was typically a few seconds.

It should be stated that to promote prompt solidification, the furnace was tapped at 1557-1560°C (2835-2840°F) to minimize the superheat in the 225-kg (500-lb) charge (solidus, 1468°C, 2675°F; liquidus, 1510°C, 2750°F). It is further understood that during solidification, the mushy zone can be estimated to be about 3.7- to 7-mm thick. The boundaries of the mushy zone are perturbed by the dumping process in that some liquid is retained between dendrites and some dendrite tips are broken off or washed away. These two perturbations tend to cancel, so that after cooling and despite neglecting the thermal expansion coefficient, the calipered wall thickness indeed provides a reasonable measure of the average solid thickness just prior to dumping, particularly if calipering is limited to the wall which is uppermost during dumping. After solidification proceeded to the point at which the top of the ingot was too constricted to allow rapid dumping, no further mechanical verification was sought.

Results and Discussion

Ultrasonic transit time measurements throughout the solidification of several ingots are shown in Fig. 7-22 as a function of time from the start of solidification. These results can be viewed as a calibration curve for the purpose of determining the constants c_x and c_y. Extrapolation of the t versus T data to $T = 0$ s yields $c_y = z/t_o$. Later, at the moment of complete solidification, the transit time t_f is obtained and used to compute $c_x = z/t_f$.

During early stages of solidification (up to $T \approx 10$ minutes) the calibration curve can be described by $t = 53.03 - 2.65\sqrt{T}$ μs.

The calipered values of wall thickness given in Fig. 7-23 show good agreement between the two techniques. Further verification of the ultrasonic technique is given by comparing with published rate equations for steel solidification, the $x/2$ versus T function (up to $T = 10$ minutes). This function can be expressed as $x/2 = -0.11 + 0.88\sqrt{T}$.

Conclusions

Based on the above results, several conclusions were drawn by Jeskey et al., 1977:

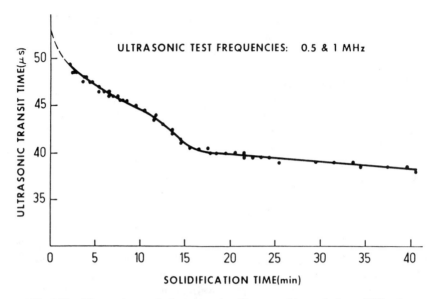

FIG. 7-22. Ultrasonic transit time through a 20-cm steel ingot during solidification.

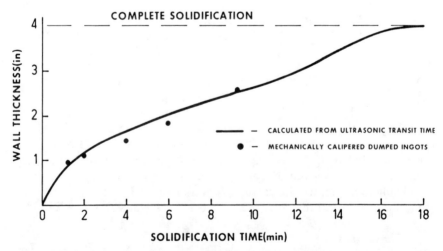

FIG. 7-23. Comparison of wall thickness calculated from ultrasonic transit time results with mechanically calipered wall measurements. Note: Most of Section 7.4 and Figs. 7-21 to 7-23 are © 1977 and reproduced with the permission of the American Foundrymen's Society.

1. Through-transmission measurements of transit time enable the extent
    of solidification to be monitored in real time, nondestructively and
    with an accuracy limitation comparable to the uncertainty inherent
    in the thickness of the solidifying ingot skin.
2. For cases similar to that reported here, the empirical sound speed
    constants c_y and c_x can be established for each situation at the be-
    ginning and end of solidification, respectively, without determining
    the detailed temperature-dependence of sound speed and without de-
    tailed knowledge of the actual temperature profile.
3. The end point of solidification can be recognized by the sudden re-
    duction in slope of the t versus T function, even if c_x and c_y are
    unknown, provided c_x and c_y differ by at least 10%.

7.5   Liquid Level and Other Fluid Interfaces

In this section, applications are broken down according to whether or not
there is any penetration of the normal pressure boundary, i.e., clamp-on
transducers versus non-clamp-on (wetted) transducers and intrusive sensors.

7.5.1   Clamp-On Transducers

Clamp-on transducers are used to measure level at one or more discrete
points, or continuously if sound speed c is known. Discrete interfaces,
as in a pipeline interface detector that senses the passage of different
gasolines, can be detected but the information probably will not be reliable
unless the system is temperature-compensated. A discrete level clamp-
on transducer representing applications where wall ringdown or opposite-
wall echo is sensed is shown in Fig. 7-24. A multiplexed application at
low g is illustrated in Fig. 7-25.

In typical industrial applications, the transducers are often clamped
onto the side of a "sight tube" which for ultrasound can be a strong steel
pipe. As an overflow detector, a transducer can be clamped under the
overflow spout or outlet.

Most often the wave mode is longitudinal, but shear (SV, vertically
polarized) and Lamb waves are sometimes advantageous when measurements are needed over a continuous range but access at the vessel bottom
is forbidden. Examples of proposed SV transducer configurations are
given in Lynnworth, 1979a and 1986. This 1986 reference also illustrates
a special "clamp-on" case, reproduced in Chapter 3, Fig. 3-39(a), wherein
a 100-kHz transducer is clamped against a thin plastic film stretched

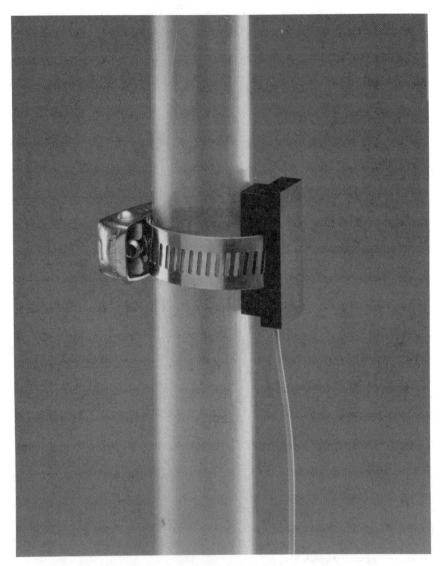

FIG. 7-24. Clamp-on discrete level application, courtesy Introtek.

across the top of a styrofoam beverage container. This is one of the few instances where a clamp-on transducer can transmit across an air path at atmospheric pressure and reflect off remote interfaces, and yield an easily-detected echo without interferences from reverberations in the wall of the container. (See also, Chapter 4, Fig. 4-9.)

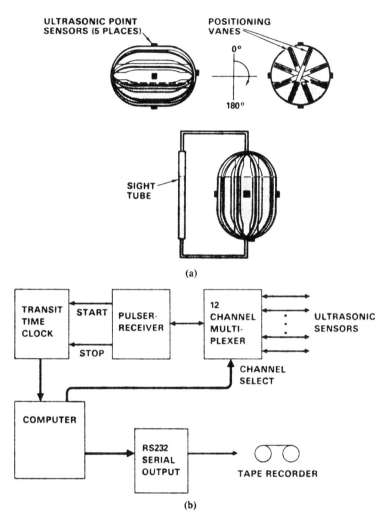

FIG. 7-25.   (a) Ultrasonic point sensor ground test setup. (b) Block diagram of 5222, mux and serial digital interface. After Jacobson et al., 1987a. © 1987 IEEE.

The common longitudinal mode ringdown method of detecting presence or absence of liquid at a particular level on the remote side of a metal wall proved impractical in the case of an automobile torque converter. The difficulty stemmed from the fact that, on the production line, the transmission oil inside the torque converter sloshed against the wall, making it difficult to reliably interpret the ringdown pattern. The wall thickness and the fluid gap were of such proportions that at normal incidence no longitudinal solution emerged. In the laboratory it *seemed* feasible to use zigzag shear waves to generate a distinct liquid-borne echo. On the production line, however, the requirement that the transducers be coupled precisely to a small protrusion on the irregular surface of the torque converter turned out to be too stringent a constraint, and so the method was abandoned. See Fig. 7-26.

7.5.2   Non-Clamp-On (Wetted) Transducers, Intrusive Sensors

Limitations of the type due to the wall ringdown as in the foregoing application are often avoidable by "eliminating" the wall and contacting the liquid directly. Vibrating elements are often used for sensing discrete levels or even distinguishing among several possible media according to their effect on the damping and resonant frequency of the detecting probe.

Time-of-flight methods can be used down to lower limits if the wall is locally removed as an echo-generating obstacle. An example of a wetted

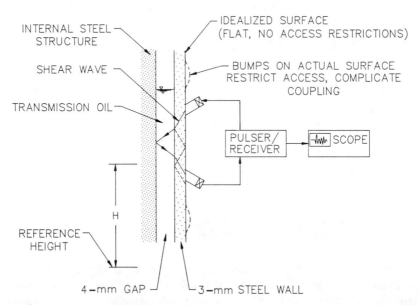

FIG. 7-26. Proposed method of detecting presence or absence of transmission oil at level *H* used obliquely incident vertically polarized shear waves. Transducers were spaced so that liquid-borne signal could be distinguished from metal-borne signal. Method worked in the laboratory on a simulator uncomplicated by bumps, but did not work on the production line. (Lynnworth, 1981, unpublished).

plug-style transducer installed in a thick-walled vessel, to measure liquid level at pressures up to 5000 psi (over 300 bar) is described as follows (Zuckerwar and Mazel, 1987).

A flat-bottom hole is drilled into a modified high-pressure plug, leaving enough wall to meet the pressure-vessel code. An ultrasonic transducer is rigidly bonded to the inner end of the hole. The plug is inserted into the pressure-vessel wall in a way that leaves the transducer exposed to the atmosphere (Fig. 7-27). This protects the transducer from high pressure and makes it easier to connect supporting electronic devices.

In operation, the transducer excites the plug flange into vibration, sending an ultrasonic pulse through the liquid to the liquid/gas interface where the

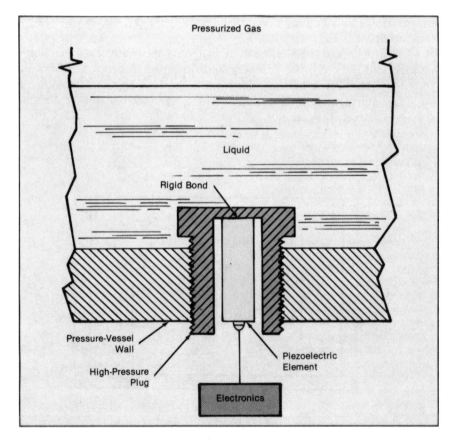

FIG. 7-27. Liquid-level-sensing piezoelectric element is mounted in a hole drilled in a high-pressure plug. The transducer is used to measure the depth of the liquid when the pressure in the vessel is high. After Zuckerwar and Mazel, 1987.

pulse is reflected. The returning pulse is intercepted by the transducer and transmitted to the supporting electronics that compute the liquid depth.

The prototype device was used in a high-temperature-structures tunnel that has cooling water pressurized to 6,000 psi (41 MN/m^2). An 11-mm hole was drilled into a 19-mm high-pressure plug, leaving at least a 3-mm wall at the inner end of the hole. A piezoelectric element having a fundamental vibration frequency of 10 MHz was bonded to the inner end of the hole. The combined piezoelectric element and plug vibrate at 258 kHz.

"Hybrid" methods of measuring liquid level, including, for example, the depth of water over a sand bed, may combine an external clamp-on transducer with an invasive beam splitter, Fig. 7-28. Alternatively, one or two immersed transducers may be used, one looking up, the other down; or one may employ two transducers, one in air, one in water, both looking down.

Measurements of transit time over an air or vapor path down to a liquid surface and back up, if c-compensated (using reference reflectors, differential paths, T-sensors or other means), provide another common avenue for interrogating the location of a fluid interface (Chapter 2, Section 2.4 and Chapter 9, Section 9.7).

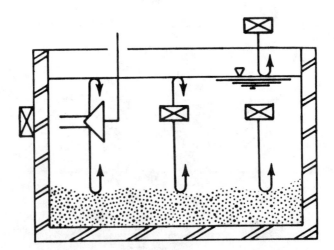

FIG. 7-28.   Left: Noninvasive transducer and invasive beam splitter proposed for measuring the depth of water above a sand bed in a laboratory flume. Center: Bidirectional immersed transducer (or pair of up- and down-looking one-way transducers) proposed for the same application. Right: Two transducers, one in air, one in water, each measure distance to nearest interface. For all three methods, the sand bed is not touched, and the water flow is hardly perturbed if intrusive elements and flow velocity are small.

7.5.2.1   Gap-Type Probes

One of the early *gap-type probes*, due to Van Valkenburg and Sansom, 1959, is shown in Fig. 7-29. This array provided for a multiplicity of discrete liquid levels to be detected. It appears that the use of gap probes has settled into applications where only one or two gaps suffice most of the time, e.g., high level and low level. Examples of commercially available gap probes appear in Fig. 7-30. Some suggested applications appear in Figs. 7-31 to 7-33.

One can think of a gap probe as a *transmission* sensor to detect presence/absence. An alternative approach might then be expected by using a *reflection* sensor, examples of which appear elsewhere in this chapter and also in Chapter 9, Section 9.9. The problem of detecting the presence of hands to be washed, while disregarding stationary objects inadvertently left in the same position, was solved by Mágori and Walker, 1987, by using a 200-kHz Doppler method. As illustrated in the bottom part of Fig. 7-30, the transducer is installed into the water tap. The transducer, as explained by Mágori and Walker,

[is] connected by a coaxial cable to a control box containing transmission-, receiving-, and echo-evaluation electronics as well as a power supply. The flow of the water is controlled by magnetic valves activated by a dc voltage of 24 V coming from the control box, which is mounted below the hand

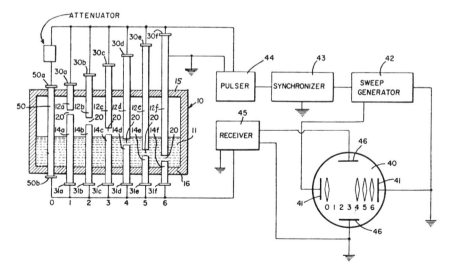

FIG. 7-29.   Array of gaps for sensing liquid level. After Van Valkenburg and Sansom, 1959.

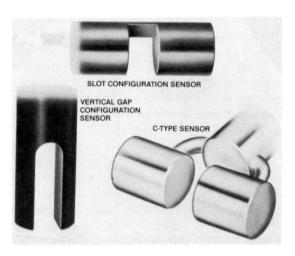

FIG. 7-30. Examples of commercially available gap probes for detecting presence or absence of liquid at a particular level, based on transmission, reflection or refraction of ultrasound, and a reflection (Doppler) system that detects the presence of (nonstationary) hands while disregarding echoes from stationary reflectors or moving targets not in a prescribed small zone. Top left: Dual-gap, high- and low-level sensor, courtesy Introtek/Magnetrol. Top right: Example of a basic slot configuration sensor for use with clear low-viscosity liquids; example of a narrow vertical gap sensor suited for detecting highly aerated as well as high solids-content liquids; and a C-type sensor with its transducers at each side of a "C" shape, for viscous or foamy liquids, courtesy National Sonics/Rosemount. Bottom left: Model 408 from Sensall/Rosemount detects interface between two immiscible liquids by utilizing an oblique interrogation technique. Bottom right: Water tap controlled by Doppler system, after Mágori and Walker, 1987, courtesy Siemens.

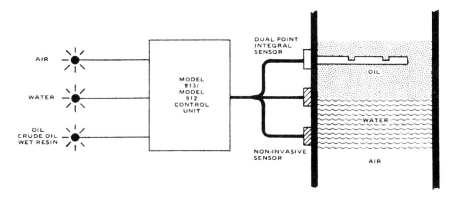

FIG. 7-31.   The Introlevel Models 912 and 913 Liquid Presence/Absence Detection Systems can be fitted with dual contact or noncontact sensors that allow the system to detect the presence of two different liquids or the interface of a liquid and settled solids in vessels or tanks. This system can be used to detect whether a pipe is filled with gas, oil or water. Other applications include the presence of water, air, or a sludge bed, or determining if the sensor is in air, water or wet resin. Separate relay contacts, audible and/or visual alarms, for each liquid are provided. Courtesy Introtek/Magnetrol.

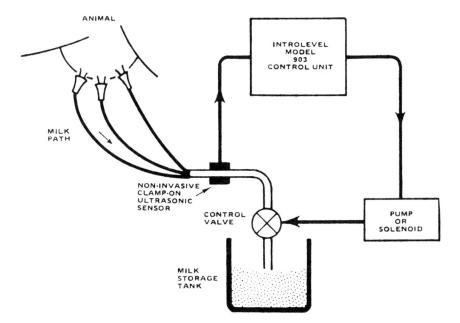

FIG. 7-32.   In the dairy industry, according to Introtek, there appears to be no efficient way to increase milk production with a conventional time control system. With such systems, the calculated milk output per cow is only an average value. The illustrated proposal suggests a means to optimize output and increase the production of milk. A basic system consists of an ultrasonic sensing device in the milk line and an electronic unit to control the pump and/or solenoid valve. The noninvasive clamp-on transducers sense milk in the tube without touching the milk. As long as milk flow is continued, the system continues milking. Courtesy Introtek/Magnetrol.

495

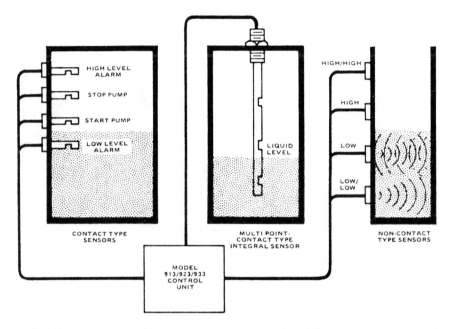

FIG. 7-33.   Diagram shows a number of multipoint level control systems for a variety of applications in or out of liquids. Courtesy Introtek/Magnetrol.

basin. The RU 200 [Siemens] transducer is aimed in a direction such that its ultrasonic beam intersects the water jet at an angle at which the echoes from the water jet do not return to the transducer directly. A short selected detection range defined by an appropriate time window is centered around this intersection between water jet and the expected position of the hands to be washed. Within the time window a Doppler-evaluation of the received signals is performed to discriminate between moving objects, as the hands which unavoidably move slightly, and stationary objects, e.g., a glass or a soap dish eventually left in the basin. [The transducer's] high directivity allows [one] to confine the region of sensitivity to a small well-defined volume. Thus the water flow is effectively interrupted as soon as the hands have been removed from this critical volume and spurious reflection from the water jet itself cannot trigger the sensor. The flow of water is initiated neither by people passing by nor by such acoustic interferences as ringing keys and the like. The sensor is not affected by dirt or water sprinkle, and the rugged open front face of the RU-transducer can be cleaned easily. These automatic water taps are used in lavatories and restrooms on highways, in restaurants, in hospitals, and medical practices.

The first RU 200 based industrial presence sensor with 200-kHz operating frequency [was] developed in 1980. Its nominal detection range of 0.2 m up to 1 m was divided into eight time windows each 10 cm wide, with one of them being selectable by a small plug arrangement. . . . All time windows before the selected one are combined to inhibit the sensor output, if an object is in the associated so called *blocking range*. Thus this ultrasonic presence sensor is active so long as an object is in the selected detection range with no other object at a shorter distance. In this way two different operation modes become possible: 1) proximity switch with a presettable detection range and 2) ultrasonic barrier with a passive reflector at a presettable distance.

7.5.2.2   Torsional Waveguide

The *torsional* waveguide discussed in Section 6.2.2 for density (ρ) applications is also applicable to liquid level applications. Liquid level may be computed either from (a) the transit time between a time reference event such as a sync pulse or a nonimmersed reference echo and the small echo at the liquid's surface, or (b) from the increase in transit time in a sensor immersed in a liquid of known ρ.

As an example of this sensor's use as a high-temperature, high-pressure water level indicator, the work of Miller et al., 1980, is presented. (Echoes at the water surface were not used in this work.) The discussion below generally follows the account by Miller et al., 1980, on the torsional sensor.

As a result of analysis of the events leading up to and contributing to the accident at Three Mile Island (TMI), the US Nuclear Regulatory Commission (NRC) issued requirements for improved instrumentation for pressurized water reactors (PWRs). One requirement is for "an unambiguous indication of inadequate core cooling." To determine whether this requirement could be met by liquid level detectors installed in the reactor vessel, the NRC sponsored the development of both heated-thermocouple and ultrasonic torsional wave sensors at Oak Ridge National Laboratory (ORNL).

Under normal operating conditions, a liquid level detector located in the reactor vessel would have to withstand temperatures to 375°C, pressures to 15.2 MPa, flowing water, and relatively intense radiation—certainly within the core, and perhaps above the core. The NRC requires that in-vessel instrumentation survive a "design-basis accident" and provide indications useful both during an excursion and for postaccident analysis. A temperature of 1260°C in the core is the most severe design-basis accident parameter postulated that the sensor must survive with continuous operation.

Manufacturers of PWRs have specified, in addition, that an in-vessel liquid level sensor must not be "event dependent;" that is, the sensor must provide an indication that can be checked, either by comparison with other plant parameters or through some form of "self-calibration." Under normal operating conditions, the reactor vessel is completely full. One conclusion from the post-TMI accident studies is that any instrument which provides no indication of liquid level other than *full* and no means for verification would run a strong chance of being disregarded. If, after months or years of indicating *full*, a sensor began to indicate some other level, the operators might conclude that the instrument was malfunctioning.

The ultrasonic torsional wave level sensor had been proposed as an instrument that would meet these requirements. The sensor element itself is a waveguide of noncircular cross section. A torsional wave can be excited in the sensor element from outside the reactor vessel by a coil that surrounds a magnetostrictive segment of the sensor element. Excitation coils can be located outside the pressure boundary, away from the severe environment inside the reactor vessel.

Depending in part on the sensor material, one of the limitations of the torsional wave sensor is that the velocity of propagation is dependent on the temperature of the sensor element. Tests were conducted at ORNL in a heated, pressurized water vessel to determine whether the temperature dependence of the velocity of the extensional wave could be applied to measure the temperature of the sensor, and whether this information could be used to remove the temperature component of the measured velocity of the torsional wave. A probe was designed and fabricated at Panametrics to allow both torsional and extensional waves to be launched on the same element, which greatly simplified the construction of the sensor. Otherwise, two probes would have been required—an extensional probe and a torsional probe. With two probes, possible temperature and other differences between the two probes could result in an erroneous measurement of level. The dual-mode probe (Lynnworth, 1970c) was constructed using two coils, one for extensional and one for torsional waves (Fig. 7-34).

Figure 7-35 is a drawing of the test vessel (pressurizer) used for testing and evaluating the probe at ORNL. The probe was inserted into the pressurizer, and the vessel was pressurized to 15.2 MPa (2200 psi). Electrical heaters on the lower half of the outside wall of the pressurizer heated the vessel and fluid to a desired temperature. In this case, the pressure containment included all of the probe wall (tubing wall) outside the pressurizer (as can be seen in Fig. 7-35). The liquid level in the pressurizer was raised by pumping cold water with a high pressure pump. For each high temperature test, the pressurizer was full at the start, and the measurements were made as the liquid level decreased. A calibration test to determine the hysteresis of the measurement was made at atmospheric pressure,

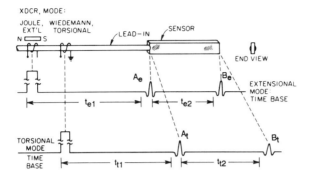

FIG. 7-34.   Four blanked intervals: t_{e1}, t_{e2}, t_{t1}, t_{t2}. After Miller et al., 1980. © 1980 IEEE.

first as the level increased and then as it decreased. The water temperature was maintained at 66°C throughout the test. The level of the water was measured by a calibrated differential pressure transmitter as a reference measurement. The total error due to nonlinearity and hysteresis at 66°C was less than 0.5% of full scale.

Figure 7-36 is a plot of error (cm) in the data taken at the various temperatures versus level (%). The greatest error was less than 7 cm.

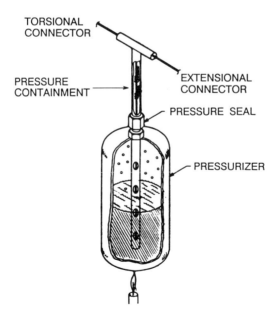

FIG. 7-35.   Pressurizer with probe. After Miller et al., 1980. © 1980 IEEE.

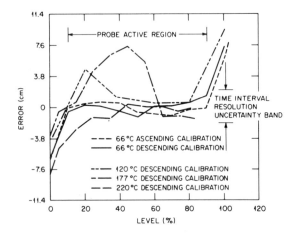

FIG. 7-36. Probe well behaved in active region. After Miller et al., 1980. © 1980 IEEE.

The method of temperature measurement derived from extensional wave transit time data resolvable to ± 50 ns, with temperature correction of the torsional time data, followed by calculation of the liquid level, results in a theoretical resolution of about 2 cm (an error of about 2.7% of span). There are several methods to improve the resolution, such as smaller time resolution, multiple transits, statistical techniques, and using sensor materials and/or cross sections exhibiting a larger change of transit time per unit of level change. (The work reported in this section, however, was conducted some nine years prior to the publication of Kim and Bau's analysis that led to the diamond cross section, a shape having some three times more sensitivity than a rectangular cross section of equal aspect ratio. See Bau et al., 1988; Kim and Bau, 1989.)

On the basis of the test results, only some of which are reproduced here, Miller et al., 1980, concluded that, in a range from 25 to 280°C, both waves in an ultrasonic extensional-torsional wave level sensor can be excited in the same probe. They further concluded that temperatures derived from the extensional wave data can be used to compensate for the temperature dependence of the torsional wave liquid level signal.

Since under normal operating conditions a pressurized water reactor is completely filled with coolant, an instrument that always indicates *full* is more likely to be disregarded if a loss of coolant accident should occur. Therefore, operators must have a way of checking the operation of the level indicator during normal reactor operation to determine that it is working properly. The ultrasonic extensional-torsional wave probe meets this need because, with a zoned probe, the output of the probe can be used to indicate not only level but to indicate temperature and density profiles as well. Correlation of these outputs with other plant sensor indications would provide a self-checking capability for the level probe. Furthermore, if the probe were located so that one zone could be confidently assumed to be completely surrounded with a medium of uniform characteristics, the probe itself would be self-calibrating, according to Miller et al., 1980.

The further possibility of using the waveguide sensor in a listening mode was suggested by Kirkpatrick and Kuzniak, 1987. They borrowed some of the ORNL equipment and continued the experiments at TVA, aiming at a design suitable for liquid level measurements in a full-size PWR for which the 15-m probes would need to be more rugged than the early experimental models made at Panametrics or subsequently at ORNL. For post-1980 developments at ORNL in this area the reader is referred to Rogers and Miller, 1982, and Dress and Miller, 1983.

7.5.2.3   Flexural Waveguide

Flexural mode tubular waveguide sensors were investigated by Dieulesaint et al., 1987a, in laboratory tests represented by Fig. 7-37. Interface reflection experiments were conducted with a partly-immersed duraluminum tube of 14-mm OD × 1-mm wall, with the transducers mounted so that incidence was from above the water. Using low-order modes with radial components, sensitivity of 1 mm was achieved, and probe lengths up to at least 10 m appeared reasonable. Temperature and viscosity effects were to be studied in subsequent experiments.

7.5.2.4   Extensional Waveguides

An extensional wave determination of liquid level H may be readily demonstrated in the laboratory by measuring the *amplitude* of the signal transmitted between two parallel waveguides (Lynnworth, 1979b). The resolution of H by this transmission method, however, is not very good. Leakage and transmission between parallel immersed waveguides may be useful in flowmeter designs (Swengel, 1955, illustrated in Chapter 3, Fig. 3-18) and in remote sensing (Motegi et al., 1987), but also may be a source of coherent noise that if unblocked may interfere with signals such as those transmitted around the sensor rings in the density profiler of Chapter 6, Fig. 6-19.

An extensional wave *reflection* method of measuring interfaces may be demonstrated by using a pair of waveguides as point-source end-fire antennas (Fig. 7-38). If the radiating end of the waveguide is about 1 cm in diameter, and if the frequency is about 100 kHz, the d/λ ratio in water is

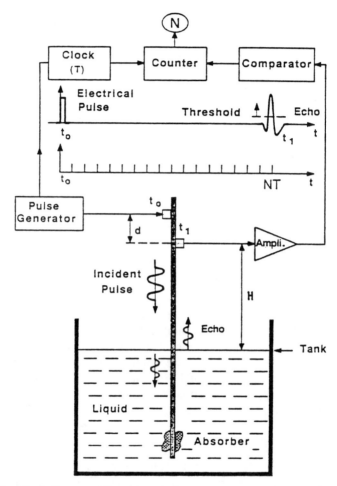

FIG. 7-37.   Block diagram of flexural mode liquid level sensor. After Dieulesaint et al., 1987a. © 1987 IEEE.

about 0.67. Since the waveguides are not very directive in this case, even if they are parallel (as in Fig. 7-38), it is easy to obtain echoes from nearby interfaces that are roughly perpendicular to them. In the laboratory this concept may be demonstrated by using wood or ice floating on or submerged under water. To block the undesired waterborne leakage, early experiments used "bubblepack" closed-pore air-cell packing sheets wrapped around each waveguide. In later acoustic shielding experiments the air (or vacuum) barrier was retained by a metal enclosure.

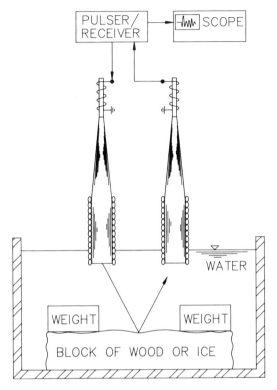

FIG. 7-38.   Reflection of 100-kHz waves from surface of a block of wood or ice held under water. Acoustic shields around each extensional waveguide block the direct paths through water.

7.6   Miscellaneous Level and Gaging Applications

As an aid in indicating the scope of interface problems addressable by ultrasonics, several special topics are collected in this section.

Level of Granular Solids

Solutions to this problem include vibrating rods and airborne time of flight. In one vibrating rod method, equipment made by Nohken consists of a vibration rod in which are mounted, near the tip of the detection tube, two piezoelectric elements, one providing vibration and the other receiving vibration.

The vibration rod is vibrated at a frequency controlled rate by the electronics. When the detection sensor tube becomes buried in the measured material, the intensity of the pipe's vibration decreases. This decrease is sensed by the receiving piezoelectric element and is then converted into an electrical signal which indicates the level.

In time of flight systems such as those made by Milltronics in 1987, air paths up to 60 m were possible. Different models have different ranges. Applications include noncontact continuous level measurements of non-solids too (sewage; liquids or slurries flowing through metering channels) but in this section attention is drawn to granular solids applications as suggested in Fig. 7-39. High-intensity transducers tend to be self-cleaning.

Problems in Fluidized Beds

Ultrasound may be expected to contribute to solving problems in this area such as measuring the rate of settlement as a function of particle param-eters (size, weight, shape), including cases where not all particles are identical. One approach would be to measure the frequency-dependence of attenuation and/or scattering as a function of level and time.

Hydrogen Slush

This cryogenic application introduces problems such as detecting the in-terface between liquid hydrogen and hydrogen slush, or measuring the vertical density profile in a dewar. The density difference between liquid and slush hydrogen is about 15%, and the density of liquid hydrogen is less than 10% of water's density. One method being considered as a pos-

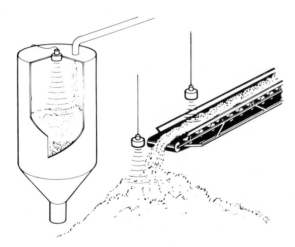

FIG. 7-39. Time of flight applications in which the level of granular solids is measured. Courtesy Milltronics.

sible solution for interface detection and/or density profiling is the torsional waveguide sensor of noncircular cross section. One torsional sensor fabricated in 1988 for hydrogen density profiling experiments in a dewar at a Colorado cryogenic test facility consisted of five horizontal SS304L rings spaced vertically about 75-mm apart. Each ring was of diamond cross section, of mean diameter ~30 mm and of aspect ratio $b/d \approx 3$. Transmitter and receiver waveguides were attached at diametrically opposed points on the ID of each ring, generally as indicated in Chapter 6, Fig. 6-19. In the initial tests planned, the magnetostrictive Remendur transducers were to be supported in air above the dewar, outside the cryogenic environment in which the temperature approaches hydrogen's triple point, ~30 K. Remendur is already known to retain its transduction properties down to $-196°C$ (Lynnworth, 1982b).

Gaging Gas or Solid Volumes Using Helmholz Resonator

Deskins et al., 1985, used the principle of the Helmholz resonator to develop a new acoustic technique to measure total body volume of premature newborns and infants. Their prototype system, termed an acoustic plethysmograph, was built and used to measure the volume of anesthetized newborn miniature pigs. Results of the animal body volume measurements compare within an average of 1.1% with body volumes measured by hydrostatic weighing.

The acoustic plethysmograph may be explained as follows. Rayleigh, 1870, modified Helmholz's original analysis to arrive at what is now the classic equation for the resonance frequency of the Helmholz resonator:

$$ f = \frac{c}{2\pi} \left(\frac{A}{V_q \left(\ell + \pi R/2 \right)} \right)^{1/2}, \qquad (7\text{-}5) $$

where c is the speed of sound in the gas, A is the cross sectional area of the opening, V_q is the volume of the resonating cavity, l is the length of the neck of the opening, and R is the radius of the opening. The derivation assumes that the wavelength is long compared to cavity dimensions. It can be seen from the above equation that the resonance frequency is inversely proportional to the square root of the cavity volume.

As Deskins et al. point out (see also, Kinsler and Frey, 1962, p. 186ff):

the long wavelength assumption needs to be satisfied to insure that the pressure field inside the resonating cavity is relatively uniform so that the pressure field, and hence the resonance frequency, is not affected by where the subject is placed inside the cavity. The ratio of wavelength to largest cavity dimension (from the neck to one corner of the cavity) is 12 in [Fig. 7-40].

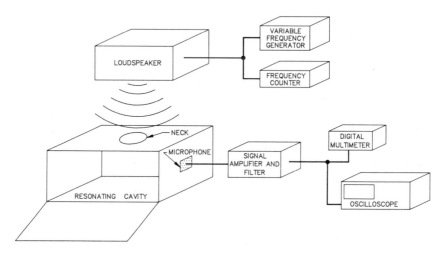

FIG. 7-40.   Schematic of the acoustic body volume measurement system, designed to measure body volumes between 0.75 and 1.5 l. Cavity dimensions are 20 × 20 × 45 cm. The diameter of the circular opening is 7.5 cm. After Deskins et al., 1985.

The empty cavity resonance frequency, on the order of 100 Hz, was found by Deskins et al. to be

very sensitive to changes in ambient temperature (~0.01 Hz/0.1°C, which compares favorably with the expected change in resonance frequency due to the dependence of the speed of sound on temperature of 0.02 Hz/0.1°C). Monitoring the empty cavity resonance frequency proved to be a convenient way to determine the temperature inside the cavity more accurately than with glass thermometers. Changes in the resonance frequency due to changes in atmospheric pressure and humidity were assumed to be insignificant compared to the changes due to temperature.

The cavity resonance was calibrated as a function of the air volume remaining, by displacing air with known volumes of water from 0.6 to 1.5 liter. For the prototype, a 1% change in the volume of the object being measured caused a 0.01% change in f, or about 0.01 Hz.

Deskins et al. concluded that the Helmholz resonator had been successfully used in a new application. The change in resonance frequency after an object has been placed inside the resonating cavity was used to measure the volume of the object. They suggest:

This new volume measuring approach might be applied in other situations where the determination of the volume of irregularly shaped objects is needed, particularly those applications where the traditional technique of hydrostatic weighing is inappropriate or undesirable.

For industrial applications, one would need to address practical limitations of a "Helmholz volume gage" such as response time, effects on accuracy due to cavity flexing, temperature and pressure pulsations or perturbations caused by introducing the test object, etc.

7.7 Cutter Vibration Monitoring Guides Mining Machine Along a Coal Seam

In a 1985 paper by Peterson et al. of MIT, a cutter vibration monitoring system is described which would map seam strata nonuniformities to permit guidance within the seam. This section closely follows that paper's explanation. In Peterson et al., a magnetostrictive commutatorless transducer is used that is compatible with conventional cutters and cutter replacement practices and is believed to be suited to the mining environment. Laboratory tests show that the system can detect bedding planes in otherwise homogeneous material. A data display system that would employ the human operator's natural pattern recognition skill and judgment is also described. Intended for remote control applications in general, the system is of direct interest for continuous excavators in coal or in any other machine excavated bedded deposit.

The interface of interest in this application is the coal seam. That is to say, one wants to sense the vertical position of a nominally horizontal seam, so that the cutting machine can be guided to follow that seam.

The selected system senses pick vibration, or more precisely, changes in pick vibration, rather than pick force, because, as configured, the vibration signal was found to be far easier to generate and transmit. Pick vibration varies with variations in strata properties. Major pick frequency correlates with dominant chip size and different strata have differing characteristic chip sizes. Even the presence of a bedding plane between identical layers of material causes a variation in pick vibration as chip formation is locally altered in the proximity of the bedding plane, according to Peterson et al., 1985.

By noting the angular position of the instrumented cutter, by using any of a number of existing instruments, and by plotting pick response versus position, a characteristic seam "signature" will be created. Repeated successive revolutions of the cutter head will create a recognizable pattern, not unlike that in seismic exploration techniques. There will be considerable noise in the pattern, since cutting generates many random responses, but, as in seismic work, the pattern of repeated, non-random variations should be visible. Also as in seismic work, there is no necessity to create an accurate measure of actual pick motion—only a record of motion that, however distorted, is repeated with each (or nearly each) pass of the pick. This, in large part, accounts for the relative ease of generating and transmitting adequate vibration data, in contrast to force data.

The actual device under development is shown schematically in Fig. 7-41. Cutter vibration is transmitted mechanically from a point on the surface of the cutter through a generally radial rod to the axis of the rotating cutterhead. At the axis the signal is transmitted from the rod, *rotating* with the cutterhead, to a *stationary* pick-up. Although there exist a variety of possible transducer concepts and transmission options, the magnetostrictive transducer described here accomplishes both functions in a simple and rugged form, according to Peterson et al. Its use in this project happened to have been suggested by the author, based upon some earlier experiments he had conducted up to 2000 rpm (Lynnworth and Bradshaw, 1971). But its implementation and evaluation were solely the work of the MIT group. For the prototype tests the rod, about 1.6-mm diameter, extended radially through a small hole drilled in the pick block, just behind the pick. The rod is spring-loaded to protrude slightly above the block surface, contacting the underside of the rear pick plug with a

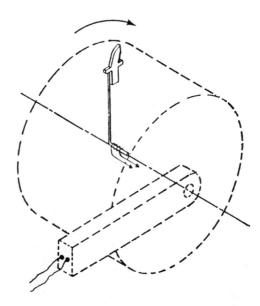

FIG. 7-41. Rotary coal mining cutter instrumented with magnetostrictive transducer to sense pick vibration signature and transmit it to a stationary coil on the outer axis. After Peterson, 1987.

light force (about 5 pounds). At the axis the rod turns 90° to coincide with the axis of rotation of the cutterhead.

As Peterson et al. explain:

> The rod passes through and rotates within the coil. Rotation of the rod, the presence of water, and even (non-magnetic) dirt do not affect the quality of the signal. The concept accomplishes, in one very simple device, transduction of the mechanical input signal to an electrical output signal, and transmission from rotating to non-rotating members. It consumes no power and generates very little power.

On the basis of laboratory tests Peterson et al., 1985, concluded:

- The magnetostrictive cutter vibration sensor appears to provide a strong signal which can be correlated to quite small variations in seam properties;
- The general concept appears to be sufficiently simple and rugged for reliable and safe operation in the mining environment.

They noted in 1985, however, that a very large gap exists between the laboratory and the mining environment. One cannot be certain that the expected patterns will be discernible above the noise until repeated cuts are made in real coal seam. Further, the ability to keep the pick mount free of dirt so that the pick is free to move slightly must be proven in the field. A staged development program was planned in which satisfactory continuous operation in a surface test facility was to be demonstrated before underground tests would be undertaken. Further details appear in the cited paper and in Peterson, 1987.

7.8 Bubbles and Particulates in Liquids

Methods for detecting discontinuities in liquids can be categorized as: passive or active; pulse or resonance; transmission or reflection; sound speed, attenuation or scattering methods; medium itself or foreign sensor (densitometer, microphone).

Bubbles in the sea, in heat transfer media (where bubbles indicate boiling), in a molten metal or in an intravenous delivery system have motivated investigators to develop a variety of bubble-detection approaches. Probably the most intuitively obvious method of detecting bubbles ultrasonically would be to utilize their reflecting properties in a reflection or transmission mode. While this simple approach works in the laboratory (Fig. 7-42), more sophisticated methods are often required in practice, especially if scatterers and sources of attenuation other than bubbles are present.

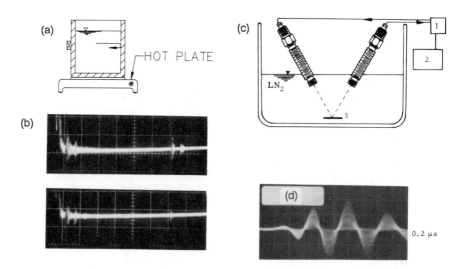

FIG. 7-42.   Water and nitrogen boiling experiments. (a) Schematic of clamp-on pulse-echo
test on vessel of ~25-cm diameter. (b) Oscillograms showing (top) measurement of echo
from opposite wall at 95°C, prior to onset of water boiling, and (bottom) measurement of
echo attenuated by boiling. After Lynnworth et al., 1971. © 1971 IEEE. (c) LN$_2$ boiling
experiment. 1–Pulser/receiver. 2–Oscilloscope. 3–Reflector creates V path. (d) Time exposure of over 100 pulses reflected during two-phase condition caused by end of warm
Remendur rod, 3-mm diameter, being placed below one leg of V path. Intensity of boiling
determines amplitude of echo. As rod cools towards -196°C, boiling diminishes and 2-MHz
signal grows. Note that two-phase condition does not appear to influence transit time. After
Nguyen and Lynnworth, 1983. © 1983 IEEE.

Starting with a bubbly air/water mixture, the fact that the air is more
compressible than water might lead one to guess that sound speed ought
to decrease monotonically as air bubbles are added to pure water. The
data of Karplus, 1958, obtained at 500 and 1000 Hz (Fig. 7-43) supports
this guess for void fractions ≤ 50%, but only if, as in Karplus' case, the
wavelength is sufficiently long compared to the bubble diameter. Karplus'
important experimental result, demonstrating 68-fold reduction in c (from
1500 m/s in "solid water" down to ~22 m/s in a 50% air void fraction
mixture) is sometimes reported without mention of the frequency. Scattering by bubbles is especially strong if the ultrasonic frequency is close
to the bubble resonant frequency. At resonance, the bubble's scattering
cross section, σ_s, defined as σ_s = total scattered power/incident intensity,
can exceed the geometrical cross sectional area, πR_o^2, by a very large
factor, e.g., 400 for a 60-kHz bubble, 5000 for a 1-kHz bubble. The dramatic multiplication of the resonant bubble's effective area for scattering

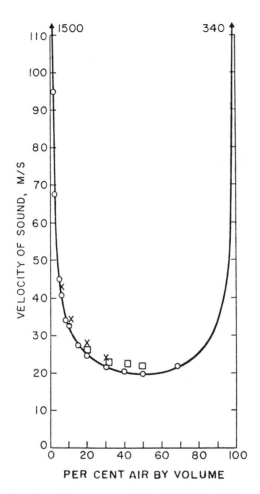

FIG. 7-43.   Velocity of sound in water containing air bubbles of diameter ≈0.01 cm. Frequency legend: ×—1000 Hz. □—500 Hz. ○—extrapolated to zero frequency. After Karplus, 1958, with permission.

is attributed to the fact that, near resonance, the bubble disturbs the phase structure of the incident wave over an area much larger than that occupied by the bubble's cross section (Medwin, 1970). Under isothermal conditions, the air bubble resonant frequency is approximately $550/d$ Hz, with d in cm. In some cases bubbles can be distinguished from solid particulates by comparing forward scattering with back scattering.

Boiling of water is detectable using a torsional density sensor, Fig. 7-44. The bottom multiple-exposure trace in the oscillogram is smeared

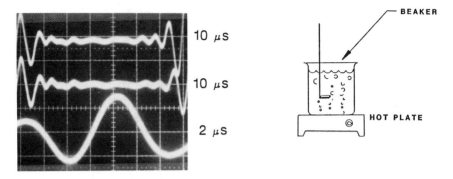

FIG. 7-44. Oscillogram for torsional sensor in air, hot water and boiling water (smeared trace). After Lynnworth, 1987.

because of the variable average density of the fluid in contact with the sensor.

Bubbles due to boiling can be detected by their noise (passive detection). Passive detection of *bubbles or boiling* under liquid sodium (Anderson and colleagues, 1971-1974) and active detection of *bubbles in blood* (Austen and Howry, 1965) or *in an intravenous solution* (Dam, 1986) provide examples of this class of applications.

Detection of *particulates in liquids* might be expected to follow the methods used in NDT to detect microstructural defects and inhomogeneities in solids. In liquid applications, however, the fluids are typically in motion and sometimes are molten metals. This adds complexity while adding new opportunities for ultrasound/measurand interaction, i.e, Doppler, range gated tracking of speckle echoes (Embree and O'Brien, 1985), phase and/or amplitude jitter, high frequency interrogation if the liquid is not attenuating, etc. The patents of Mansfield, 1981, and Abts, 1986, underlie the discussion below on particulate sensing applications in liquids.

Regarding high temperature *acoustic sensors for boiling detection*, Gavin and Anderson, 1971, stated that:

> Early detection of boiling in a fuel channel of a liquid metal fast breeder reactor may offer a means of safely shutting down the reactor before core damage occurs from loss of sufficient cooling. One method is to provide an acoustical sensing system [that] is capable of discriminating noise produced by boiling from normal reactor noise. The acoustic sensors can be located on top of acoustic wave guides which extend from just above the reactor core to outside the reactor tank; or they can be immersed in the coolant near the reactor core.

In the former case, the sensors would be subjected only to atmospheric conditions outside the reactor tank and would be easily recalibrated or replaced. Sensors for this type of service are commercially available. The problems of this system are associated with the acoustic waveguide, which would consist of either a solid rod or a liquid-metal-filled tube of uniform cross section extending into the coolant above the core. These problems include transmission losses, echoes, and narrow frequency response.

The use of immersed sensors offers advantages of more frequency bandwidth at high and low frequencies. They may discriminate more readily the acoustic emissions from boiling. To be useful for this application the sensor must be capable of producing a reliable output signal under gamma and neutron irradiation at temperatures up to 650°C.

Because such sensors are not commercially available, Argonne National Laboratory (ANL) developed a rugged microphone for use as an immersed acoustic monitor in liquid-metal fast breeder reactors (LMFBRs). These microphones have been tested extensively in water, in air and in sodium at 650°C. They have been successfully installed and employed as acoustic monitors in several operating liquid-metal systems, according to a later report (Gavin and Anderson, 1974).

The design that had been employed in the construction of about twenty-five microphones by 1975 is shown in Fig. 7-45. This ANL design is

based on the symmetric-drive concept in which two crystals in the form of circular plates are arranged with faces of like polarity facing a central electrode in a housing between flexible diaphragms. A change of pressure in the liquid in which the unit is immersed produces an electrical charge at the surfaces of the crystals; this charge is transmitted to the readout instrumentation via the high-temperature coaxial cable attached to the sensor housing.

To maintain sufficient electrical resistivity in the lithium niobate crystals, all microphones intended for operation at temperatures in excess of 1000°F [538°C] for periods of more than one month require a case vent consisting of a small diameter tube (1.6-mm OD × 0.5-mm ID) connected to an external source of oxygen maintained at <1 atm abs.

These microphones have a sensitivity of −160 dB referenced to 1 pC/μPa (1 μPa = 10^{-5} μbar). Frequency response from a low-frequency cutoff of the charge amplifier (typically 10 Hz) is flat to 10 kHz, then varies within a −5 to +10 dB [band] to a fast roll-off point at 90 kHz.

Typical units exhibit no loss of sensitivity after nine months exposure at 650°C in the ANL air-furnace calibration facility. One unit was in good condition after three months' exposure to sodium at 650°C in the ANL facility for acoustic calibration of transducers.

A microphone placed just under the surface of the sodium in the test vessel of the core components test loop (CCTL) operated satisfactorily for

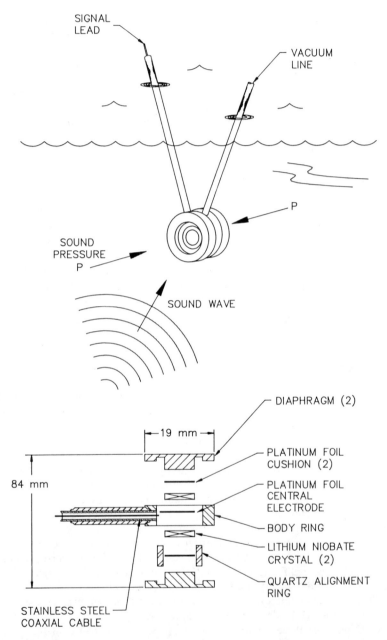

FIG. 7-45.   Top: High temperature acoustic transducer for operation in liquid sodium. Bottom: Assembly of ANL high temperature acoustic transducer. After Anderson et al., 1972.

120 days at 600°C. During a series of tests to determine the limits of detection of gas bubbles in the system, the sound produced by argon gas injected at 2 cm³/minute into the 525-gpm CCTL flow was readily audible when the amplified signal from the microphone was monitored by means of a loud-speaker.

A microphone placed in the plenum of the heat transfer simulation loop (HTSL) provided an audible indication of the onset of boiling during a series of sodium superheat tests. Signals from the microphone and from a waveguide were recorded for future analysis.

One ANL microphone together with a waveguide was installed in the primary sodium tank of EBR-II (Experimental Breeder Reactor II). Background-noise spectra were recorded at various reactor operating conditions. For additional details, see Anderson et al., 1972.

Regarding detection of *bubbles in blood* being delivered intravenously, consider the following application, as described by Powers (1987, priv. comm.) of Haemonetics, using ultrasonic equipment designed by Dam, 1986.

Haemonetics manufactures several types of automated blood processing equipment. In this area, the determination of air or fluid in blood tubing is critical for both safety and machine function reasons. In order to fill the need for a reliable and effective method of air detection, Haemonetics has utilized ultrasonic technology exclusively. They have found well designed ultrasonic systems to be effective in dealing with foam, small air bubbles, and segmented air and fluid moving at high velocities. See Fig. 7-46.

Some industrial applications of clamp-on transducers for detecting voids in liquids are suggested in Fig. 7-47.

Ultrasonic Inspection of Molten Aluminum

During the production of aluminum castings, quality control engineers and foundry managers are concerned about the quality of their product. Whether the end product is a structural casting for critical applications or a high quality commercial casting, plant personnel require some means of evaluating product quality and determining that the product meets the specifications and requirements of their customers.

During casting, according to Mansfield, 1981, molten metal quality can be determined ultrasonically to provide an immediate evaluation of the casting (ingot) before it is processed by using the Mansfield molten metal monitor (4M™) system. The probe material underlying this system is Ti, specially prepared to promote wetting by molten Al (Fig. 7-48.)

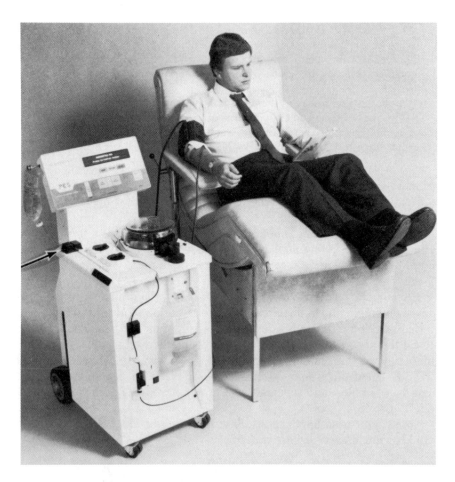

FIG. 7-46.   Plasma collection system uses ultrasonics to detect foam, small air bubbles and segmented air and fluids. Arrow points to ultrasonic module designed by Dam, 1986. Illustration courtesy Haemonetics.

According to Mansfield and Bradshaw, 1985:

The primary benefits of the 4M™ System can be summarized as follows:

1. The ultrasonic quality of metal is determined during the casting process. This saves time and expense by identifying poor quality metal before submitting the material to the final processing stages.
2. The presence of inclusions and non-metallic particles is determined, and these defects can be directly related to locations in the solidified ingots. Defective areas can be removed or [relegated] to less critical applications.

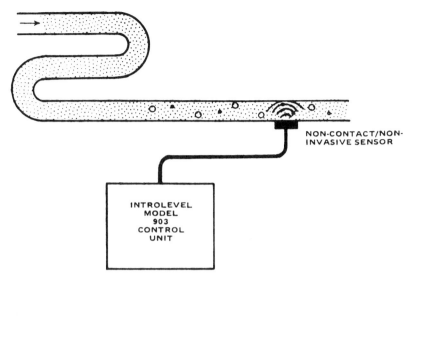

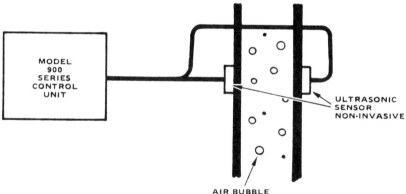

FIG. 7-47. Top: Detection of bubbles in a refrigeration system during charging. Bottom: Detection of air bubbles in a tube of diameter 3 to 25 mm using external transducers. Courtesy Introtek/Magnetrol.

3. For the first time, a method is available for on-line evaluation of cast house processes such as filtering, fluxing, and holding time.

4. The automatic calibration and probe/transducer evaluation features are very important. The system can be operated by plant personnel with minimal ultrasonic training and the inspection results are independent

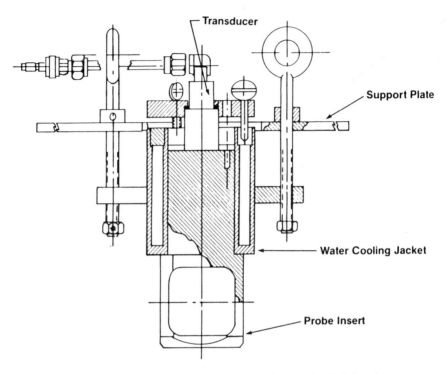

FIG. 7-48.   Mansfield probe as used in Reynolds Metals 4M™ System.

of operator variations and, therefore, *consistent* with each 4M™ System.

System . . . functions and computations are implemented in software. This provides for flexibility and adaptability, since new functions or controls can be supported by modifications to the existing software without redesign of hardware components.

Oil Droplet Detection in Secondary Oil Recovery

According to Abts and Dahl, 1986:

In secondary oil recovery systems, oil is recovered by pumping water into the ground through a secondary hole near the oil deposit. The water seeps through the ground, and as it is under pressure, the water forces the oil to the surface through a primary drill hole. This oil flow, however, will contain oil, water and solid particles. Most of the oil is separated from the rest of the flow at the surface, and the residual liquid is recirculated back into the ground through the secondary drill hole. The use of this residual water

substantially reduces the amount of fresh or sea water which must be con-
tinually added while the system is in operation. [See Fig. 7-49.]

There is an important drawback to this method. Despite efforts to separate the oil and water at the surface, some oil will be carried back into the
ground by the recirculated water. While the oil droplets do not adversely
affect the system, other than by possibly confusing detectors trying to detect
solid particles in the recirculated flow, the droplets could, if of sufficient
quantity or size, put a great deal of oil back into the ground thereby defeating
the purpose of the secondary oil recovery system to some degree.

[Abts and Dahl, 1986] discovered that discontinuities, particularly oil
droplets, in a recirculatory flow for an oil recovery system can be identified
and their concentration and size determined by using an ultrasonic transmitter to send pulses of ultrasonic energy into the flow whereby forward-
scattering of the pulses from the droplets can be detected and measured to
determine the concentration of oil in the flow.

In the preferred embodiment, a first transducer is mounted on a pipe
carrying the recirculating water flow. A second transducer is also mounted
on the pipe at an angle of at least 90° to the first. The ultrasonic pulses into
the flow from the first transducer are forward-scattered by any oil droplets,
but not by solid particles . . . Forward-scattering is detected by the second
transducer and counted. The number of such counts per unit volume determines the oil concentration, and the threshold level for the counter also
gives an indication of minimum droplet size.

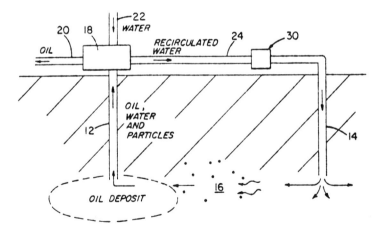

FIG. 7-49.   Diagram represents a secondary oil recovery system that comprises a primary
drill hole 12 which extends into the oil deposit and a secondary drill hole 14, which is nearby.
The lower end of secondary drill hole 14 is separated from the oil deposit by a ground seepage
area 16. The surface end of the primary drill hole 12 is connected to a separator unit 18.
Separator unit 18 also has an oil output pipe 20, a water input pipe 22 and a recirculated
water pipe 24. A detector 30 is disposed in the recirculated water pipe 24. After Abts and
Dahl, 1986.

In another preferred embodiment, the arrangement of the first and second transducers is the same, but the second transducer is used to measure the amount of ultrasonic absorption . . . of the transmitted ultrasonic pulse. If oil droplets are present, their contribution to the forward-scattered energy may be interpreted in terms of oil concentration.

An example of a focused transducer arrangement contained in the Abts and Dahl patent appears in Fig. 7-50.

On-Line Wear Particle Monitoring

As reported by Whitesel et al., 1986:

During normal machine operation, friction causes microscopic particles to wear away from the surfaces of bearings, gears, cylinders, pistons and other machine parts. By successfully correlating wear-particle size-concentration with machinery condition, engineers can optimize maintenance scheduling for most plant equipment as well as accurately diagnose incipient failure, simply by periodically sampling oil or hydraulic fluid.

Although wear-particle monitoring has vastly improved machinery diagnostics, none of the current methods, such as spectrography, chip detection, or ferrography, has been able to provide both qualitative and quantitative information on-line. One developing technology involving ultrasonic measurement provides this potential for lubrication and hydraulic systems. The instrument, an ultrasonic particle sensor (UPS), has been successfully used to count and distinguish wear particles from air bubbles inside the lubrication system of a bearing-failure test machine. An additional advantage of the ultrasonic technique is its ability to detect wear over a large size

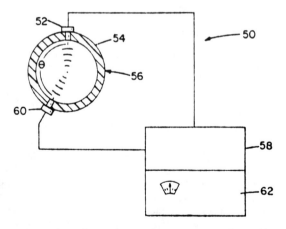

FIG. 7-50.   Focused transducer forward scattering arrangement due to Abts and Dahl, 1986.

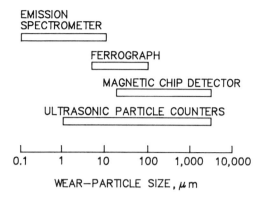

FIG. 7-51.   Comparison of wear-particle size ranges covered by various methods, after Whitesel et al., 1986.

range, practically encompassing other methods practiced [up to 1986 (Fig. 7-51)].

Tests in an oil recirculation loop, initially conducted with only two transducers, disclosed air/oil ambiguities and air bubble/particle ambiguities. These were resolved by increasing to three transducers and placing them equidistantly around the pipe, at locations near those predicted from scattering theory (40° in theory; 60° in practice). See Fig. 7-52.

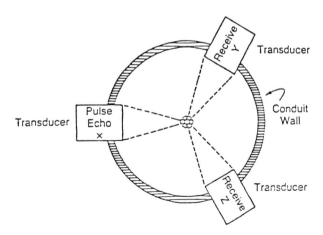

FIG. 7-52.   Detection at 60-deg scattering angle proved more beneficial because crosstalk is reduced and three-transducer configurations are possible, in study reported by Whitesel et al., 1986.

On the basis of further tests in the oil recirculation loop containing entrained air, glass beads and steel wear particles, Whitesel et al. concluded:

> The ultrasonic pulse-echo and scattering measurements can be used to detect particles and air bubbles in oil. For 5-MHz, focused, wetted transducers, the following performance characteristics were observed:
> - Particle and air-bubble size limits range from three to about 2000 μm.
> - The upper size limit could be extended . . . by using wider-beam transducers.
> - Minimum detectable contaminant concentration could be extended . . . by using wider-beam transducers.
> - Minimum detectable contaminant concentration is about 0.01 ppm.
> - Maximum measurable contaminant concentration is estimated at about 0.05%, but depends on material composition and size distribution.
>
> The UPS can be applied . . . where . . . immiscible materials need to be detected and counted in small quantities. Potential applications include detection of particles, air, or water in oil, and particles, air, or oil in water.

In a 1987 patent, Riebel proposed the simultaneous measurement of solids concentration and particle size distribution by using a plurality of frequencies whose wavelengths spanned the expected sizes of particles. In one example, the number of frequencies equals the number of size intervals into which the distribution is divided. By measuring absorption versus frequency, a number of simultaneous equations are obtained, the solutions yielding the sought concentration and size distribution.

Real-Time Particle Monitoring in an Ultrapure Water System

An increasing number of ultrapure water facilities are making use of continuous monitoring not only for ionic contamination but also for particulate contamination and organics. Motivation stems from economic considerations such as improving product yields in semiconductor fabrication facilities, maintaining sterility in pharmaceutical applications, and assuring the integrity of filtration operations in the production of process water for beverage and food applications. Gaucher, 1984, considers as an important reason for monitoring for particles at various points in the process, the ability both to assure consistent product water quality and to anticipate potential upset conditions. Specifically, continuous monitoring provides continuous real-time particle contamination data with a diagnostic capability unmatched by periodic sampling. Illustrative of this diagnostic capability, Gaucher refers to an application where counts at a monitoring point increased exponentially over a period of several days, then suddenly collapsed. Almost immediately, counts started slowly rising again, accelerated, then mysteriously collapsed again a few days later. The Facilities Manager recognized the phenomenon as a buildup of bacteria, rising exponentially until the food supply was exhausted, then rebuilding after the population collapsed. In another application reviewed by Gaucher, 1984, what appeared to be an erroneous high count was ultimately traced to a small pipe break on the suction side of a booster pump, causing cavitation and bubbles downstream. In the last example that we take from Gaucher, 1984, a semiconductor facility's DI water deteriorated after the final filters had been replaced and rinsed. This was later traced to the maintenance operator who had inadvertently replaced 0.2 μm filters with 2.0 μm filters.

Related background information and further discussions of bubble and particle detection using Micro Pure, Rexnord and later Monitek Technologies equipment manufactured in the 1980s and based on or related to the work of Abts and Dahl, 1986, appear in the following references: Abts, 1982; Behrman and Larson, 1987; Foote, 1985; Luppens et al., 1986; Massimino et al., 1983; Morin and Gaucher, 1986; Norvelle, 1984; Tolliver and Schroeder, 1983; Vogler and Groetsch, 1989.

Reporting on studies conducted at Yale, Roy, 1987, describes his

acoustic scattering technique for determining the compressibility and density of individual particles having diameters on the order of 10 μm, modeled as fluid spheres. Ultrasonic tone bursts of 2-μs duration and 30-MHz center frequency scatter from individual particles as they traverse the focal region of two confocally positioned transducers. One transducer acts as a receiver while the other both transmits and receives acoustic signals. The resulting scattered bursts are detected at 90° and at 180° (backscattered). Using either the long wavelength (Rayleigh) or the weak scatterer (Born) approximations, it was possible to determine the compressibility and density of the particle provided one possesses *a priori* knowledge of the particle size and the host properties. The detected scattered signals were digitized and stored in computer memory. With this information one can compute the mean compressibility and density averaged over a population of particles (typically 1000 particles) or display histograms of scattered amplitude statistics.

Sawada and Kitamori, 1988, describe the principle and applications of laser-induced breakdown acoustic spectroscopy to detect and count ultrafine particles in ultrapure water.

Gaseous Air in Pulp Suspensions

In the pulp suspensions used in papermaking machines, entrained air can cause difficulties in the papermaking process. The most severe consequences of disturbances caused by entrained air include decreased wet-web strength, retarded drainage and losses in pumping capacity. In other words, a high level of entrained air in a pulp suspension can limit the production capacity of a paper machine.

As explained by Karras et al., 1988, the amount of entrained air in a pulp suspension is controlled through the use of mechanical deaerators and chemical defoamers. Defoamers are surface-active chemicals that operate by reducing the stability of air bubbles and foam, causing them to collapse and coalesce. While necessary, defoamer chemicals are expensive, and using too much can reduce the tensile strength of paper and cause formation of deposits on the paper machine. The economical use of defoamers has been limited by the lack of an on-line device for measuring air content.

In 1988, Karras et al. reported a laboratory observation involving ultrasonic attenuation in a pulp suspension. This observation may be the basis for an improved way to control the feed of chemical defoamer. In the laboratory experiments, large air bubbles were observed to have only a small effect on ultrasonic attenuation, while entrained air produced a much larger attenuation. The entrained air apparently becomes attached to the fiber flocs in the suspension and therefore strongly scatters sound. According to Karras et al., the inertia of loose bubbles is too low for agitation by acoustic waves, and so they do not scatter sound as effectively as entrained air.

Based on attenuation measurements as a function of air void fraction (roughly in the 0.03 to 0.15% range) in pine pulp suspensions at different levels of Schopper-Riegler freeness, and field tests in several paper machines, Karras et al., 1988, concluded that when a given volume of air in a pulp suspension is dispersed into small bubbles, these bubbles attach themselves to fibers in the suspension, especially at network junctions. The suspension retains these small bubbles for relatively long periods of time, even under continuous stirring. These entrained air bubbles substantially increase the ultrasonic attenuation. It appears that the quantity of air in a pulp suspension can be monitored continuously by measuring the attenuation.

Karras et al. found that ultrasonic attenuation responds specifically to the presence of dispersed and entrained air bubbles. To the extent that attenuation is relatively insensitive to the presence of large, loose bubbles, attenuation may be the preferred and nearly-ideal control signal for detecting the harmful fraction of suspension-borne air. With a meter controlling the feed of chemical defoamer, a mill could maintain the air content at low levels without excessive consumption of chemical defoamers. The savings in chemical consumption would easily justify the cost of such a meter, Karras et al. concluded.

7.9 Position Sensing and Communication for Robotics, Machines, Tools, Fingers, Nails

In this section we include: echo ranging in air as applied to robotics; sensing of the position of parts of a machines such as carriage travel, or the axis of a lathe relative to a cutting tool; sensing of the position of an operator's finger on the curved screen of a computer terminal; location of a pointer on a map; continuity test to locate a hidden nail causing a protrusion in vinyl floor tile.

7.9.1 Robotics, Ranging and Remote Control by Acoustic Telemetry

Robotics is a large and fast-growing field. Attempts to summarize in this section even the narrower subject of robotic position sensing must necessarily be incomplete. To obtain some perspective on this topic, we borrow the following remarks from Schoenwald's 1985 review:

> Robotics is an acknowledged critical component of present and future manufacturing methods. The competitive drive for still higher productivity is pressing robotics technology to provide machine intelligence capable of accurate, real-time adaptive behavior. Sensing is considered a major component of machine intelligence. For manufacturing, repair and maintenance, a sense of touch, proximity and depth perception, traditionally considered uniquely human activities, are among the key requirements for any machine to emulate—or surpass. Acoustic and ultrasonic based sensors are demonstrating a capacity to provide perceptual information for range, object recognition, tactile force and shape determination. They also complement machine vision. Numerous systems make use of acoustics/ultrasonics and/or rely on the piezoelectric properties of transduction sensors. The device physics, data acquisition, signal processing and feature extraction capabilities are always interrelated.

Commenting on the future of acoustics in robotics, Schoenwald noted that

> the existence of any obstacle to implementing acoustics in robots and intelligent machines was not in the technology of acoustics by itself. Sensor and front end signal processing electronics can be very compact, and in some cases [had already been demonstrated by 1985]. The time delays involved in pulse-echo ranging methods are compatible with the 20-40 ms frame intervals generally considered appropriate for real-time control for working volumes where the robot arm extension is on the order of two

meters or less. Tactile sensors that are matrix addressable are also amenable to signal processing components that can satisfy these timing requirements.

The technology gap is much more apparent in the computer architecture and communications between components of a robotic system, e.g., sensors/preprocessors, real-time feature extractors and manipulator controllers, concurrent simulator and predictive modeler, off-line process planner and symbolic process controller. Generally speaking, acoustic sensors and the means for robotics to take advantage of them is not a bottleneck for automation science. What is required is a more homogeneous environment for signal processing, information extraction, inferencing and control.

For further details the reader is referred to Shoenwald's 1985 review, and to the body of literature from recent international symposia or journals on robotics.

In this section, the purpose is to identify the field of robotics as one likely to remain important for ultrasonics, and to mention applications that have already received considerable attention. Such applications include:

- Determining range or proximity of objects, as used in mobile robot navigation, robot arm control, micro-distance (e.g., range resolution of 0.2 μm in air at $f = 8.4$ MHz), and prosthetic aids.

- Determining shape of objects, i.e., acoustic imaging.

- Tactile sensing.

- Underwater applications analogous to those in air, especially searching, manipulating objects, assisting divers; underwater military applications for unmanned undersea vehicles including monitoring the vehicle's location, avoiding obstacles, locating targets, detecting threats such as mines, vessels or torpedoes, and imaging (Shaker, 1988).

A robotic excavator (REX) developed at the Carnegie-Mellon Robotics Institute in 1987 had an air time of flight sensing system (Section 9.7) that constructs a surface depth model to guide its air-jet tooling actions and thereby recognize, for example, non-erodable objects like large rocks. Figure 7-53 shows the Carnegie-Mellon Terregator (terrestrial navigator) navigating in a coal mine using a cluster of polymer transducers.

Apart from robotics, ranging in air (or water) can also aid human subjects directly. As an example, a multitransducer headset system due to Davison, 1987, enables a quadriplegic to control the cursor of a computer display by head motion (Coleman, 1985).

FIG. 7-53.   Terrestrial navigator in coal mine employs cluster of polymer transducers. Illustration courtesy W. Whittaker, Carnegie-Mellon Robotics Institute (priv. comm., 1988).

Remote Control by Acoustic Telemetry

Remote control of underwater equipment, including subsea communications over distances of 11 km, was reported in 1988 by Franceschini et al. and Trett. Franceschini et al. described tests in the Ionian Sea of a prototype acoustic system designed to control or monitor up to fifteen subsea wells in water depths of 1 km (Fig. 7-54). Transmission consists of 500 W pulses having a center frequency of 22 kHz, with 10° beamwidth. Trett's 1988 report, reviewed by Snyder, 1988, deals with undersea communication between a floating production system and remote subsea wells in the 1984 to 1988 period including ranges of 11 km. (See also, Urban, 1984.)

7.9.2   Wiedemann Linear Displacement Transducer

In a Temposonics™ linear displacement transducer, a torsional strain pulse is induced in a magnetostrictive tube by the momentary interaction of two magnetic fields (Wiedemann effect). One of these fields emanates from a permanent magnet which passes along the outside of the tube. The

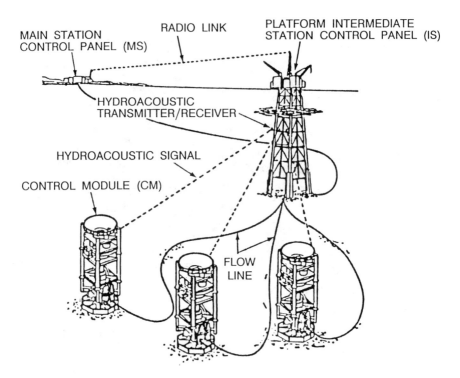

FIG. 7-54. Subsea wells acoustic control system (SWACS) designed to control or monitor up to 15 subsea wells. Hydroacoustic transmission links platform and wells. Radio link provides communication from onshore base to platform. After Franceschini et al., 1988. © 1988 Ocean Industry.

other is produced by a current pulse launched along a wire inside the tube. The interaction between the two fields produces a strain pulse which travels at the torsional velocity down the tube that serves as a waveguide, and is detected by a coil arrangement at the end of the device. See Fig. 7-55.

The Temposonics™ sensor tube is constructed of a magnetostrictive alloy that is processed to make its modulus of elasticity largely independent of temperature and aging effects. It is mounted under tension in a ~9.5 mm stainless steel tube for additional protection. Damping materials are clamped around the tube's ends to absorb the strain pulses and to prevent distortions resulting from their reflection.

The electric pulse-carrying conductor is an insulated wire coaxially threaded through the ~0.5-mm diameter × ~0.08-mm wall tube. This circuit is electrically isolated from the tube and the strain pulses. The

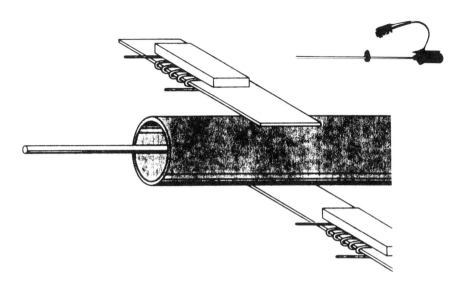

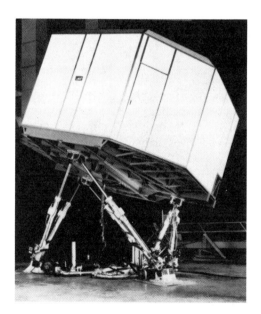

FIG. 7-55. Top: Diagram for explaining Temposonics™ Wiedemann transducer. Bottom: Application of linear displacement transducer in multi-axis tester. Source: MTS Sensor Division.

torsional strain pulse is mode converted by metallic tapes welded perpendicular to the tube.

The sensing coils generate the output signals and also provide common mode noise rejection from external electrical noise sources. The mode converter design responds only to torsional strain pulses on the tube; longitudinal pulses caused by shock or vibration are not detected.

Among the specifications for the Temposonics™ linear displacement measuring system:

- Repeatability, better than 0.01% of full stroke, with minimum full stroke of ±25 μm

- Linearity, ±0.05% of full stroke, with minimum full stroke of 50 μm

- Resolution, from 2.5 μm

- Displacement range, 9 m

- Temperature coefficient, ~5 ppm/°C (rod only)

One of its *limitations* is that probes must be installed correctly to prevent ferrous metal in the environment from shunting transducer magnets. *Applications* include:

- Control of position set points in a 220-ton plastic injection molding machine;

- Precision position and velocity control in a linear electrohydraulic servo actuator. The Temposonics™ unit is wholly enclosed in the cylinder to protect it from damage in the industrial environment;

- The linear transducer, combined with a hydraulic control package enabled elimination of all limit switches and stops heavy duty press rams within 25 μm;

- Determining the position of the knee on a heavy duty sawmill carriage, used to produce high quality lumber. In this system, the Wiedemann effect transducer acts in a closed-loop with the hydraulic positioning cylinder to accurately and consistently position the log before each pass through the saws. Each of the three knees on this carriage can be independently positioned, allowing the optimum cutting pattern for each log and maximum yield of lumber;

- The area of multi-axis motion control systems, as illustrated in Fig. 7-55, opens up numerous potential applications for the Wiedemann effect transducer. Link Flight Simulation Division of the Singer

Company uses the transducer in motion systems for flight simu-
lators. Six transducers are used in the six-degree-of-freedom platform motion system which is hydraulic actuator driven and uses a
computer controller. The motion control system replaces a rotary
potentiometer with racks and pinions, for a significant increase in
longevity and reliability.

7.9.3   Tool Datum

A machine tool datum monitoring system due to Tehon and Roberts, 1985,
of GE incorporates an ultrasonic detector with sensitivity to machine
vibrations in the 50–kHz frequency region. It detects abnormal cutting
operation such as chatter due to chipping of the tool bit. See Fig. 7-56.
   For the purposes of automated machine control, each cutting operation
must be measured from accurately known position references, requiring
recalibration of position for each change in the cutting tool surface.
   The datum, described in the paper by Tehon and Roberts,

   provides the means for precise location of that surface. It is an accurately
   machined steel bar, square in cross section, vibrating with an amplitude of
   the order of a few microinches and a frequency of 50 kHz. It is mounted
   so that its surface lies on the reference axis of the machine. When a cutting
   tool is brought up to that surface, at the instant of contact 50–kHz vibrations

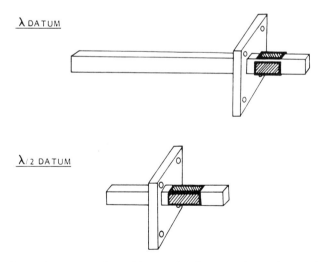

FIG. 7-56.   Datum bar construction, with mounting flanges and ceramic transducer chips.
After Tehon and Roberts, 1985. © 1985 IEEE.

are carried through the tool into the machine, and hence to the ultrasonic detector, indicating that the cutting surface is within microinches of the reference axis. Since the datum is square, two accurately located faces can be used to locate an X-axis and a Y-axis, individually.

Datum Construction.   Each bar is square in cross section, ~6 mm × 6 mm. Its mounting flange, a 25.4-mm square plate used for bolting the datum into the lathe, is 3 mm thick, and is centered about a nodal section of bar length. The bar and flange, machined from one piece of steel, are given a thin nickel plating for protection against corrosion. The nickel also provides a hard, wear-resistant working surface.

Each transducer is a PZT-4 ceramic chip 0.5 mm thick, measuring ~6 × 12.7 mm, poled in thickness between surface electrodes. Four chips are mounted on the faces of the datum bar, adjacent to the flange in a region of high vibration strain, as shown in [Fig. 7-56]. Solder and conducting epoxy have both been used satisfactorily for bonding; the datum tends to exhibit much higher resonant Q with solder bonds.

In order to reduce the number and intensity of spurious resonances, the geometry is symmetrical about the length of the bar, purposely minimizing any tendency to excite flexural vibrations.

The ultrasonic Datum Bar was reported by Tehon and Roberts, 1985,

to provide a precise reference for locating cutting edges of tools in an automated lathe system. It has the advantage that each tool is the probe used for measuring its location in the machine coordinate system.

(In passing, it is noted that worn tool detectors have been designed by other investigators, e.g., Begin, 1985, or Bischoff et al., 1987, based on passively sensing abnormal vibrations.)

7.9.4   Finger Position on a Curved Screen

Since the early days of ultrasonic NDT using surface waves (Firestone and Ling, 1945) the existence and location of the beam of ultrasound on the surface of the part was often verified by touching the surface and observing the corresponding damping.

Conversely, as shown by Adler and Desmares, 1985, 1986, one of the ways that the location of one's finger (and finger pressure) on a computer screen can be determined is to recognize a dip in the received SAW signal, when suitable reflective arrays are combined with two transducers.

The principle may be explained with reference to Fig. 7-57.

Figure 7-57 (top) illustrates the operation of the horizontal coordinate. Reflector strips tilted at 45° are placed along the top and bottom edges. A short pulse launched along the top edge from left to right traverses those reflector

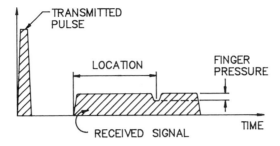

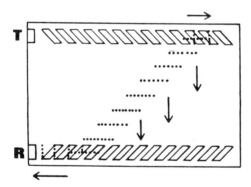

FIG. 7-57.   Top: Principle of operation. Bottom: Input pulse and output signal. After Adler and Desmares, 1986. © 1986 IEEE.

strips and is partially re-directed downward by each of the strips. It travels across the panel via many parallel vertical paths, is then partially intercepted and once more re-directed by the reflector strips along the bottom edge, and finally arrives at the bottom left corner. By now it has acquired a long rectangular shape, with each point in time corresponding to a specific vertical path across the panel. A finger touch causes an amplitude dip whose timing and depth indicate location and finger pressure (Fig. 7-57, bottom).

On curved surfaces, SAWs travel along great circles. Great circles do not intersect at right angles. Thus the two right-angle turns which work for a flat panel are wrong for a curved one. (See Chapter 2, Fig. 2-55.)

Adler and Desmares, 1986, describe several solutions.

The first places the two arrays on great circles but modifies reflector orientation and spacing to achieve the required reflection angles. A second solution uses arrays with a transverse phase velocity gradient which forces the SAW to follow a parallel rather than a great circle, thus restoring the right angle turns. A third solution takes advantage of the waveguide action of the reflector arrays to achieve a similar result.

7.9.5   Map Coordinate Resolver

Consider the problem of measuring the coordinates of a point designated on a map having an area of 1 m^2. One solution, due to Dieulesaint (1987, priv. comm.) and colleagues, is based on the following principle:

The abscissa X of a pointer is deduced from the periodic measurement of the time necessary for an acoustic pulse to propagate from a reference to the pointer, $X = cNT$, where c is the propagation velocity of the pulse and N is the number of beeps of a clock of period T. The ordinate Y is calculated in a similar way, a time T_o later: $Y = cMT$.

To implement this principle, one may choose for the acoustic pulse a symmetrical Lamb wave propagating in a duraluminum plate. This pulse is generated by piezoelectric bars bonded onto the edges of the plate. It is detected by a set of piezoelectric disks placed in the feet of a tripod (Fig. 7-58). The coordinates of the center of the tripod are calculated from the signals coming from the three disks. The map indicating the "objects" whose coordinates have to be sensed is placed on the duraluminum plate. The objects are successively pointed at by the tripod with the aid of an aiming part. The tripod, displaced by hand, is pressed slightly on the map. The effects of dispersion and of temperature changes are taken into account by the computer.

Specifications of the Prototype Used for Military Applications:

- Duraluminum plate. Dimensions: 1200 × 1200 × 1.5 mm^3. Useful area: 1000 × 1000 mm^2. This plate rests on cylindrical hard-rubber blocks carried by a honeycomb structure.

- Resolution: 0.1 mm.

- Precision: 0.5 mm.

- Measurement frequency: 100 points/second.

- Computer: HP 9816.

- Transducers: 15 piezoelectric bars (74 × 3 × 1.5 mm) for each side.

- Electrical pulse applied to each row of transducers: 200 V.

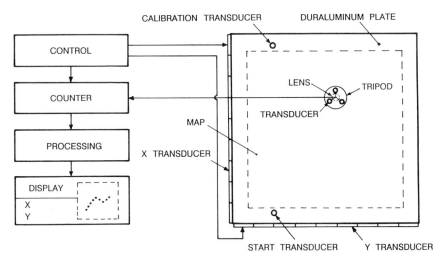

FIG. 7-58.   Diagram of operation of the map coordinate resolver. The tripod provided with a lens and a reticule is placed on the map, just above a chosen point. The coordinates of this point are deduced from the time-delayed signals induced in the tripod transducers by acoustic waves launched periodically along X and along Y. Source: Dieulesaint, 1987, priv. comm.

- Electrical pulse rise time: 200 ns.
- Clock frequency: 60 MHz.

This prototype has passed mechanical, thermal, electromagnetic radiation and sensitivity tests according to French Mil Specs. Three important characteristics provided by this particular ultrasonic approach are:

- Simple and rugged structure
- Insensitivity to electromagnetic disturbances
- Very low electromagnetic radiation level

Further details appear in Dieulesaint et al., 1984, 1987b.

7.9.6   Continuity Test Locates Protruding Flooring Nail Under Vinyl Tile

A few days after adhesive-backed vinyl tile was installed over a newly-laid plywood floor in the author's kitchen, three bumps appeared in two adjacent tiles. The size and uniformity of the bumps, ~6-mm diameter x

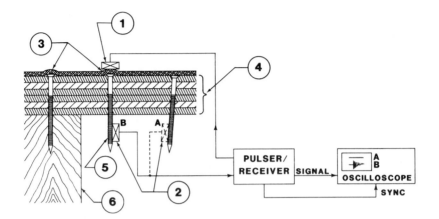

FIG. 7-59.   Continuity test finds the flooring nail that is responsible for bump in vinyl tile. Legend: 1–Transmitting transducer. 2–Receiving transducer. 3–Bumps in vinyl floor tile. 4–New plywood floor nailed to old plywood subfloor. 5–Ribbed flooring nail. 6–Floor joist. The transmitting transducer (1) is coupled to a bump. The receiving transducer (2) is coupled to the side of different nails until a signal is received (location B, oscilloscope trace B), identifying nail (5) as the culprit.

$\leq\frac{1}{2}$-mm high, suggested that they were due to the heads of the new flooring nails working their way up out of the plywood as the floor flexed. If the position of the offending nails could be determined beneath the kitchen in the basement from among the $100+$ nails whose points penetrated through the subfloor, it was thought that they could be pulled down tight and clinched, thereby undoing the bump permanently.

Using two glycol-coupled 100-kHz transducers in the through-transmission mode turned out to be an easy way to positively identify a nail whose point was accessible and whose head caused a bump. Actually only one of the three was located by the method shown in Fig. 7-59, but this was sufficient because once it was found the location of the other two neighboring nails was determined visually to be in the floor joists. As a result of this ultrasonic continuity test, it became clear that simply clinching every accessible nail that could be clinched would not have eliminated the three bumps. In other words, the test identified the problem with precision and thereby provided a basis for selecting a remedy from among several repair options.

8. Elastic Moduli Applications

In this chapter we generally limit discussion to isotropic elastic solids, we disregard differences between static and dynamic moduli (isothermal versus adiabatic elastic constants, respectively) and, except for one application, we disregard nonlinear effects. More general treatments appear in the literature (Mason, 1958; Green, 1973). This latter reference contains a comprehensive yet concise survey of the theoretical literature on this topic up to 1970. Elasticity (sound velocity) measurements at *high pressure* are reviewed by Anderson and Liebermann, 1968.

It is convenient to divide the applications into two categories, according to whether the specimen's cross sectional dimensions are large (bulk wave) or small (guided wave) compared to wavelength λ. By "bulk specimens," we mean specimens in which only bulk wave (longitudinal, transverse shear) propagation is to be considered. In "slender specimens," we shall be concerned primarily with the propagation of extensional and torsional waves. Equations relating Young's modulus E, shear modulus G and Poisson's ratio σ to sound velocities appear in Chapter 3.

Generally speaking, one chooses ultrasonic methods of measuring moduli over mechanical alternatives for reasons including:

High accuracy—for a representative specimen, accuracy is ultimately limited only by errors in path length and time interval or frequency

- Speed of measurement, including specimen preparation and testing
- Ease of overcoming constraints imposed by small specimen size
- Ultrasonic strains are within the linear elastic region
- Brittle materials can be measured despite small strains
- E, G and σ can be measured while specimen is under uniaxial or hydrostatic load
- Data needed at temperature extremes, especially high temperature
- Possibility of simultaneously testing a number of specimens using multiplexed transducers
- Low cost per data point
- Can be applied to irregular shapes and to materials while they are being processed
- Measurements of one or more sound speeds provides a rapid measure of product uniformity, for quality control

Precautions include avoiding errors due to faulty length or time measurements, and errors due to perturbing the measurand. A faulty length can be caused by non-ideal joints or too thick a coupling layer applied to the specimen. Faulty timing can be caused by using an uncalibrated time base; by selecting a half-cycle of the wrong polarity; or by selecting the wrong cycle in a pulse train. Timing errors also can occur because of multipaths in the specimen, e.g., spurious sidewall contributions, and in dispersive materials, failure to take into account the frequency or bandwidth of the measured signal. The measurand can be perturbed by overheating when making the joint; by high-stress or cold-work fabrication methods; by allowing couplant to be absorbed into a porous specimen. Chapter 3 or the abundant literature on this topic suggest means of avoiding these potential errors, consistent with the accuracy sought.

8.1   Bulk Specimens

Most ultrasonic measurements of elastic constants up to ~1985 were not aimed at "process control" but rather at determining the physical properties of a single crystal or a polycrystalline specimen. However, the increasing number of papers on ultrasonic monitoring of curing in polymers, and on evaluating green-state ceramic powder during compaction (e.g., Jones and Blessing, 1987) suggests a growing interest in the process control aspect of measuring elastic properties. In some cases (e.g., Gieske, 1978-1985), high-temperature elasticity or sound speed data are needed in part to interpret subsequent reentry data obtained with an ultrasonic ablation gage, or as calibration data in support of ultrasonic thermometers (Papadakis et al., 1974). Folds, 1972, measured the longitudinal velocity c_L and its temperature dependence in sixteen plastics, two elastomers and three foams by clamping the specimens between 500-kHz transducers. If an automatic intervalometer or sound velocimeter is available it is possible to measure the moduli in a large number of materials relatively quickly (Zacharias et al., 1974).

In many cases, the longitudinal and shear velocities are obtained in small cubes or other cylindrical samples of minimum dimension on the order of 10 mm. Contact pulse-echo measurements are typically obtained at a convenient high frequency that is not excessively attenuated. For many materials this means $f = 1$ to 10 MHz. For smaller samples $f > 10$ MHz may be required. For composites it may be necessary to test at $f < 1$ MHz. In the work of Nakano and Nagai, 1988, E and G of a machinable ceramic were measured up to 800°C with a microcomputer-assisted method that was found to have a time resolution of ± 0.1 ns and accuracy better than 3 ns.

To cite a current example, GTE uses ultrasonically determined modulus data to evaluate their research efforts on silicon nitride based composites (Katsoulakos, 1987, priv. comm.). In their work, E, G and σ were determined using shear and longitudinal transducers and a pulse-echo-overlap intervalometer system. Young's modulus of some of the materials tested ranged from 1×10^6 to 90×10^6 psi (1 psi = 69,000 dynes/cm^2). See Tables 8-1 and 8-2.

Table 8-1. Young's Modulus Measured for Selected Materials at Room Temperature (Katsoulakos, 1987, priv. comm.)

Material	Young's Modulus (10^6 psi)
Graphite	1.22
Cast iron	19.31
Nickel	30.40
Si$_3$N$_4$	43.00
Al$_2$O$_3$	53.66
SiC (NL203)	65.50
WC-6 w/o Co	89.00–91.00

Table 8-2.   Young's Modulus of Si_3N_4
            Composites (Katsoulakos, 1987,
            priv. comm.)

Material	Young's Modulus (10^6 psi)
Si_3N_4 + 10 v/o TiC	43.75
Si_3N_4 + 20 v/o TiC	45.91
Si_3N_4 + 30 v/o TiC	47.88

The measured Young's modulus of silicon nitride based composites shows that the elastic properties of these materials can be tailored by composite design. Increasing additions of a TiC–dispersed phase linearly increase Young's modulus (Table 8-2), and the observed increases fall within the theoretical upper and lower bounds (Paul, 1960; Fig. 8-1).

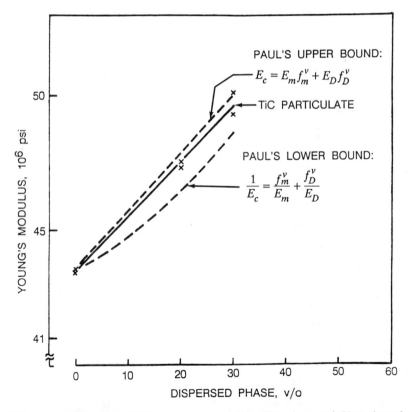

FIG. 8-1.   Additions of a TiC dispersed phase result in a linear increase in Young's modulus. Source: Katsoulakos, 1987, priv. comm. Illustration courtesy GTE.

Viscoelastic Specimen

Resonance techniques are available to measure elastic moduli, including cases where the viscous losses are not negligible. A method that in some respects is similar to that used by Köster, 1948, for testing metals up to high temperatures, is utilized in equipment supplied by Nametre.

To measure the viscoelasticity of a solid sample, e.g., a bar 1.03 cm × 4.60 cm × 19.31 cm, the bar is supported on a transducer like that shown at the lower left of Fig. 8-2. At either end of the bar and opposite the driver and detector coils, tiny expendable steel pole pieces are fas-

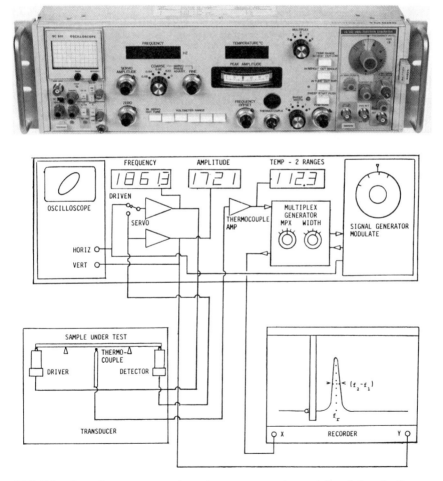

FIG. 8-2. Acoustic resonance equipment measures complex modulus of viscoelastic specimen. Illustration courtesy Nametre.

tened. With the equipment illustrated, explains Fitzgerald, 1987, priv. comm., the frequency is swept manually up to 50 kHz.

The fundamental resonance mode for the bar appears as the peak plotted on the X-Y recorder. Resonance is also indicated by the ellipse on the oscilloscope. The complex modulus $E^* = E' + iE''$, where E' is the storage modulus and E'' is the loss modulus.

8.2   Slender Specimens

To indicate the broad applicability of the pulse-echo method using approximately 100-kHz extensional and torsional waves, Table 8-3 lists typical applications in a variety of materials and in different geometries (but all of diameter D or other cross sectional dimensions small compared to wavelength), and over a range of test conditions. (In testing *textile* fibers much lower frequencies are normally used, down to 5 kHz, as in H.M. Morgan's equipment shown in Chap. 2, Fig. 2-57. The papers by Seferis and Samuels, 1979, Hussain et al., 1984 and Ryan and Postle, 1981 are

Table 8-3.   Examples of Moduli Measurements Using Thin-Line Ultrasonics and Ultrasonic Frequencies on the Order of 100 kHz

Glass rods and fibers
Boron-tungsten filaments
Silicon-carbide whiskers
Graphite-fiber yarn
Foam plastic
Sapphire single-crystal fibers
Mo, Re, W wires, to their melting points (2610, 3180, 3410°C, respectively)
Glassy carbon, graphite rods
Noble metal wires, ranging up to 2200°C
Kanthal in air to 1000°C
Iridium in air to 2000°C
Indium wire, cryogenic to room temperature
Paper—moist versus dry
Grass blades, leaves
Textile yarn
Wood: toothpick-size specimen
Plastic rods, ~3-mm diameter
Be rods, ~8-mm diameter
Measurements on wires self-heated electrically to incandescence
Simultaneous E and G
Switching to two or more specimens in furnace
Specimens of circular, square or other cross sections
Lead-in and (if used) lead-out members coupled to specimen by using honey, epoxy, solder, brazing alloy, welding or pressure

among the many on this equipment and method published in the past twenty years. On the other hand, the glass transition point of thin visco-elastic polymer films was measured at 158 MHz by Ballantine, 1988.)

The following paragraphs, mainly extracted from Lynnworth, 1973b, briefly present results in a few specific applications. We proceed from a very slender specimen, $D \approx 16$ µm, up to $D \approx 6.25$ mm. In its earliest form the measurement utilized two axially-separated coils (Fig. 8-3), due to Bell, 1957. This arrangement was later simplified by combining transmit and receive functions in one coil placed at the end (Lynnworth and Spencer, 1972). The two-coil arrangement has the advantage that each coil can be optimized for its function. The one-coil design is easier to use, and mainly for this reason appears to have become the most commonly used magnetostrictive transducer configuration with respect to thin-line moduli studies (Fig. 8-4). In passing, we may note that even if two coils are used they can both be wound over the end of one waveguide for pulse-echo measurements. At the end of this chapter, we shall explain a two-coil arrangement motivated by a pressure-coupled through-transmission method, where each coil is at the end of its respective waveguide.

Glass Fiber, $D \approx 16$ µm

Bacon, 1971, compares static versus dynamicYoung's moduli measurements at room temperature for round glass fibers mechanically drawn to diameters in the range 16 to 42 µm (0.62 to 1.66 $\times$ 10^{-3} in.) The static measurements were obtained on an Instron CRE tester at 5 mm/minute. For one composition the average of sixty static measurements averaged 15.93 $\times$ 10^6 psi. For this same material, the average of sixteen dynamic measurements, obtained using a Panatherm® 5010 ultrasonic instrument, averaged 15.94 $\times$ 10^6 psi, with much less scatter than for the static data. The averages agree to better than 0.1%. (The ultrasonic instrument automatically resolved time to ± 0.1 µs.)Bacon's comparison of static versus dynamic moduli for eleven different types of mechanically drawn glass fibers is given in Table 8-4.

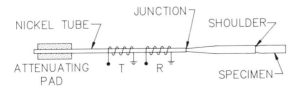

FIG. 8-3.   Diagram of waveguide used initially to measure the velocity of sound in metals at high temperature, and later nonmetals too. After Bell, 1957.

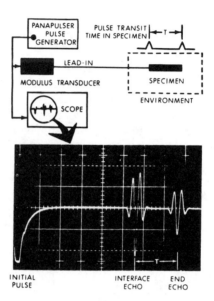

FIG. 8-4. Functional block diagram of basic equipment used to measure elastic moduli of slender specimens and a typical oscillogram showing a time interval T between joint and end echoes. Like polarity corresponds to a specimen of acoustic impedance less than that of the lead-in. The "modulus transducer" represents a coil surrounding the end of a magnetostrictive waveguide segment.

Table 8-4. Comparison of Static Versus Dynamic Determinations of Young's Modulus for a Group of Mechanically Drawn Glass Fibers Tested at UARL (United Aircraft Research Labortory). After Bacon, 1971

Glass Number	Density (g/cm^3)	Static Test (Mpsi)	Dynamic Test (Mpsi)
Owens-Corning "S"	2.48	—	12.6
UARL 320	2.929	18.6	16.2
321	3.632	17.4	18.2
331	3.629	18.3	19.8
344	3.390	18.3	18.6
348	3.651	—	17.6
367	3.531	17.5	18.3
368	3.623	18.8	18.5
370	3.629	17.7	17.8
371	3.566	18.4	18.1
405	3.726	—	18.4

Regarding ultrasonic measurement details, we should note that, since the specimens were so slender, it was necessary to insert a tapered impedance-matching section between the lead-in and specimen. This tapered section, ~1-m long, is made of glass, typically twice the Remendur lead-in diameter at the lead-in connection, and twice the specimen diameter at the specimen connection. The connections are butt joints, held together with a thermosetting adhesive such as beeswax. (Technique due to Fowler, ca. 1970, and reported in Lynnworth et al., 1977. See also, Krause et al., 1979.)

Glass Rod, $D \approx 1$ mm

Duchateau et al., 1970, 1971, used the thin-line technique to measure elasticity and physical properties of glass rods of diameters near 1 mm from room temperature up to elevated temperatures, ~700°C. They used both extensional and torsional waves, measured transit times with the same type of instrument as Bacon, and interpreted velocities and Poisson's ratio σ in terms of glass composition, structure, phase and viscosity. Table 8-5 summarizes the glass types, temperature ranges and salient observations reported by these investigators.

Table 8-5. Summary of Observations in Several Glasses at Elevated Temperatures, after Duchateau et al., 1970, 1971

Glass Type	Max. Temp., °C	Salient Observations
Soda-lime	600	c_e, c_t slopes indicate two distinct domains on either side of the transformation temperature. These correspond to expansion coefficients. Extending the tangents of the low- and high-temperature branches of the c_e versus T curves, they intersect near 550°C, where log η = 13.3, coinciding with the middle of the transformation zone (η = viscosity coefficient). Structural stabilization observed at 450°C. Poisson's ratio is minimum at transformation temperature.
Borosilicate	700	c_e, c_t increase with temperature at low temperature. c_e, c_t decrease beyond transformation temperature. When c_e is maximum, log η = 14.5. Phase separation was seen. Poisson's ratio showed large transformation domain. Three domains identified: velocity slightly increasing, slightly decreasing, rapidly decreasing as temperature increases. For a borosilicate of low expansion coefficient, Poisson's ratio, when plotted versus temperature, appears to indicate a number of singular regions below 600°C; its rapid increase from 0.20 at 600°C to about 0.28 near 700°C marks the beginning of the transformation zone.
Alumina-silicate	700	Effect of heat treatment (2 hr at 890°C) was noted.

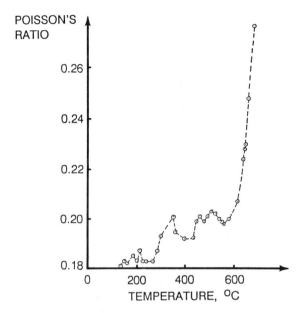

FIG. 8-5. Poisson's ratio versus temperature for a borosilicate glass of low expansion coefficient. After Duchateau et al., 1970, 1971.

It is clear that from thin-line measurements, one can not only compute elastic moduli, Poisson's ratio (Fig. 8-5) and compressibility, but also, as Duchateau et al. conclude, one can see evidence of the transformation zones, for determination of several characteristic viscosity points, and one can observe the duration of structural relaxation under diverse conditions.

Titanium Sheet, D = 1.5 × 3 mm

Thin-line ultrasonic measurements in titanium specimens of rectangular cross section, 1.5- × 3- × 7.5-mm long, demonstrate the utility of extensional waves in testing specimens of noncircular cross section. Moreover, by comparing extensional wave results with independent results using sonic resonance and also using strain gage techniques, the validity of the extensional wave technique is reinforced. Furthermore, this application takes the extensional wave method beyond the constraint of isotropic materials.

Specimens of titanium alloy sheet were sheared at 0°, 45° and 90° to the rolling direction. The as-sheared specimens, without further machining, were cemented to the end of a length of 0.5-mm diameter wire of Remendur magnetostrictive alloy.

A BC D E F

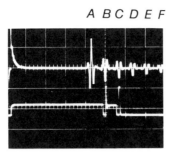

FIG. 8-6.   Oscillogram for titanium specimen of rectangular cross section. Top trace: receiver monitor shows echoes A, B, . . ., F. Bottom trace: selector monitor shows selection of echo pair B, C for automatic measurements of time interval. After Fowler, 1969. © 1969 ASM International.

Fowler, 1969, used the same type of intervalometer as did Bacon to measure the transit times. In the oscillogram, Fig. 8-6, the upper trace shows the reverberation of the pulse within the sample. The lower trace shows pedestals indicating the pair of echoes B and C selected from among these reverberations. Depending on the signal-to-noise ratio, higher precision is sometimes achieved by selecting echo pairs such as B, D or B, E representing multiple transits.

The Young's modulus of other specimens of the same material was measured independently by A.G. Martin, AMMRC, Watertown, Massachusetts, by sonic resonance and strain gage techniques. Table 8-6 compares the results.

While resonance and strain gage techniques require time-consuming, careful specimen preparation, the pulse-echo thin-line measurements

Table 8-6.   Comparison of Independent Determinations of Young's Modulus (Millions of Psi) in Rolled Titanium Sheet. After Fowler, 1969

Orientation to Rolling Direction	Sonic Resonance	Strain Gage	Extensional Wave Velocity
Parallel	16.1	16.1	16.1
		16.8	16.2
		16.07	
45°	16.8	17.1	16.9
		17.7	16.8
		16.78	
Perpendicular	19.7	19.6	19.6
		19.7	19.5
		19.67	

quickly yielded essentially the same modulus values in as-sheared specimens. For example, with the pulse technique, six samples were measured in less that thirty minutes including time for bonding.

Shear modulus too can be determined in specimens of rectangular or other noncircular cross section using the thin line technique and torsional waves, but one must take into account the noncircular cross section. For shape factor analyses, see Spinner and Valore, 1958; Tefft and Spinner, 1961; Bau, 1986; Kim and Bau, 1989. The shape factor for a few special cases is given in Chapter 3, Eqs. (3-10) to (3-13).

Binary Fe-Based Alloys, D = 3.2 mm

Speich et al., 1972, determined elastic constants in circular cross section, centerless-ground specimens of dimensions 3.2-mm diameter × 63.5- to 76.2-mm long, from liquid nitrogen temperature, nearly −200°C, up to + 200°C. Speich also measured E and G in iron and iron-chromium alloys from ambient to 1400°C.

The Young's modulus and shear modulus of isotropic Fe and binary Fe-C, Fe-Co, Fe-Cr, Fe-Ir, Fe-Mn, Fe-Ni, Fe-Pt, Fe-Re, Fe-Rh, and Fe-Ru alloys were determined as functions of composition (0-10 at. %) and

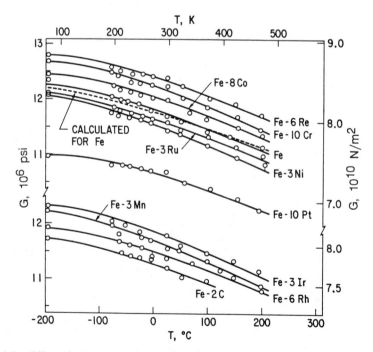

FIG. 8-7. Effect of temperature on Young's modulus of Fe and Fe-base solid solutions. After Speich et al., 1972. © 1972 ASM/TMS.

temperature (77-473 K) by the thin-line technique (~120-kHz elastic waves).

Speich's electronics essentially consisted of the basic equipment shown in Fig. 8-4, plus a time mark generator to calibrate the oscilloscope.

Time interval was averaged over five reflections within the specimens. Poisson's ratio σ and the bulk modulus K were computed from E and G. Speich found that the rates of change of E and G with composition ($\Delta E/\Delta C$ and $\Delta G/\Delta C$) depend on the change of lattice parameter with composition and upon the position of the element in the periodic table. Both negative and positive values of $\Delta E/\Delta C$ and $\Delta G/\Delta C$ were observed. These values are generally different so that σ may increase or decrease with composition. Comparable changes in the value of K/G also occur. These changes cannot be used to predict the effects of alloying additions on the toughness of Fe. The temperature dependence of E and G of the alloys is similar to that of Fe, decreasing in a nonlinear manner from 77 to 273 K and linearly from 273 to 473 K (Figs. 8-7, 8-8).

In Speich's work to 1400°C, the magnetic anomaly and the $\alpha \rightarrow \gamma \rightarrow \delta$ phase transformations were clearly observed. (Compare with high-temperature Fe data plotted in Chapter 2, Fig. 2–62, due to Hub, 1963).

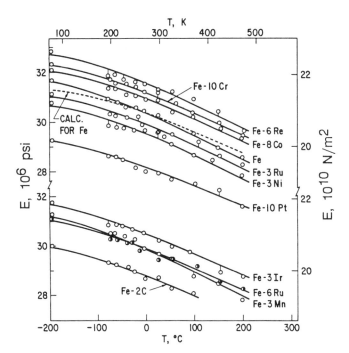

FIG. 8-8. Effect of temperature on shear modulus of Fe and Fe-base solid solutions. After Speich et al., 1972. © 1972 ASM/TMS.

Tantalum–Base Alloys, D = 6.4 mm

E, G and σ were measured in the Ta base alloy T111 and Astar 811-C by Sheffler and Doble, 1972. Specimen dimensions were 6.4-mm diameter by 100-mm long.

For these tests, the ultrahigh vacuum was not to be jeopardized by the transducer. Figure 8-9 illustrates the test arrangement. Results in the T111 alloy are presented in Table 8-7. Transit times were measured automatically to ± 100 ns with an ultrasonic intervalometer, as in Bacon, 1971.

Noble Metals, Thermocouple Alloys, Materials Studies

Papadakis et al., 1974, measured sound speed and attenuation in a number of refractory waveguide specimens, the materials being those commonly used in high temperature thermocouples and sheaths. One of the motivations for conducting these measurements to over 2000°C was to establish a database for the design and future use of these materials as temperature sensors according to the methods of Chapter 5, Section 5.2.1. Figures 8-10 and 8-11 are plots of Young's moduli in Mo and Re to over 2000°C.

In a series of studies between 1973 and 1985, Krause and colleagues at Bell Laboratories used the thin-line technique to investigate the large *anharmonicity* of amorphous and crystalline phases of a Pd-Si alloy (Testardi et al., 1973), stress dependence of E (1975; see Fig. 8-12), correlation between E and *thermal properties* of metallic glasses (Chen and Krause, 1977), *deviations from linearity* in elongation tests of fiberoptic waveguide material (1979), *anomalies* in Pd-Si glasses (Chen et al., 1980), *effects of*

Table 8-7. T-111 Physical Property Data Measured at High
Vacuum. After Sheffler and Doble, 1972

Temperature		Modulus of Elasticity (Mpsi)		Poisson's Ratio
°F	°C	Longitudinal	Shear	
75	24	26.5	10.6	.26
500	260	25.4	10.2	.25
1000	538	24.4	9.9	.24
1500	816	23.5	9.5	.24
1800	982	23.1	9.3	.24
2000	1093	22.9	9.1	.25
2200	1204	22.6	9.0	.26
2400	1316	22.5	8.8	.28
2600	1427	22.3	8.5	.30

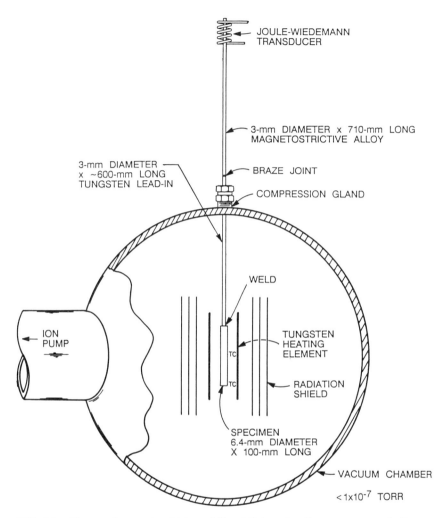

FIG. 8-9. Vacuum furnace in which moduli and Poisson's ratio were measured ultrasonically to over 1400°C by Sheffler and Doble, 1972, 1973. Method allowed investigators to reach high temperatures without contaminating the vacuum, and was easier to use than alternative extensometer methods available at that time, according to Doble (1987, priv. comm.).

quench rate on E (Chen et al., 1983), as well as E and other measurements in icosahedral $Al_{86}Mn_{14}$ (Chen et al., 1985).

McLellan, Yoshihara and their colleagues and students at Rice University, over roughly the same time period, used the thin-line method to study the variation of E in austenite with *carbon concentration* (Arnoult

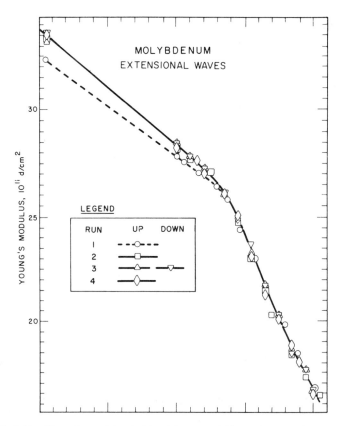

FIG. 8-10. Young's modulus in Mo. After Papadakis et al., 1974. © 1974 AIP.

and McLellan, 1975), *T-dependence* of *E* and *G* in pure Ni, Pt and Mo (1977), and later, in polycrystalline Nb, Ta and Va (1979) and in Pd single crystals (1987).

Grossman, 1984, used the method to measure, in effect, *E* averaged over the composite cross section of a corroding zircaloy wire, demonstrating a capability as an *in-situ corrosion monitor*. Corrosion film thicknesses <0.1μm were measured.

One potential application to be considered is the special case where the waveguide material is magnetostrictive. Here it becomes especially easy to monitor the sound speed even while the waveguide (wire) is in motion. The measurement may be accomplished by using one transducer coil plus reflective rollers or by using two coils spaced apart a known distance *L* (Fig. 8-13). Using the reflection or transmission "flowmeter" principles of Chapter 4, it will be understood that the wire's speed *V* can

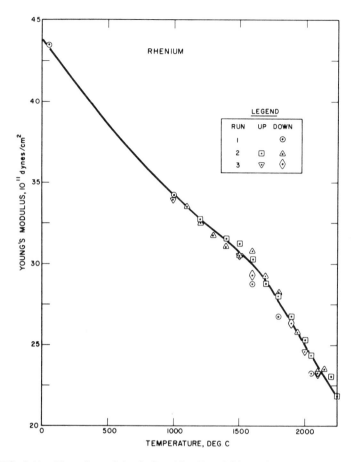

FIG. 8-11.   Young's modulus in Re. After Papadakis et al., 1974. © 1974 AIP.

be measured, not just its elastic properties. At V = 50 m/s, for a wire in which the sound speed c = 5000 m/s, if the coil to coil spacing L = 1 m, the contrapropagation transit time is $\Delta t = 2LV/c^2 = (2)(1)(50)/25 \cdot 10^6) = 4$ μs. In contrast to fluid flow problems, the entire wire cross section moves at the same speed, so there is no need for "profile correction." The length X of wire passing through the inspection station during time interval τ is given by the integral of $V(t)dt$ evaluated from $t = 0$ to $t = \tau$. In contrast to a determination of X inferred from a weight measurement, the foregoing ultrasonic determination would be independent of the wire's density or cross sectional dimensions. (Compare with speed and tension measurements in paper sheet by Jartti and Luukkala, 1977 and Marttinen and Luukkala, 1985, Section 9.3, respectively.)

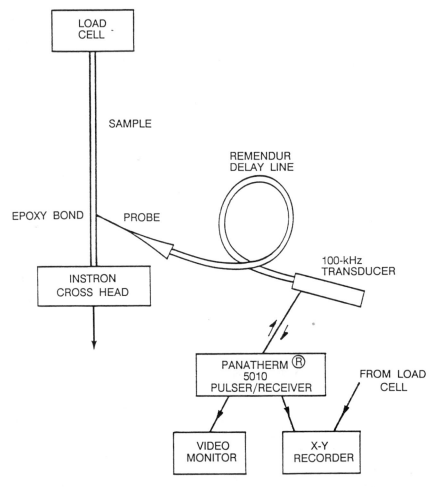

FIG. 8-12. Diagram of the experimental apparatus used to measure the force-elongation dependence of glass fibers. After Krause et al., 1979. See also, Fox, 1985.

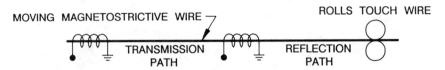

FIG. 8-13. Proposed pulse-echo and through-transmission inspection methods for a moving wire that is magnetostrictive. Wave speed and wire speed may both be determined by these methods while the wire is in motion.

Pressure Coupling

In Chapter 3, Section 3.3, dry pressure coupling is mentioned in connection with testing *bulk* specimens at high temperature (Carnevale et al., 1964; Figs. 3-21, 3-22). Pressure coupling is applicable to *slender* specimens too, if they are rigid. Pressure coupling offers advantages at ordinary temperature as well as at high or low temperature. With pressure coupling, in order to test slender specimens easily it is helpful, although not essential, if there is access to both ends of the specimen and if some simple alignment fixture is utilized such as a pair of nesting counterbored tubes (Fig. 8-14). For modulus testing at room temperature the fixture may be made of plastic, whereas for high temperature testing, graphite, ceramic or other refractory material will be preferred. Pressure may be applied manually, perhaps assisted by levers or springs, or automatically. Pressure may be applied continuously or only momentarily when a reading is desired, analogous to its 1964 use as referred to above, where elastic moduli were measured in bulk specimens at elevated temperatures using high pressure and momentary contact.

Crecraft, 1964, found that for dry pressure coupling to bulk steel having an ordinary machined surface roughness, pressures up to the order of 1 kbar (100 MPa) were required in order to *maximize* transmission, but pressure of only 5% of the maximum yielded detectable signals. Smooth flat surfaces can be coupled at even lower pressures. Translating this to a waveguide of 1.6-mm diameter, forces of ~2.5 lb ought to suffice, i.e., the weight of a ~1-kg mass. If the lead-in and lead-out are sharpened to a point, even less force is required for coupling, as Fowler, 1971, demonstrated when he developed an acoustic emission test set. One should limit the pressure so as to not deform the specimen. For nonrigid specimens, a resilient contact pad between the waveguides and the specimen may be considered. See also, Lacey, 1955, Bradfield, 1957 or Sproule, 1961, regarding point contact coupling.

Limitations

Dispersive Materials. The analysis and procedures presented above are applicable to elastic solids. In anelastic or viscoelastic dispersive materials, where sound speed depends on the test frequency, one should use a narrowband pulse (gaussian-, $\cos^2$- or rectangular-enveloped rf burst, for example), spectral analysis (e.g. Sachse and Kim, 1987) or other means such that the velocities and the computed moduli can be specified at a particular frequency. In acrylic and several other plastics, for example, it is observed that the dynamic E, G and σ determined at ultrasonic loading rates equivalent to ~100 kHz are substantially different from the values

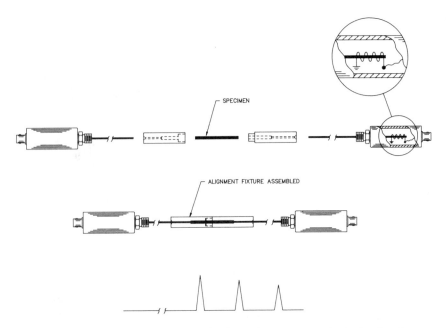

FIG. 8-14. Nesting tubes make it easy to align and pressure-couple, by hand or by mechanical means, slender waveguides of ~1- to 2-mm diameter to a slender specimen. The method is useful when executing elastic moduli measurements with extensional or torsional pulse techniques at room temperature, or at temperature extremes when the specimen material or the temperature make it hard to find a suitable bonding or coupling agent. As long as the transmission coefficients are not too large at each end of the specimen, the transmitted signal contains echoes as represented in simplified form in the sketch. The time between two successive echoes equals the round trip transit time in the specimen, eliminating the need to subtract delays in the lead-in and lead-out buffer waveguides from the total through-transmission transit time. A photograph of a commercially available transducer that may be used for these measurements appears in Chapter 3, Fig. 3-73.

obtained in the more usual tensile test. (See also, earlier remarks in this chapter regarding resonance techniques for measuring a viscoelastic specimen; Fig. 8-2.)

Slenderness.   At a given frequency f, if one seeks to increase D up to the maximum allowable specimen diameter, the required D/λ limit applies first to extensional waves, and later to torsional waves. For example (Mason, 1958), one may usually operate with extensional waves for diameters up to $D/\lambda_E = 0.2$, and with torsional waves, up to $D/\lambda_T = 1.2$. Note that λ_E and λ_T correspond to c_e/f and c_t/f, where f is, strictly speaking, the highest frequency contained in the pulse spectrum. That is to say, the specimen's slenderness is to be compared to the *shortest* wavelength in the pulse's spectrum. Since λ_{min} occurs at f_{max} one is led to ask: What is f_{max}? Referring to Fig. 8-6, if the time between two successive zero crossings in a given echo is denoted $T_z/2$, such that the rise time $t_r \approx T_z/4$, then $f_{max} \approx 1/2t_r \approx 2/T_z$. For torsional measurements in non-circular cross sections, a useful guideline is: Keep $\lambda_T > 10b$ where $b = $ largest cross sectional dimension.

The error in Young's modulus E calculated from ρc^2 increases as d/λ increases from zero, unless the proper correction is applied. For d/λ small compared to unity, E exceeds ρc^2, according to Eq. (3-6) in Chapter 3, Section 3.1.1 by about $2[\sigma\pi(a/\lambda)]^2$ or $2[\sigma\pi(d/2\lambda)]^2$. As a numerical example, according to Eq. (3-6), the error is less than 2% if $a/\lambda < 0.1/\sigma\pi$. Taking into account the above slenderness requirements on d/λ for propagation at $c_e \approx (E/\rho)^{1/2}$, and also the conclusion of Tu et al., 1955, that d/λ should exceed 5 for propagation of longitudinal waves, it is clear that compressional wave sound speeds are easiest to interpret in terms of Young's modulus for diameters either very small or very large compared to wavelength, i.e., $<1/5$ or >5. In process control applications where *relative* measurements of moduli suffice, relaxing the d/λ requirements may be justified. In some applications it may be convenient to test **bulk** materials by using **slender** waveguides to introduce and detect the sound (compare point-source, point-receiver measurements of Sachse and Kim, 1987). This situation would develop in Figure 8-14 if the specimen diameter were allowed to increase without limit.

9. Other Parameters— Special Topics

9.1 Viscosity and Polymerization

Of the three types of acoustic energy loss mechanisms, viscous, thermal, and molecular exchanges, this section treats applications involving only the first. However, the section also includes cases where changes in the amplitude of a reflected or transmitted signal indicates curing of a resin, even though such amplitude changes are not directly or necessarily due to a changing dissipation mechanism. For example, during cure, changes occur in impedance Z and, as a corollary, changes occur in leakage (coupling) out of or reflection back into a waveguide or other probe.

Stokes, 1845, is associated with the viscous part of the now-classic Stokes-Kirchhoff equation accounting for sound attenuation in fluids:

$$\alpha = \frac{\omega^2}{2\rho c^3} \frac{4\eta}{3} + \frac{K_{th}(\gamma - 1)}{C_p} \tag{9-1}$$

where α = attenuation coefficient, ω = radian frequency, ρ = density, c = sound speed, η = absolute viscosity, K_{th} = coefficient of thermal conductivity, γ = specific heat ratio and C_p = specific heat at constant pressure (Kinsler and Frey, 1962, p. 226).

559

Apart from any theoretical or empirical correlation between η and K_{th}, Eq. (9-1) suggests that α is at least a function of η, even if not in exact proportion to it. Accordingly, a number of interrogating means may be devised to convert viscous losses to an α-related parameter. Examples include (see Chapter 2): shear wave complex reflection coefficient (McSkimin, 1960; McSkimin and Andreatch, 1967; Cohen-Tenoudji et al., 1987), prf or power needed to sustain a particular amplitude in an extensional mode waveguide [Roth and Rich, 1953; Bendix' and later CE's (Combustion Engineering) Ultra-Viscoson] or in a torsionally-driven sensor element (Fitzgerald and Matusik, 1986; Nametre's Viscoliner; amplitude of a signal transmitted through a curing composite (Hinrichs and Thuen, 1985) or through a waveguide embedded therein (Sofer and Hauser, 1952; Papadakis, 1974; Harrold and Sanjana, 1987). A *clamp-on* pulse-echo measurement qualitatively showing the difference in attenuation of two fluids of very different viscosities (0.5 versus 100 poise) has been demonstrated in a 3-inch schedule 80 SS pipe (Lynnworth et al., 1971). Speake et al., 1974, reported on measurement of the cure of resins by ultrasonic techniques. Examples of other work on ultrasonic viscometry include: use of quartz tuning fork by Fisch et al., 1976, to measure η in the vicinity of the gas-liquid critical point with sample height hardly more than 1 mm, thereby avoiding ρ-gradient errors; use of strip delay line by Knauss et al., 1973. Other references of interest include: Harrison and Barlow, 1981, and Fisch et al., 1976. But as Kim and Bau, 1989, point out, many of the earlier "viscosity" sensors actually respond not to η alone but to the square root of the density·viscosity product, $\sqrt{\rho \cdot \eta}$.

Apart from amplitude effects, viscous drag may be detected in some instances by its retarding effect on a guided wave, analogous to mass loading. The retarding effect was used by Hunston et al., 1972, in their studies of viscoelastic constants with a *transverse shear* mode strip operated in the 1- to 7-MHz range. Later, in 1977, by using 100-kHz *torsional* pulses, the retarding effect was demonstrated by the author by immersing a rectangular cross section fused silica sensor into glycerine at temperatures from 6 to 90°C (η = ~10 to ~0.1 poise, respectively). At 6°C the fractional reduction in sound speed attributed to η was 21% (Lynnworth, 1977, p. 31). The retarding effect of a viscous fluid on a torsional guided wave is maximized if the sensor has a small polar moment of inertia and large surface area (Bau et al., 1988; Kim and Bau, 1989). Using a thin-walled 2.3-mm diameter threaded aluminum tube, Kim (1988, priv. comm.) obtained the torsional propagation data plotted in Fig. 9-1. For these early experiments, Kim used mixtures of glycerine and water at room temperature to vary the density-viscosity product. Depending on the sensor design and the range of density and viscosity in the fluid, sound

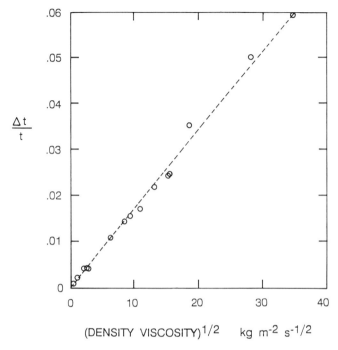

FIG. 9-1.   Fractional change in torsional transit time, $\Delta t/t$, versus the square root of the density-viscosity product, for a threaded Al tube having a thin wall. After Kim (1988, priv. comm.)

speed or attenuation could provide the better measure of viscosity. To the extent that the effects of ρ and η can be separated, a torsional sensor can be devised as a *kinematic viscosity* (v) probe, where $v = \eta/\rho$. In Bau et al., 1988, one of the configurations proposed for this purpose consists of a dual sensor (Fig. 9-2). One part has a diamond cross section responsive to ρ and to the square root of $\eta\rho$. Another part is a threaded tube responsive essentially to the viscous effect only. Hence, one has two equations and two unknowns, which when solved yield η and ρ. In Kim and Bau, 1989, this concept is explained using a circular cross section responsive to $\sqrt{\rho \cdot \eta}$ and a rectangular cross section responsive to both ρ and a $\sqrt{\rho \cdot \eta}$ term.

Hybrid Viscometer

Analogous to acoustomechanical flowmeters such as a rotameter whose float position is measured ultrasonically (Chapter 2, Fig. 2-1, entry 10), one can imagine a hybrid viscometer in which ultrasound measures the

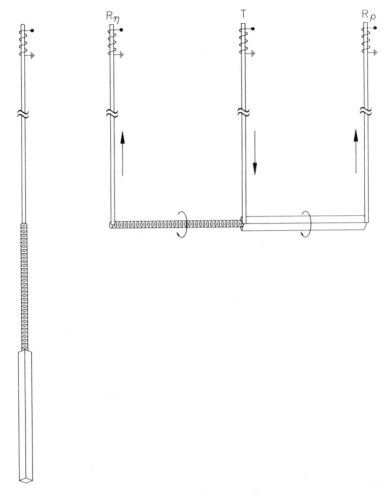

FIG. 9-2. Example of a two-zone torsional sensor in which the first sensing zone (e.g., a smooth or threaded tube) responds only or primarily to a viscous term proportional to $\sqrt{\rho\eta}$ and the second sensing zone (having a noncircular cross section such as a rectangle or diamond) responds to both ρ and a viscous term proportional to $\sqrt{\rho\eta}$. After Bau et al., 1988.

level of a liquid of known density falling at a rate determined by a constricted passageway and the liquid's viscosity, or the successive *positions* or *velocity* of a body falling through a liquid of known density. The falling body may be a sphere, cylinder or needle. (Park and Irvine, 1984, describe a *non*ultrasonic falling needle method.)

Oscillation Viscometer

As reported by Fitzgerald and Matusik, 1986, Nametre's Vibrating Sphere Viscometer, although designed originally for laboratory applications, later found its main use for in-line process control. Its advantages over some other in-line viscometers include: (1) no rotating parts in contact with the liquid, (2) the viscometer mounts directly into pipelines or tanks, (3) it operates even where flow is fast and turbulent, (4) its response to viscosity changes is almost instantaneous, (5) it is strong and impact resistant, and (6) it works at high temperatures and pressures. (See also, Fitzgerald et al., 1988a.)

Principle

The principle of operation is analogous to some other acoustic viscometers. When a vibrating rod is inserted into a liquid, the vibrations decay faster in a high-viscosity liquid than in a low-viscosity fluid. The rate of decay is a measure of the viscous resistance of the liquid and the density, ρ.

In the Nametre Oscillation Viscometer, the viscous resistance of the liquid is measured by the electrical power required to maintain a constant amplitude of vibration. The configuration of the transducer lets the tip be oscillated in a torsional mode by means of an inner shaft hermetically sealed from the liquid. According to Fitzgerald and Matusik, this construction improves accuracy over a wide range of viscosities, and results in a strong, reliable measuring tip.

Figure 9-3 shows the elements of the in-line oscillation-viscometer sensor. A strong but compliant stainless-steel sheath extends below the metal support plate to which it is welded. Magnetic driver and magnetic detector elements are mounted above the support plate onto a crossbar attached to a rigid central shaft, the bottom of which is welded to the free end of the sheath. The vibrating sensor is attached to the free end of the sheath, where the amplitude of vibration is maximum. The amplitude, less than 1 μm, can not be felt or seen. The driver coil causes the crossarm to oscillate at the resonance frequency of the transducer, 750 Hz. The detector coil detects the vibration and, by closed-loop circuitry, maintains a constant amplitude. When the tip is immersed, the viscous resistance of the liquid is overcome exactly by an increase in power.

A variety of SS316 sensor shapes are available (Fig. 9-3), including designs for temperature up to 343°C and for high pressure. These cover η from 0.1 to 200,000 cP or 10^1 to 2×10^6 cP. Applications mentioned in the cited article include: blending of motor oils, polymers, non-Newtonian papermill black liquor and mineral slurries.

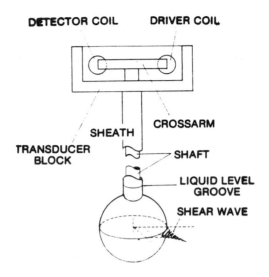

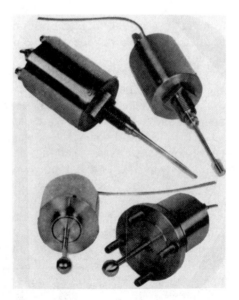

FIG. 9-3.   Top: Basic elements of Nametre's vibrating sphere viscometer. Bottom: Four in-line configurations for the oscillation viscometer available in 1987. Illustrations courtesy Nametre. See also, Fitzgerald et al., 1988a.

Polymerization Monitored by Embedded Waveguide

In ultrasonic thermometry (Chapter 5) approaches were divided into "medium itself" versus "foreign sensor." Ultrasonic viscosity/polymerization sensing can be similarly divided. One member of the "foreign sensor" category is an embedded waveguide technique that Harrold and Sanjana, 1987, of Westinghouse developed to monitor, *in situ*, the cure of thermosetting materials during manufacture, and subsequently, to monitor internal stresses, strains and emissions during the material lifetime. In their words:

> An acoustic waveguide is embedded within the thermosetting material as it cures. Monitoring the attenuation and transit time of ultrasound transmitted through the waveguide provides information on the curing process. Subsequently, the waveguide can be used to monitor impact damage and externally imposed stresses. Relative advantages of the technique include the ability to interrogate large volumes of the material, compatibility with host material and the potential for lifetime monitoring of material properties.

Experiments with the cure of 3501-6 epoxy resin and AS4/3501-6 graphite-epoxy prepeg are described in the cited 1987 paper:

> The waveguide, which is a solid rod, is embedded within the curing material which is either in an oven or a press. External to the curing environment, at both ends of the waveguide are bonded piezoceramic sensors which are used to transmit and receive ultrasound. The change in the peak amplitude and transit time of the signal through the waveguide are measured.

Harrold and Sanjana found these parameters to be very sensitive

> to changes in the viscosity and rigidity of the surrounding medium which take place during cure. Experiments with two different waveguides, a polymer-glass composite and a metal waveguide, indicate a strong attenuation of the signal with increasing viscosity and a strong minimum in amplitude at or near gelation. After gelation, the signal attenuation decreases as the rigidity of the polymer increases with cure in the solid state. Preliminary results indicate that a rapid increase in transit time (slower acoustic wave velocity) accompanies the liquid-to-solid transition (gelation).

Figure 9-4 outlines one approach Harrold and Sanjana used to monitor the cure of resins and composites.

> An acoustic waveguide of 1.5-mm diameter polyester-fiberglass is centrally located inside a mold when a resin casting is involved, or in between sheets of prepeg when laminates are involved. The waveguide then passes through acoustic, heat and vacuum isolation seals depending on whether an oven,

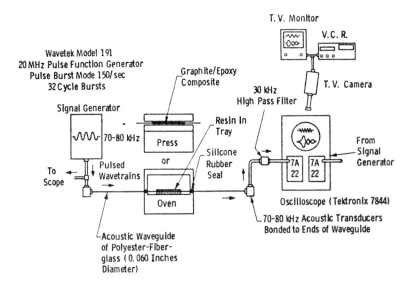

FIG. 9-4. Schematic of instrumentation and systems used for acoustic waveguide cure monitoring of resin and graphite/epoxy composite. After Harrold and Sanjana, 1987. © 1987 Plenum.

press or autoclave is used for curing, and acoustic transmitters/receivers are bonded to the waveguide terminations which are in an ordinary non-hostile environment.

The procedure is then to transmit pulsed acoustic wavetrains (~70 kHz) through the waveguide and oscillographically record the wave transit time and peak value of received signal throughout the cure cycle. A TV camera and VCR are used to record the signals on the oscilloscope screen so that rapidly changing signals can be analyzed later.

(Compare with Hansman and Kirby, 1986; Section 7.3.2.3).

Resin and Prepeg Results

In Fig. 9-5, the log of the received signal is plotted for 3501-6 epoxy resin together with the cure temperature versus time. The waveguide system is sensitive to density changes in the surrounding media, and as the resin softens and flows, changes are observed in the attenuation of the waveguided signal.

As the resin begins to polymerize and harden, the magnitude of the transmitted acoustic signal falls rapidly (4 orders of magnitude) to reach a minimum at gelation (liquid-to-solid transition) that occurs at about 170 minutes into the cure cycle. The signal then rises (3 orders of magnitude) to a new

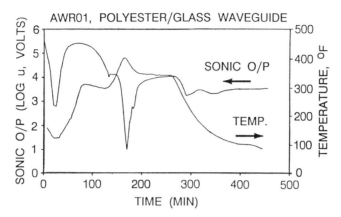

FIG. 9-5. Acoustic cure monitoring of 3501-6 resin using a polyester-fiberglass waveguide. After Harrold and Sanjana, 1987. © 1987 Plenum.

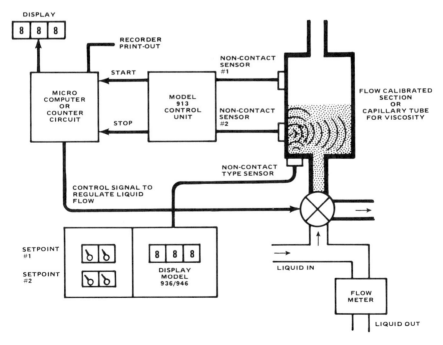

FIG. 9-6. Viscosity measurement converted to a liquid level measurement. Illustration courtesy Introtek/Magnetrol.

level which does not change greatly throughout the remainder of the cure cycle. As these experimental data were obtained using a 1.5-mm diameter polyester-fiberglass waveguide which may possibly have been influenced by the curing temperature (180°C), it was decided to also try a nichrome (80 Ni, 20 Cr) waveguide.

The data for the experiment repeated with the nichrome waveguide yielded an attenuation pattern similar to that obtained with the non-metallic waveguide. Experiments with AS4/3501-6 graphite-epoxy prepeg, for a polyester-fiberglass waveguide and a nichrome waveguide, give essentially the same results as those found for the resin alone. Throughout all these experiments, the acoustic wave transit time (wave velocity) was measured, and generally, near gelation, a large increase in transit time (slower wave velocity) occurs. This phenomenon is being investigated further by Harrold and Sanjana, 1987. (See also, Sun and Winfree, 1987.)

A viscosity measurement can sometimes be converted to a flow or liquid level measurement. This is suggested in Fig. 9-6. Note that in this simplified drawing the level is shown horizontal. A non-planar surface would complicate the interpretation.

9.2   Pressure

As indicated in Chapter 2, pressure P in gas, liquid or solid media can be detected as a consequence of its influence on sound propagation. As liquid examples of the "medium itself" method, if there were no other variables present, one could interpret studies of sound speed c versus P as calibrations for water (Litovitz and Carnevale, 1955; Del Grosso, 1973), CCl_4 and silicone fluids (McSkimin, 1957), in 1-propanol (Hagelberg, 1970) or petroleum products (Korycki et al., 1979). A solid example of an ultrasonic P gage is due to Hoechli et al., 1972, of IBM. Reference is also made to ultrasonic tension meters. An example of equipment that responds to bolt stress and strain is shown in Fig. 9-7.

Regarding the "foreign sensor" approach, experimental SAW devices are reported from time to time (Hauden, 1987). But apparently, the most widely used ultrasonic pressure sensor in the past twenty-odd years is that based on the resonance frequency of a quartz crystal. The main applications for the Hewlett-Packard quartz pressure gage (Chapter 2, Fig. 2-81) are probably "downhole" to measure quickly yet accurately the high pressures in wells. A probe designed for this application is shown in Fig. 9-8. Reasons for selecting an ultrasonic device, namely the quartz P gage, will be evident in the following applications, according to information supplied by Hewlett-Packard.

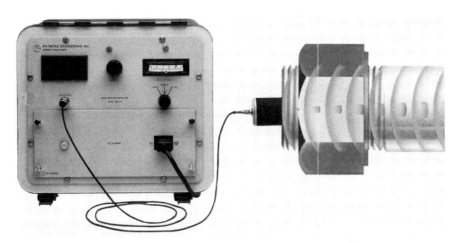

FIG. 9-7.   Raymond Model PDX 633 extensometer measures bolt tension or pre-load. Operator attaches the transducer to the end of the bolt. Using a single knob, operator adjusts
zero prior to tightening. Bolt is then tightened and instrument reads change in length of bolt,
which is proportional to pre-load. Courtesy Raymond Engineering.

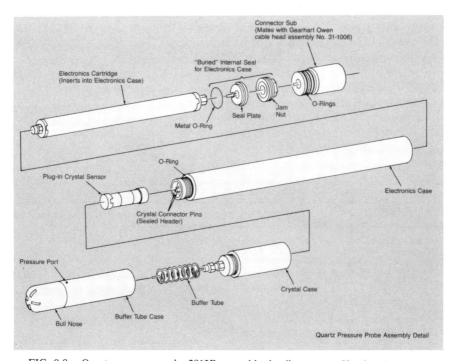

FIG. 9-8.   Quartz pressure probe 2813B assembly detail, courtesy Hewlett-Packard.

(As a more general introduction to the problems in *downhole* instrumentation, see Havira, 1986, Carson and Wolfenbarger, 1986, or the *Proc. Technical Review "Advances in Geothermal Reservoir Technology—Research in Progress"* (1988); see also, Section 9.10.)

Interference testing has long been accepted as the most accurate pressure test for determining permeability and other important *downhole reservoir properties*. Until the advent of the quartz P gage, however, few interference tests were run because long test times and small pressure changes made these tests uneconomical.

The quartz gage makes interference testing economical by significantly reducing test time. Pressure transients travel at a velocity that is determined only by reservoir properties. The quartz gage, however, detects pressure changes sooner because it resolves pressure to 0.01 psi (69 Pa) at any pressure up to 12,000 psi (82.7 MPa). This ultra-high resolution enables interference tests to be completed in a fraction of the time required by nonultrasonic gages. For example, consider three reservoirs with the properties listed in Table 9-1.

Table 9-2 compares the time to run an interference test on each reservoir with a conventional gage (1 psi resolution) and a quartz pressure gage (0.01 psi resolution) at the observation well.

As shown in Table 9-2, it would take 72,000 hours or over eight years to test Reservoir C with a conventional gage. The quartz gage would require only one day. According to Hewlett-Packard,

> *New wells* frequently are in remote locations far away from major pipelines or storage facilities. This lack of storage necessitates low flow rates for initial pressure tests. Low flow rates create only small pressure changes.
>
> *Data Retrieval Time.* The initial tests on new wells are run to determine if a well should be completed. If the well is to be completed, the extent of

Table 9-1. Examples of Reservoir Properties. Source: Hewlett-Packard

	Reservoir A	Reservoir B	Reservoir C*
Permeability k, md	15	10	10
Thickness h, ft	50	50	50
Porosity ϕ	0.1	0.2	0.2
Compressibility C, psi^{-1}	5×10^{-6}	10×10^{-6}	10×10^{-6}
Viscosity μ, cp	1.0	0.5	0.02
Formation Volume Factor B, rb/stb	1.1	1.1	0.0025
Change in Flow Rate q	150 b/d	100 b/d	2 mcf/d
Distance Between Wells r, ft	1500	660	2000

* Reservoir C is a gas reservoir.

Table 9-2. Comparison of Time Required to Conduct an Interference Test. Source: Hewlett-Packard

	Reservoir A	Reservoir B	Reservoir C
Conventional Gage 1 psi Resolution	480 hours	670 hours	72,000 hours
Quartz Pressure Gage 0.01 psi Resolution	113 hours	66 hours	24 hours

the wellbore damage also needs to be determined. These decisions need to be made as quickly as possible because of the large amount of capital equipment and labor tied up at the wellhead.

With conventional bottom-hole recording devices, complete pressure data is usually received a day or more after the test has been completed. However, the quartz pressure gage has recorded all the data even before it is pulled out of the well. Therefore, decisions can be made about well completion a day or more sooner. As a result, use of equipment and labor is maximized.

Test Design Aid. The quartz gage reduces the effect of test design error. Pressure tests are generally based on assumed values for permeability, porosity or other reservoir parameters. If the assumed reservoir properties differ sufficiently from the real reservoir properties, the test may be ruined. For example, if a buildup test is designed with an assumed permeability far greater than the actual permeability, then the test would be too short. The test would be ended before pseudo-steady state was reached and the test results would be inconclusive.

When the quartz gage is used, this problem does not occur. Horner plots, for example, are derived from the pressure data recorded at the surface. From these plots it would be determined that the test should be lengthened. The surface recording feature allows the test to be redesigned during the test. This is particularly important in interference testing because values of directional permeability and other non-homogeneous properties are difficult to approximate.

Housed in a 36.5-mm OD case made of 17-4PH stainless steel, the probe can withstand pressures in excess of 12,000 psi (~83 MPa). It can be operated in flowing gas or liquid wells. Mechanical vibration reportedly has no effect on its performance.

The 2813 case contains a quartz crystal pressure-sensing oscillator and a reference oscillator that is protected from applied pressure. The frequency of the sensor oscillator varies with pressure and is subtracted from the reference oscillator frequency. The resulting difference frequency, which is a function of pressure, is transmitted up the cable to the signal processor on the surface.

In the 1980s, several quartz resonator designs emerged that differed from that used in HP's pressure gage. Examples referred to in a 1988 paper by Wearn and Paros include the latter's 1983 patented configurations. The Paroscientific double-ended tuning fork resonator, Fig. 9-9, senses pressure by responding to the strain of a bellows, diaphragm or Bourdon tube. When internal pressure is applied to the Bourdon tube, the crystal is subjected to tension, raising its frequency. The small adjustable balance weights tend to eliminate inertial forces and torques, yielding the desirable low sensitivity to linear acceleration and vibration. The internal vacuum eliminates air damping and contamination, thereby maintaining high Q and a stable pressure reference. The torsional tuning fork shown in the diagram is necessary for temperature compensation (Chapter 5, Table 5-7), accomplished with a multiplexed counter. Applications and test data are reported by Busse, 1987, Busse and Wearn, 1988, Wearn and Paros, 1988, and Miller, 1988.

A vibrating cylinder pressure standard designed to measure P independent of gas density ρ, and to reduce package size and weight (including electronics) to <1 liter and <1 kg is described by Love and Juanarena, 1988. The device, illustrated in Fig. 9-10, is intended for wind tunnel (especially transonic wind tunnel) and in-flight applications. One of these applications is air speed measurement. Electronics includes a 10-MHz clock to measure resonant frequency (~4 to 5 kHz), and an EPROM in which fifteen calibration constants are burned. These enable the characteristic equation to be solved in a way that corrects for zero offset, sensitivity, nonlinearity and thermal errors. The air speed measurement

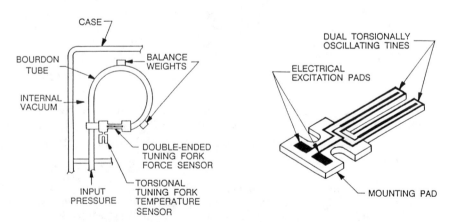

FIG. 9-9.   Tuning fork resonator senses pressure by responding to pressure-induced strain. Temperature-sensitive quartz resonator provides temperature-compensation. Reference: Wearn and Paros, 1988. Courtesy Paroscientific.

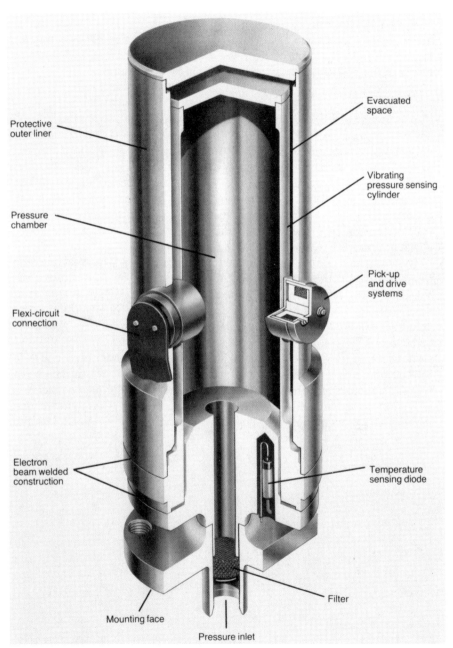

FIG. 9-10.   Vibrating cylinder pressure sensor, courtesy SONIX. Reference: Love and Juanarena, 1988.

illustrates the "hybrid" category (Chapter 2, entry 12). That is, a pressure-dropping geometry such as a venturi or orifice generates ΔP as a square-law (nonlinear) function of flow velocity, and then ΔP is measured in terms of the resonant frequencies of the pressure sensors.

9.3   Paper Tension

An acoustic, noncontacting instrument to measure tension in a moving paper web has been described by Marttinen and Luukkala, 1985. Tension is one of the most important measurements in the process of papermaking. The acoustic solution is based on the velocity of membrane waves being dependent on the tension in the membrane, i.e., proportional to the square root of tension, provided the air load is taken into account, or can be neglected for relative (not absolute) measurements.

The Marttinen and Luukkala method illustrated in Fig. 9-11 may be thought of as the dual of the transit time (contrapropagation) flowmeter. To determine tension, measurements are averaged "upstream" and "downstream" in order to eliminate the contribution due to the paper's velocity, which is not negligibly small compared to the membrane wave velocity.

Measurements at a full-size installation at a slitter-winder provided data for comparisons between the tension profile and the other profiles: basis weight, moisture and thickness profiles measured at the paper machine. It is found that if the latter three profiles correlate strongly with one another the correlation will usually be strong with the tension profile

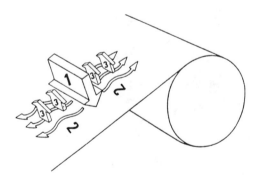

FIG. 9-11.   Paper tension measuring principle utilizes a wave burst (1) applied to the membrane, creating a membrane wave that (2) begins to propagate in the direction of tension. Its propagation velocity is monitored by microphones (3). After Marttinen and Luukkala, 1985. © 1985 IEEE.

(although negative in sign since wet paper corresponds to lower tension). In practice, however, the tension profile cannot be predicted from the other profiles and tension may have peaks that have no corresponding peaks or valleys in any of the three other profiles.

A combined graph of these four profiles is shown in Fig. 9-12. The correlation between these profiles is not very high.

Marttinen and Luukkala note that correlation is most clearly noticed

> between the tension and basis weight profiles. One should also pay attention to a peak in the tension profile at the position two meters from the left. The peak is surrounded by slack areas on both sides. The paper maker would suspect that the origin of the fault may be found in the beginning of the process in the spread of the mass on the Fourdinier wire. The mass has been spread unevenly, less in the middle of the peak and more on both sides of it. This is also seen in the basis weight profile. The disturbance also causes changes in the fiber orientation that . . . magnifies the disturbance in the tension profile.

The tension profile provides information to the paper maker that is not less important than moisture, basis weight and thickness profiles, which are routine in the paper mill. There is no question of maintaining quality and productivity without them. Tension, the fourth member in this group

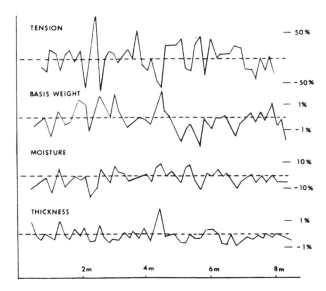

FIG. 9-12.   Profiles of tension, basis weight, moisture and thickness versus position across the paper. After Marttinen and Luukkala, 1985. © 1985 IEEE.

since 1983, seems to be reliably measurable by acoustics, according to Marttinen and Luukkala, 1985.

9.4   Gas Composition

The equation for sound speed c in an ideal gas of molecular weight M, specific heat ratio γ (constant pressure/constant volume) and absolute temperature T is

$$c = (\gamma RT/M)^{1/2} \qquad (9\text{-}2)$$

where R = gas constant. This suggests that M, or composition, can be determined from measurements related to c. Such measurements, in principle, can be time-of-flight, cavity resonance (Tallman, 1978), fluidic (Pacanowski et al., 1986), etc.

An ultrasonic trace gas analyzer manufactured by Tracor utilizes a measurement cell that has mainly been used in laboratory gas chromatography (GC) since its introduction in the 1960s. Background references include: Noble et al., 1964, Grice and David, 1969, and Skogerboe and Yeung, 1984.

For the Tracor ultrasonic detector cell, Fig. 9-13, the phase shift $\Delta\phi$ between transducers is given by Eq. (9-3), where subscript c means carrier gas, subscript x, the analyte, s is the distance between the transducers, f is the frequency of the sound wave, n is the mole fraction, M is the molecular weight and C_p is the molar heat capacity at constant pressure:

$$\Delta\phi = 180 sfn_x \left[\frac{M_c}{RT\gamma_c} \right]^{1/2} \left[\left(\frac{M_x}{M_c} - 1 \right) + \frac{C_{px}}{C_{pc}} \left(\frac{\gamma_c}{\gamma_x} - 1 \right) \right]. \qquad (9\text{-}3)$$

To arrive at (Eq. 9-3), one assumes that both gases are ideal, the gas mixture is homogeneous, and the mole fraction n is small. These conditions are always satisfied in GC, according to Skogerboe and Yeung.

In their 1984 work,

The response of the ultrasonic gas chromatographic detector is used to determine the absolute weight of an unknown analyte without a calibration curve. The method does not require that any physical properties of the analyte be known. The procedure for determining the weight concentration of an unknown requires that its response be measured in two carrier gases with different molecular weights but equal molar heat capacities, such as He and Kr. A detectability of 1 pg of hexane was found using He as the carrier gas.

We turn our attention now to the detector. Within the detector are two independent cells, each containing a pair of gold-plated ceramic transducers set up as a transmitter and receiver (Fig. 9-13). In operation they produce a phase difference that is translated by electronics into an output signal which can be integrated or recorded.

When a gas sample is introduced, changes in the propagation are displayed as chromatographic peaks showing the exact amounts of the various compounds of interest within the matrix.

> The predictable response is directly related to the molecular weight of the detected species in relation to a chosen carrier gas. A wide variety of carrier gases may be selected including: hydrogen, helium, nitrogen, argon and oxygen. Many analyses are simplified by using the same carrier gas as in the sample matrix, so that only impurities in the sample are detected with no response to the matrix gas.

Applications in gas chromatography include low ppm analysis, differential analyses, testing ultrapure gases for research and semiconductor manufacturing, including flammable or corrosive gases not easily handled by nonultrasonic GCs, according to Tracor. As an indication of sensitivity, the minimum detectable level for ethane with He carrier is 1 ng.

One application that emerged in the latter 1980s is the measurement of the molecular weight of binary gas mixtures encountered in metalorganic chemical vapor deposition (MOCVD), a process for producing compound semiconductor thin films (Thrush et al., 1987; Stagg, 1988; Rask, 1988, priv. comm.; Weyburne, 1988, priv. comm.). Since the specific heat ratio γ varies with the composition of these gases, it is not sufficient to merely measure their sound speed and temperature when determining M.

Another application in binary gas mixtures evolved at the Stanford Linear Accelerator Center, as part of the Cherenkov Ring Imaging De-

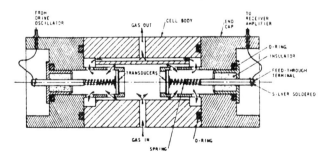

FIG. 9-13.   Ultrasonic detector cell. The internal volume between transducers is about 180 microliters. Courtesy Tracor.

tector. Hallewell, 1988, and Hallewell et al., 1988, reported the detection
of <1% fluctuations in the relative concentration of the constituent gases
having γ from 1 to 1.67, some of which were not ideal gases. In this work
the transmitted pulse was eight cycles of a 45-kHz burst.

Sound velocity experiments in five gases (N_2, CH_4, C_2H_6, C_4H_{10} and
C_5F_{12}) were compared to the theoretical predictions of three equations
of state: ideal gas, Van der Waals equation, and a "generalized" equation
of state. The last gave best agreement with experiments (discrepancy 0.2%
or less, in four cases) with 4.2% discrepancy in the C_5F_{12} prediction attributed to uncertainties in available thermodynamic data for that uncom-
mon gas.

Redding, 1978, proposed geometries such as those in Fig. 9-14 along
with low-cost electronics based on coherent modulation, for measuring
the presence and concentration of a hazardous gas. Mechlenburg, 1985,
developed equipment specifically designed to measure ethylene oxide
in air (Fig. 9-15) or other gas mixtures. Thirty years earlier, interest in
measuring respiratory gases is represented by the work of Kniazuk and
Prediger, 1955.

Water vapor in hot air, and steam quality, enthalpy and mass flow rate
measurements represent potential applications wherein information derived from c (at known temperature and pressure) is of value per se, and
of further value when combined with flow velocity V. These cases are
analogous to the flare gas application (Chapter 4, Section 4.2.2.2) where
c is correlated with molecular weight, and when combined with V and
other data (T, P) provides a basis for computing the flare gas mass flow
rate. In air, sound propagation for various conditions may be computed
by using program listings given in Kleppe, 1989.

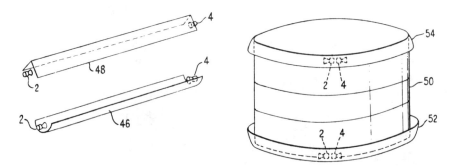

FIG. 9-14. Proposed methods for measuring the presence and perhaps concentration of a
hazardous gas having a density substantially different from air. Collectors are intended to
avoid dilution due to convection, diffusion or wind while also providing some wave guidance
between the transducers 2 and 4. After Redding, 1978.

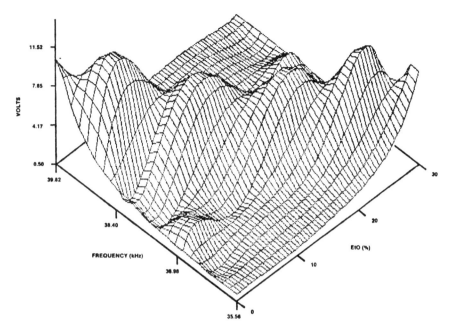

FIG. 9-15.   This surface is a three-dimensional, empirical plot of percent EtO in air versus frequency and amplitude. Different gas mixtures would have different surfaces similar to this one, but characteristic of that particular gas mixture. After Mechlenburg, 1985. © 1985 Helmers Publ., Inc.

9.5   Electrokinetic Sonic Amplitude

Electrokinetic Sonic Amplitude (ESA) is an electroacoustic method for determining the electrical charge on colloidal particles suspended in a liquid. According to information from Matec Instruments (Cannon, 1988, priv. comm.):

> The study of colloids and colloidal phenomena encompasses a vast array of materials, processes and physical properties. Colloids are usually defined as systems containing very small particles ranging in size from 1 to 1000 nm. Common examples include ceramic and clay slurries, milk, blood, pigments, pharmaceuticals, inks and dyes. Many of the important properties of colloids are determined directly or indirectly by the presence of charge on the particles. The charge can arise from a variety of mechanisms such as ionization of surface chemical groups, absorption of ions, or unequal dissolution of ions from an ionic crystal lattice. The charge at the surface of a particle is balanced by an equal and opposite charge of ions distributed as a diffuse cloud around the particle. The combined region of surface and

diffuse charge is known as the "double layer." The electrical potential at or near the beginning of the diffuse layer is termed the zeta potential and determines the electrostatic interaction forces between the particles (Hunter, 1981). Important colloid properties such as suspension rheology and the stability of the particles towards coagulation are governed by the zeta potential.

Until recently, the measurement and control of zeta potential in industrial processes have been very difficult because traditional measurement techniques are limited to extremely dilute suspensions. Microelectrophoresis has been the standard procedure for determining zeta potential. In this method, the velocity of individual particles is measured optically, under the influence of a known dc or low frequency electric field. Zeta potential can then be calculated from the measured electrophoretic mobility, u, of the particles, defined as the particle velocity per unit electric field. The major drawback of microelectrophoresis is that the sample must be extremely dilute (typically <100 ppm) to carry out the measurement. This makes sampling and control of the zeta potential of suspensions at process concentration difficult and time consuming. Matec Instruments has developed a new method, Electrokinetic Sonic Amplitude (ESA), that measures the electrophoretic mobility of particles directly in concentrated suspensions.

The ESA Measurement Technique. The Matec ESA 8000 System measurement technique is based upon a set of electroacoustic phenomena that arise because of the interaction of alternating electric and pressure fields in a colloidal suspension. When an alternating electric field is applied across a colloidal suspension with a pair of electrodes, the particles undergo oscillatory accelerations because of the electrical forces acting on the charged particles. If there is a density difference between the particles and the liquid, an acoustic wave is developed between the electrodes because of a net momentum flux at the boundaries. This effect was discovered at Matec and has been termed the ESA of the colloid (Oja et al., 1985). ESA is the pressure amplitude per unit electric field generated by the colloid.

When sound waves are passed through a suspension of charged particles, a reciprocal effect occurs. A density difference between the particles and liquid leads to relative motion between the particles and liquid. The relative motion creates a periodic polarization of the electrical double layer around each particle resulting in an alternating dipole at the frequency of the sound wave. The alternating dipoles sum to a potential that can be detected by placing a pair of electrodes in the sound field. This effect is termed the *Ultrasonic Vibration Potential* or UVP and was first predicted for electrolyte solutions by Debye in 1933. In 1938, Rutgers and Hermans pointed out that the effect also would be present in colloidal suspensions. The UVP of colloids is often referred to as the *Colloidal Vibration Potential* or CVP. CVP is measured in units of volts per unit velocity amplitude of the applied acoustic field, or volts per m/s. A detailed theory for CVP was first given by Enderby in 1951. Extensive experimental studies of the UVP in electrolytes have been carried out by Yeager and coworkers (Zana and Yeager, 1982). Recently, a new instrument has been described that uses cw methods for measuring the CVP in colloids (Marlowe et al., 1988). O'Brien, 1988, has developed a general theoretical treatment of electroacoustic effects in colloids and has derived a reciprocal relation linking the ESA and CVP.

Relation of Electroacoustic Effects to the Electrophoretic Mobility.   The magnitudes of the ESA and CVP effects are both proportional to the electrophoretic mobility of the particles. The mobility in this case, however, is the dynamic or AC mobility of the particles:

$$ESA(\omega) = P/E = c\Delta\rho\Phi G_f u_d(\omega) \tag{9-3}$$

and

$$CVP(\omega) = \Delta\Psi/U_0 = [c\Delta\rho\Phi G_f u_d(\omega)]/K^* \tag{9-4}$$

where the following notation is used:

ω angular frequency
P pressure amplitude of sound wave
E amplitude of applied electric field
c sound velocity in the suspension
$\Delta\rho$ density difference between the particles and liquid
ϕ volume fraction of particles
G_f geometrical factor for electrode geometry
Ψ potential difference measured at electrodes
U_o velocity amplitude of the sound wave
K^* high frequency conductivity of the suspension
u_d dynamic or high frequency electrophoretic mobility.

For the case of a parallel plate electrode geometry, a reciprocal relation exists between the ESA and UVP effects (O'Brien, 1988), namely, ESA/K^* = CVP. For a given mobility, the relative magnitude of the two electroacoustic effects depends upon the complex conductivity of the suspension. The CVP is greater at low conductivities and the ESA is greater at high conductivities. *In most instances, the ESA measurement offers a distinct advantage over CVP because it is directly proportional to the mobility and does not require knowledge of the high frequency conductivity of the suspension to compute the mobility.* This is especially true for suspensions of particles in low dielectric constant organic media where the complex conductivity is difficult to measure accurately and has a strong capacitive component.

Determination of the Zeta Potential from the Dynamic Mobility.   The dynamic mobility determined from the electroacoustic measurements differs from the low frequency or dc mobility determined by microelectrophoresis because of particle inertia effects. The particle velocities in an alternating field generally will be out of phase with the applied field. The higher the frequency of the applied field, the more the particle will lag the field. Because of particle inertia, the dynamic mobility is a function of frequency, particle size and particle density. For the case of spherical particles with thin double layers and low zeta potential, the following formula can be applied (O'Brien, 1988):

$$u_d = \frac{\epsilon\zeta}{\eta} G (\omega a^2/\nu)$$

(9-5)

where

$$G(\alpha) = (1 - \frac{1}{9} i\alpha (3 + 2\Delta\rho/\rho)/\{1 + (1 - i)\sqrt{\alpha/2}\})^{-1}$$

(9-6)

where:

α particle radius
ϵ dielectric permittivity of the suspension
ν kinematic viscosity of the liquid (η/ρ)
η viscosity of the liquid
ζ zeta potential

The formula for the dynamic mobility given in Eq. (9-5) above is identical to the well known Smoluchowski equation (Hunter, 1981) for the dc electrophoretic mobility except for the $G(\alpha)$ inertial term. Figure 9-16 compares the dc electrophoretic mobility calculated from ESA measurements with actual microelectrophoresis measurements of the mobility for a ceramic aluminum oxide suspension. Mobilities were measured as a function of pH and compared. Inorganic metal oxides like aluminum oxide have surface charges that are determined by the pH of the suspension. The ESA measurements were carried out at a particle concentration of 1.8 volume percent, and the microelectrophoresis measurements were made at approximately 10 ppm. The two measurements are in excellent agreement with one another.

Concentrated Suspensions. The theory used to derive Eq. (9-5) assumes that the electroacoustic effects vary linearly with particle volume fraction. This assumption is not valid in highly concentrated suspensions where the effect of particle interactions must be taken into account. Both hydrodynamic and electrical double layer interactions lead to a nonlinear dependence on volume fraction that results in lower dynamic mobility than predicted by Eq. (9-5). The particle interactions tend to retard particle motion for a given applied field strength. The volume fraction where nonlinear behavior begins varies with the colloidal system under investigation, but a good rule of thumb for thin double layer systems is approximately 10 volume percent. A detailed theory that models both particle-particle interactions and the propagation of sound waves in highly concentrated suspensions was under development in 1988 (O'Brien, priv. comm., 1988).

ESA measurements are still useful at high-volume fractions, even if a theory for calculating the zeta potential is not available. The dynamic mobility obtained from the ESA can still be used to assess the *relative* change

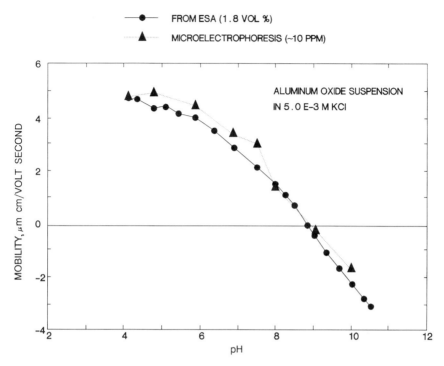

FIG. 9-16. ESA method compared with conventional microelectrophoresis. Measurements were made in an aluminum oxide suspension as a function of pH. Source: Matec.

in charge on the particles, as a function of the dosage of chemical additives, or to monitor the change in charge versus process conditions. In cases where the zeta potential must be known, electroacoustic measurements at 5 to 10 volume percent concentrations offer significant advantages over traditional methods that require dilution to the ppm range.

Matec ESA 8000 Measurement System Description. The ESA 8000 is a computer controlled instrument system for measuring the ESA of colloids in the MHz region. Pulsed ultrasonic methods with phase-sensitive detection are used to measure the magnitude and phase angle of the electroacoustic signals. Figure 9-17 is a block diagram of the standard system configuration. An electroacoustic probe-based sensor is incorporated in the SSP-1 ESA Sample Cell Assembly that also houses sensors for pH, conductivity, and temperature. The system is fitted with a digital titrator for carrying out automated potentiometric, conductometric, and generic volumetric titrations with simultaneous measurement of the ESA for determination of the sample zeta potential. An optional electroacoustic flow-through sensor is available for measurements on high viscosity samples that must be pumped or for in-line sampling of process lines. The probe-based

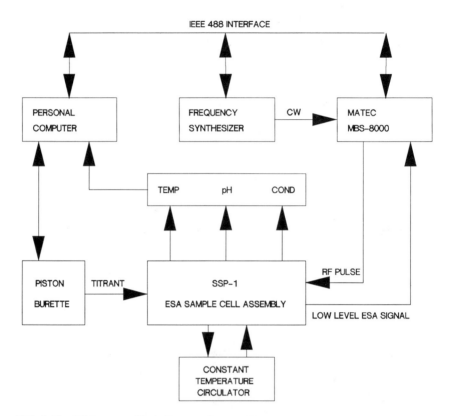

FIG. 9-17.   ESA system block diagram. Source: Matec.

sensor is not dedicated to the titration vessel and can be located remote from the system electronics.

Details of the signal processing and electroacoustic sensor design are given elsewhere (Cannon, 1989). A brief description will be presented here. Figure 9-18 is a schematic representation of the electroacoustic sensor and the primary signal processing blocks. The sensor consists of an acoustic delay line (buffer rod) that has a piezoelectric transducer mounted on one end and a solid gold electrode laminated to the other end. A second gold planar electrode is positioned parallel to the buffer rod electrode. The electrode spacing is an odd multiple of one half the acoustic wavelength in the colloid (typically $3\lambda/2$). The colloid under investigation fills the region between the gold electrodes. In an ESA measurement, an rf pulse from a gated amplifier is applied to the electrodes and generates a standing acoustic wave between the electrodes. The pulse duration is long enough to achieve a steady-state amplitude but is shorter than the acoustic propagation delay

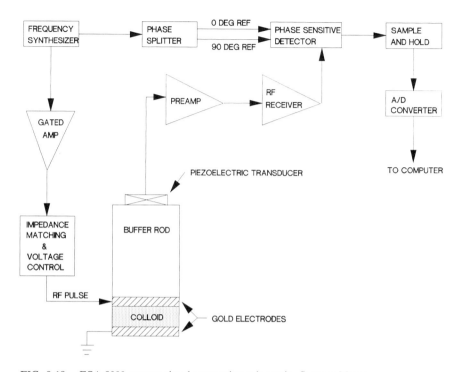

FIG. 9-18.   ESA 8000 system signal processing schematic. Source: Matec.

through the buffer rod. The acoustic wave generated between the electrodes launches a sound pulse down the buffer rod that is detected by the piezoelectric transducer. The buffer rod provides an acoustic delay that allows the low-level electroacoustic signal to be separated in time from the electromagnetic crosstalk caused by the energizing rf pulse. After preamplification, a dual quadrature phase sensitive detection scheme is used to determine the amplitude and phase of the transducer signal. The analog signal information is then processed digitally to compute the dynamic mobility and zeta potential of the colloid.

Applications.   The ESA measurement technique has been successfully applied to a wide range of materials and processes. The instrument is being used for both fundamental research in colloid science and in many industrial process applications. Some of the areas in which the instrument is being used include:

1.   Scale-up of flocculation processes
2.   Monitoring catalyst production
3.   Preparing coal-water slurries
4.   Pigment characterization and dispersion

5. Emulsion and inverse emulsion characterization
6. Preparation of magnetic tape slurries
7. Screening ceramic raw materials
8. Evaluation of pharmaceutical products
9. Dispersing clay and other mineral slurries
10. Extraction of oil from tar sands
11. Latex production
12. Paper coating formulation
13. Basic electrical double-layer research

For a discussion of related work at the University of Maine and at Pen Kem on the electrokinetic characterization of concentrated colloidal slurries, see Pendse and Marlow, 1988.

9.6   Multiparameter Sensing

The possibility of sensing two or more measurands in a given "application" by simultaneous or sequential interrogation by ultrasound is a recurring theme in this book. For example, see Table 3-17 in Section 3.9, or see Section 5.1.1.3 where sound speed yields the temperature T at which a gas' transport properties are determined from the attenuation coefficient. Besides using the sound speed and the absorption of ultrasonic waves, the designer also has wave options, namely, longitudinal, shear, Rayleigh, Lamb, Love, Sezawa, Stonely, extensional, torsional, . . . each of which preferentially responds to a selected parameter in the "right" circumstances. In addition there are sensor material and shape options. In some multizone multiparameter probes (e.g., Lynnworth, 1970 and 1971) Al and Cu segments were proposed as being differently responsive to temperature and nuclear fluence. In the 1977 and 1980 torsional sensor designs discussed in Chapter 6, extensional and torsional modes distinguish between density and temperature. Earlier in this chapter, methods were indicated for separating viscosity from density. If magnetostrictive 100-kHz and longitudinal 1-MHz transducers are connected electrically in parallel, electrical filtering can be used to sort out their respective responses. A mass flowmeter (Section 9.10) in which the number of electrical lead wires must be minimized can use this technique to advantage. Electrical filtering was also used by Silvermetz and Adams, 1981, in a gap-type liquid level probe, wherein the low-frequency radial mode was used in a self-test mode. In the normal operating or running mode, high-pass filtering blocked the radial mode acoustic short circuit, and allowed the detection circuitry to concentrate on the high frequency thickness mode longitudinal wave transmitted across the gap when an acoustically-transmissive fluid filled that gap.

One example of a gap-type impedometer probe intended for downhole measurements of several parameters is shown in Fig. 9-19, due to Birchak and Lygas, 1986. This probe is designed to be lowered down into an oil or gas well to measure the properties of the drilling mud or whatever fluid may be down there. A piezoelectric transducer *36* at the top is covered with sound absorbing material *28*. Except for a void *26*, which provides a reference echo, the probe is solid and homogeneous, but it has a slot *30* which fills with the fluid to be examined. The transducer emits ultrasonic pulses, and receives echoes from the three surfaces *25*, *32*, and *34*. The time difference between the last two is a measure of the speed of sound, and the ratio of their intensities is a measure of the attenuation through the fluid. The magnitude of the echo from surface *32*, relative to the reference echo, is a measure of the specific acoustic impedance of the fluid in the slot. According to Birchak and Lygas, these three quantities would enable a computer to calculate the density, compressibility and viscosity of the fluid. The validity of such calculations would depend in part on the validity of the assumptions or approximations underlying the analysis of a device such as this in the intended application.

Two transducer designs that include within one housing separate piezoelectric elements for low and high frequency operation, e.g., 0.1 and 1 MHz, are shown in Fig. 9-20 (top). Such designs are intended for applications where the fluid may sometimes be transmissive to the higher frequency (preferred for better time resolution), but at other times, due

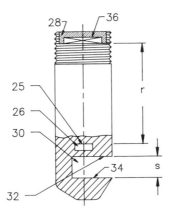

FIG. 9-19. Downhole gap probe proposed by Birchak and Lygas, 1986, for measuring several parameters using only one device.

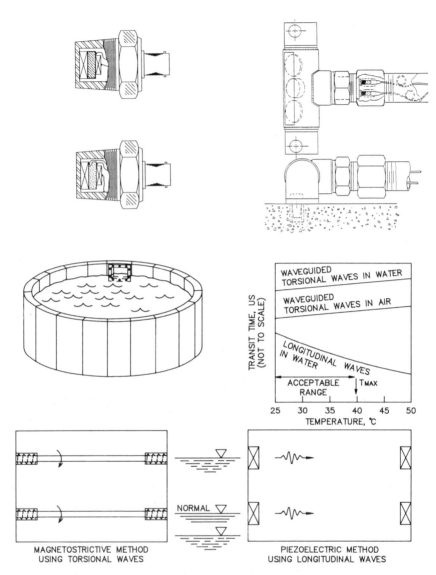

FIG. 9-20. Top: Examples of transducer housings containing separate piezoelectric elements resonant at widely different frequencies such as 0.1 and 1 MHz. Bottom: Examples of skimmer-mounted magnetostrictive (0.1-MHz) and piezoelectric (~1-MHz) approaches to sensing the water level and temperature in a hot tub, to make sure that the water is not too hot, not too high and not too low. As shown in the graph, the transit time for torsional waves in each waveguide indicates the presence or absence of water at the level of that waveguide, as well as the temperature. The piezoelectric longitudinal transducers operate as gap-type presence/absence sensors, and when water completes the acoustic path, the transit time indicates the temperature of the water. Illustration based on concept that originated at Racquetball of Toms River (R. Lynnworth, priv. comm., 1988).

588

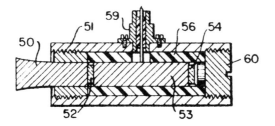

FIG. 9-21.   Probe containing longitudinal and shear wave elements. After Lynnworth and Carnevale, 1967.

to multiphase conditions, only the lower frequency would be able to penetrate.

Control of both temperature and water level in a hot tub provides an example of a two-parameter problem that is addressable with 0.1-MHz torsional/extensional waveguide technology as well as with ~1-MHz unbounded longitudinal wave technology. See Fig. 20, bottom.

An earlier two-element probe, intended for red-hot steel applications where the propagation of both longitudinal and shear modes is of interest, is shown in Fig. 9-21. Miller et al., 1980, used extensional and torsional modes in the same waveguide sensor to separate temperature from liquid level (Section 7.5.2.2). Magnetostrictive transducer coil designs afford interesting opportunities for sensing temperature. Options include: measuring coil resistance; measuring resonant frequency of an intentionally mismatched transducer stub (impulse-induced resonance); utilizing thermoelectric materials for the lead wires (Shepard, in Lynnworth et al., 1971).

In the prototype flowcell of Fig. 9-22 (Lynnworth et al., 1986), three ultrasonic modes are used (longitudinal, extensional, torsional). From these measurements the following information might be derived:

- Flow velocity V

- Sound speed c

- Attenuation coefficient α

- Density ρ

- Mass flowrate M

- Temperature T

- Viscosity η

- Composition of a mixture

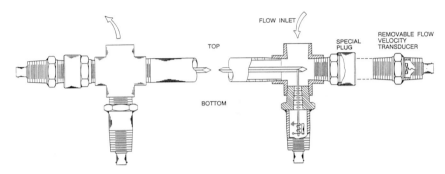

FIG. 9-22.   Ultrasonic mass flowmeter contains density and flow velocity transducers. If installed at the bottom of a loop the densitometer can also indicate whether the liquid in the line is a hydrocarbon of low density, or water. In that case the torsional sensor is both a hydrometer and hygrometer.

A mass flowmeter similar to that shown in Fig. 9-22 is currently under evaluation in a chemical plant in a recycling area. Installed at the bottom of a loop, in which low density hydrocarbons are normally present, but in which water may be present under upset conditions, the flowcell provides, through the densitometer, a means of determining when the liquid in the path has a density of about one rather than the normal value of about 0.66 g/cm³. (In some cases this function could be achieved indirectly, from the sound velocity.) In this particular design the transducers for the contrapropagation flow measurement are not installed as part of the pressure boundary but are removably coupled to the Pan-Adapta® plugs. This allows the transducers to be changed during an experimental period, without draining the pipeline.

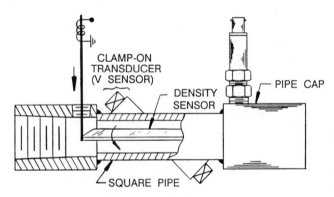

FIG. 9-23.   Piezoelectric V sensors are external. Magnetostrictively-driven torsional ρ sensor is internal but not very obstructive, in this proposed in-line flowcell designed to measure mass flow rate while minimizing pressure drop.

The two-parameter mass flowmeter of Fig. 9-23 measures ρ and V in an in-line configuration having less pressure drop than the offset design of Fig. 9-22.

9.7 Polymer Membrane Transducers

Polymer membrane transducers are flexible electroacoustic transducers in which the electroacoustic action occurs by a piezoelectric or electrostatic mechanism. Typically the element is very thin compared to the wavelength in itself or in the adjacent medium into which it radiates. Some of the technical characteristics common to both piezoelectric and electrostatic polymer membrane transducers include: broad bandwidth; better match to gases than possible with ceramic transducers; fast ringdown after excitation. Aside from these and other technical features, and despite drawbacks such as need for bias voltage (>100V in some electrostatic designs) and insufficient ruggedness for some applications, the ready acceptance of these devices within the last five years or so in $\sim 10^5$ industrial applications is attributable to

- Low cost

- Availability in easy-to-use piezoelectric sheets or in a packaged, easy-to-use nonpiezoelectric (electrostatic) form

- Electronic support also available: low-cost chips, components and circuit boards

- Packaged electrostatic type proven in $>10^6$ echo-ranging autofocus cameras

- Do-it-yourself low cost kits available from Pennwalt and Polaroid

Readers interested in background information and technical details on how to mount these transducers, attach leads in the case of Pennwalt's piezoelectric polymer, PVDF, or copolymer, VF2-VF3, avoid crosstalk in PVDF arrays, or modify standard Polaroid circuit boards (which already contain transmitter, receiver and signal processing) are advised to consult the growing body of polymer transducer technical literature of the past twenty years, or data from the manufacturers, e.g., Pennwalt and Polaroid (Ultrasonic Components Group). As a start, the following references may be helpful: Fukada, 1968; Biber et al., 1980; Lerch and Sessler, 1980; Muggli, 1980; Wilson et al., 1982; Pennwalt's Kynar Piezo Film Technical Manual, 1983; Kaneko et al., 1985; Kirby and Paglia, 1985; Leibson, 1985; Campbell, 1986; Carlisle, 1986; Chatigny and Robb, 1986;

Higuchi et al., 1986; Squire, 1986; Fiorillo et al., 1987; Gallantree and Smith, 1987; Lewin and Schafer, 1987; Smith and Dunhill, 1987; Xu et al., 1987. Current papers are typically found in the same journals or annual symposia proceedings in which the above papers appeared.

As one reviews the growth of ultrasonics, one observes step functions which correspond in many cases to transducer materials or transducer technology breakthroughs, or transducer discoveries. Starting with the Curies' discovery of piezoelectricity over a century ago, consider the impact of materials like barium titanate, lead zirconate titanate, lithium niobate, special cuts in quartz, and recently, PVDF and Kapton-based transducers. Housed ceramic (nonpolymer) transducers available from Massa (see Section 9.9; see also, Schoenwald et al., 1987), Siemens, NDT equipment manufacturers, or others, often at relatively low cost, will probably retain or improve their position in many application areas because of their ruggedness, tolerance to difficult environments, or other characteristics, and their success being well known over many years of "field trials." But if one considers the consequences of new materials being readily available to designers at a cost low enough to allow a variety of high-volume applications to be addressed competitively for the first time, and usable with μP-based equipment, the implication is that by the early 1990s one may witness another step function in the growth of ultrasonic measurements. It is beyond the scope of this section to explain the membrane transducers in detail. The main technical limitations still evident in 1987 were: lack of ruggedness (in that a strong solid housing would appear to defeat one of the main advantages of polymer transducers, low acoustic impedance), sensitivity to fouling, and temperature. To overcome some of the fouling and temperature limits the transducers may be mounted remotely using a tubular waveguide to direct ultrasound into and/or receive ultrasound from the hostile environment. As will be seen below, in some tactile sensors a soft rubber layer protects the membrane from direct exposure to a part being touched by a robot's finger. It is not unreasonable to suppose that, if the rubber were a quarter wave thick, the transducer and impedance matching layer could be housed in an all-metal housing whose front window was thin compared to wavelength. Ti and stainless steel sealed designs of this type have been in industrial use in flare gas and natural gas applications since about 1982, where the transducer is a PZT-type (Lynnworth et al., 1981; Lynnworth et al., 1984; Smalling et al., 1984). No doubt other methods of improving ruggedness will emerge, as incentives are ample. While most of the *active* applications to be indicated below involve transmission into a fluid (air or water, primarily), *passive* applications include monitoring of body functions (e.g., breathing) by contact with the skin or perhaps through cloth or clothing; tremor drug research involving a small animal in a cage that incorporates PVDF film under its floor (*Sensors*, September 1988, p. 58) and monitoring the response of a piezoelectric transducer, as in the sandwich construction proposed by Redding, 1985. PVDF polymer and VF2-VF3 copolymer piezoelectric films have been incorporated into sealed assemblies for NDT applications. The outermost face may be a soft, resilient material to couple to solids which may not necessarily be very rigid.

The polymer transducer applications to be listed or discussed below are selected because they illustrate a technical advantage of ultrasonics over a competing technology (e.g., optics or IR), because they are technically instructive or illustrative of a basic principle or limitation, or because they indicate the order of magnitude of the number of units needed, which is one measure of the importance of an application.

9.7.1   Piezoelectric Film—PVDF

Available from several manufacturers in Europe, Japan and the US, polyvinylidene fluoride (PVDF) is a long-chain semicrystalline polymer containing repeating units of CH_2-CF_2. The material has excellent resistance to stress fatigue, abrasion, and cold flow. It is lightweight ($\rho = 1.78$ g/cm^3), transparent and flexible. Its longitudinal wave speed in the thickness direction is 2200 m/s, leading to a characteristic acoustic impedance greater than water's by a factor of only 2.6. PVDF is readily fabricated in continuous sheets or complex shapes. It is supplied polarized and metallized. (The copolymer VF2-VF3 can be molded into curved shapes; repoling is not required.)

To gain familiarity with PVDF, one may utilize the available literature (Chatigny and Robb, 1986, for example) or obtain a sample technical manual, or kit, if available from the manufacturer. A list of some of PVDF's properties is given in Table 9-3.

In a 1987 index to the literature on piezoelectric polymer films (Pennwalt, 1987), the following categories are represented: acoustics, boats (prevention of marine fouling), electromechanical devices, flow, force transducers, hydrophones, keyboards, microphones, medical (apnea, osteogen transducers), music, NDT, optical, pyroelectric, power generation, robotics, speakers, and other general or miscellaneous items. Table 9-4 is a corresponding tabulation of applications categorized by Chatigny and Robb, 1986.

Aside from tactile sensors and ranging in fluids, as applied to robotics and perhaps other situations, and flow, the above list might not appear strongly oriented towards process control applications. Indirectly, how-

Table 9-3. PVDF Properties, Partial List. Source: Pennwalt

Broadband; flat to 40 MHz at 28-μm thickness, for example.
No ringing.
Low Z, 3.9×10^6 kg/m^{2}s (thickness direction).
No evidence of radial mode.
Form factor: flexible, conformable, large areas possible.
Uniform.
Electrode patterns may be small, allowing many discrete elements on one polymer carrier.
Stackable (acoustically in series, electrically in parallel).
Mean lifetime before failure better than many ceramics.
Relative permittivity $= 12$.
Dielectric strength 75V/μm; usable to 30V/μm.
$d_{31} = 23 \times 10^{-12}$ C/N (poor transmitter compared to PZT).
$d_{33} = -33 \times 10^{-12}$ C/N
$g_{31} = 216 \times 10^{-3}$ Vm/N (good receiver compared to PZT).
$g_{33} = -339 \times 10^{-3}$ Vm/N
$k_{31} = 12\%$ at 1 kHz, electromechanical coupling coefficient.
$k_{33} = 29\%$, low compared to 50% for ceramic—but PVDF can be used at much higher
electrical fields, leading to higher mechanical output per unit volume.
Maximum operating temperature $= 100°$C.

---

ever, the use of PVDF to test and calibrate ultrasonic transducers touches on a number of industrial process control applications. Also, as reported by Carlisle, 1986, ac excitation of a polymer film can be used to sense loads and liquid level. In the method reported, two piezo films are attached to a member whose natural frequency changes under load. One film, ac powered, induces vibrations into the load-sensing member. The other film acts as a receiver, detecting the frequency of the vibrations. If this type sensor were small enough, the resonances would fall in the ultrasonic domain, justifying their inclusion here. In a liquid level gage concept, again one film excites a sensor, this time tubular, into resonance. Somewhat analogous to the vibration sensors in the liquid level section (7.5.2), contact with a liquid changes the resonant frequency, which is detected by the second film.

Examples of housed piezoelectric polymer transducers are illustrated in Figs. 9-24 to 9-27. The receiving transducer design in Fig. 9-24 incorporates acoustically matched backing materials, which absorb acoustic energy entering the front of the probe, thereby suppressing undesired internal acoustic reflections. The rubber rod in the center serves three purposes. It is (1) an electrical conductor to carry the output voltage to the rf connector; (2) a mechanical support for the 1-mm diameter PVDF disk that is held in place against the front window by this rubber "spring," and (3) an acoustically absorbent material to attenuate acoustic energy entering the housing. The device is well shielded from electromagnetic

Table 9-4. List of Piezo Film Applications According to Market-Product Segments. After Chatigny and Robb, 1986. © 1986 Helmers Publ., Inc.

Military-Government

- Hydrophones: Hull mounted sonar systems, Sonobuoys, Towed cable arrays, Geophysics systems
- Fuses: Impact and thermal types
- Security Systems: Perimeter control
- Aerospace: Research on space particles
- Energy Generation: Ocean wave, Wind power
- Acoustic Emission: Nondestructive testing

Computer (Input-Output)

- Keyboards: Force transducer arrays
- Printers: Impact and ink jet
- X-Y Coordinates: CRT, Digitizers, Mouse
- Copiers: Switches and Counters

Industrial

- Security and Energy Management: Passive IR detectors, Vidicon, Vibration
- Robotics: Tactile sensors
- Load Cells: Consumer bathroom scales, Industrial scales
- Impact/Switch: Industrial, Coin sensor, Glass panel breakage, Traffic monitors
- Pressure Regulators: Pumps
- Flowmeters: Air conditioning ducts
- Automotive: Engine knock monitors, Tire pressure gage, Fuel injection control, Fuel tank level gage, Driving control switches

Instrumentation

- Machinery Monitor: Vibration, Noise
- Micropositioners: Manipulation devices
- Adaptive Optics: Deformable mirrors
- Range Finders: Pulse-echo transducers
- Displays: Large area shutter elements

Medical

- Apnea Monitors: Body wrap, Crib-mattress
- Blood Pressure: Cuff
- IV Fluid Monitors: Flow and level, Drop counter
- Pacemaker: Handicapped switch, Gait, Startle
- Ultrasound: Imaging systems
- Instant Thermometer: Pyroelectric response
- Osteogenesis: Bone and wound healing

Telecommunications

- Microphones: Various models
- Speakers: Handset and squawk-box
- Keypads
- Tone Generators

Consumer

- Musical Instrument Pick-ups: Custom models for OEM and after market
- Sports Equipment: Martial arts, foul line, touch pad jogger shoe
- Toys-Games: Various models
- Acoustic Speakers: Headset, Tweeter, Automotive
- Switches: Musical keyboards

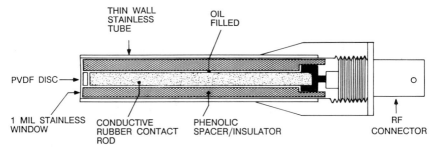

FIG. 9-24. Internal structure of a PVDF microprobe with an acoustically matched backing. This transducer is designed for high fidelity detection of short pulses up to 10 MHz. After Wilson et al., 1979, 1982. Illustration courtesy Raytheon.

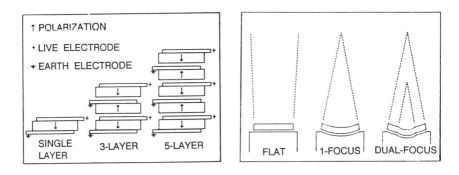

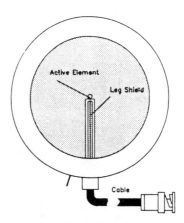

FIG. 9-25. Stacked, focused and shielded designs due to Gallantree and Smith, 1987, and Smith and Dunhill, 1987. © 1987 IEEE. (Compare stacked structures with folded multilayer structures of Chen et al., 1978.)

FIG. 9-26. NDT pulse-echo transducer, 10-mm diameter, courtesy Marconi Research Centre.

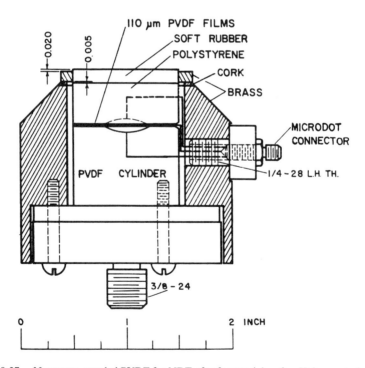

FIG. 9-27. Neoprene-coupled PVDF for NDT of soft materials, after Habeger et al., 1988.

interference; the exterior housing is entirely metallic. The thin (25-μm) stainless steel at the front of the device serves as an acoustic window, and also as the electrical ground. During assembly, just prior to sealing, the device is oil-filled to eliminate any air bubbles. A discussion of the probe's design and performance appears in Wilson et al., 1979, 1982.

9.7.2   Electrostatic Module

Available from Polaroid since the early 1980s, this type of polymer film transducer is variously termed a capacitive type, electret or electrostatic type. As supplied in 1987, the housed device enclosed a gold-plated Kapton element stretched over a grooved backing member. Details are given in Fig. 9-28. As seen in the photograph, printed circuit boards hardly larger than the transducer were also available in 1987. These boards can supply the bias voltage and also the transmitter, detector and signal processing functions similar to those used in Polaroid's autoranging cameras. Apart from camera applications, sales of the Polaroid transducer were reported to be millions of units by the end of 1988. Electronic tape measures comprised much of the 1988 consumer market. The industrial market was hundreds of thousands of units in 1988 (J. O'Brien, priv. comm., 1989).

Since this transducer and circuits evolved at Polaroid (in conjunction with Texas Instruments) specifically for autoranging of a camera in air it is not surprising that the process control applications at least initially would follow the same well traveled path. It may not be immediately obvious, however, how many diverse applications exist under the heading "ranging." The following list of ranging-related applications and companies supplying particular instruments is by no means complete, but may serve to indicate the scope of such applications, which is our present purpose. Only a few shall be illustrated.

- Vision Assist, an ultrasonic automotive proximity detection system designed to assist drivers in the operation of their vehicles by providing both distance and directional data of obstacles in the vehicle path and/or immediate vicinity—especially for parking and lane change maneuvers. In potential applications reported by Ford, sensors mounted in the bumpers activate an instrument panel display to alert the driver of objects within 2 m of the front and 6 m of the rear of the vehicle. This display shows the distance to the nearest front or rear object both digitally (in ~30-mm increments) and through the progressive illumination of green, yellow and red bar segments which indicate the distance as well as the direction of the object relative to the vehicle. Additional sensors in the rear view

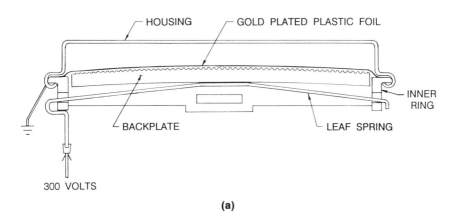

(a)

(b)

FIG. 9-28. (a) Transducer assembly. After Biber et al., 1980. (b) Transducers and various protective covers and a small printed circuit board. The general function of the ultrasonic ranging board is as follows. It generates the 300V peak-to-peak, 50-kHz signal for driving the transducer; contains the variable gain, variable-Q amplifier for receiving the returning echo; and it provides the two outputs (start of transmit and echo received flag) for measuring the elapsed time between pulses to determine distance. Courtesy Polaroid. (O'Brien, 1987; Jackman, 1987, priv. comm.)

mirror housings provide lane change assistance by monitoring the space in the adjacent lane just to the rear of the vehicle - the so-called "blind spot." If a vehicle is in this area when the turn signal is activated, a red warning light in the mirror housing illuminates and an audible warning sounds. (Evans, 1987, priv. comm.) See Fig. 9-29.

- Retro-Guard™, by Dal-Star Research, Inc., is intended as a retrofit device for cars and trucks to prevent back-up collisions. Aero-Guard is used in some Federal Express vans, to prevent backing into aircraft waiting to be loaded. See Fig. 9-30. When in operation, Retro-Guard™ warns with an audio signal. Starting from ~2 m, as one approaches the obstacle, the audio signal will increase both in frequency and volume. At 30 cm it will become a continuous audio tone, corresponding to the digital readout on a display. Retro-Guard™ is designed to aid in rear maneuvers where visual contact

FIG. 9-29. Electret polymer transducer mounted in rear view mirror assists driver in lane change maneuver. Illustration courtesy Ford Motor Company Design Center (Evans, 1987, priv. comm.)

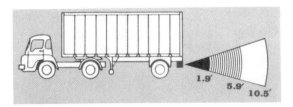
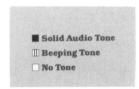

FIG. 9-30.   Electrostatic transducer in rear end of truck warns driver of hidden obstacle, and indicates distance by an audible signal. Courtesy Dal-Star Research, 1987.

is not possible or where precision backing-up maneuvers are necessary.

• Patient distance, to provide data to an instrument for automatically setting the optimum radiographic technique. (See Fig. 9-31; Kleinman, 1983).

• Tapeless measurer measures the distance between opposite walls. (Calculation Industries; Etec; Exact Technologies; Measurement Specialties; others.)

• Ink level—water level in weirs and flumes—liquid level in bins. (Toltec Electronic Systems, Inc.; Electronic Sensors, Inc.)

• Distance/thickness measurements such as gaging flat stock from an extruder, calendar or mill, positioning end effectors and paint nozzles, locating belt edges. (Xecutek Corp.)

• Loop control. (Waddington Electronics, Inc. and Echo Controls, Inc.) See Fig. 9-32. (*Loop* controllers are sometimes reported as

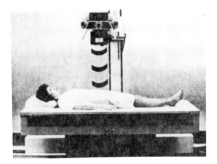
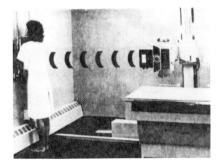

FIG. 9-31.   Echo ranging combined with radiography, for automatic setting up of optimum technique. After Kleinman, 1983. Illustration courtesy Bennett X-Ray Corp.

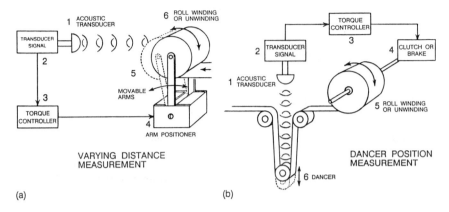

FIG. 9-32.   Measurement of distance to rollers in motion. Courtesy Echo Controls, Inc.

tension controllers; compare, however, with direct measure of tension as described in Section 9.3.)

- Finger location in front of a terminal screen, as in Fig. 9-33. (Contaq Technologies. Compare with surface wave method in Section 7.9.4.)

- Ranging and interfacing to a personal computer (Fig. 9-34).

- Forklift carrier uses proximity sensor to identify locations of open wire mesh baskets (reportedly not practical with photoelectric sensors, based on tests) and avoids collisions with previously loaded baskets (Fig. 9-35).

- Snow height gage.

9.7.3   Composite Transducers

In attempts to secure some of the advantages of a low-characteristic-impedance transducer material such as found in polymer transducers, yet still retain some of the advantages of ceramic transducers, several investigators have chosen the path of the composite. In some cases the acoustic impedance, bandwidth, and radiation patterns of the composite transducer can be controlled in a manner beyond what is possible in single-phase materials. Examples of structures under investigation in the mid-1980s are illustrated in Fig. 9-36.

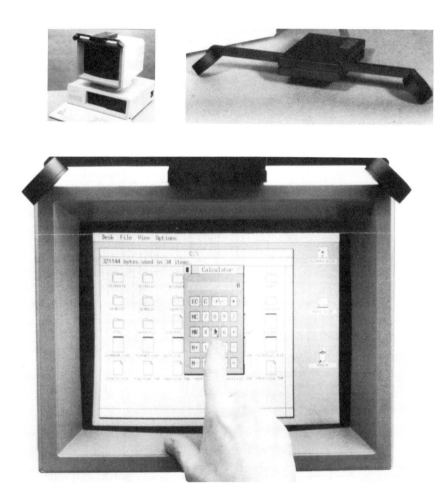

FIG. 9-33.   Three views of ScreenMouse™, an ultrasonic pointing accessory from Contaq Technologies. (Boehm, 1987, priv. comm.)

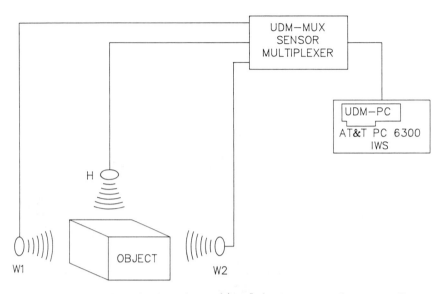

FIG. 9-34. Three-dimensional ranging, and interfacing to a personal computer. Source: Contaq Technologies Corp.

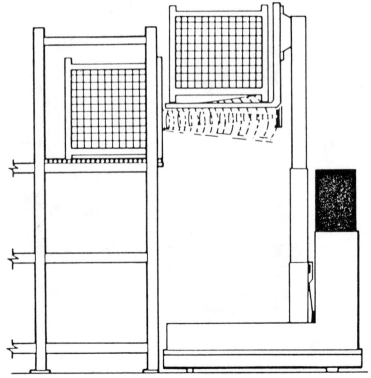

FIG. 9-35. Agastat proximity sensor installed in a Volvo forklift carrier, for use with open mesh wire baskets. © 1987 Helmers Publishing Corp.

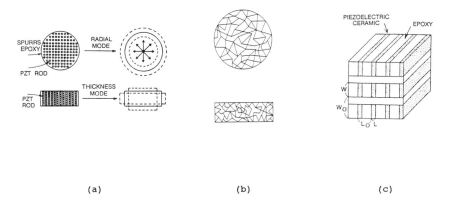

FIG. 9-36.   Examples of composite transducers. (a) Rods in epoxy. After Gururaja et al., 1985. © 1985, IEEE. (b) Fractured ceramic "repaired" by epoxy, described by Montero de Espinosa et al., 1986. (c) Stacked layered structure. After Hashimoto and Yamaguchi, 1986. © 1986 IEEE.

9.8   Ultrasonic Control of Ultrasonic Processes

Potential users of ultrasonic measuring equipment for industrial applications frequently ask the sensor/instrument manufacturer if ambient noise and vibration are likely to frustrate the measuring system. This is a justifiable concern, especially if the noise spectrum extends up into the ultrasonic domain where the interrogating frequency is located. Ultrasonic noise is likely to be present in high-energy processes, e.g., combustion, and is certainly present if the process itself is in the high-energy ultrasonic category, e.g., cleaning (McQueen, 1988). In some cases the noise indicates a situation that may block the ultrasonic wave, e.g., cavitation.

The designer has several options to avoid errors due to noise. One option is to sense the measurand using frequencies far removed from the noise. This usually means high interrogation frequencies, yet not so high that attenuation becomes a problem. Other options include narrowband filtering, and correlation detection (Nuspl et al., 1986) perhaps aided by chirping (Mylvaganam, 1987–1989) or coded transmitted waveforms (Jacobson et al., 1987b). Signal averaging tends to suppress noise that is not transmitted in synchronism with the interrogating wave. As we learn more about signal processing used by bats, dolphins, whales, or bat-evading moths, more sophisticated solutions may emerge. It is assumed above that the noise itself is to be eliminated. But there are ample examples where noise is "the signal" and provides useful information on leaks, incipient boiling, combustion, or failure of a part.

As examples of the "averaging" approach to noise suppression, reference is made to Section 7.7 on tracking a coal seam based on cutter vibration (Peterson et al., 1985) and to Section 7.3.2.4 on ablation gaging (McGunigle and Jennings, 1975). We may also make note of simple waveguide experiments such as illustrated in Fig. 9-37, wherein the noise due to a doorbell buzzer pounding against a waveguide, or the noise associated with spinning a waveguide at the maximum rpm of a hand-held electric drill, does not interfere with obtaining clear signals or echoes on an oscilloscope.

In the next two examples, the noise generated by high intensity ultrasonic processes does not prevent successful process control by ultrasound because in one instance, the tuning of a homogenizer is based on producing the loudest possible noise at the receiver monitoring crystal; in the other case body concretions (kidney stones) are fragmented by a shock wave, but the ultrasonic localizing transducer provides position information prior to initiation of the shock wave.

9.8.1   Homogenization, Mixing

In a paper on applications of high-intensity ultrasound to organic chemistry, Davidson et al., 1987, begin with the well known observation that the application of ultrasound to organic liquids under appropriate conditions can cause cavitation and microstreaming. In addition, Davidson et al. point out, the presence of solid objects such as metal particles in the liquids leads to intense perturbation at the solid-liquid interface, causing erosion at the surface. These properties are believed to be of considerable potential for the little-exploited application of high-intensity ultra-

 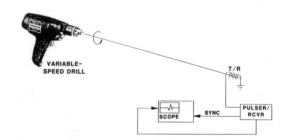

FIG. 9-37.   Waveguide experiments illustrate averaging out of mechanical noise despite pounding of a doorbell buzzer or the rapid spinning of a waveguide by an electric drill at maximum rpm. After Lynnworth et al., 1969, and Lynnworth and Bradshaw, 1971, respectively.

sound to organic chemistry on a major scale. In recent (pre-1987) years, laboratory scale organic syntheses have been carried out in which the application of ultrasound was crucial to the success of the reaction, they report. See Fig. 9-38.

In a paper describing the use of high-intensity ultrasound for mixing, dispersing and homogenizing, Booth, 1986, explains the need for "tuning" the machine. Tuning consists of adjusting the jet stream configuration by altering the gap distance between an inlet orifice and a blade, moving "adjusters" near the blade, or introducing a resonant block, or presumably other means. As Booth explains, if a crystal pick-up is mounted in the vicinity of the blade and connected to a meter then the device can be adjusted to yield maximum acoustic intensity. Since materials vary in viscosity, density and other parameters, the ultrasonic pick-up is perhaps the most reliable and most general solution available for controlling the high-intensity ultrasonic process. In this example the "noise" is the signal.

9.8.2   Kidney Stone Pulverizer

In an invention due to Makofski et al., 1986, a device for the noninvasive fragmentation of body concretions includes an integral ultrasonic locating and positioning means whereby the concretion is localized with ultrasonics and the device is positioned in response to ultrasonically-derived information.

Many attempts have been made to develop a simple and effective noninvasive treatment of kidney stones, one of the most common body concretions (Flachenecker, 1987). According to Makofski et al.,

The advent of high-speed physics and the development of a method of generating shockwaves by an underwater spark gap led to a method of non-

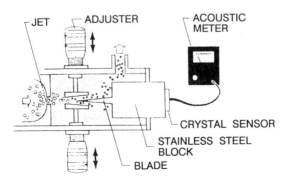

FIG. 9-38.   Layout of Ultra Son ultrasonic homogenizing head. After Davidson et al., 1987.

invasive fragmentation of body concretions. One such device for the non-invasive fragmentation of kidney stones includes a large bath in which the patient is immersed, crossed X-ray beams for the localization of the stone and an underwater spark gap for the generation of high energy shockwaves which are focused at the kidney stone.

To avoid the disadvantages of such a system, Makofski et al. propose a different solution, as follows. Figure 9-39, as they explain it,

is a cross sectional view of main housing 24 and shows the positioning of the main housing 24 in relation to a concretion such as a kidney stone, represented at 40, in a human kidney, represented at 42. The patient's skin is represented at 44 and a portion of body tissue is represented at 46. The main housing 24 comprises a closed space 48 filled with a fluid with acoustical properties essentially similar to the acoustical properties of body tissue. Such a fluid could be water or a saline solution of water.

One portion of enclosed space 48 is bounded by a reflector surface 50 with a first focus f_1, represented at 52 and a second focus, f_2, represented at 54. The reflector surface of the preferred embodiment is described by

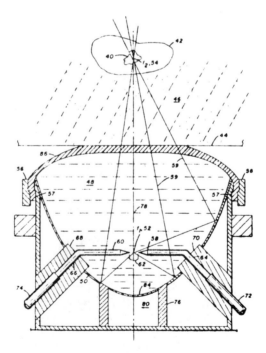

FIG. 9-39. Apparatus for noninvasive fragmentation of body concretions such as kidney stones. After Makofski et al., 1986.

an ellipsoid of revolution. The remaining portion of enclosed space 48 is bounded by a flexible membrane 86.

An ultrasonic transducer 76 is made integral with main housing 24 and is positioned at an end of the reflector surface 50. The ultrasonic transducer 76 is positioned so that an axis of the transducer is coincident with a line, indicated at 78, extending through the first and second focus, f_1 and f_2, of the ellipsoid of revolution. The transducer 76 is rotatable around the line 78 for at least an angle of ± 90 degrees. The transducer 76 includes a radiating and detecting element 80 which radiates an acoustical signal and detects reflected portions of the acoustical signal. A flexible membrane 84 provides an interface between the radiating and detecting element 80 and the space 48. The transducer is caused to rotate by the doctor or technician until a body concretion is indicated on a visual display. The doctor or technician moves the main housing in the first, second and third direction as discussed above until the concretion is located at the second focus of reflector 50. The third direction is perpendicular to both the first and second directions and is essentially perpendicular to the patient's body. Once the concretion is visually indicated as being at the second focal point f_2, a shock wave is initiated by the doctor and the process is repeated until the concretion is fragmented.

9.9 Bowling Alley Pinsetter

An automatic ultrasonic scoring system that was developed and mass produced by Massa Products Corporation for use by the AMF Bowling Companies, Inc. to locate the positions of all standing pins, no matter where they slide over the pindeck, is described by Massa, 1987.

Among the advantages of ultrasonics over optical systems, Massa cites: easier ranging; no blockage of the field of view by standing pins; not sensitive to dirt on pins, nor to highly reflective background walls or other surfaces.

As explained by Massa, 1987:

> Each ultrasonic system is designed to independently control two bowling lanes and consists of two ultrasonic arrays connected to a common Z80-based electronic control board. An ultrasonic array is mounted under each pinspotter machine above and in front of the number one pin, and 559 mm above the pindeck. See Fig. 9-40.
>
> Each array contains 14 ultrasonic transducers precisely positioned so that the entire pindeck is insonified by 14 accurate ultrasonic beams. The array bar is a rigid aluminum extrusion with 14 precision cavities machined into the front edge to hold the transducers. The axis of each cavity is machined at a different angle in both the horizontal and vertical planes so that the entire pindeck will be insonified by the 14 separate sound pulses.

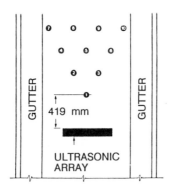

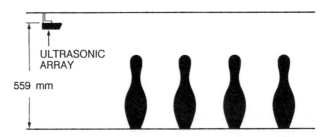

FIG. 9-40.   Location of array of 150-kHz transducers shown above and in front of ten standing pins in bowling alley. After Massa, 1987. © 1987 Helmers Publishing, Inc.

A circuit board containing both the amplifier and detection circuits is also located in the array bar. This allows the signals to be processed inside a completely shielded enclosure. All of the signals going in and out of the grounded array housing are digital, eliminating any possibility of picking up the transient electromagnetic noise that can occur on low-level analog signal leads.

The sealed, ~13-mm diameter transducers, Model TR-2404, are driven with a 130-V_{pp} tone burst and operate at a resonant frequency of 150 kHz ±450 Hz. The beamwidth is 12°, with side lobes suppressed to a negligibly small intensity.

The ultrasonic pinsetting and scoring system self-calibrates for temperature or humidity-dependent changes in sound speed based on echoes from pins five and nine each time all the pins are set up. These reference echoes normally occur about 5.5 and 7.25 ms after the transmit pulse has fired. After calibration for c, operation proceeds as follows, according to Massa, 1987:

When the bowler rolls the first ball, it knocks down several pins and then hits a switch at the back of the lane. This switch sends a signal to the control electronics board indicating a ball has just been bowled. The computer waits 1.7 s to allow for late falling pins, and then sequentially pulses each of the 14 transducer channels in the array. Each channel sends a digital signal to the control electronics whenever any echoes are received. Since the computer knows in which direction each of the narrow-beam transducers is pointing, it is able to calculate the exact range and bearing of the reflected echoes and, therefore, the exact location of each standing pin.

It takes 0.3 s to accumulate the data from the 14 channels and determine which pins are standing. For increased reliability, all areas of the pindeck are insonified by at least two transducers in the array. In addition, all the transducers are sequentially pulsed a second time. A pin must be detected during both detection sequences in order to be called standing.

The control electronics then send the standing pin information to the graphics computer, which displays on a screen in proper scoring format the number of pins knocked down. In addition, the positions of the standing pins are displayed, together with an indication of where the second ball should be rolled in order to obtain a spare. Meanwhile, the pinspotter is sweeping the deadwood from the pindeck and replacing the standing pins. The bowler then rolls the next ball, and the sequence is repeated, except that all pins are swept and ten new pins are placed on the pindeck. In this manner, the score is automatically updated and continuously displayed to the bowler.

An additional benefit occurs when a strike is bowled. The control electronics recognize the strike and instantly signal the pinspotter to sweep the pindeck and drop ten new pins, rather than waste time by first going through its normal cycle of trying to pick up any pins that might be left standing.

Massa has been manufacturing large quantities of ultrasonic scoring systems for several years and successful installations have been made in over 24,000 bowling lanes. The system speeds up bowling games by as much as 25%. The AccuSonic Automatic Scoring System pays for itself in a relatively short period of time because of the additional number of games that can be played during a given period of time.

9.10   Downhole Flowmeter

The problem of measuring geothermal, oil, or gas flows downhole requires the development of novel sensor and associated electronics that can survive for hours in the downhole environment to 250°C in many cases and in the 300 to 400°C range in some cases; pressures to 15,000 psi (100 MPa); operable with a downhole data logger (''slickline'' tool) or sometimes with only one monoconductor power and data line; preferred tool diameter not more than 43-mm OD for oil and gas and about 80-mm OD for many geothermal cases. Furthermore, the tool should be able to measure *total flow,* and also *flow profile* around itself in vertical and in deviated (non-vertical) wells. Measurements are needed in cased and uncased boreholes. The tool design should anticipate single- and two-phase flow; low and high velocity flow; aggressive fluid; rough handling. The ideal downhole flow tool would provide reliable data not just on flow velocity but also on density, mass flow rate, composition, and quality and enthalpy in geothermal applications, all at low cost, of course. If a tool could achieve most or all of these presently unavailable characteristics, it could be used to assess oil, gas and geothermal resources more precisely. As a leak detector the tool could be used to inspect casings and joints.

Two approaches will be considered here: internal and external. In the *internal* approach, flow is collected and funneled by a diverter basket through the flow tool. In the *external* approach, the flow tool is centralized in the wellbore and the *annular flow* outside the tool is measured. In the author's laboratory both internal and external approaches were investigated using experimental tools similar to those shown in Figs. 9-41 and 9-42. Data uncorrected for profile or end effects are plotted in Fig. 9-43. To obtain the data plotted in this figure, the internal approach used an axial path ~356-mm long, while the external approach measured flow in the annulus between a 42-mm OD tool and a nearly-concentric pipe having an ID = 154 mm (Lynnworth et al., 1989a).

For further background see Carson and Wolfenbarger, 1986, DiBello et al., 1987, McBane et al., 1988, Vos and Vogel, 1988, and the *Proc. Technical Review, "Advances in Geothermal Reservoir Technology— Research in Progress,"* 1988.

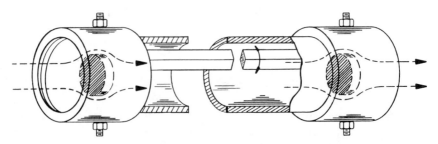

FIG. 9-41. Laboratory model of a restricted-OD flowmeter, where flow is measured *inside* the "flow tool," similar to design in Lynnworth et al., 1986. The torsional mode ρ sensor having a diamond-like cross section was first analyzed and fabricated by Bau and Kim at the University of Pennsylvania.

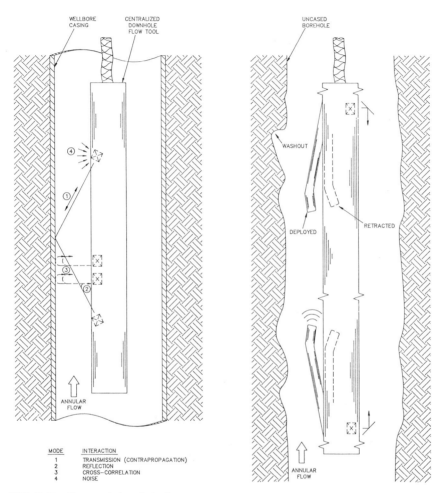

FIG. 9-42.   Proposed downhole flowmeter concepts for measuring flow *outside* the tool, in the annulus between the tool and the wellbore or wellbore casing. Left: No moving parts, but operation depends on centralizing the tool within a generally concentric, specularly reflective wellbore. Right: Transducers or reflectors are deployed downhole in this design that is intended for measuring flow in the annular region even when the boundaries of the hole are irregular, as may be caused by washout. In 1988, work began in the author's laboratory on the design, construction and testing of transducers to meet the high-temperature, high-pressure requirements of this downhole geothermal application, in a DOE-sponsored Small Business Innovation Research program (Lynnworth, 1988d).

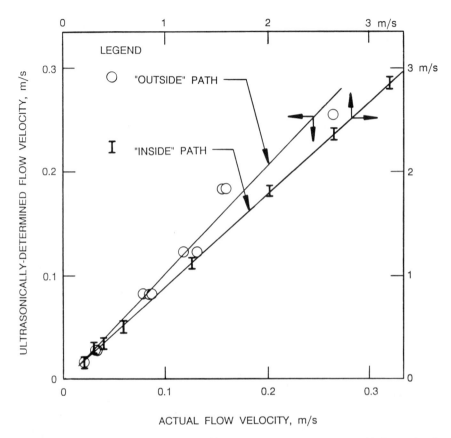

FIG. 9-43.   Laboratory data obtained with water at room temperature with *internal* and *external* ultrasonic flow tool experimental designs. The *internal* flow tool, similar to that in Fig. 9-41, was tested in a flow loop where all the water flow was confined to an axial path inside a 1.25-inch schedule 40 SS pipe section. Passageways represented by the dashed lines in Fig. 9-41 conveyed the water around the piezoelectric transducers at each end of the axial path. The *external* flow tool was tested inside a 6-inch schedule 40 Ni-plated steel pipe using an annular zigzag path.

9.11   Food Processing and Food Systems

In a 1988 review of applications of ultrasound to food systems, Javanaud's sixty-three literature citations include work dating back to 1956. (See also, Upchurch et al., 1985; Fitzgerald et al., 1988b.) Table 9-5, from Javanaud's review, categorizes ultrasonic methods according to their being based on velocity, attenuation, specular reflectance or scattering. As expected, many applications in food technology require techniques already developed to solve problems in biomedical and NDT areas. Interpretation

Table 9-5. Ultrasonic Methods Used to Evaluate Foods. After Javanaud, 1988. © 1988 Butterworths

System	Property
Velocity	
Cattle, pigs and sheep	% solid fat
Fruit juices	% fruit flesh or % sugar
Coffee (in water)	% grains
Wine	% alcohol and % solids
Milk	% fat and % solids
Emulsions	% oil
Oils	% solid fat
Yeast Slurry	% solid
Ice cream	moisture content, structure
Ice/water mixture in meat	% ice
Fruit (various)	ripeness
Eggs (white and yolk)	age
Egg shells	thickness
Cheese	crack detection
Biscuits	crispness
Attenuation	
Ice/water mixture in meat	% ice
Orange juice	stability
Fruit (apples, cantaloupe)	ripeness
Potatoes	age
Eggs (white and yolk)	age
Specular reflectance	
Orange skin	smoothness
Tomato skin	cracks
Husked sweetcorn	defects
Scattering	
Fish	lipid content
Beef	quality grade through marbling

ranges from relatively straightforward in a pure liquid or solid, to difficult or presently impossible in multiphase systems. But even in multiphase systems, velocity changes correlate well with concentration changes. Javanaud's review does not deal with "industrial"-type process control measurements in food processing such as flow, temperature, liquid level, distance ranging or counting, for which ultrasound is sometimes the method of choice.

Coupling to the object (e.g., a dry biscuit or an egg) or subject (e.g., an animal) presents interesting problems sometimes solved the same way as in industry [soft-tipped probe (compare with Hastings et al., 1961)] and sometimes by a different way [molten paraffin (compare Section 3.3)].

Javanaud envisages future developments including on-line applications of scattering, and parallels related to advances in medical ultrasonics.

10 Historical Notes and Anecdotes

Histories of acoustics and ultrasonics such as those recounted by Beyer, 1966, Herrick, 1977, and Graff, 1981 provide interesting perspectives to both newcomers as well as veterans in the field of ultrasonic measurements. One of the more difficult tasks for anyone wishing to be a historian of ultrasonic technology is to determine which inventor or author was "first" with respect to an improvement or breakthrough. First to "conceive" is not as easy to define or document as first to file, patent or publish. Accordingly, references in this book to early work generally should be interpreted as a guide to a history of the subject, and not necessarily a definitive judgment of who was first.

The author is of the opinion that the motivating circumstances surrounding the invention of early methods or apparatus for measuring flow, liquid level, temperature, etc., might be instructive or interesting to today's students or workers in the field who may be using said inventions or improvements thereof, or who themselves may be contemplating improvements. To determine the history, need and circumstances in a few cases of special interest to the author, inquiries were sent to several workers in ultrasonics whose work conducted before 1970 is still recognized in, or has had an influence upon, today's methods and equipment. The responses of Mason, Swengel, Greenspan, Kritz, Van Valkenburg, Welkowitz and Joy are presented below in that order, with minor editing in some cases.

10.1   Use of High Amplitude Strains in Studying Wear and Ultrasonic Fatigue in Metals*

High amplitude strains were first produced at Bell Telephone Laboratories by using a barium titanate driver attached to an exponential brass horn which amplified the motion in the ratio of the large diameter to the small diameter. A pick-up from the transducer was fed back to the driving electrode through a limiting amplifier which controlled the amplitude of the motion.

At that time, work was being done on a wire spring relay which is a relay with 60 simultaneous contacts attached to 60 wires. It was desired to have a 40-year life which corresponded to about a billion operations. At first the wires were straight, but for a billion operations the plastic card which moved the contacts caused wear in the cards which actuated the relay so that it lost its ability to make contacts with the stationary contacts.

In order to study the wear, a relay wire was mounted in the end of the horn and this was pressed against a pivoted arm that allowed a definite force to be applied to the wire. The wear in cubic mil-inches for a billion vibrations was plotted against the length of the stroke. For the vibration frequency of 21.5 kHz, this requires about 13 hours for each point. If the displacement of the wire can be held down to 0.1 mil-inches the wear is very small. This was accomplished by putting a double bend in the wire. A negative displacement at the card is given by the flexure motion while a positive motion is caused by the hinge motion. The two together gave a small enough motion at the card to meet the 40-year objective of the relay wear.

Since I had a device for producing high strains in materials at a high frequency, I became interested in studying the internal friction and fatigue in metals. The ultrasonic frequency allowed the fatigue to be studied in a much smaller time than was possible with the ordinary fatigue systems. A specimen was used similar to the one shown in Fig. 10-1. This gave a constant strain over the whole length of the small section and the combination could be tuned to the same frequency as the transducer and horn. Furthermore, by measuring the ratio of the pick-up voltage to the applied voltage the internal friction could be measured while the strain in the sample can be determined by measuring the pick-up voltage.

Using this device a number of metals were measured. For brass the internal friction was independent of the amplitude up to strains of 10^{-4}

---

* Contributed by Warren P. Mason

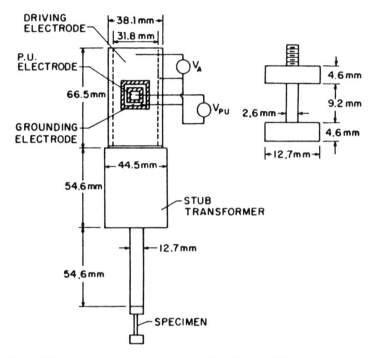

FIG. 10-1.   Ultrasonic generator and specimen. After Mason, 1981.

after which slip bands begin to form at strains of 3×10^{-4}. These results were confirmed by photographs of polished surfaces. The first effect was isolated slip bands beginning at strains of 3×10^{-4}. Next, slip bands degenerate into microcracks while internal friction increases rapidly and the microcracks join to form a fatigue failure. The internal friction increases asymptotically at the fatigue point as does the change of the elastic modulus.

At the age of 65, I retired from Bell Telephone Laboratories and then became an adjunct professor with the civil engineering department and later the metallurgical department of Columbia University. I continued to study ultrasonic fatigue in metals. The first effort was to improve the driving mechanism. Work was being done on the alloy 90% Ti, 6% Al, 4% V for possible use in supersonic airplanes and that was one of the materials that I examined. Measurements show that the internal friction is low and the linearity extends up to strains of 5×10^{-3}.

With this improved material, Fig. 10-1 shows the type of driving mechanism and specimen shape that was finally used. The barium titanate was replaced by a lead titanate-lead zirconate composition (PZT-4) which was more sensitive and would take a larger strain value. The exponential horn was replaced by a stub transformer which gave a transformer ratio proportional to the square of the large diameter to the small diameter. The specimen shape produced another factor of 5 in the strain in the small section. All together linear strains of 6×10^{-3} could be produced in the small section of the specimen if the material could take it.

Many materials were studied. Shortly after this investigation, considerable work on ultrasonic fatigue was done by a number of companies and universities. The first international conference on ultrasonic fatigue was sponsored by the Engineering Foundation. It was held in 1981. Over 50 papers on various phases of fatigue and corrosion fatigue by authors from many countries were presented. On account of my early work the conference was dedicated to me.

(For a more complete summary of the late W.P. Mason's scientific achievements, the reader is referred to Thurston, 1988.)

10.2   Early Flowmeters

An understanding of the ideas underlying both Doppler and transit time (Δt) flowmeters existed in the last century, over 100 years ago. The classic demonstration in 1844 of the frequency shift predicted by Doppler, 1842, for *light* waves emanating from a receding star utilized a constant-frequency *sound* source (a renowned trumpeter, on a moving train) and a reputable judge of frequency. This 1844 demonstration led to controversy at the time, but the validity of the observations and conclusions today would not be disputed, regarding the increase or decrease in received pitch as the train approached and then departed from the listener (White, 1982.)

Regarding the transit time method, in Rayleigh's *The Theory of Sound*, 1877, he notes on p. 2 that, in order to measure c in air correctly, one needs to measure with the wind and against it, to eliminate errors. We may assume that Rayleigh did not think it necessary in that introductory passage to write down the $c + V$, $c - V$ equations for transit time in the two directions, from which V could be computed.

Among the early patents associated with these two basic methods of acoustic flow measurement, Herrick, 1977 and others frequently cite those of Chilowsky (1932, for Doppler) and Rütten (filed and published in 1928, issued in 1931, for transit time.) See also, the review by McCullough and Graeper, 1979.

10.2.1 Swengel's Contributions

The earliest "practical" implementation of these two types appears to be the transit time of flight instrument of the late R.C. Swengel, who achieved initial results in water around 1947 or 1948, according to his notes and Herrick's review. He is author or co-author of several documents published between 1950 and 1956 that report his early work on ultrasonic flowmeters. Swengel appears to be the first to have *rapidly* interchanged the alternating directions of interrogation by electronic switching means. Others apparently recognized the need to do this by 1943 (F.V. Hunt, 1945, p. 42, brought to the author's attention by Swengel), but the means of switching, and the required rate of switching, are not discussed. Greenspan (1986, priv. comm.) brought to the author's attention the 1948/1949 suggestion by Hanson that the sing-around method could be applied to flow measurement (Ref. 6 in Greenspan and Tschiegg, 1961; see Section 10.3). Details on how that might be accomplished, however, were not published until the 1950s, e.g., Garman and Droz, 1954; Kritz, 1959.

Swengel also brought to the author's attention, the 1948 patent application by Ono, granted in 1950, in which *simultaneous* excitation of both transducers is proposed. The author was able to obtain limited information about Ono's work only through an intermediary (T. Matsushima, 1978, priv. comm.). According to Matsushima, Ono worked on his invention "on his desk," and tried to build it under Japanese government support from a branch that was disorganized just after World War II. Having lost his sponsor, Ono abandoned the invention. (A diagram from his patent appears on p. 434 of the author's 1979 review of ultrasonic flowmeters.)

Swengel's second major contribution to ultrasonic flowmetering is his recognition of the need to area-average the flow profile, and his solution using a sheet of ultrasound to interrogate 100% of the cross section of a large rectangular duct. Swengel's interchanging switch and area-averaging *waveguide antenna* transducers are shown in Fig. 10-2. His area-averaging *piezoelectric* transducers are shown in Chapter 4, Fig. 4-74.

One factor that contributed to Swengel's solutions, but not mentioned in the usual technical literature, was his background as a ham radio operator and inventor. Another factor was that he was unemployed shortly before being asked if he could solve the penstock flow problem. Had he been fully occupied with other tasks the rapid switching and area averaging solutions might not have been invented and demonstrated until much later.

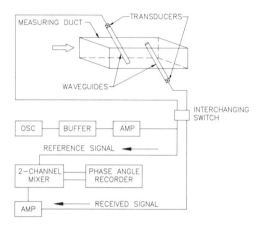

FIG. 10-2. Swengel's contributions to rapid switching upstream/downstream and area-averaging (1950-1956). Note pioneering use of "interchanging switch" enabling the same path to be utilized sequentially for upstream and downstream measurements, and also oblique interrogation of full cross-sectional area using "line" sources.

10.3   The Sing-Around Velocimeter†*

The earliest mention we have found of the sing-around principle is in the US patent [3] filed in 1937 and granted in 1943 to F.H. Shepard, Jr., of RCA. As mentioned in a FIAT review [11], Freund and Hiedemann filed a German patent application in 1940. Shortly thereafter, in 1941, W. Kock, then of the Baldwin Piano Company, filed for a similar US patent; this was granted in 1946 [4]. Huntgren and Hallman [12] discussed possible applications to radar (they used the term "ring-around") in 1947; in the same year, M.J. Larsen [5] filed for a US patent, granted in 1949, covering a sing-around echo-location system for the blind. In 1948, R.D. Holbrook [13], working at Brown University, made what so far as we know was the first serious laboratory application of the principle; this was for measurement of small changes in the speed of sound in solids. It was in 1948, also, that R.L. Hanson [6] read his paper, based on Kock's patent, in

* Contributed by Martin Greenspan and Carl E. Tschiegg
† A history of the sing-around velocimeter was presented in 1961 by Greenspan and Tschiegg. It is reproduced in this section with permission, essentially as they wrote it. (See also, Section 3.5.4.1.) In a 1986 "update," Greenspan (priv. comm.) remarked that in 1970 there were about six manufacturers of sing-around equipment. In that same private communication, Greenspan described his contribution, and Tschiegg's, as an engineering one. Their first sing-around instrument was orders of magnitude better than any they knew of at the time, so far as stability, accuracy, and precision are concerned. See Chapter 3, Fig. 3-54. See also, *J. Acoust. Soc. Am.* **83** (4), p. 1700 (April 1988).

which the term "sing-around" was coined; and at the same meeting, W.E. Kock and F.K. Harvey demonstrated a system using loud speakers in air. This demonstration and Hanson's paper introduced the subject to us. Barrett and Suomi [7] in 1949 experimented with a balloon-borne sing-around device for the measurement of air temperature. They used a thyratron pulser and a 16-in. path. The electrical time delay was 34 μs. Holbrook's work at Brown was continued by Cedrone and Curran [14], who by 1954 had produced an instrument employing a pulse-modulated 10-MHz carrier with an accuracy in liquids of about 0.1%. A much simpler instrument, utilizing video pulses and good to about 1% was described in 1956 by Ficken and Hiedemann [8].

Our own work began in 1952 and by the end of the year a prototype model of high stability, described in NBS Report 2702, January 2, 1953, was in operation. The first operational model had a straight path and quartz transducers, and was field tested in June, 1953. This instrument was used by the Chesapeake Bay Institute for several years. All succeeding vacuum-tube velocimeters, of which more than a dozen were built, had ceramic transducers and a singly bent path with a reflector of hard rubber or perforated metal. These were described in 1957 [9] but were first announced in 1955 [15]. The transistorized version was developed in 1957 to meet the need for a deep-sea instrument. At the present time (August, 1961), 65 of these instruments have been manufactured and we know of current invitations to bid on 61 more. Of the 65, three were made by NBS, two by US Coast and Geodetic Survey, two by the Woods Hole Oceanographic Institution, and the remainder by three different commercial manufacturers. Two instruments are in England, two in Norway, one is at the Saclant ASW Research Center in Italy; most of the rest are owned by various naval or oceanographic installations in the United States.

In addition, about 25 instruments are being made with the same sound head but a different timing mechanism. There are also three instruments with a 10-cm path for 1.5-V operation.

In 1958, A. Lutsch of the NPL of the Union of South Africa reported on an instrument [16] similar to that of Cedrone and Curran [14], but of much higher accuracy.

References

1. M. Greenspan and C.E.Tschiegg, "Effect of Dissolved Air on the Speed of Sound in Water," *J. Acoust. Soc. Am.*, **28**, 501 (1956).
2. M. Greenspan, C.E. Tschiegg and F.R. Breckenridge, "Temperature Coefficient of the Speed of Sound in Water Near the Turning Point," *J. Acoust. Soc. Am.*, **28**, 500 (1956).

3. F.H. Shepard, Jr., U.S. Patent No. 2,333,688 (November 9, 1943).
4. W.E. Kock, U.S. Patent No. 2,400,309 (May 14, 1946).
5. M.J. Larsen, U.S. Patent No. 2,580,560 (January 1, 1952).
6. R.L. Hanson, "Applications of the Acoustic Sing-Around Circuit," *J. Acoust. Soc. Am.*, **21**, 60–61 (1949).
7. E.W. Barrett and V.E. Suomi, "Preliminary Report on Temperature Measurement by Sonic Means," *J. Meterol.*, **6**, 273–276 (1949).
8. G.W. Ficken, Jr., and E.A. Hiedemann, "Simple Form of the Sing-Around Method for the Determination of Sound Velocities," *J. Acoust. Soc. Am.*, **28**, 921–923 (1956).
9. M. Greenspan and C.E. Tschiegg, "Sing-Around Ultrasonic Velocimeter for Liquids," *Rev. Sci. Inst.*, **28**, 897–901 (1957).
10. M. Greenspan and C.E. Tschiegg, "Speed of Sound in Water by a Direct Method," *J. Research NBS*, **59**, 249–254 (1957).
11. E. Hiedemann, *FIAT Rev. Ger. Sci.*, 1939–1946, Part 1, 178 (1947).
12. R.D. Huntgren and L.B. Hallman, "The Theory and Application of the Radar Beacon," *Proc. Inst. Radio Engrs.*, **35**, 716–730 (1947).
13. R.D. Holbrook, "A Pulse Method for Measuring Small Changes in Ultrasonic Velocity in Solids with Temperature," *J. Acoust. Soc. Am.*, **20**, 590 (1948).
14. N.P. Cedrone and D.R. Curran, "Electronic Pulse Method for Measuring the Velocity of Sound in Liquids and Solids," *J. Acoust. Soc. Am.*, **26**, 963–966 (1954).
15. *Tech. News Bull. NBS* **39**, 89 (1955).
16. A. Lutsch, "An Apparatus for Measuring and Recording the Velocity of Sound and Temperature Versus Depth in Sea Water," *Acustica*, **8**, 387–391 (1958).

10.4 Oscillating Loops, Density and Profile Compensation in Round Pipes*

In the early 1940s as a young graduate electrical engineer, serving in the armed forces, I was somewhat frustrated by not being able to practice my profession. During a short period of military leave spent in the company of a chemical engineering colleague at his parents' farm, we were discussing his problems in process control. When I boastfully claimed that electronic devices would ultimately take the major role in providing measurement and control solutions, he challenged me to come up with a flowmeter that would satisfy the criteria he voiced. They were as follows:

- Smooth-bore pipe with minimal pressure drop.

- Calibration independent of fluid type.

- Accuracy better than 1% of full scale.

- Wide dynamic range, i.e., very low to very high flow rates in the same instrument.

* Contributed by Jack Kritz

This problem intrigued me and stimulated my thinking during both my remaining time in the military service and afterward as I became an active engineer in the radio-electronics industry.

I soon realized that the only practical form of radiant energy capable of being transmitted through a variety of liquids and gases was sound. This provided the direction of my thinking that ultimately resulted in my first and subsequent patent applications. Initially my thoughts were involved with the measurement of the difference between upstream and downstream time of flight—via phase-measurement of the exciting carrier. The trouble with the time-difference measurement was that the result gave calibration values that were a function of the speed of sound in the fluid, thus violating the above second criterion of the ideal instrument. It should be remembered that in this time period, the advent of semiconductors and the ability to perform calculations with small hardware had not yet made its appearance.

The equations show that what is needed is the difference in the reciprocals of the upstream and downstream time periods. Thus the idea of the oscillating loops, as I called them, was born. Reciprocals of the periods were obviously frequencies. Furthermore, the process of taking the difference with absolute accuracy was already established in the radio art as "heterodyning." According to my notes, a first drawing of the invention was made on March 2, 1944.

I was given the opportunity to actively exploit these ideas after joining The W.L. Maxson Corporation in New York City in 1949. The item of first priority insisted upon by the patent attorneys was the "reduction to practice" of the oscillating loop.

This was accomplished with a quick and simple device using a small microphone and radio loudspeaker mounted at either end of a cardboard tube and set into feedback oscillation via an audio amplifier. An electric fan provided the variable speed air stream. Change in frequency of the loop was displayed via Lissajous patterns on an oscilloscope.

The big boost in developing a practical and successful ultrasonic flowmeter was the acquisition of government research funds from the Wright Air Development Center in 1951. This program was aimed at the development of a meter for use in the in-flight refueling boom of the KC-135 tanker. Here the flow rates were high (1500 gal/minute in a 4-inch line). The fluids to be measured ran the gamut of jet fuels JP-1, JP-3, and JP-4 as well as grades 80, 91, 100, and 115 gasolines. The instrument was required to measure the mass flow rate where the maximum was of the order of 10,000 pounds per minute (76 kg/s).

The challenges offered by this program were severe. The temperature environment precluded the use of the then only available piezoelectric ceramic, barium titanate. Quartz was the material of choice. There were no published values for the propagation velocities or absorption coefficients of the fluids. The program started with the measurement of these characteristics, and was immediately faced with designing test setups including the handling of highly inflammable liquids.

Mass flow rate determination required development of an ultrasonic densitometer. The densitometer used the measurement of the acoustic impedance (ρc) via a sensing quartz crystal and performed a computation to generate the mass flow rate as the output. Figure 10-3 is a photograph of the prototype 4-inch pipe section. One set of dual transducers is seen at the bottom. Four transducers in near-coincident paths were used to provide upstream and downstream directions.

Figure 10-4 shows a breadboard pipe section mounted in a flow test stand. It is interesting to note that the flow test stand was filled with a hydrocarbon dry cleaning fluid of high flash point that successfully simulated the acoustic characteristics of the jet fuels and gasolines. This

FIG. 10-3.   Prototype 4-inch pipe section for use aboard the KC-135 in-flight refueling tanker. Illustration courtesy J. Kritz.

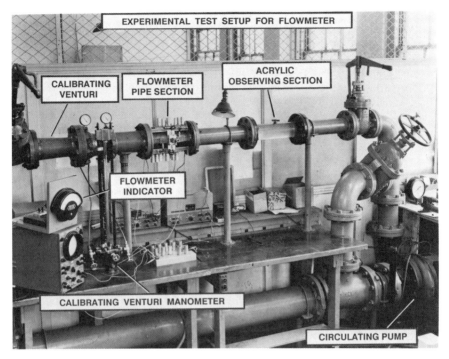

FIG. 10-4. Test stand used with simulated fuel and a venturi reference flowmeter. The breadboard ultrasonic flow section is shown mounted in the stand, and was first tested ca. June 3, 1952, according to Kritz's notes (1986, priv. comm.) Illustration courtesy J. Kritz.

arrangement surprisingly enough was approved by the New York City Fire Department for use in a Manhattan office building. Questions have always been asked of me as to whether it ever sprung a leak. The answer is "YES." The only significant result of this incident was that for a short time we had the cleanest manufacturing floor in Manhattan.

The prototype unit was delivered in 1955. The successful tests led to a production contract for the KC-135 tanker by the Boeing Aircraft Corporation. This was later cancelled at the insistence of the government. The reason advanced was that they had no confidence in the performance of electronic equipment using radio tubes, especially those mounted on a section of pipe in the refueling boom. Ironically, many years later, it was disclosed that Boeing was using the prototype unit in their boom tower to test and calibrate the mechanical flowmeters delivered on the production contract.

In 1955, an agreement was reached between The W.L. Maxson Corporation and Fisher & Porter, manufacturers of flowmeters and other process control instruments, for the development of commercial units. It was during this period that I attacked the problems of flow profile to enhance the accuracy of measurement. The results of these early studies were published in 1955. Unfortunately the work was terminated by both companies when I chose to change my employment.

10.5   Remote and Noninvasive Sensors*

As a result of Sperry Products' work in ultrasonic NDT, in the 1950s, my colleagues and I were asked from time to time if we could measure liquid level in nuclear or fossil fueled plants, preferably from the outside of a pressure boundary. These inquiries led us to consider or to adapt methods used in NDT, such as longitudinal waves in rods, angle beam zigzag shear waves in rods or vessel walls, and Rayleigh (surface) waves, to liquid level measurements. These led us to various designs including those shown in Fig. 7-29 (gap sensors) and Fig. 3-18c (Rayleigh and zigzag shear wave probes). Illustrations of test facilities at the Boiler and Turbine Laboratory of the Philadelphia Naval Base, experimental results at high pressure and up to 300°C, obtained in a 1960-1962 program, and a bibliography of ultrasonic liquid level work in the 1950s is contained in a Sperry Products' report†. Work conducted at that time showed that while ultrasound was unable to distinguish water level when the density and sound speed in the liquid and vapor hardly differ, the simpler cases could certainly be solved by the proposed method.

One of the interesting liquid level patents from that era is that of Mongan, 1961, filed in 1955.

Regarding the above–mentioned zigzag (vertically polarized) shear wave probe, if to that probe were added another shear wave, horizontally polarized, then that probe ought to be able to additionally function as a viscosity (or viscosity·density) sensor.

10.6   Ultrasonic Beam Deflection Flow Measurements**

The development of the beam deflection ultrasonic flowmeter was part of an overall activity at Gulton Industries in the 1950s and 1960s on ap-

---

* Contributed by H.E. Van Valkenburg
† Sperry Products, TR-084C, for Navy Department, Bureau of Ships, Electronics Division, Contract NObsr-77613, Index No. NE141402 (1962).
** Contributed by Walter Welkowitz, Department of Electrical Engineering, Rutgers University

plications of ultrasonics. This particular work resulted in a number of patents assigned to Gulton Industries:

2874568	Petermann	February 4, 1959
2959054	Welkowitz	November 8, 1960
3020759	Welkowitz	February 13, 1962
3178940	Dahlke & Welkowitz	April 20, 1965
3204456	Welkowitz	September 7, 1965
3236098	Dahlke & Welkowitz	February 22, 1966

Clearly, the major people involved besides myself were Drs. Dahlke and Petermann.

The fundamental problem related to this system is that it is an amplitude system and therefore is inherently less accurate than frequency based systems. Most of the problems, however, were state-of-the-art problems related to transducers and electronics. These problems are somewhat obviated by recent developments in these fields. A brief write-up of this system appears in Welkowitz and Deutsch, *Biomedical Instruments: Theory and Design*, 1976, and in Dahlke and Welkowitz, 1960. A system block diagram and an example of beam drift test data obtained on a pipe are given in Chapter 4, Figs. 4-71 and 4-72, respectively.

[Comment by author: Despite the interest expressed in this method in the 1950s and early 1960s, the *beam drift method* has *not* been an important method of measuring flow in industry. But in a mid-1980 *transit time* ultrasonic flowmeter designed to measure flare gas flow rate up to *high* velocity, $\gtrsim 100$ m/s, the placement and orientation of the transducers was intended to take the drift into account (Mylvaganam et al., 1985, 1987; Mylvaganam, 1987, 1989; see Section 4.2.2.2). These high velocities, despite their infrequent occurrences, nevertheless represent the major oil loss in some offshore flares. Depending on the flare gas molecular weight, and hence the sound speed c, velocities over 100 m/s correspond to Mach numbers V/c which can exceed 30%. Therefore the drift, which is usually negligible in industrial transit time flowmeters, is no longer insignificant.

Section 4.2.2.3 describes a particle-laden mass flow meter due to Hamade, 1982 and Leffert, 1987, 1988, that senses beam drift to determine the flow velocity of air carrying pulverized coal, and interprets attenuation in terms of coal mass loading. In Hamade's work, the receiving transducer was moved in a direction parallel to the air flow until the received signal was maximized. From the receiving transducer's position downstream of the transmitting transducer, the flow velocity was computed. In the simplest form of this measurement, the tangent of the angle of beam drift equals the Mach number.]

10.7   Early Recollections of the J-Tec Vortex Sensor Invention*

J-Tec was originally an oceanographic sensor house and one of the needs evident was for a more modern ocean current sensor. A form of a paddle wheel sensor had been used for decades. In the summer of 1969 Russ Colton and I set about to see if we could come up with some better way to measure this. Russ had previously worked with aircraft design and mentioned the wing tip vortices which are a problem on aircraft but occur at a rate proportional to the speed. We decided to see if we could make use of this phenomenon. Our first sensor used a flat plate about ½-inch wide with a second plate extending downstream behind the vortex "generator." Strain gages were put on the second plate and sure enough we could detect vortices. However, as we increased the velocity, the output frequency came up to the resonant frequency of the plate and stayed there. We tried changing the characteristics of the plate but were never able to get a wide range with any reasonable sensitivity. Therefore we decided to get rid of the second plate and instead measure the rotation force imparted on the vortex generator by vortices. Not having much equipment (and even less money) I took the cartridge out of an old phonograph and connected it to the vortex strut. Again we had very limited success. The problems in both these methods were later solved by Fisher-Porter and Yokogawa.

I had earlier built some ultrasonic pingers used as fish tags for tracking salmon and one day suggested that maybe we could see a Doppler shift from the vortices. The transducers were ring type, so to create a narrow beam we put these in a small metal box with a hole. These didn't work very well under water so we tried them in air and recorded a beautiful output signal over the full range that we could get from the electric fan which was our source of flowing air.

Looking for support, we took our data to one of the Navy Labs and while they didn't have any development funds, they did agree to buy a prototype for measuring wind speed on a carrier. This order for $1,000 started us in the vortex business using ultrasonics.

We then found that the same technique worked in water and proceeded to build an ocean current meter. I like to say that the unit was very successful—we captured almost the total current meter market . . . maybe 10 units per year.

Not finding any support for a current meter, we went back to air measurement and found a believer in the Air Force at Wright-Paterson AFB.

* Contributed by Robert D. Joy, J-Tec

We built several versions for measuring the speed of helicopters. Still being an oceanographic company, we tried to sell the current meter concept to the National Data Buoy Program but they didn't want anything protruding from the hull of their buoy. However, they said that they did want a no-moving-parts anemometer. They awarded us a contract and we were in the gas flow business.

One other recollection that might be of interest occurred in 1972 when we decided that the sensor might be used to measure breath flow and began working with the Bourns Life Sciences group. Russ and I were both research oriented and when we delivered the first prototype, it was tacked together with a lot of modifications. A week later Bourns called and said that they had installed the unit on a respirator and it was being used in a hospital to keep a patient alive. All I could think about was the wires I had flexed and cold solder joints that were probably in the unit. The unit didn't fail. The design proved to be very successful with Bourns (later this Division was sold to Bear Medical Co.) selling at least 30,000 of these devices.

The total number of ultrasonic vortex meters that have been sold is hard to determine exactly. I can break them down into four categories: (1) Spirometers built by Bear Medical (formerly Bourns Life Sciences Division); (2) automobile air intake meters built by Mitsubishi; (3) cross-wind sensors for use on military tanks; and (4) the industrial or commercial sensors.

Mitsubishi uses the ultrasonic units on all the cars they sell which have turbo-chargers. By the end of 1987, this represented 206,000 on cars exported from Japan. The cross-wind sensor used for fire control purposes on military tanks amounts to some 3,500 units.

The industrial and commercial units probably amount to 10,000 units, so in total there have been more than 250,000 ultrasonic vortex meters sold between 1970 and 1987.

References

ANL/NBS, Hot Steel NDT Method Announced by Argonne, *Metal Progress*, p. 24 (March 1987).

ANSI/ASME MFC-5M-1985, *Measurement of Liquid Flow in Closed Conduits Using Transit-Time Ultrasonic Flowmeters*, available from Amer. Soc. Mech. Engrs., 345 E. 47th St., NY 10017 (1985).

ARINC (Aeronautical Radio, Inc.) Airlines Electronic Engineering Committee, AEEC Letter 86-036/FQS-01 (March 25, 1986).

ASNT, *Fifth Annual NDT Buyer's Guide: Manufacturers' and Distributors' Listings for NDT Equipment and Supplies* **45** (6), pp. 641–714 (June 1987).

ASTM, STP 505, *Acoustic Emission* (1972).

R.A. Abbott, Quality Concepts and Statistical Methods, pp. 401–406 in: *ASM* (Amer. Soc. Metals) *Metals Handbook*, 8th Edition, Vol. **11**, Nondestructive Inspection and Quality Control, ASM (1976).

W.E. Abbotts, Methods and Apparatus for Measuring the Densities of Fluids, U.S. Patent No. 3,648,512 (March 14, 1972); Vibrating Fluid Density Meter, *Instrum. Techn.* **19** (7), p. 66 (July 1972).

L.R. Abts, The Ultrasonic Detection of Microparticles, Microbubbles, and Microemboli in Flowing Liquids, Doctorate Thesis, Brown University (June 1982).

L.R. Abts and P.H. Dahl, Ultrasonic Determination of Component Concentrations in Multi-Component Fluids, U.S. Patent No. 4,580,444 (April 8, 1986).

J.A. Achenbach, *Wave Propagation in Elastic Solids*, North Holland Pub. Co. (1975).

R.C. Addison, Jr., L.J. Graham, R.S. Linebarger and B.R. Tittmann, Synthesis of an Ultrasonic Array Using Laser-Based Techniques, pp. 1109-1113, *1987 Ultrasonics Symp. Proc.*, IEEE (1987).

R. Adler and P.J. Desmares, An Economical Touch Panel Using SAW Absorption, pp. 499–502, *1985 Ultrasonics Symp. Proc.,* IEEE (1985).

R. Adler and P.J. Desmares, SAW Touch Systems on Spherically Curved Panels, pp. 289–292 in: *1986 Ultrasonics Symp. Proc.*, IEEE (1986).

J. Agar, Measuring of Fluid Density, U.S. Patent 3,763,692 (October 9, 1973).

N.S. Ageeva, Ultrasonic Method for Measuring the Height of the Fluid Level in a Vessel by Means of Flexural Oscillation of a Thin Elastic Strip, *Sov. Phys.—Acoustics* **6** (1) pp. 116–117 (Jan.-Mar. 1960).

N. Ahmad, Surface Acoustic Wave Flow Sensor, pp. 483–485 in: *1985 Ultrasonics Symp. Proc.*, IEEE (1985).

N. Ahmad, A Microprocessor-Based SAW Resonator Temperature Sensor, pp. 347–351, *Proc. Sensors Expo 1987*, Helmers Publ. (1987).

M. Alagar and V. Krishnasamy, Ultrasonic Properties of Tetraalkoxysilanes, *Ultrasonics* **25** (5), pp. 283–287 (1987)..

C.P. Albertson and H.E. Van Valkenburg, Ultrasonic Detection, Sizing and Counting of Sub-sieve Particles, pp. 147–149, *Proc. Fourth International Conference on Nondestructive Testing*, Butterworths, London (1964).

E.E. Aldridge, A Study of the Ultrasonic Micrometer, *IEEE Trans. Sonics and Ultrasonics* **SU-14** (2), pp. 89–99 (April 1967).

G.A. Alers, R.A. Chesebrough and D.T. MacLaughlan, Application of Surface Skimming SH Waves to Stress and Texture Measurement in Steel, in D.O. Thompson and D.E. Chimenti (ed.), *Rev. Progress in Quantitative NDE*, Plenum Press (1988).

G.A. Alers, D.T. MacLaughlan and L.R. Burns, Electromagnetic Acoustic Transducer, U.S. Patent No. 4,777,824 (Oct. 18, 1988).

C. Alquie and J. Lewiner, A New Method for Studying Piezoelectric Materials, *Revue Phys. Appl.* **20**, pp. 395–402 (1985).

O.L. Anderson and R.C. Liebermann, Sound Velocities in Rocks and Minerals: Experimental Methods, Extrapolations to Very High Pressures, and Results, pp. 329–472 in: W.P. Mason (ed.) *Physical Acoustics* Vol. 4, Part B, Academic Press (1968).

T.T. Anderson, A.P. Gavin, J.R. Karvinen, C.C. Price and K.J. Reimann, Detecting Acoustic Emission in Large Liquid Metal Cooled Fast Breeder Reactors, *Acoustic Emission, ASTM STP* **505**, American Society for Testing and Materials, pp. 250–269 (1972).

W.L. Anderson, Y. Motiwala and M.S. Toth, Detection of Subsurface Stress in Metals Using Ultrasound and Time-Varying Temperature Gradients, pp. 531–533, *1986 Ultrasonics Symp. Proc.*, IEEE (1986).

K.W. Andrews, M. Druce and J.E. Russell, Use of Ultrasonics for the Examination of Hot Billets and for Cleanliness Assessment, *Metals Techn.* **1**, pp. 94–106 (February 1974).

K.W. Andrews, Chap 10, pp. 411–436 in: J. Szilard (ed.), *Ultrasonic Testing*, John Wiley and Sons Ltd., London (1982).

Anon., Ultrasonic Sensor Solves Wire Mesh Transparency Problems, *Sensors* **4** (7), pp. 34–37 (July 1987).

J.H. Apfel, Acoustic Thermometry, *Rev. Sci. Instrum.* **33** (4), pp. 428–430 (1962).

G.C. Aprilesi, G. De Cicco and A. Taroni, A Microprocessor-Based, Three Axes, Ultrasonic Anemometer, pp. 295–298 in: *1983 Ultrasonics Symp. Proc.*, IEEE (1983).

H. Araki and Y. Matsunaga, Ultrasonic Flow Meter, U.S. Patent No. 4,014,211 (March 29, 1977).

A.E. Arave, IN-1441, An Ultrasonic Void Fraction Detector Using Compressional Stress Waves in a Wire Helix (Oct. 1970).

A.E. Arave, IN-1442, An Ultrasonic Liquid Level Detector Using Shear Wave Attenuation in a Bar (Nov. 1970).

A.E. Arave, F.E. Panisko, and J. A. Christensen, Idaho Nucl. Corp. Rpt. ANCR-1091, High-Temperature Ultrasonic Thermometer In- Reactor Fuel Rod Centerline Temperature Test Results (1972).

A.E. Arave, E. Fickas and W. Shurtliff, Instrumentation in the Aerospace Industry—Vol. **24**, *Proc. 24th International Instrumentation Symposium*, pp. 609–620, ISA (1978).

A.E. Arave, Ultrasonic Densitometer Development, presented at NRC Instrumentation Review Group Meeting, Silver Spring, Maryland (July 24–26, 1979).

D.L. Arenberg, Ultrasonic Solid Delay Lines, *J. Acoust. Soc. Am.* **20** (1), pp. 1–26 (1948).

P.E. Armstrong, J.M. Dickinson and H.L. Brown, Temperature Dependence of the Elastic Stiffness Coefficients of Niobium (Columbium), *Trans. Met. Soc. AIME* **236**, pp. 1404–1408 (1966).

W. J. Arnoult, Problems in the Deformation of Metals: 1. Vacancy Formation Free Energy; 2. Dislocation Attack Frequency; 3. Microstrain in I-Easy Glide Region, PhD dissertation, Rice Univ. (1973).

W.J. Arnoult and R.B. McLellan, Variation of the Young's Modulus of Austenite with Carbon Concentration, *Acta Metall.* **23** pp. 51–56 (1975).

M. Aronson (ed.), Open-Channel Flowmeters, *Measurements & Control* **21** (6), pp. 218–223 (Dec. 1986).

H. Asada and M. Yamamoto, Ultrasonic Flowmeter for Exhaust Gases in Stacks, pp. 227–231, *Proceedings of the IMEKO Symposium on Flow Measurement and Control in Industry*, Tokyo (November 1979).

E.A. Ash and C.R. Hill (ed.), *Acoustical Imaging* **12**, Plenum Press, New York and London (1982).

A. Ashrafzadeh, J.Y. Cheung and K.J. Dormer, Analysis of Velocity Estimation Error for a Multidimensional Doppler Ultrasound System, *IEEE Trans. UFFC* **35** (5), pp. 536–544 (Sept. 1988).

K.N. Astill, *Elementary Experiments in Mechanical Engineering*, esp. pp. 105–111 by R. Greif; NSF Grant GY-2467; Tufts University (Feb. 1971).

A. Atalar and H. Köymen, Generation of Focused Surface Waves with a Solid Wedge, pp. 681–684 in: *1987 Ultrasonics Symp. Proc.*, IEEE (1987).

P. Atkinson, An Ultrasonic Fluctuation Velocimeter, *Ultrasonics* **13** (6), pp. 275–278 (1975).

P. Atkinson and J.P. Woodcock, *Doppler Ultrasound and Its Use in Clinical Measurement*, Academic Press, New York (1982).

J. Attal and C.F. Quate, Investigation of Some Low Ultrasonic Absorption Liquids, *J. Acoust. Soc. Am.* **59** (1), pp. 69–73 (January 1976).

J.D. Aussel, A. Le Brun and J.C. Baboux, Generating Acoustic Waves by Laser: Theoretical and Experimental Study of the Emission Source, *Ultrasonics* **26** (5), pp. 245–255 (Sept. 1988).

W.G. Austen and D.H. Howry, Ultrasound as a Method to Detect Bubbles or Particulate Matter in the Arterial Line During Cardiopulmonary Bypass, *J. Surgical Research* **V** (6), pp. 283–284 (June 1965).

S. Ayter, Focusing Surface Waves Using Conical Transducers, pp. 301–304 in: *1987 Ultrasonics Symp. Proc.*, IEEE (1987).

M. Azimi and A.C. Kak, An Analytical Study of Doppler Ultrasound Systems, *Ultrasonic Imaging* **7** (1), pp. 1–48 (1985).

BHRA (British Hydromechanical Research Association, Cranfield, UK), Multiphase Update, bimonthly newsletter starting September 1987.

T. Baba, *Jpn. J. Waterworks Assoc.* No. 4, pp. 21–31 (1964).

O.I. Babikov, *Ultrasonics and Its Industrial Applications*, Consultants Bureau, Plenum Press (1960).

J.F. Bacon, The Kinetics of Crystallization of Molten Binary and Ternary Oxide Systems and Their Application to the Origination of High Modulus Glass Fibers, UARL Rpt. K910939–4, NASA Contracts NASW-1301, 2013 (January 31, 1971).

J.A. Bailey and A. Dula, Acoustic Technique for Use in Some Solidification Rate Studies. *Rev. Sci. Instrum.* **38** (4), pp. 535–538 (1967).

D.W. Baker, F.K. Foster and R.E. Daigle, Doppler Principles and Techniques, pp. 161–287 in F.J. Fry, *Ultrasound: Its Application in Medicine and Biology*, **I**, Elsevier (1978).

D.S. Ballantine, Jr., Use of SAW Devices to Monitor Visco-Elastic Properties of Materials, pp. 559–562, *1988 Ultrasonics Symp. Proc.*, IEEE (1988).

A.S. Bandes, Leak Detection Through Airborne Ultrasound, *Sensors* **4** (11), pp. 26–33 (November 1987).

X.J. Bao, W. Burkhard, V.V. Varadan and V.K. Varadan, SAW Temperature Sensor and Remote Reading System, pp. 583–586, *1987 Ultrasonics Symp. Proc.*, IEEE (1987).

Y. Bar-Cohen and D.E. Chimenti, NDE of Effects in Composites Using Leaky Lamb Waves, pp. 202–208 in: *Proc. 15th Symp. on NDE*, D.W. Moore and G.E. Matzkanin (ed.), SwRI, San Antonio, TX (1985).

Y. Bar-Cohen and D.E. Chimenti, NDE of Composite Laminates by Leaky Lamb Waves, pp. 1199–1206 in: D.O. Thompson and D.E. Chimenti (eds.), *Review of Progress in Quantitative NDE* **5B** (Plenum Press, 1986).

Y. Bar-Cohen, Ultrasonic NDE of Composites—A Review, pp. 187–201 in: *Solid Mechanics Research for Quantitative NDE*, J.D. Achenbach and Y. Rajapakse (ed.) Martinus Nijhoff, Boston (1987).

Y. Bar-Cohen, Nondestructive Characterization of Defects in Multilayered Media Using Ultrasonic Backscattering, presented to Ultrasonic International 1987, Kensington Town Hall, London, U.K. (July 6–8, 1987).

A.R. Barber, K.E. Kneidel, C.S. Fitzgerald and L.C. Lynnworth, Ultrasonic Temperature Profiling System for Determining Critical Heat Flux in Nonuniformly Heated Tube Bundles, *J. Heat Transfer* **101**, pp. 622–627 (Nov. 1979).

A. Barone and J.A. Gallego Juarez, Flexural Vibrating Free-edge Plates with Stepped Thicknesses for Generating High Directional Ultrasonic Radiation, *J. Acoust. Soc. Am.*, p. 953 (1971).

M.F. Barsky, D.K. Linder and R.O. Claus, Robot Gripper Control System Demonstrating PVDF Piezoelectric Sensors, pp. 545–548, *1986 Ultrasonics Symp. Proc.*, IEEE (1986).

N.K. Batra, H.H. Chaskelis and P.P. Delsanto, Ultrasonic Characterization of Highly Attenuative Fiber-Reinforced Composites, pp. 975–978, *1985 Ultrasonic Symp. Proc.*, IEEE (1985).

H.H. Bau, Torsional Wave Sensor—A Theory, *ASME J. Appl. Mech.* **53** (4), pp. 846–848 (Dec. 1986).

H.H. Bau, J.O. Kim, L.C. Lynnworth, and T.H. Nguyen, Improved Torsional Wave Sensor and System, U.S. pat. pending (Oct. 14, 1988).

J. Baumoel, Clamp-On Transit-Time Ultrasonic Flowmeters, *Measurements & Control* **18** (3), pp. 186–194 (June 1984).

R.E. Beatty, Jr., Boundary Layer Attenuation of Higher Order Modes in Rectangular and Circular Tubes, *J. Acoust. Soc. Am.* **22** (6), pp. 850–854 (1950).

M.S. Beck and A. Plaskowski, *Cross Correlation Flowmeters-Their Design and Application*, Hilger, Techno House, UK (1987).

J.D. Begin, Worn Tool Detector Utilizing Normalized Vibration Signals, U.S. Patent No. 4,514,797 (April 30, 1985).

C. Behrman and J. Larson, On-Line Ultrasonic Particle Monitoring of Brewing Operations, MBAA Technical Quarterly **24** (2), pp. 72–76 (1987).

J.F.W. Bell, The Velocity of Sound in Metals at High Temperatures, *Phil. Mag.* **2**, pp. 1113–1120 (1957).

J.F.W. Bell, Dynamic and Static Elasticities of Solids, *Nature* **181** (4619), p. 1330 (1958).

J.F.W. Bell, Some Acoustic Effects at Phase Changes, Paper J52 in *Proceedings of the Fourth International Congress on Acoustics*, Copenhagen, 1962 (Organization Committee of the 4th ICA and Harlang and Toksvig, Copenhagen, 1963), Pt. 1.

J.F.W. Bell and G.E.R. Mobsby, UK Patent 54,647 (1966).

J.F.W. Bell, B.P. Doyle and B.S. Smith, An Instrument for the Measurement of Acoustic Pulse Velocity and Attenuation in a Solid Probe, *J. Sci. Instrum.* **43** (1), 28–31 (1966).

J.F.W. Bell, An Ultrasonic Thermometer, Abstract, *Ultrasonics* **5** (4), p. 265 (Oct. 1967).

J.F.W. Bell, UK Patent 4,934 (1969).

J.F.W. Bell, Ultrasonic Thermometry Using Resonance Techniques, pp. 709–713 in: H.H. Plumb (ed.-in-chief), *Temperature—Its Measurement and Control in Science and Industry* **4**, ISA (1972).

J.F.W. Bell, A.C. Johnson and J.C.K. Sharp, Pulse-Echo Method of Investigating the Properties of Mechanical Resonators, *J. Acoust. Soc. Am.* **57** (5), pp. 1085–1093 (May 1975).

A. Benjaminson and F. Rowland, The Development of the Quartz Resonator As A Digital Temperature Sensor with a Precision of 1 x 10^{-4}, pp. 701–708 in: H.H. Plumb (ed.-in-chief), *Temperature—Its Measurement and Control in Science and Industry* **4**, ISA (1972).

K.D. Bennett, S.J. Hanna and R.O. Claus, Monitoring of Strain in Layered Media Using Clad Rod Acoustic Waveguides, pp. 1064–1067, *1985 Ultrasonics Symp. Proc.*, IEEE (1985).

L. Bergmann, *Der Ultraschall*, 6th ed., Hirzel, Stuttgart (1954). See also, *J. Acoust. Soc. Am.* **32** (4), p. 515 (1960).

R.T. Beyer, Nonlinear Acoustics, pp. 231-264 in: W.P. Mason (ed.), *Physical Acoustics* Vol. **II**, Part B, Academic Press (1965).

R.T. Beyer and S.V. Letcher, *Physical Ultrasonics*, Chap. 6, p. 161 ff., Academic Press (1969).

A.B. Bhatia, *Ultrasonic Absorption*, Oxford Univ. Press, London (1967).

C. Biber, S. Ellin, E. Shenk and J. Stempeck, The Polaroid Ultrasonic Ranging System, Audio Engineering Soc. Preprint 1696(A-8), (1980).

J.R. Birchak and E. Lygas, Acoustic Device for Measuring Fluid Properties, U.S. Patent No. 4,571,693 (February 18, 1986).

G.I. Birger and N.I. Brahznikov, *Ultrasonic Flow Meters*, Moscow: Metallurgiya, in Russian (1964). Reviewed by I.N. Kanevskii, *Sov. Phys.-Acoustics* **12** (4) 441–442 (1967).

B. Bischoff, S. Ramalingam and W.P. Robbins, Toolbit Mounted Thin Film Zinc Oxide Sensors for Process Control in Lathe and Milling Machine Applications, pp. 605–609, *1987 Ultrasonics Symp. Proc.*, IEEE (1987).

W.K. Blake, *Mechanics of Flow-Induced Sound and Vibration*, Vol. 1—General Concepts and Elementary Sources, 358 pp.; Vol. 2—*Complex Flow-Structure Interactions*, 465 pp., Academic Press (1986).

G.V. Blessing and D.R. Flynn, New Technique for Evaluating Installation of Home Insulation, *NBS Research Reports*, NBS Spec. Publ. 735, p. 30 (December 1987); Acoustic Evaluation of Thermal Insulation, U.S. Pat. 4,672,851 (June 16, 1987).

J. Blitz, *Fundamentals of Ultrasonics*, 2nd ed., Plenum Press (1967).

O. Bonnefous and P. Pesqué, Time Domain Formulation of Pulse-Doppler Ultrasound and Blood Velocity Estimation by Cross Correlation, *Ultrasonic Imaging* **8**, pp. 73–85 (1986).

O. Bonnefous, Measurement of the Complete (3-D) Velocity Vector of Blood Flow, pp. 795–799, *1988 Ultrasonics Symp. Proc.*, IEEE (1988).

S.E. Booth, Ultrasonics as a Method of Mixing, Dispersion and Homogenisation, *Paint and Resin*, pp. 17, 18, 24 (December 1986).

D. Boyd and P. Sperline, Noncontact Temperature Measurements of Hot Steel Bodies Using an Electromagnetic Acoustic Transducer (EMAT), pp. 1669–1676 in: D. O. Thompson and D.E. Chimenti (ed.), *Review of Progress in Quantitative NDE*, Plenum Press (1988).

D.M. Boyd, B.D. Droney, P.D. Sperline, J.F. Jackson and J.R. Cook, In-Plant Demonstration of High-Temperature EMAT System on Continuous Caster Strand, in: *Review of Progress in Quantitative NDE*, Plenum Press (1989).

R. Boyle, *New Experiments Physico-Mechanical, Touching the Air,* Miles Fleshner, London, 3rd ed., Expt. 27, p. 103(1682).

J.G. Brace, T.S. Sanfelippo and S.G. Joshi, Flow Sensing Using Surface Acoustic Waves, Paper N6, Abstr., *IEEE Trans. UFFC* **35**, *1987 Ultrasonics Symp. Proc.*, IEEE (1987).

G. Bradfield, Brit. Patent 766,981 (Jan. 30, 1957).

G. Bradfield, *Elastic Properties of Solids, Use in Industry of Elasticity Measurements in Metals With the Help of Mechanical Vibrations,* Notes on Applied Science, No. 30 (1964).

J.E. Bradshaw and N.E. Pedersen, Acoustic Flowmeter with Envelope Midpoint Tracking, U.S. Patent No. 4,480,485 (Nov. 6, 1984).

J.H. Bradshaw, C. Carey and E.H. Carnevale, Research Study and Experimental Program to Determine the Transport Properties of High Temperature and High Density Gases, Arnold Engineering Development Center, Air Force Systems Command Report AEDC-TR-71-191 (September 1971).

J.H. Bradshaw, A Method for the Experimental Determination of the Coefficients of Viscosity and Thermal Conductivity of Cases by Means of an Acoustic Cavity Resonator, Doctorate Thesis, Department of Physics, Boston College (1972).

M. Braeuel, F. Nadeau, and M.M. Bayoumi, An Acoustic Method for the Detection of Defects in the Nozzle of Plasma Cutting Torches, *IEEE Trans. UFFC* **UFFC-34** (2), pp. 259–262 (March 1987).

H. Braun, Flow Measurement by Correlation—An Alternative Method to Transit Time Correlation, ACTA IMEKO 88, ISA (1988).

N.I. Brazhnikov, Method of and Device for Controlling Gas-Liquid or Liquid-Liquid Interface in Monolayer Reservoirs, U.S. Patent 4,118,983 (Oct. 10, 1978).

M.A. Breazeale, Physics and Engineering Principles of Nonlinear Acoustics, *IEEE Trans. UFFC* **34** (4), p. 430 (July 1987).

L.M. Brekhovsikh, *Waves in Layered Media, Applied Mathematics and Mechanics* Vol. 16, Academic Press (1980).

J.A. Brennan and A. Takano, A Preliminary Report on the Evaluation of Selected Ultrasonic and Gyroscopic Flowmeters at Cryogenic Temperatures, *Proc. 9th Int'l. Cryogenic Engineering Conference,* pp. 655–658 (1982).

M. C. Brenner and J.J. Fitzgibbon, Surface Acoustic Wave Touch Panel System, U.S. Patent 4,644,100 (Feb. 17, 1987).

C.F. Brockelsby, J.S. Palfreeman and R.W. Gibson, *Ultrasonic Delay Lines,* London Iliffe Books Ltd., London (1963). See also, S. Davidson, Wire and Strip Delay Lines, *Ultrasonics* **3** (3), pp. 136–146 (July-Sept. 1965).

G.J. Broekgaarden and H. Lammerse, Ultrasonic Gas Flow Measurements in Reflection Mode in Underground Pipelines, pp. 793–807 in: *Proc. AGA Int'l Symposium on Fluid Flow* (November 1986).

E. Brookner, *Radar Technology,* 432 pp., Artech (1977).

L.J. Brooks, R.G. Castile, G.M. Glass, N.T. Griscom, M.E.B. Whol, and J.J. Fredberg, Reproducibility and Accuracy of Airway Area by Acoustic Reflection, *Amer. Phys. Soc., Respirat. Environ. Exercise Physiol.* **57** (3), pp. 777–787 (1984).

G.H. Broomfield, The Effects of Temperature and Irradiation on Piezoelectric Acoustic Transducers and Materials, UKAEA Harwell Rpt. AERE R11942 (December 1985).

H.L. Brown and P.E. Armstrong, Young's Modulus Measurements Above 2000°C, *Rev. Sci. Instrum.* **34** (6), pp. 636–639 (1963).

M.K. Brown, Controlling a Robot with Sonic and Ultrasonic Means, pp. 553–562, *1987 Ultrasonics Symp. Proc.,* IEEE (1987).

R. Brown, Piezo Film for Monitoring Vibration, *Sensors* **5** (1), pp. 20–26 (January 1988).

R.F. Bruner, Phase Modulation, Ultrasonic Flowmeter, U.S. Patent No. 4,528,857 (July 16, 1985).

V. Brusasco, K.C. Beck, M. Crawford and K. Rehder, Resonant Amplification of Delivered Volume During High-Frequency Ventilation, *J. Appl. Physiol.* **60** (3), pp. 885–892 (1986).

K.G. Budden, *The Wave-Guide Mode Theory of Wave Propagation*, Prentice-Hall, Englewood Cliffs, N.J. (1961).

C. Buess, P. Pietsch, W. Guggenbuhl and E.A. Koller, Design and Construction of a Pulsed Ultrasonic Air Flowmeter, *IEEE Trans. Biomed. Eng'g.* **BME-33** (8), pp. 768–774 (August 1986).

S.F. Burch and J.T. Burton, Ultrasonic Synthetic Aperture Focusing Using Planar Pulse-Echo Transducers, *Ultrasonics* **22** (6), pp. 275–281 (November 1984).

S.F. Burch, An Amplitude Correlation and Differencing Method for the Monitoring of Flaws in Repeat Ultrasonic Inspections, *Ultrasonics* **23** (6), pp. 246–252 (November 1985).

C.P. Burger, T.D. Dudderar, J.A. Gilbert and J.A. Smith, The Use of Fiber Optic Interferometry to Sense Ultrasonic Waves, pp. 209–215 in: *Proc. 34th Int'l. Instrum. Symp.*, ISA (1988).

L.R. Burns, D.T. MacLauchlan and G.A. Alers, Compact EMAT Receiver for Ultrasonic Testing at Elevated Temperatures, pp. 1677–1683 in: D.O. Thompson and D.E. Chimenti (Ed.), *Review of Progress in Quantitative NDE*, **7B** Plenum Press (1988).

D.W. Busse, Quartz Transducers for Precision Under Pressure, *Mechanical Engineering* **109** (5), pp. 52–56 (May 1987).

D.W. Busse and R.B. Wearn, Intelligent Digital Pressure Transmitters for Aerospace Applications, *Measurements & Control*, Vol. **22** (1), pp. 162–164 (February 1988).

D.W. Busse, An Intelligent Addressable Pressure Transmitter, pp. 311–315 in: *Proc. 34th Int'l. Instrum. Symp.*, ISA (1988).

B. Butler, S.B. Palmer and G.J. Primavesi, Techniques for the Generation of Ultrasound for Extended Periods at High Temperatures, *Ultrasonics* **17** (6), pp. 249–254 (1979).

C.F. Buynak and R.L. Crane, A Novel Acoustic Coupling Device Using Permeable Membrane, *Materials Eval.* **45**, pp. 743–746 (June 1987).

C. Cachard, G. Gimenez and D. Vray, Ultrasonic Doppler Device for Measurement of Time-Dependent and Space-Dependent Flow Speed, pp. 901–904, *1988 Ultrasonics Symp. Proc.*, IEEE (1988).

W.G. Cady, *Piezoelectricity*, McGraw-Hill, NY (1946).

J. Callerame, R.H. Tancrell and D.T. Wilson, Comparison of Ceramic and Polymer Transducers for Medical Imaging, pp. 117–121, *1978 Ultrasonics Symp. Proc.*, IEEE (1978).

L. Camp, *Underwater Acoustics*, Wiley-Interscience, NY (1970).

D. Campbell, Ultrasonic Noncontact Dimensional Measurement, *Sensors* **3** (7), pp. 37–43 (July 1986).

M.A. Campbell and A. McNab, A Novel Instrument for the Control of a Phased Array for NDE, pp. 994–997, *1985 Ultrasonics Symp. Proc.*, IEEE (1985).

D.W. Cannon, Electro-acoustic Effects in Electrolytes and Colloidal Suspensions, PhD Dissertation, Brown University, Rhode Island (1989).

C.A. Carey, E.H. Carnevale, S. Uva and T. Marshall, Experimental Determination of Gas Properties at High Temperatures and/or Pressures, AEDC-TR-69-78 (March 1969).

W.M. Carey, Acoustic-Ultrasonic Leak Location on Long Pipes, *IEEE Trans. Sonics and Ultrasonics* **SU-26** (2), 148 [Abstract C-3] (March 1979).

B.H. Carlisle, Piezoelectric Plastics Promise New Sensors, *Machine Design* **58** (25), pp. 105–110 (October 23, 1986).

G.A. Carlson, W.H. Sullivan and H.G. Plein, Application of Ultrasonic Thermometry in LMFBR Safety Research, pp. 24–28, *1977 Ultrasonics Symp. Proc.*, IEEE (1977).

E.H. Carnevale, J.M. Yos and H.L. Poss, Ultrasonic Temperature Determination in a Plasma, AVCO RAD-TR-61-13 (1961); E.H. Carnevale, H.L. Poss and J.M. Yos, Ultrasonic Temperature Determinations in a Plasma, pp. 959–967 in C.M. Herzfeld (ed.), *Temperature—Its Measurement and Control in Science and Industry*, Reinhold (1962).

E.H. Carnevale, L.C. Lynnworth and G.S. Larson, Ultrasonic Measurement of Elastic Moduli at Elevated Temperatures, Using Momentary Contact, *J. Acoust. Soc. Am.* **36** (9), pp. 1678–1684 (Sept. 1964); Abstr., *ibid.* **35**, 1883 (1963).

E.H. Carnevale, G. Larson, L.C. Lynnworth, C. Carey, M. Panaro and T. Marshall, Experimental Determination of Transport Properties of High Temperature Gases, NASA CR-789 (June 1967a).

E.H. Carnevale, pp. 73–103 in: G.G. Mannella (ed.), *Aerospace Measurement Techniques*, NASA SP-132 (1967).

E.H. Carnevale, L.C. Lynnworth and G.S. Larson, Ultrasonic Determination of Transport Properties of Monatomic Gases at High Temperatures, *J. Chem. Phys.* **46** (8), pp. 3040–3047 (April 15, 1967b).

E.H. Carnevale and L.C. Lynnworth, Ultrasonic Measurement Apparatus, U.S. Patent No. 3,315,520 (April 25, 1967).

E.H. Carnevale, S. Wolnik, G. Larson, C. Carey and G. Wares, Simultaneous Ultrasonic and Line Reversal Temperature Determination in a Shock Tube, *The Physics of Fluids* **10** (7), pp. 1459–1467 (July 1967c).

E.F. Carome and J.M. Witting, Theory of Attenuation in Cylindrical and Rectangular Waveguides, *J. Acoust. Soc. Am.* **33** (2), pp. 187–197 (Feb. 1961).

P.H. Carr, Reflection of Gigacycle-per-Second Ultrasonic Waves from an Optical-Contact Bond, *J. Acoust. Soc. Am.* **37** (5), pp. 927–928 (May 1965).

J.E. Carrington and H.R. Martin, Performance of an Ultrasonic Flowmeter Under Pulsating Flow, pp. 92–97 in: *Symposium on the Measurement of Pulsating Flow Proc.,* University of Surrey, Guildford, Surrey, England (April 2–3, 1970).

J.E. Carrington, An Investigation of the Use of Ultrasonics for Measuring Unsteady Flow with Particular Reference to Use with High Pressure Hydraulic Oils, PhD Thesis, Queen's Univ., Belfast (1976).

J.E. Carrington and D. McCloy, The Use of Ultrasonics for Measuring Unsteady Flow in Oil Hydraulic Systems, 5th International Fluid Power Symposium, University of Durham (Sept. 13–15, 1978).

C.C. Carson and F.M. Wolfenbarger, Development of Slickline Logging Tools for Very High-Temperature Applications: Soc. Petr. Engrs. Paper SPE-15606, presented at the 61st Annual Technical Conference and Exhibition of the SPE at New Orleans, LA (October 5–8).

J.M. Cassanto and C.R. Droms, Re-Entry Vehicle Flight Test Pressure Measurements (Steady-State and Fluctuating): An Overview Progress Report, *ISA Transactions* **17** (3), pp. 11–27 (1978).

G. Cataland, M. Edlow and H.H. Plumb, The Determination of Absolute Temperatures from Sound Velocity Measurements, pp. 129–132 in: Charles M. Herzfeld (Editor-in-chief), *Temperature—Its Measurement and Control in Science and Industry*, Vol. **3**, Part 1, *Basic Concepts, Standards and Methods* (Edited by F.G. Brickwedde), Reinhold, NY (1962).

D. Cathignol, Z. Trawinski and J.Y. Chapelon, Full Range Pseudorandom Doppler Flowmeter Using Serial Processing, pp. 889–892, *1987 Ultrasonics Symp. Proc.*, IEEE (1987).

D. Censor and V.L. Newhouse, Theory of Ultrasound Doppler-Spectra Velocimetry for Arbitrary Beam and Flow Configurations, pp. 923–928, *1986 Ultrasonics Symp. Proc.*, IEEE (1986).

D. Censor, V.L. Newhouse, T. Vontz and H.V. Ortega, Theory of Ultrasound Doppler-Spectra Velocimetry for Arbitrary Beam and Flow Configurations, *IEEE Transactions on Biomedical Engineering* **BME-35** (9), pp. 740–751 (Sept. 1988).

J.R. Chamuel, Ultrasonic Aircraft Ice Detector Using Flexural Waves, U.S. Patent 4,461,178 (24 July 1984).

K.H. Chan, H.L. Bertoni and D.A. Davids, Effect of Higher Modes of a Layered Substrate on V(z) in Acoustic Microscopy, pp. 795–799, *1987 Ultrasonics Symp. Proc.,* IEEE (1987).

J.Y. Chapelon, D. Cathignol, V.L. Newhouse and P.M. Shankar, A Double Frequency Doppler Technique for Bubble Size Measurement, pp. 885–888, *1987 Ultrasonics Symp. Proc.,* IEEE (1987).

J.Y. Chapelon, V.L. Newhouse, C. Cathignol and P.M. Shankar, Bubble Detection and Sizing with a Double Frequency Doppler System, *Ultrasonics* **26** (3), pp. 148–154 (May 1988).

G.T. Chase and L.J. Galbiati, ISA Paper 17.18-4-65 (1965).

J.V. Chatigny and L.E. Robb, Piezo Film Sensors, *Sensors* **3** (5), pp. 6–18 (May 1986).

M. Chávez, V. Sosa and R. Tsumura, Speed of Sound in Saturated Pure Water, *J. Acoust. Soc. Am.* **77** (2), pp. 420–423 (February 1985).

J.N.C. Chen, Ph.D. Thesis, Boston College (1975); with E.P. Papadakis, E.H. Carnevale and C.A. Carey, High Temperature Attenuation and Modulus Measurements, pp. 530–533 in: *1974 Ultrasonics Symp. Proc.,* IEEE (1974).

C.-T. Chen and F.J. Millero, Speed of Sound in Seawater at High Pressures, *J. Acoust. Soc. Am.,* Vol. **62** (5), pp. 1129–1135 (November 1977).

H.S. Chen and J.T. Krause, Correlation Between Young's Modulus and Thermal Properties of Metallic Glasses, *Scripta Metallurgica* **11**, pp. 761–764 (1977).

H.S. Chen, J.T. Krause, K. Shirakawa and T. Masumoto, On the Anomalies in Density, Young's Modulus and Glass Temperature of Pd-Si Glasses, *J. of Non-Crystalline Solids* **41**, pp. 79-88 (1980).

H.S. Chen, J.T. Krause, A. Inoue and T. Masumoto, The Effect of Quench Rate on the Young's Modulus of Fe-, Co-, Ni- and Pd-Based Amorphous Alloys, *Scripta Metallurgica* **17**, pp. 1413–1414 (1983).

H.S. Chen, C.H. Chen, A. Inoue and J.T. Krause, Density, Young's Modulus, Specific Heat, and Stability of Icosahedral $Al_{86}Mn_{14}$, *Physical Review B* **32** (4), pp. 1940–1944 (15 August 1985).

W.H. Chen and J.L. Deng, Ultrasonic Nondestructive Testing Using Barker Code Pulse Compression Techniques, *Ultrasonics* **26** (1), pp. 23–26 (January 1988).

W.H. Chen, H.J. Shaw, D.G. Weinstein and L.T. Zitelli, PVF_2 Transducers for NDE, pp. 780–783 in: *1978 Ultrasonics Symp. Proc.,* IEEE (1978).

N.P. Cheremisinoff and P.N. Cheremisinoff, *Instrumentation for Process Flow Engineering*, Technomic Publishing Co., Inc., Lancaster, PA 17604, 234 pages (1987).

C.I. Chessell, Three-Dimensional Acoustic-Ray Tracing in an Inhomogeneous Anisotropic Atmosphere using Hamilton's Equations, *J. Acoust. Soc. Am.* **53** (1), pp. 83–87 (1973).

C. Chilowsky and P. Langevin, Production of Submarine Signals and the Location of Submarine Objects, U.S. Patent No. 1,471,547 (Oct. 23, 1923).

C. Chilowski, Method and Means for the Observation and Measurement of the Speed of a Vessel by Directed Beams of Ultra-Audible Waves, U.S. Patent No. 1,864,638 (June 28, 1932).

D.E. Chimenti and Y. Bar-Cohen, Signal Analysis of Leaky Lamb Wave Spectra for NDE of Composites, pp. 1028–1031 in: *1985 Ultrasonics Symp. Proc.,* IEEE (1985).

Y.C. Cho and E.J. Rice, High-Frequency Sound Propagation in Spatially Varying Mean Flow, *J. Acoust. Soc. Am.* **70** (3), pp. 860–865 (1981).

C.-H. Chou, B.T. Khuri-Yakub and G.S. Kino, Lens Design for Acoustic Microscopy, *IEEE Trans. UFFC* **35** (4), pp. 464–469 (July 1988).

D.A. Christensen, *Ultrasonic Bioinstrumentation*, Wiley (1988).

P. Cielo, *Optical Techiques for Industrial Inspection*, Academic Press (1988).

A.E. Clark, Magnetostrictive Rare Earth Fe_2 Compounds, *Ferromagnetic Materials* Vol. 1, pp. 531–589 (North-Holland Pub. Co., 1980).

A.E. Clark and H.T. Savage, Magnetostriction of Rare Earth-Fe_2 Compounds Under Compressive Stress, *Proceedings of the International Magnetics Conference*, Kyoto, Japan (1982).

A.E. Clark, H.T. Savage and M.L. Spano, Intermag. Conference, Hamburg, Germany (April 1984).

A.V. Clark, Jr. and J.C. Moulder, Residual Stress Determination in Aluminum Using Electromagnetic Acoustic Transducers, *Ultrasonics* **23** (6), pp. 253–259 (November 1985).

K.E. Clark, *VLSI/VHSIC Package Test Development*, RADC-TR-86-94 (Dec. 1986).

L.R. Clarke, C-H. Chou and B.T. Khuri-Yakub, Acoustic Evaluation of Grinding Damage in Ceramic Materials, pp. 979–982, *1985 Ultrasonic Symp. Proc.*, IEEE (1985).

C.A.E. Clay, Practical Experience with a Multi-Beam System, presented in Cranfield Institute of Technology Short Course on Electromagnetic and Ultrasonic Flowmeters (1985).

M.R. Coates, T. Lam and J.D. Byles, Total Air Temperature Measurement, *Sensors* **6** (3), pp 17–22 (March 1989).

R.S.C. Cobbold, P.H. Velting and K.W. Johnson, Influence of Beam Profile and Degree of Insonation on the C.W. Doppler Ultrasound Spectrum and the Mean Velocity, *IEEE Trans. SU* **SU-30** (6), pp. 364–370 (1983).

F. Cohen-Tenoudji, W.J. Pardee, B.R. Tittmann, L.A. Ahlberg and R.K. Elsley, A Shear Wave Rheology Sensor, *IEEE Trans. UFFC* **UFFC-34** (2), pp. 263–269 (March 1987).

K. Coleman, Device Aids User's Eyes and Head to Move Cursor, *Design News* **41** (18), p. 42 (September 23, 1985).

J. Companion, J.S. Heyman, J.R. Oleson and M. Engler, Ultrasonic Hyperthermia Temperature Monitoring Technique, pp. 977–980, *1986 Ultrasonics Symp. Proc.*, IEEE (1986).

D.M. Considine, ed., *Process Instruments and Control Handbook*, McGraw-Hill Book Co., New York, NY 10020 (1985).

G.R. Cooper and V.L. Newhouse, Ultrasonic Random Signal Doppler Flow Measurement System, U.S. Patent 3,940,731 (February 24, 1976).

H. Cox, Approximate Ray Angle Diagram, *J. Acoust. Soc. Am.* **61** (2), pp. 353–359 (February 1977).

M.T. Covington, Method and Apparatus for Detecting a Break or Other Occurrence in a Pipeline Containing Gas Under Pressure, U.S. Patent No. 3,903,729 (September 9, 1975); Pipeline Rupture Detection and Controls, ASME paper no. 78-Pet-54 (1978).

Crane Co. Engr. Div., *Flow of Fluids Through Valves, Fittings and Pipe*, Techn. Paper 410, Crane, New York (1980).

G.M. Crean, M.G. Somekh, A. Golanski and J.C. Oberlin, The Influence of Thin Film Microstructure on Surface Acoustic Wave Velocity, pp. 843–848 in: *1987 Ultrasonics Symp. Proc.*, IEEE (1987).

D.I. Crecraft, Launching Ultrasonic Shear Waves Into Solids at Normal Incidence by Pressure Coupling, *J. Sound Vib.* **1** (4) pp. 381–387 (Oct. 1964); The Measurement of Applied and Residual Stresses in Metals Using Ultrasonic Waves, *J. Sound Vib.* **5** (1), pp. 173–192 (1967). See also Królikowski, J. Szczepek and Z. Witczak, Ultrasonic Investigation of Contact Between Solids Under High Hydrostatic Pressure, *Ultrasonics* **27** (1), pp. 45–49 (Jan. 1989).

R. Culver, A Preliminary Assessment of the Use of Quartz Thermometers for the Determination of Pump Efficiency In-Situ, *J. Inst. Eng., Australia*, pp. 13–21 (Jan.-Feb., 1968).

J. and P. Curie, Développement, par Pression, de l'Électricité Polaire dans les Cristaux Hémièdres à Faces Inclinées, pp. 294–295; Sur l'Électricité Polaire dans les Cristaux Hémièdres à Faces Inclinées, pp. 383–386, in *Compt. Rend.* **91** (1880).

G. Curtis, A Broadband Polymeric Foil Transducer, *Ultrasonics* **12** (4), pp. 148–154 (July 1974).

G. Curtis, Wave Propagation Techniques in Determining the Dynamic Elastic Properties of Wires and Fibres, Chapter 14 in: J. Szilard (ed.), *Ultrasonic Testing*, Wiley (1982).

G. Curtis, P. Lloyd and K. Allen, Correlation of the Shear Strength of a Napkin-Ring Adhesive Joint and the Dynamic Elastic Stiffness and the Torsional Wave Damping, in: K. Allen (Ed.), *Adhesion* **6**, Applied Science Pub. (1982).

H. Dahlke and W. Welkowitz, A New Ultrasonic Flowmeter for Industry, *ISA J.* **7** (10), pp. 60–63 (October 1960).

N. Dam, Ultrasonic Level Monitoring, *Measurements & Control* **18** (5), pp. 169–171 (October 1984).

N. Dam, Non-Contacting Liquid Level Detection System, U.S. Patent No. 4,630,245 (December 16, 1986).

B.J. Davis and P.P. Lele, An Acoustic Phase Shift Technique for the Non-Invasive Measurement of Temperature Changes in Tissues, pp. 921–924, *1985 Ultrasonics Symp. Proc.*, IEEE (1985).

R.S. Davidson, A. Safdar, J.D. Spencer and B. Robinson, Applications of Ultrasound to Organic Chemistry, *Ultrasonics* **25** (1), pp. 35–39 (1987).

K.K. Davison, Apparatus and Method for Controlling a Cursor on a Computer Display, U.S. Patent No. 4,682,159 (July 21, 1987).

K.R. Dawber and M. Sinclair, An 'Acoustic' Anemometer Based on Turbulent Pressure Fluctuations, *J. Phys. E* **10** (11), pp. 1112–1114 (1977).

C.K. Day and R.W. Smith, Under-Sodium Viewing, pp. 191–194 in: *1973 Ultrasonics Symp. Proc.*, IEEE (1973).

V.A. Del Grosso and E.M. Spurlock, The Feasibility of Using Wholly External Ultrasonics to Measure Fluid Flow Within Thick-Walled Metal Pipes, 40 pp., *NRL 4967* (Nov. 12, 1957); AD149409.

V.A. Del Grosso and C.W. Mader, Speed of Sound in Pure Water, *J. Acoust. Soc. Am.*, **52** (5) Part 2, pp. 1442–1446 (1972).

V.A. Del Grosso, Tables of the Speed of Sound in Open Ocean Water (with Mediterranean Sea and Red Sea Applicability), *J. Acoust. Soc. Am.*, **53** (5), pp. 1384–1401 (May 1973).

M. de Billy and G. Quentin, Experimental Investigation of Reflection Coefficients for Lossy Liquid-Solid-Liquid Systems, *Ultrasonics* **22** (6), pp. 249–252 (November 1984).

P. Debye, A Method for the Determination of the Mass of Electrolyte Ions, *J. Chem. Phys.* **1**, pp. 13–16 (1933).

G. De Cicco, B. Morten, M. Prudenziati, A. Taroni and C. Canali, A 250 kHz Piezoelectric Transducer for Operation in Air: Application to Distance and Wind Velocity Measurements, pp. 321–324, *1982 Ultrasonics Symp. Proc.*, IEEE (1982).

M.O. Deighton, A.B. Gillespie, R.B. Pike and R.D. Watkins, Mode Conversion of Rayleigh and Lamb Waves to Compression Waves at a Metal-Metal Interface, *Ultrasonics* **19** (6), pp. 249–258 (November 1981).

M. Deka, Air Coupled Ultrasonic Transducer for NDE, pp. 543–546, *1987 Ultrasonics Symp. Proc.*, IEEE (1987).

J. Delsing, A New Velocity Algorithm for Sing-Around-Type Flowmeters, *IEEE Trans. UFFC* **34** (4), pp. 431–436 (July 1987).

T. Derenzini and A. Giacomini, *Ric. Sci.* **13**, 27 and 242 (1942).

W. Derham, Experimenta et observationes de soni motu aliisque as id attinentibus, *Phil. Trans. Roy. Soc.* (London) **26**, pp. 1–35 (1708).

W.G. Deskins, D.C. Winter, H. Sheng and C. Garza, Use of a Resonating Cavity to Measure Body Volume, *J. Acoust. Soc. Am.* **77** (2), pp. 786–758 (February 1985).

C.S. Desilets, J.D. Fraser and G.S. Kino, The Design of Efficient Broad-Band Piezoelectric Transducers, *IEEE Trans. Sonics and Ultrasonics* **SU-25** (3), pp. 115–125 (May 1978).

M.J. Diamond and R.H. Lutch, Ultrasonic Measurement of Material Nodularity, U.S. Patent No. 3,603,136 (September 7, 1971).

D.B. Dianov, Ultrasonic Radiation Through Plane-Parallel Layers, *Sov. Phys.—Acoustics* **5**, pp. 30–37 (1959).

E.G. DiBello, J.L. McWilliams, K.R. Haack, G.E. Pax and S.S. Waterbury, Downhole Flowmeter Final Report, BDM (October 1987).

E. Dieulesaint and D. Royer, *Elastic Waves in Solids—Applications to Signal Processing*, Translated by A. Bastin and M. Motz, Wiley (1980). Originally published under the title *Ondes Élastiques dans les Solides*, Paris (1974).

E. Dieulesaint, D. Royer and A. Billman, French Patent No. 2–575–281 (21 December 1984).

E. Dieulesaint, D. Royer, O. Legras and F. Boubenider, A Guided Acoustic Wave Liquid Level Sensor, pp. 569–572 in: *1987 Ultrasonics Symp. Proc.*, IEEE (1987a).

E. Dieulesaint, D. Royer, A. Chaabi and B. Formery, Lamb Wave Graphic Tablet, *Electronics Letters* **23**, pp. 982 984 (10 September 1987b).

J.L. Dion and A.D. Jacob, Ultrasonic Intensity and Phase Imaging at 3.6 MHz Using Liquid Crystal Conversion, *IEEE Trans. UFFC* **34** (5), pp. 550–557 (September 1987).

J.L. Dion and A. Barwicz, Practical Ultrasonic Spectrometric Measurement of Solution Concentrations by a Tracking Technique, pp. 529–532, *1988 Ultrasonics Symp. Proc.*, IEEE (1988).

R.C. Dixon, *Spread Spectrum Systems*, Second Edition, Wiley, New York (1984).

B.B. Djordjevic and S.C. Traugott, Ultrasonic Liquid Jet Probe, U.S. patent no. 4,507,969 (April 2, 1985).

J.C. Doppler, On the Colored Light of Double Stars and Some Other Heavenly Bodies, delivered to Royal Bohemian Society of Learning (1842).

A.E. Douglass, *Climatic Cycles and Tree-Growth*, Vol. 1, Carnegie Institute of Washington Publ. 289 (1919); Vol. 2 (1928); Vol. 3 (1936).

E. Drescher-Krasicka, J.N. Meder and A.V. Granato, A Teflon Pressure-Transducer-Specimen Bond for Ultrasonic Measurements Over a Wide Temperature Range, *Ultrasonics* **23** (6), pp. 281–282 (November 1985).

W.B. Dress and G.N. Miller, An Ultrasonic Level and Temperature Sensor for Power Reactor Applications, *First Proc. Amer. Nucl. Soc. Nucl. Thermal Hydraulics 1983 Winter Meeting*, pp. 240–247 (1983).

B.E. Droney and T.J. Pfeiffer, Ultrasonic Inpection of Hot Steel Blooms to Detect Internal Pipe, *Materials Eval.* **38** (6), pp. 31–36 (June 1980).

C.J. Drost, Volume Flow Measurement System, U.S. Patent No. 4,227,407 (October 14, 1980).

P.S. Dubbelday and K.M. Rittenmyer, Shear Modulus Determination of Foamed Aluminum and Elastomers, pp. 1052–1055, *1985 Ultrasonics Symp. Proc.*, IEEE (1985).

J. Duchateau, C. Georgeon and M. Schneegans, Application des Ultra-sons aux Mesures de Températures et à l'Étude des Caracteristiques d'Élasticité des Verres en Fonction de la Température, in: *Proc. Congrès U.S.C.V.*, Stresa (1970).

J. Duchateau, C. Georgeon and M. Schneegans, Étude des Caracteristiques de Propagation des Ultra-sons dans les Milieux Vitreux. Applications aux Mesures de Quelques Propriétés Physiques des Verres, in: Section A 1.5, pp. 573–594, *Proc. Congrès U.S.C.V.*, Versailles (1971).

H.L. Dunegan, High Temperature Dynamic Modulus Measurements by Use of Ultrasonics, *Materials Eval.* **22**, pp. 266–272 (1964).

W. Durgin and D. Roberti, Rotational Flow in Low Gravity, NASA/USRA University Advance Design Program, Third Annual Summer Conf., WPI, Appendix B, Washington, D.C. (June 1987).

EDO Western Corp., *Instruction Manual, Model 429 Doppler Current Meter*, Rpt. No. 13056 Rev. A (November 4, 1970); *Instruction Manual, Model 519 Doppler Flowmeter*, Rpt. No. 13098 (July 6, 1970).

EDO Corp., Flowmeter Monitors Oil, Chemicals, Other Liquids, *Chem. Engineering* **81** (25), p. 54 (Nov. 25, 1974).

EPRI, Acoustic Leak Detection section in *Proc. Third EPRI Conference on Incipient Failure Detection in Power Plants* (March 1987); see esp. papers by J.E. Coulter, T.P. Sherlock, J.R. Scheibel and S. Gehl; by M. Famiglietti and S. Ghia; by R.S. Evans, J.R. Scheibel, S.P. D'Alessio and R.W. Yavorsky; by M.H. Lind.

C.F. Eck, Acoustical Flow Meter, U.S. Patent No. 4,028,938 (June 14, 1977).

C.F. Eck and H.H. Hill, Digital Knotmeter and Log, U.S. Patent No. 3,729,993 (May 1, 1973).

I. Edler and C.H. Hertz, The Use of Ultrasonic Reflectoscope for the Continuous Recording of the Movements of Heart Walls, *Kungl. Fysiografiska Sällskapets* I Lund Fördhandlingar Band **24** (5), pp. 40–48 (1954).

E.P. EerNisse, R.W. Ward and R.B. Wiggins, Survey of Quartz Bulk Resonator Sensor Technologies, *IEEE Trans. UFFC* **35** (3), pp. 323–330 (May 1988).

N.G. Einspruch, Ultrasonics—A Solid State Research Tool, pp. 104–133, *Proc. Symp. Physics and Nondestructive Testing*, SwRI (1963).

N.G. Einspruch, Generation of Circularly Polarized Transverse Elastic Waves, *J. Acoust. Soc. Am.* **36** (5), pp. 971–972 (May 1964).

P.J.M. Elands, P.A.O.G. Korting, F. Dijkstra, and T. Wijchers, Combustion of Polyethlene in a Solid Fuel Ramjet—A Comparison of Computational and Experimental Results, AIAA '88 paper AIAA-88-3043, *Proc. AIAA/ASME/SAE/ASEE 24th Joint Propulsion Conference* (1988).

P.M. Embree and W.D. O'Brien, Jr., The Accurate Ultrasonic Measurement of the Volume Flow of Blood by Time Domain Correlation, pp. 963–966, *1985 Ultrasonics Symp. Proc.*, IEEE (1985).

J.A. Enderby, On Electrical Effects Due to Sound Waves in Colloidal Suspensions, *Proc. Roy. Soc.* (London) **A207**, pp. 329–342 (1951).

H.E. Engan, B.Y. Kim, J.N. Blake and H.J. Shaw, Optical Frequency Shifting in Two-Mode Optical Fibers by Flexural Acoustic Waves, pp. 435–438, *1986 Ultrasonics Symp. Proc.*, IEEE (1986).

R.H. Engler, D.W. Schmidt, W.J. Wagner and B. Weitenmeier, Ultrasonic Method for Flow Field Measurements in Wind Tunnel Tests, *J. Acoust. Soc. Am.* **71**, pp. 42–50 (1982).

D.C. Erdman, Apparatus for Coupling Ultrasonic Waves, U.S. Patent No. 2,751,783 (June 26, 1956).

R.K. Erf, ed., *Holographic Nondestructive Testing*, Academic Press (1974).

G.P. Erickson and J.C. Graber, Jr., Ultrasonic Flowmeters for Hydroelectric Plants, *Mech. Eng.* **105** (11), pp. 84–88 (November 1983).

W.M. Ewing, W.S. Jardetsky and F. Press, *Elastic Waves in Layered Media*, McGraw-Hill, N.Y. (1957).

R.W. Faas, Analysis of the Relationship Between Acoustic Reflectivity and Sediment Porosity, *Geophysics* **34** (4), pp. 546–553 (Aug. 1969).

P. Fairhurst (ed.) *Proc. of the 3rd International Conf. on Multi-Phase Flow*, The Hague, Netherlands, 435 pages, BHRA (May 1987).

S.S. Fam, L.C. Lynnworth, and E.H. Carnevale, Ultrasonic Thermometry, *Instrument and Control Systems* **42** (10), pp. 107–110 (Oct. 1969).

S.S. Fam, Acoustic Sensing System, U.S. Patent No. 3,540,279 (Nov. 17, 1970).

C. Farrow, Supersonic Testing of Hot Articles, Republic Steel Corp., U.S. Patent No. 2,697,936 (December 1954).

M.F. Feil and E.M. Zacharias, Jr., The Determination of Yeast Slurry Consistency and Wort Plato by Sonic Solution Analysis, *Brewer's Digest* **46** (11), pp. 76, 78–80 (Nov. 1971).

M.P. Felix, Distortion of Short-Duration Stress Pulses Propagating in Solids and Liquids, *J. Acoust. Soc. Am.* **58** (3), pp. 626–629 (September 1975).

R.J. Ferraro and R.B. McLellan, Temperature Dependence of the Young's Modulus and Shear Modulus of Pure Nickel, Platinum, and Molybdenum, *Met. Trans.* A, Vol. **8A**, 1563–1565 (Oct. 1977); Diffusion of Oxygen and Nitrogen in Niobium, *Mater. Sci. Eng.* Vol. 33, pp. 113–116 (1978); Diffusivity of Heavy Interstitials in bcc Metals, *ibid.*, Vol. **39**, pp. 47–56 (1979).

R.J. Ferraro and R.B. McLellan, High Temperature Elastic Properties of Polycrystalline Niobium, Tantalum, and Vanadium, *Met. Trans. A.*, Vol. **10A**, 1699-1702 (Nov. 1979).

M.E. Field, Development of Ultrasonic Thermometry for Application in LMFBR Safety Research, pp. 450–455, *1984 Ultrasonics Symp. Proc.*, IEEE (1984).

M.E. Field, Development of Ultrasonic Thermometry for High-Temperature High-Resolution Temperature Profiling Applications in LMFBR Safety Research, Sandia Rpt. SAND 84-1341 (May 1986).

L. Filipczynski, Z. Pawlowski and J. Wehr, *Ultrasonic Methods of Testing Materials*, Butterworths (1966).

A. Fiorillo, P. Dario, J. Van der Spiegel, C. Domenici and J. Foo, Spinned P(VDF-TrFE) Copolymer Layer for a Silicon Piezoelectric Integrated Ultrasonic Transducer, pp. 667–670, *1987 Ultrasonics Symp. Proc.*, IEEE (1987).

F.A. Firestone and D.S. Ling, Jr., University of Michigan, *Reflection and Refraction of Supersonic Waves, The Raybender*, p. 7 (March 1945).

F.A. Firestone, Method of Supersonic Inspection, U.S. Patent No. 2,592,134 (June 28, 1945).

M.R. Fisch, R.P. Moeller and E.F. Carome, Improved Acoustic Viscosimeter Technique, *J. Acoust. Soc. Am.* **3** (3), pp. 623–625 (September 1976).

R.E. Fishbacher, The Ultrasonic Flowmeter, *Trans. S. Instrum. Technol.* **11**, pp. 114–119 (1959).

R.E. Fishbacher, Flowmeters, U.S. Patent No. 3,097,526 (July 16, 1963).

S.G. Fisher and P.G. Spink, pp. 139–159 in: C.G. Clayton, ed., *Modern Developments in Flow Measurement*, Peregrinus (1972).

J.V. Fitzgerald and F.J. Matusik, Oscillation Viscometer, A Viscometer for Many Purposes, *Measurements & Control* **20** (3), pp. 175–179 (June 1986).

J.V. Fitzgerald, F.J. Matusik, D.W. Nelson and J.L. Schrag, Apparatus and Method for Determining the Viscoelasticity of Liquids, U.S. Patent No. 4,754,640 (July 5, 1988a).

J.V. Fitzgerald, F.J. Matusik and T.M. Walsh, New Method and Instrument for Rheology of Dough, *Cereal Foods World* **33** (11), pp. 908–912 (Nov. 1988b).

G. Flachenecker, Ultrasonic and Focused Shock-Wave Lithotrites: A Revolution in the Treatment of Bladder-, Kidney- and Ureter-Stones, pp. 1001–1010, *1987 Ultrasonics Symp. Proc.*, IEEE (1987).

R.S. Flemons, Velocity Measurement System with Synchronized Demodulation, U.S. Patent no. 3,858,446 (Jan. 7, 1975).

D.L. Folds, Experimental Determination of Ultrasonic Wave Velocities in Plastics, Elastomers, and Syntactic Foam as a Function of Temperature, *J. Acoust. Soc. Am.* **52** (1, Part 2), pp. 426–427 (1972).

K.G. Foote, Ultrasonic Particulate Sensing, U.S. Patent No. 4,527,420 (July 9, 1985).

G.A. Forster and H.B. Karplus, Ultrasonic Flowmeter for LMFBR Applications, Presented at the Specialists' Meeting on "Sodium Flow Measurements in Large LMFBR Pipes," held at Interatom, Bergisch Gladbach, FRG (1980).

G.A. Forster, H.B. Karplus, and T.P. Mulcahey, ANL-84-46, Multipath Ultrasonic Flow Measurements in Water (June 1984).

R.B. Foster, J.A. Brewer, S.R. Montgomery and D.O. Pederson, High Temperature Acoustic Bond Compatible with Fluoride Fluorites, *J. Acoust. Soc. Am.* **73** (1), pp. 352–354 (January 1983).

J.M. Fouke and K.P. Strohl, Effect of Position and Lung Volume on Upper Airway Geometry, *J. Appl. Physiol.* **63** (1), pp. 375–380 (1987).

K.A. Fowler and L.C. Lynnworth, Ultrasonic Measurements of Temperature Using Extensional and Torsional Waves, *Proc. 6th Symp. Temp. Measurement Soc.*, pp. 191–208, Hawthorne, CA (April 1969).

K.A. Fowler, A New Way to Measure Elastic Moduli, *Metal Progress* **95** (6), p. 21 (June 1969).

K.A. Fowler and L.C. Lynnworth, Ultrasonic Techniques for High Temperature Measurements and Nondestructive Inspection, *Proc. Sixth International Conference on Nondestructive Testing,* Hannover, Germany, Report L2, pp. 15–26 (June 1–5, 1970).

K.A. Fowler, Acoustic Emission Simulation Test Set, *MTRSA* **11** (3), pp. 35–36 (March 1971a).

K.A. Fowler, Ultrasonic Sensing System, U.S. Patent No. 3,595,069 (July 27, 1971b).

K.A. Fowler and E.P. Papadakis, Observation and Analysis of Simulated Ultrasonic Acoustic Emission Waves in Plates and Complex Structures, pp. 222–237 *Acoustic Emission*, ASTM Special Techn. Publ. 505, American Society for Testing and Materials (1972).

W.W. Fowlis, Liquid Metal Flow Measurements Using An Ultrasonic Doppler Velocimeter, *Nature Phys. Sci.* **242**, pp. 12–13 (March 5, 1973).

A. Fox, Mechanical Testing Activities at AT&T Bell Laboratories, *ASTM Standardization News*, pp. 37–41 (October 1985).

F.E. Fox, S.R. Curley and G.S. Larson, Phase Velocity and Absorption Measurements in Water Containing Air Bubbles, *J. Acoust. Soc. Am.* **27** (3), pp. 534–539 (May 1955).

J.D. Fox, B.T. Khuri-Yakub and G.S. Kino, High Frequency Acoustic Wave Measurements in Air, pp. 581–584, *1983 Ultrasonics Symp. Proc.*, IEEE (1983).

J.D. Fox, B.T. Khuri-Yakub and G.S. Kino, Excitation and Detection of 8 MHz Waves in Air, pp. 475–479, *1984 Ultrasonics Symp. Proc.*, IEEE (1984).

G. Franceschini, C. Chimisso and C. Del Lago, Acoustic Control System Operates Subsea Well, *Ocean Industry* **23** (5), pp. 17–22 (May 1988).

H. Frankenberger, B. Grohs and E. Häusler, Ultrasonic Dynamic Air Flowmeter, pp. 682–683, *1974 Ultrasonics Symp. Proc.*, IEEE (1974).

W. Franz, *Z. Naturforsch.* A9, 705 (1954).

J. Fraser, B.T. Khuri-Yakub and G.S. Kino, The Design of Efficient Broadband Wedge Transducers, *Appl. Phys. Lett.* **32** (11), pp. 698–700 (June 1, 1978).

J.R. Frederick, A Study of the Elastic Properties of Various Solids by Means of Ultrasonic Pulse Techniques, PhD Thesis, University of Michigan (1947).

J.R. Frederick, Ultrasonic Measurement of the Elastic Properties of Polycrystalline Materials at High and Low Temperatures (Abstr.) *J. Acoust. Soc. Am.* **20**, p. 586 (1948).

J.R. Frederick, *Ultrasonic Engineering*, Wiley, N.Y. (1965).

C.L. Friant, G.B. Groff, D.C. Nagle, B.E. Buddemeyer and S.L. Van Doren, Ultrasonic Characterization of SiC Fiber-Reinforced Glass Matrix Composites, pp. 380–387, *Nondestructive Testing of High-Performance Ceramics*, Am. Cer. Soc. (1987).

H.C. Fritts, *Tree Rings and Climate*, Academic Press (1976).

R.J. Fritz, *ASME J. Eng. Ind.* **94** 167–173 (Feb. 1972).

H.M. Frost, Electromagnetic-Ultrasound Transducer: Principles, Practice and Applications, 179–276 in: W.P. Mason and R.N. Thurston (ed.), *Physical Acoustics* **14** Academic Press (1979).

W.J. Fry and F. Dunn, Ultrasonic Intensity Gain by Composite Transducers, *J. Acoust. Soc. Am.* **34**, (2), pp. 188–192 (1962).

E. Fukada, Piezoelectricity in Polymers and Biological Materials, *Ultrasonics* **6** (4), pp. 229–234 (1968).

J.J. Galkowski, The Design of Two Thermometry Systems for Use in Hyperthermia, Master's Thesis, University of Illinois (1983).

J.J. Galkowski, Copper Vapor Laser Acoustic Thermometry System, U.S. Patent No. 4,697,270 (Sept. 29, 1987).

H.R. Gallantree and M.R. Smith, A Comparison of Polymer Pulse-Echo Transducers, pp. 761–764, *1987 Ultrasonics Symp. Proc.*, IEEE (1987).

L. Ganlin, L. Saihua, C. Meijuan and W. Binxiu, The Effect of Pressure on Ultrasonic Testing Sensitivity in Hot Steel Blooms by Pressure Contact Method, Abstract Z5, p. 121, in: *Program and Abstracts, IEEE 1988 Ultrasonics Symp.*, IEEE (1988a).

L. Ganlin, H. Xianmao, C. Meijuan, Z. Minghuei, W. Binxiu, Y. Songzhang, Z. Shenggao and J. Weiping, An Ultrasonic Inspection System to Detect Internal Pipes in Hot Steel Bloom, *Chinese J. Acoust.* **7** (4), pp. 339–346 (1988b).

R.L. Garman, M.E. Droz, and J.W. Gray, Supersonic Flow Meter, U.S. Patent No. 2,669,121 (Feb. 16, 1954).

R.H. Garwood, Noncontact Presence Sensing: What Works Where and Why, *Sensors* **5** (8), pp. 19–29 (August 1988).

D. Gaucher, Real-Time Particle Monitoring in An Ultrapure Water System, *Ultrapure Water* **1** (2), pp. 26–28 (Sept./Oct. 1984).

A.P. Gavin and T.T. Anderson, High-Temperature Acoustic Sensors for Boiling Detection, *IEEE Trans. Nucl. Sci.* **NS-18** (1), pp. 340–344 (February 1971).

A.P. Gavin and T.T. Anderson, Design and Application of Immersible Microphones for LMFBR Acoustic Monitoring, *ANS Trans.* **19**, p. 335 (1974).

I.L. Gelles, Optical-Fiber Ultrasonic Delay Lines, *J. Acoust. Soc. Am.* **39** (6), pp. 1111–1119 (1966).

I.L. Gelles, Ultrasonic-Pulse Propagation through Films, Foils, Fibers, and Whiskers, *J. Acoust. Soc. Am.* **40** (1), pp. 138–147 (1966).

G.L. Gibby, An Ultrasonic Temperature Measurement System, Master's Thesis, University of Illinois (1980).

J.H. Gieske, Ultrasonic Shear Wave Velocity in CMT Graphite Up to 6500°F, Sandia Laboratories Report SAND78–1448 (September 1978).

J.H. Gieske, Ultrasonic Shear Wave Velocity in CLF/CMT Graphite from Room Temperature to 2000°F, Sandia Laboratories Report SAND80–1850, (November 1980).

J.H. Gieske, Ultrasonic Longitudinal Wave Velocity in Carbon-Carbon Pitch and Pitch/M2 Shape Stable Nosetip Material from 70°F to 5400°F, Sandia Laboratories Report SAND85–0853 (June 1985).

J.H. Gieske, G.J. Hochrein and G.C. Stoker, Ultrasonic Measurement of Reentry Vehicle Antenna Window Recession, pp. 595–606, *Proc. 33rd Int'l Instrum. Symp.*, ISA (May 1987).

W. Gilbert, *De Magnate* (1600). See P.H. Sydenham, *Measuring Instruments: Tools of Knowledge and Control*, Part 1 of the IEE History of Technology Series (19) or *Meas. & Control* **21** (6) cover (Dec. 1986).

T.P. Gill, *The Doppler Effect: An Introduction to the Theory of the Effect*, Academic Press (1965).

A.B. Gillespie, M.O. Deighton, R.B. Pike and R.D. Watkins, A New Ultrasonic Technique for the Measurement of Liquid Level, *Ultrasonics* **20** (1), pp. 13–17 (January 1982).

D. Ginesi, A Review of Insertion Type Flowmetering Devices, *ISA Transactions* **26** (4), pp. 1–8 (1987).

A. Goldstein, B-Scan Transducer Peak Frequency Measurement by Diffraction Grating Spectroscopy, *J. Ultrasound Medicine* **1** pp. 53–66, esp. p 57 (March 1982).

J.H. Goll and B.A. Auld, *IEEE Trans. Sonics and Ultrasonics* **SU-22** (1), pp. 52–53 (1965).

G.E. Goode and R. Lewis, A Momentary-Contact System for Ultrasonic Testing of Steel at Temperatures up to 1200°C, *Non-Destructive Testing* **8** (6), pp. 313–319 (1975).

P. Goodman, Maximum Surface Temperature Measurement by Means of Kryptonates, pp. 749–754 in H.H. Plumb (1972).

R. Gopal, J.R. Smith and G.V. Rao, Acoustic Monitoring Instrumentation for Pressurized Water Reactors, *ISA Transactions* **17** (3), pp. 71–80 (1978).

Y. Gorfu and G. Hayward, A Hardware Digital Correlation Scheme for Ultrasonic Non-Destructive Testing in Explosive Environments, pp. 573–576, *1986 Ultrasonics Symp. Proc.*, IEEE (1986).

J. C. Graber, Ultrasonic Meters, pp. 41–58, in: *Recent Developments in the Custody Transfer Measurement of Crude Oil*, Conference Transcript, Oyez (November 1982).

K.F. Graff, A History of Ultrasonics, pp. 1–97 in: W.P. Mason and R.N. Thurston (ed.), *Physical Acoustics* Vol. XV, Academic Press (1981).

D.E. Gray (Ed.), *Am. Inst. of Physics Handbook*, McGraw-Hill (1957).

T.A. Gray, Utilization of Ultrasonic NDE to Modify Product Designs to Guarantee Inspectability, p. 155, *Program and Abstracts, 1988 Ultrasonics Symp.*, IEEE (1988).

R.E. Green, Jr., Ultrasonic Investigation of Mechanical Properties, in: *Treatise on Materials Science and Technology*, Vol. 3, Academic Press, NY (1973).

R.G. Green, H.K. Kwam, M.S. Beck and R. John, A New Low-Cost Instrument for Solids Flow Measurement, *J. Power and Bulk Solids Tech.*, **3** pp. 20–25 (1979).

S.F. Green, An Acoustic Technique for Rapid Temperature Distribution Measurement, *J. Acoust. Soc. Am.* **77** (2), pp. 759–763 (February 1985).

S.G. Green and A.U. Woodham, Rapid Furnace Temperature Distribution Measurement by Sonic Pyrometry, Central Electricity Generating Board, Marchood Engineering Laboratories, Marchood, Southampton, England (1983).

M. Greenspan and C.E. Tschiegg, A Sing-Around Velocimeter for Measuring the Speed of Sound in the Sea, in: *Underwater Acoustics*, pp. 87–101, Plenum Press, New York (1962).

M. Greenspan, Acoustic Transmission Line: Some Impedance Properties, *J. Acoust. Soc. Am.* **52** (1, Part 2), pp. 455–458 (1972). See also, Lea and Fozooni (1985).

H.W. Grice and D.J. David, Performance and Applications of an Ultrasonic Detector for Gas Chromatography, *J. Chrom. Sci.* **7** pp. 239–247 (April 1969).

R.J. Grossman, A High Resolution In Situ Ultrasonic Corrosion Monitor, *IEEE Trans. Sonics and Ultrasonics* **SU-31** (1) pp. 25–31 (Jan. 1984).

M.V. Grund and L.C. Lynnworth, Step-Type Gamma Ray Ablation Gage(S), Avco RAD Doc. 60-1627 (14 Oct. 1960).

T.R. Gururaja, W.A. Schulze, L.E. Cross, R.E. Newnham, B.A. Auld and Y.J. Wang, Piezoelectric Composite Materials for Ultrasonic Transducer Applications. Part I: Resonant Modes of Vibration of PZT Rod-Polymer Composites, *IEEE Trans. Sonics and Ultrasonics* **SU-32** (4), pp. 481–498 (July 1985).

J. Gutterman, Dual Frequency Acoustic Fluid Flow Method and Apparatus, U.S. Patent 4,527,432 (July 9, 1985a); Method and Apparatus for Measuring Fluid Flow, U.S. Patent No. 4,527,433 (July 9, 1985b); Reflective Acoustic Fluid Flow Meter, U.S. Patent No. 4,555,951 (December 3, 1985c).

C.C.H. Guyott and P. Cawley, The Measurement of Through Thickness Plate Vibration

Using a Pulsed Ultrasonic Transducer, *J. Acoust. Soc. Am* **83** (2), pp. 623–631 (February 1988).

C.C. Habeger, W.A. Wink and M.L Van Zummeren, Using Neoprene-Faced, PVDF Transducers to Couple Ultrasound into Solids, *J. Acoust. Soc. Am.* **84** (4) pp. 1388–1396 (October 1988).

M.P. Hagelberg, Ultrasonic-Velocity Measurements and *B/A* for 1–Propanol at Pressures to 10,000 kg/cm^2, *J. Acoust. Soc. Am.* **47** (1) Part 2, pp. 158–162 (1970).

H.E. Hagy, High Precision Photoelastic and Ultrasonic Techniques for Determining Absolute and Differential Thermal Expansion of Titania-Silica Glasses, *Appl. Optics* **12** (7), pp. 1440–1446 (July 1973).

H.E. Hagy and W.D. Shirkey, Determining Absolute Thermal Expansion of Titania-Silica Glasses: A Refined Ultrasonic Method, *Appl. Optics* **14** (9), pp. 2099–2103 (1975).

J.M. Hale, Ultrasonic Density Measurement for Process Control, *Ultrasonics* **26** (6), pp. 356–357 (Nov. 1988).

S.G. Hallett, J.H. Gieske and P.D. Walkington, Measurement of Linear-Shaped Charge Jet Penetration Using an Ultrasonic Pulse Echo Technique, Sandia Laboratories Report *SAND85–2812* (April 1986).

G. Hallewell, G. Crawford, D. McShurley, G. Oxoby and R. Reif, A Sonar-Based Technique for the Ratiometric Determination of Binary Gas Mixtures, *Nuclear Instruments and Methods in Physics Research* A264, pp. 219–234 (1988).

G.D. Hallewell, A Sound Method for Measuring Gas Concentrations, *Research & Development* **30** (9) pp. 98–101 (Sept. 1988).

J. Halttunen and E. Luntta, Effect of Velocity Profile on Ultrasonic Flowmeters, *ACTA IMEKO 88*, ISA (1988).

T.A. Hamade, PhD Dissertation, *Ultrasound Attenuation in Pipe Flow of Turbulent Gas and Suspended Solids*, Wayne State University (1982).

D.L. Hammond, C.A. Adams, and P. Schmidt, A Linear, Quartz-Crystal, Temperature Sensing Element, *Trans. Instr. Soc. Am.*, **4**, pp. 349–354 (October 1965).

S.M. Handley, M.S. Hughes, J.G. Miller and E.I. Madras, Characterization of Porosity in Graphite/Epoxy Composite Laminates with Polar Backscatter and Frequency Dependent Attenuation, pp. 827–830, *1987 Ultrasonics Symp. Proc.*, IEEE (1987).

R.J. Hansman, Jr. and M.S. Kirby, Measurement of Ice Accretion Using Ultrasonic Pulse-Echo Techniques, *J. Aircraft* **22** (6), pp. 530–535 (June 1985).

R.J. Hansman, Jr. and M.S. Kirby, Measurement of Ice Growth During Simulated and Natural Icing Conditions Using Ultrasonic Pulse-Echo Techniques, *J. Aircraft* **23** (6) pp. 492–498 (June 1986).

E.N. Haran, Gasdichtemessung Mittels Akustischer Methode (Gas Density Measurement by an Acoustic Method), Armament Development Authority, Haifa, *Isr Technisches Messen* TM **50** (2), pp. 43–48 (Feb. 1983).

E.N. Haran, Acoustic Gas Density Measuring Cell, *Rev. Sci. Instrum.* **59** (9), pp. 2059–2062 (1988).

E. Härkönen, J. Tornberg and M. Karras, Ultrasonic Probe Characterizing the Pulp Suspension, pp. 870–873, *1983 Ultrasonics Symp. Proc.*, IEEE (1983).

G. Harrison and A.J. Barlow, Dynamic Viscosity Measurement, pp. 137–178 in: P.D. Edmonds (ed.), *Methods of Experimental Physics* **19**, Academic Press (1981).

R.T. Harrold and Z.N. Sanjana, Non-Destructive Evaluation of the Curing of Resin and Prepreg Using an Acoustic Waveguide Sensor, in: D.O. Thompson and D.E. Chimenti (ed.), *Review of Progress in Quantitative Nondestructive Evaluation*, Vol. **6B**, pp. 1277–1285, Plenum Publ. Co. (1987).

Z. Hashin, The Elastic Moduli of Heterogeneous Materials, *J. Appl. Mech.* **29** (1), pp. 143–150 (1962).

K.Y. Hashimoto and M. Yamaguchi, Elastic, Piezoelectric and Dielectric Properties of Composite Materials, pp. 697–702, *1986 Ultrasonics Symp. Proc.*, IEEE (1986).

D.P.H. Hasselman, On the Porosity Dependence of the Elastic Moduli of Polycrystalline Refractory Materials, *J. Amer. Cer. Soc.* **45** (9), pp. 452–453 (Sept. 1962).

D. Hassler, W. Härer, G. Temme, E. Schmidt, P. Wegener and P. Krämmer, Degradation of Image Quality by Sound Velocity Fluctuations and Its Dependence of the Aperture Size, pp. 935–938, *1987 Ultrasonics Symp. Proc.*, IEEE (1987).

C.H. Hastings, S.A. LoPilato and L.C. Lynnworth, Ultrasonic Inspection of Reinforced Plastics and Resin-Ceramic Composites, *Nondestructive Testing* **19** (5), pp. 340–346, esp. p. 345 (1961).

D. Hauden, Miniaturized Bulk and Surface Acoustic Wave Quartz Oscillators Used as Sensors, *IEEE Trans. UFFC* **34** (2), pp. 253–258 (March 1987).

R.M. Havira, Ultrasonic Techniques in Oil Well Logging, pp. 563–571, *1986 Ultrasonics Symp. Proc.*, IEEE (1986).

A.T.J. Hayward, *Flowmeters*, MacMillan-London and Halstead-NY (1979).

G. Hayward and Y. Gorfu, A Digital Hardware Correlation System for Fast Ultrasonic Data Acquisition in Peak Power Limited Applications, *IEEE Trans. UFFC* **35** (6) pp. 800–808 (Nov. 1988).

T. Hazen, How to Predict Bearing Failure Well in Advance, *Engineer's Digest* **15** (7), p. 68 (July 1987).

J. Heiserman, Acoustic Measurements in Superfluid Helium, pp. 414–453 in: P.D. Edmonds (ed.) *Methods of Experimental Physics* **19**, Academic Press (1981).

R.F. Hellbaum and H.D. Garner, Sensor Requirements for Hypersonic Flight Instrumentation, *Sensors* **5** (3), pp. 34–40 (March 1988).

J. Hermans, Charged Colloidal Particles in an Ultrasonic Field, *Phil. Mag.* **25**, pp. 426–438 (1938).

J.F. Herrick, History of Early Ultrasonic Blood Flowmeters, *Med. Instrum.* **11** (3), pp. 144–148 (May-June 1977).

K.F. Herzfeld and T.A. Litovitz, *Absorption and Dispersion of Ultrasonic Waves*, Academic Press (1959).

W.B. Hess, R.C. Swengel and S.K. Waldorf, An Ultrasonic Method for Measuring Water Velocity, AIEE Miscellaneous Paper 50–214 (October 1950). See also, Measuring Water Velocity by an Ultrasonic Method, *Elec. Eng'g* **69** (11), p. 983 (Nov. 1950).

P.L.M. Heydemann, A Fringe-Counting Pulsed Ultrasonic Interferometer, *Rev. Sci. Instrum.* **42** (7), pp. 983–986 (July 1971).

J.S. Heyman, D.R. Dietz and J.G. Miller, A Non-Doppler Ultrasonic Monitor for Particles in Flowing Liquids, pp. 561–564, *1975 Ultrasonics Symp. Proc.*, IEEE (1975).

J. Heyman, A Self-Exciting Ultrasonic Reflection CW Instrument pp. 113–116, *1976 Ultrasonics Symp. Proc.*, IEEE (1976).

R. Hickling and S.P. Marin, The Use of Ultrasonics for Gauging and Proximity Sensing in Air, *J. Acoust. Soc. Am.* **79** (4), pp. 1151–1160 (April 1986).

R. Hickling and S.P. Marin, Ultrasonic Gauging and Proximity Sensing in Air, pp. 453–477, *Sensors Expo Proc. 1987*, Helmers Publ. (1987).

K. Higuchi, K. Suzuki and H. Tanigawa, Ultrasonic Phased Array Transducer for Acoustic Imaging in Air, pp. 559–561, *1986 Ultrasonics Symp. Proc.*, IEEE (1986).

C.R. Hill (Ed.), *Physical Principles of Medical Ultrasonics*, Ellis Horwood Ltd., Chichester, UK (Halsted Press, a division of John Wiley and Sons), 495 pp. (1986).

J.E. Hill and A.L. Ruoff, Velocity of Sound Measurements in Liquid Metals, *Rev. Sci. Instrum.* **36** (10), pp. 1465–1472 (October 1965).

K. Hirama, T. Ohshima, and Y. Tomikawa, Wireless Temperature Sensing Using Quartz Sensor for Hyperthermia, pp. 917–920, *1985 Ultrasonics Symp. Proc.*, IEEE (1985).

U.T. Hoechli, F. Mueller and H. Nievergelt, Ultrasonic Pressure Gauge for Static Pressure, *Ultrasonics* **10** (6), page 243 (November 1972).

R.B. Holt, H.I. Smith and M.S. Gussenhoven, Research on Optical Contact Bonding, Air Force Cambridge Research Lab. Final Report No. AFCRL-66-649 (30 May 1966).

J.P. Hou and H. Van de Vaart, Mass Sensitivity of Plate Modes in Surface Acoustic Wave Devices and Their Potential as Chemical Sensors, pp. 573–578, *1987 Ultrasonics Symp. Proc.*, IEEE (1987).

J. Hou, B.H.T. Chai, and D. Badding, High Temperature Characteristics of Fresnoite Crystal Resonators, *1988 Ultrasonics Symp. Program and Abstracts*, pp. 48–49, IEEE (1988).

N.N. Hsu, *Exptl. Mech.* **14** (5), pp. 169–176 (May 1974).

J.K. Hu, Q.L. Zhang and D.A. Hutchins, Directional Characteristics of Electromagnetic Acoustic Transducers, *Ultrasonics* **26** (1), pp. 5–13 (January 1988.)

D.R. Hub, Measurements of Velocity and Attenuation of Sound in Iron Up to the Melting Point, Paper J51 in *Proceedings of the Fourth International Congress on Acoustics*, Copenhagen (1963).

D.S. Hughes, W. L. Pondrum and R.L. Mims, Transmission of Elastic Pulses in Metal Rods, *Phys. Rev.* **75**, pp. 1552–1556 (1949).

R. Hughes and A.V. Clark, Jr., Characteristics of the Reflection of SV Waves from Submicron Thickness Liquid Layers, *Air Force/DARPA Review of Progress in Quantitative NDE*, Univ. of Colorado, Boulder, CO, 2–7 August (1981).

R.J. Hunter, *Zeta Potential in Colloid Science*, Academic Press (1981).

G.F.S. Hussain, K.R.K. Iyer and N.B. Patril, Sonic Modulus of Cotton Yarn and its Relationship with Recovery Parameters, *Textile Research Journal* **54** (11), pp. 761–765 (November 1984).

D.A. Hutchins, J. Hu, K. Lundgren, A Comparison of Laser and EMAT Techniques for Noncontact Ultrasonics, *Materials Eval.* **44**, pp. 1244–1253 (September 1986).

D.A. Hutchins, R.P. Young, R. Stoner, D. Jansen, J.K. Hu and Q.L. Zhang, Non-Contact Ultrasonic Tomography of Metal Cylinders, pp. 1037–1040, *1987 Ultrasonics Symp. Proc.*, IEEE (1987).

D. A. Hutchins, Ultrasonic Generation by Pulsed Lasers, pp. 21–123 in: W.P. Mason and R.N. Thurston (ed.), *Physical Acoustics* **18** (1988).

Y. Ikenaga, H. Matsumoto, S. Takahashi, Y. Ozaki, M. Oda, T. Tomoda, and M. Tanaka, Development of an Ultrasonic Flowmeter for Pwrs, *Trans. Am. Nucl. Soc.* **45**, pp. 557–558 (1983).

U. Ingard and V.K. Singhal, Upstream and Downstream Sound Radiation into a Moving Fluid, *J. Acoust. Soc. Am.* **54** (5), pp. 1343–1346 (1973).

F.A. Inkley, D.C. Walden and D.J. Scott, Flow Characteristics of Vortex Flowmeters, *Meas. Control* **13**, pp. 166–169 (1980).

G.L. Innes, Use of Edge-Tone Resonators as Gas Temperature Sensing Devices, pp. 689–700 in: H.H. Plumb (1972).

T. Inoue, T. Nada, T. Miyama, K. Sugiuchi and S. Takahashi, Low-Frequency Flextensional Piezoelectric Transmitter with Displacement Amplifier, pp. 765–770 in: *1987 Ultrasonics Symp. Proc.*, IEEE (1987).

A. Ishimaru, *Wave Propagation and Scattering in Random Media, Vol. 2, Multiple Scattering, Turbulence, Rough Surfaces and Remote Sensing*, Academic Press (1978).

S.A. Jacobson, P.N. Denbeigh and D.E.H. Naude, A New Method for the Demodulation of Ultrasonic Signals for Cross-Correlation Flowmeters, *Ultrasonics* **23** (3), pp. 128–132 (May 1985).

S.A. Jacobson, J.M. Korba, L.C. Lynnworth, T.H. Nguyen, G.F. Orton and A.J. Orazietti, Low-Gravity Sensing of Liquid/Vapor Interface and Transient Liquid Flow, *IEEE Trans. UFFC* **34** (2), pp. 212–224 (March 1987a).

S.A. Jacobson, L.C. Lynnworth and J.M. Korba, Differential Correlation Analyzer, U.S. patent application serial no. 103,066 (Sept. 30, 1987b). U.S. Patent No. 4,787,252 (Nov. 29, 1988).

M. Jaeschke and H.M. Hinze, Using Densitometers in Gas Metering, *Hydrocarbon Processing* **66** (6), pp. 37–41 (June 1987).

P. Jartti and M. Luukkala, Ultrasonic Method for Web Speed Measurement, *Tappi* **60** (11), p. 167 (1977).

C. Javanaud, Applications of Ultrasound to Food Systems, *Ultrasonics* **26** (3), pp. 117–123 (May 1988).

C.K. Jen, Similarities and Differences Between Fiber Acoustics and Fiber Optics, pp. 1128–1133, *1985 Ultrasonics Symp. Proc.*, IEEE (1985).

C.K. Jen, Acoustic Fibers, pp. 443–454, *1987 Ultrasonics Symp. Proc.*, IEEE (1987).

C.K. Jen, K. Sreenivas and M. Sayer, Ultrasonic Transducers for Simultaneous Generation of Longitudinal and Shear Waves, *J. Acoust. Soc. Am.* **84** (1), pp. 26–29 (July 1988).

G.V. Jeskey, L.C. Lynnworth and K.A. Fowler, An Ultrasonic Transmission Technique for Real Time Monitoring of Steel Solidification, *AFS International Cast Metals J.* **2** (4), pp. 26–30 (Dec. 1977).

G.V. Jeskey, priv. comm. (March 11, 1987).

C.P. Jethwa, M. Kaveh, G.R. Cooper and F. Saggio, Blood Flow Measurements Using Ultrasonic Pulsed Random Signal Doppler System, *IEEE Trans. Sonics and Ultrasonics* **SU-22** (1), pp. 1–11 (January 1975).

S.A. Johnson, J.F. Greenleaf, M. Tanaka and G. Flandro, pp. 335–359 in: Natl. Bur. Stand. (U.S.), *Spec. Publ. 484* (1977a).

S.A. Johnson, D.A. Christensen, C.C. Johnson, J.F. Greenleaf and B. Rajagopalan, Non-Intrusive Measurement of Microwave and Ultrasound-Induced Hyperthermia by Acoustic Temperature Tomography, pp. 977–982, *1977 Ultrasonics Symp. Proc.*, IEEE (1977b).

M.P. Jones and G.V. Blessing, The Dynamic Poisson's Ratio of a Ceramic Powder During Compaction, pp. 587–590, *1987 Ultrasonics Symp. Proc.*, IEEE (1987).

S.G. Joshi, Flow Sensor Using Surface Acoustic Waves, pp. 555–558, *1988 Ultrasonics Symp. Proc.*, IEEE (1988).

J.P. Joule, On the Effects of Magnetism Upon the Dimensions of Iron and Steel Bars, *Phil. Mag.* (III) **30**, 76 (1847).

R.D. Joy and R.F. Colton, Sonic Velocity Sensing, U.S. Patent 3,680,375 (August 1, 1972).

R.D. Joy, R.J. Mahany, G.A. Thorne and R.F. Colton, Method and Apparatus for Determining Fluid Density and Mass Flow, U.S. Patent 4,240,299 (December 23, 1980).

R.D. Joy, Ultrasonic Vortex Flowmeters, pp. 219–230 in: *Proc.30th International Instrumentation Symp.*, ISA (1984).

R.D. Joy, Vortex Flow Meter Frequency Adjustment, U.S. Patent 4,437,349 (March 20, 1984).

R.D. Joy, et al., Final Report, U.S. Army Contract DAMD17-85-C-5044 (1985).

J.C. Kaimal and D.A. Haugen, An Acoustic Doppler Sounder for Measuring Wind Profiles in the Lower Boundary Layer, *J. Appl. Meteor.* **16**, pp. 1298–1305 (1977).

J.C. Kaimal and J.E. Gaynor, The Boulder Atmospheric Observatory, *J. Climate Appl. Meteor.* **22**, pp. 863–880 (1983).

A.C. Kak and M. Slaney, *Principles of Computerized Tomographic Imaging*, IEEE (1988).

H.P. Kalmus, Electronic Flowmeter System, *The Review of Scientific Instruments* **25** (3), pp. 201–206 (March 1954).

H.P. Kalmus, A.L. Hedrich and D.R. Pardue, The Acoustic Flowmeter Using Electronic Switching, *IRE Trans. PGUE* **1** (1), pp. 49–62 (June 1954); reprinted in *IEEE Trans. Sonics and Ultras.* **SU-31** (6), pp. 49–62 (Nov. 1984).

H.P. Kalmus, New Ultrasonic Liquid Level Gage, *Rev. Sci. Instrum.* **36** (10), pp. 1432–1435 (October 1965).

B.A. Kalugin and I.G. Mikhailov, A New Ultrasonic Method for Measuring the Elastic Properties of Solids at High Temperatures, *Akust. zh.* **7** (2), pp. 195–200 (1961) [*Sov. Phys.—Acoustics* **7** (2), pp. 154–158. See also, Ultrasonic Method of Measuring the Elastic Moduli of Metals to 3000°K, *Sov. Phys.—Acoustics* **12** (1), pp. 91–92 (1966).]

N. Kaneko, N. Nakamura, M. Koyama, S. Saitoh and H. Honda, Polymeric Piezoelectric Ultrasonic Probe, European Patent No. 0 186 096 (Dec. 17, 1985).

H.-G. Kang, J.-N. Zhang and S.-B. Zou, A Method of Demodulating Doppler Frequency Shift for Pulsed Ultrasonic Doppler Blood Flow Imaging, pp. 893–895, *1988 Ultrasonics Symp. Proc.*, IEEE (1988).

H.B. Karplus, The Velocity of Sound in A Liquid Containing Gas Bubbles, Armour Res. Found., ITT Rpt. COO-248, page 11 (June 11, 1958); with J.M. Clinch, Sound Propagation in Two-Phase Fluids, *J. Acoust. Soc. Am.* **36** (5), p. 1040 (May 1964).

H.B. Karplus and M.F. Tupper, Transmission of Sound Across Dry Metallic Interfaces, paper NN17, *J. Acoust. Soc. Am.* **58** Suppl. p. S82 (Fall 1975).

H.B. Karplus, High Temperature Shear-Mode Transducers for LMFBR Ultrasonic Flowmeters, Argonne National Laboratory Tech. Memo. ANL-CT-77-6 (August, 1977).

H.B. Karplus and A.C. Raptis, Flow Measurement of Dense Slurries Using the Sonic Doppler Principle, pp. 291–295, *1978 Ultrasonics Symp. Proc.*, IEEE (1978).

H.B. Karplus, A.C. Raptis and D.R. Canfield, Development and Testing of High Temperature Acoustic Doppler Flowmeter, ANL Report No. ANL/FE-81-64 (September 1980).

H.B. Karplus, A.C. Raptis and D.R. Canfield, Development and Testing of High-Temperature Acoustic Doppler Flowmeter, Argonne National Laboratory, Fossil Energy Report ANL/FE-81-64 (September 1981).

M. Karras, E. Härkönen, J. Tornberg and O. Hirsimäki, Pulp Suspension Flow Measurement Using Ultrasonics and Correlation, pp. 915–918, *1982 Ultrasonics Symp. Proc.*, IEEE (1982).

M. Karras, J. Tornberg and E. Härkönen, The Ultrasonic Inferential Mass Flowmeter for Solids Carried By the Pulp Suspension, International Conference on Flow Measurement, August 20–23, 1985, Melbourne, Australia, pp. 177–179 (1985).

M. Karras, T. Pietikäinen, H. Kortelainen and J. Tornberg, Ultrasonic Measurement of Gaseous Air in Pulp Suspensions, *Tappi J.* **71** (1), pp. 65–69 (January 1988).

E.N. Kaufman, Resonance Testing, *Meas. & Control* **22** (2), pp. 145–147 (April 1987).

W. Kaule, Magnetostrictive Ultrasonic Testing of Materials, pp. 291–294 in: *Proc. Fourth International Conf. Nondestructive Testing*, Butterworths, London (1964).

R.P. Keech, The KPC Multichannel Correlation Signal Processor for Velocity Measurement, *Trans. Inst. MC* **4** (1), pp. 43–52 (1982).

L.W. Kessler, P.R. Palermo and A. Korpel, Practical High Resolution Acoustic Microscopy, in: *Acoustical Holography* **4** pp. 51–71, Plenum Press, New York (1972), ed. by Glen Wade.

L.W. Kessler, Acoustic Microscopy – An Industrial View, pp. 725–728, *1988 Ultrasonics Symp. Proc.*, IEEE (1988).

B.T. Khuri-Yakub, J.H. Kim, C.-H. Chou, P. Parent and G.S. Kino, A New Design for Air Transducers, pp. 503–506, *1988 Ultrasonics Symp. Proc.*, IEEE (1988).

S. Kikkawa, T. Yamaguchi, K. Tanishita, and M. Sugawara, Spectral Broadening in Ultrasonic Doppler Flowmeters Due to Unsteady Flow, *IEEE Trans. Biomed. Eng.* **BME-34** (5), pp. 388–391 (May 1987).

J.O. Kim and H.H. Bau, A Torsional Wave Sensor For the Measurement of Density and

Density-Related Fluid Characteristics Such as the Level of Liquids and the Solid Particle Concentration in Suspensions, Abstract EN4, *Bull. Amer. Phys. Soc.* **31**, p. 1744 (Nov. 1986).

J.O. Kim, Y.-Z. Wang and H.H. Bau, Effects of Adjacent Fluid Viscosity on the Transmission of Torsional Stress Waves in Waveguides (1988, priv. comm.)

J.O. Kim and H.H. Bau, On Line, Real Time Densimeter—Theory and Optimization, *J. Acoust. Soc. Am.* **85** (1), pp. 432–439 (Jan.1989).

G.S. Kino, J. Fanton and B.T. Khuri-Yakub, Optical Measurements of Acoustic and Photoacoustic Effects, pp. 505–514, *1986 Ultrasonics Symp. Proc.*, IEEE (1986).

G.S. Kino, *Acoustic Waves*, Prentice-Hall (1987).

L.E. Kinsler and A.R. Frey, *Fundamentals of Acoustics*, 2nd Ed., Wiley (1962).

J.B. Kirby and R. Paglia, Ultrasonic Transducer For Use in a Corrosive/Abrasive Environment, European Patent Application No.85101849.9 (September 11, 1985).

M.S. Kirby and R.J. Hansmann, Jr., Method and Apparatus for Measurement of Ice Thickness Employing Ultrasonic Pulse Echo Technique, U.S. Patent No. 4,628,736 (16 December 1986).

A. Kircher, *Musurgia Universalis*, Francisci Corbelletti, Rome (1650).

J. F. Kirkpatrick and W.C. Kuzniak, Development of Large-Scale Acoustic Waveguides for Liquid-Level Measurement, *Trans. Amer. Nucl. Soc.*, pp. 719–721 (1987).

A.H. Kits van Heyningen, Solid State Doppler Wind Sensor, Abstr., SBIR Topic No. N85–181 (1987).

A. Sh. Kiyasbeili et al., *Ultrasonic Frequency-Time Flowmeters and Counters* (in Russian), Mashinostroenie, Moscow (1984).

B. Kleinman, Automated Setting of Technic Factors for X-Ray Examinations, U.S. Patent No. 4,403,337 (September 6, 1983).

P. Kleinschmidt and V. Mágori, Ultrasonic Robotic-Sensors for Exact Short Range Distance Measurement and Object Identification, pp. 457–462, *1985 Ultrasonics Symp. Proc.*, IEEE (1985).

R.A. Kline and D. Hashemi, Using Interface Waves for Monitoring Fatigue Damage Development in Adhesively Bonded Joints, pp. 1032–1035, *1985 Ultrasonics Symp. Proc.*, IEEE (1985).

J.A. Kleppe, *Engineering Applications of Acoustics*, 250 pp., Artech (1989).

C. Knapp, Geschwindigkeits- und Mengenmessung strömender Flüssigkeiten mittels Ultraschalls, Prom.-Nr. 2795, PhD thesis, ETH (1958); *VDI-Ber.* **86**, pp. 65–71 (1964).

C.J. Knauss, D. Leppo and R.R. Myers, Viscoelastic Measurement of Polybutenes and Low Viscosity Liquids Using Ultrasonic Strip Delay Lines, *J. Polymer Sci.*, Symposium No. 43, pp. 179–186, Wiley (1973).

K.E. Kneidel, Advances in Multizone Ultrasonic Thermometry Used to Detect Critical Heat Flux, *IEEE Trans. Sonics and Ultrasonics* **SU-29** (3), pp. 152–156 (May 1982).

M. Kniazuk and F.R. Prediger, Sonic Gas Analyzer for Measuring Respiratory Gases, Paper No. 55-9-2, 5 pages, ISA (1955).

L. Knopoff, On Rayleigh Wave Velocities, *Bull. Seismol. Soc. Am.* **42**, pp. 307–308 (1952).

L. Knopoff et al., 2nd Annual Report, *Seismic Scattering Project*, Chapter 12, Institute of Geophysics, UCLA (April 1957). See also, Fig. 6 in R.L. Morris et al., The Use of Compressional and Shear Acoustic Amplitudes for the Location of Fractures, AIME Paper No. SPE-723 (Oct. 1963).

M. Knuuttila and P. Hiismäki, An Acoustic Method for High Precision Gas Flow Measurements, abstr., *Ultrasonics* **25** (6) p. 382 (November 1987).

Y. Kobori, Ultrasonic Gas Flow Meter, GF Series, pp. 779–792, *Proc. AGA Int'l Symp. on Fluid Flow* (November 1986).

D.R. Koehler, Quartz-Resonator Pressure Gauges: Temperature Performance, SAND-81-0130, 22 pp. (September 1981).

R.A. Kolano and R.O. Rowlands, Doppler Shift in a Medium with Velocity Gradients (abstr. G17), *J. Acoust. Soc. Am.* **62**, Suppl. 1 (1977).

I.A. Kolmakov and A.G. Safin, Measurement of Turbulent Flow Rate by Ultrasonic Means, *Russian Ultrasonics* **17** (4), pp. 171–174 (1987).

H. Kolsky, *Stress Waves in Solids*, Dover, 1963; *J. Sound Vib.* **1** (1), pp. 88–110 (1964).

R. Kompfner, M. Chodorow and R.A. Lemons, Method of and Apparatus for Acoustic Imaging, U.S. Patent No. 4,012,950 (March 22, 1977).

A. Korpel, L.W. Kessler and P.R. Palermo, Acoustic Microscope Operating at 100 MHz, *Nature* **232** (5306), pp. 110–111 (July 9, 1971).

P.J. Kortbeek, M.J. P. Muringer, N.J. Trappeniers and S.N. Biswas, Apparatus for Sound Velocity Measurements in Gases Up to 10 kbar: Experimental Data for Argon, *Rev. Sci. Instrum.* **56** (6), pp. 1269–1273 (June 1985).

J. Korycki, Z. Kozlowski, W.H. Szachnowski and B.J. Wislicki, Ultrasound Velocity Measurement and Determination of Elasticity Constants of Liquids as a Function of Temperature and Pressure, *Ultrasonics*, pp. 166–174 (July 1979).

W. Köster, Die Temperaturabhängigkeit des Elastizitätsmoduls reiner Metalle (Temperature Dependence of the Elastic Moduli of Pure Metals), *Z. f. Metallkunde* **39** pp. 1–9, 9–12 (1948), translated by M.W. Buxton, R&DB, Windscale Works, UKAEA (Sept. 1956).

A.H. Kou, W.R. Peickert, E.E. Polenske and M.G. Busby, A Pulsed Phase Measurement Ultrasonic Flowmeter for Medical Gases, *Annals of Biomedical Engineering* **12**, pp. 263–280 (1984).

P. Krämmer and D. Hassler, Measurement of Spatial Time-of-Flight Fluctuations of Ultrasound Pulses Passing Through Inhomogeneous Layers, pp. 939–942, *1987 Ultrasonics Symp. Proc.*, IEEE (1987).

J.T. Krause, Gold-Indium Bond for Measurement of Ultrasonic Properties in Solids at High Temperatures, *J. Appl. Phys* **39**II pp. 5334–5335 (1968).

J.T. Krause and H.S. Chen, Uniaxial Stress Dependence of Young's Modulus By An Ultrasonic Technique, *Proc. of the Second Int'l Conf. on Rapidly Quenched Metals*, pp. 425–430, MIT (1975).

J.T. Krause, L.R. Testardi and R.N. Thurston, Deviations From Linearity in the Depennce of Elongation Upon Force For Fibres of Simple Glass Formers and of Glass Optical Lightguides, *Physics and Chemistry of Glasses* **20**, No. 6, pp. 135–139 (December 1979).

J.K. Krause, P.R. Swinehart, and J.R. Bergen, Temperature Sensors for Cryogenic Applications, *Sensors* **5** (2), pp. 18–24 (February 1988).

J. and H. Krautkrämer, *Ultrasonic Testing of Materials*, 3rd Ed., Springer, New York (1983).

K. Kristoffersen, Time-Domain Estimation of the Center Frequency and Spread of Doppler Spectra in Diagnostic Ultrasound, *IEEE Trans. UFFC* **35** (4), pp. 484–497 (July 1988); ibid (6) pp. 685–700 (Nov. 1988).

J. Kritz, An UltraSonic Flowmeter for Liquids, ISA Paper No. 55-16-3, pp. 1–6 (1955a).

J. Kritz, Ultrasonic Flowmeter, *Instruments & Automation* **28**, pp. 1912–1913 (Nov 1955).

J. Kritz, Acoustic Velocity Measuring System, U.S. Patent No. 2,826,912 (March 18, 1958).

J. Kritz, Continuously Indicating Electroacoustic Densitometer, U.S. Patent 2,869,357 (January 20, 1959).

J. Królikowski, J. Szczepek and Z. Witczak, Ultrasonic Investigation of Contact Between Solids Under High Hydrostatic Pressure, *Ultrasonics* **27** (1), pp. 45–49 (1989).

S.J. Krupski, Hot Wall Thickness Variation Measurement System, WVT-QA-7901, Watervliet Arsenal Technical Report, AMCMS No. 5397.OM.6350 (May 1979).

R. Kuc, K. Haghkerdar and M. O'Donnell, Presence of Cepstral Peak in Random Reflected Ultrasound Signals, *Ultrasonic Imaging* **8** (3), pp. 196–212 (July 1986).

G.J. Kühn and A. Lutsch, Elastic Wave Mode Conversion at a Solid-Solid Boundary with Transverse Slip, *J. Acoust. Soc. Am.* **33** (7), pp. 949–954 (1961).

D.S. Kupperman and R.N. Lanham, Ultrasonic Shear Wave Couplant, U.S. Patent No. 4,559,827 (December 24, 1985).

D.S. Kupperman, T.N. Claytor, T. Mathieson and D. Prine, Leak-Detection Technology for Reactor Primary Systems, *Nuclear Safety* **28** (2), pp. 191–198 (April-June 1987).

W. Kurz and B. Lux, *Arch. Eisenhüttenwes.* **39** (7), pp. 521–530 (July 1968).

W. Kurz and B. Lux, Ultrasonic Testing of Hot Semi-Finished Steel Products at Approximately 1000°C, *Arch. Eisenhüttenwes.* **42** (2), pp. 99–105 (1974), Brutcher Translation HB 9221.

M.D. Kyser and W.H. Vander Heyden, Ultrasonic Contrapropagation Clamp-On Flowmeter and Measurements in Partly Full Pipes, pp. 192–199, *Proc. of the 30th International Instrumentation Symposium*, ISA (1984).

M.D. Kyser, E.F. Martinez and M.R. Stuart, Solar-Powered Transit Time Ultrasonic Flowmeters, pp. 689–692, *Proc. 34th Int'l. Instrum. Symp.*, ISA (1988).

H.K. Kytomaa, Cooldown Simulation for CIT PF Coils—Test Plan Outline for Central Solenoid (April 30, 1987); priv. comm. (1987).

S.A. Lacey, Brit. Patent 732,083 (June 15, 1955).

A. Laenen, The Use of Acoustic Velocity Meters in the Measurement of Streamflow, pp. 201–208, *Proc. of the 30th International Instrumentation Symp.*, ISA (1984).

D.B. Lake, Flowmeters, U.S. Patent No. 3,050,997 (Aug. 28, 1962).

J.M. LaPlant and D.J. Flood, *Cryo.* **12** (3), p. 234 (June 1972).

C.T. Lancée, J. Souquet, H. Ohigashi and N. Bom, Ferroelectric Ceramics Versus Polymer Piezoelectric Materials, *Ultrasonics* **23** (3), pp. 138–142 (May 1985).

V.M. Lantux, Contact Material for Ultrasonic Defectoscopy of Metals, Russian Patent No. 1,193,576A (November 23, 1985).

S.E. Larsen, F.W. Weller and J.A. Businger, A Phase-Locked Loop Continuous Wave Sonic Anemometer-Thermometer, *J. App. Meteorology*, **18** (4), pp. 562–568 (April 1979).

G.S. Larson, R. N. Lawson and L.C. Lynnworth, Upper Atmospheric Sonic Thermometry, *ISA Trans.* **5** (3), pp. 233–241 (July 1966).

R.H. Latiff and N.F. Fiore, Ultrasonic Attenuation and Velocity in Two-Phase Microstructures, *J. Acoust. Soc. Am.* **57** (6), Part II, pp. 1441–1447 (June 1975).

A. Latuszek, A New Method of SAW Generation on the Surface of Arbitrary Solid Materials, pp. 213–215, *1986 Ultrasonics Symp. Proc.*, IEEE (1986).

F. Laville, G. Abbott and M. Miller, A Rainfall Measurement Device Using Underwater Acoustic Signals, Phase I Final Report, DOE Contract No. DE-AC-01-87-ER80478 (December 1987).

P. Lavocat, Ultrasonic Flow Measurement in Liquid Helium, pp. 577–581, in: G. and I. Klipping (ed.), *Proc. 11th Int'l. Cryogenic Engineering Conference*, Butterworths (1986).

M.J. Lea and P. Fozooni, The Transverse Acoustic Impedance of an Inhomogeneous Viscous Fluid, *Ultrasonics* **23** (3), pp. 133–137 (May 1985).

H. Lechner, Ultrasonic Flow Metering Based on Transit Time Differentials Which Are Insensitive to Flow Profile, *J. Acoust. Soc. Am.* **74** (3), pp. 955–959 (September 1983).

S. Lees, Data Reduction from Critical Angle Reflection Measurements, *Ultrasonics* **13** (5), pp. 213–215 (September 1975).

C.B. Leffert, Airborne Coal Mass Flowmeter, priv. comm. (December 17, 1987).

C.B. Leffert, and L.H. Weisman, Ultrasonic Instrument to Measure the Gas Velocity and/ or the Solids Loading in a Flowing Gas Stream, U.S. Patent No. 4,726,235 (Feb. 23, 1988).

P.J. Legendre and G.T. Chase, Jr., The Operational Performance of Reentry Vehicle Heatshield Thermodynamic Instrumentation, *ISA Trans.* **13** (2) pp. 132–141 (1974).

S.H. Leibson, Reduced Costs Help Ultrasonic Sensors Compete in Position-Detection Systems, EDN Special Issue, pp. 75–82 (November 28, 1985).

R.A. Lemons and C.F. Quate, A Scanning Acoustic Microscope, pp. 18–21, *1973 Ultrasonics Symp. Proc.*, IEEE (1973).

R.A. Lemons and C.F. Quate, Advances in Mechanically Scanned Acoustic Microscopy, pp. 41–44, *1974 Ultrasonics Symp. Proc.*, IEEE (1974).

L.M.A. Lenihan, Mersenne and Gassendi. An Early Chapter in the History of Sound, *Acustica* **2**, pp. 96–99 (1951).

L.M.A. Lenihan, The Velocity of Sound in Air, *Acustica* **2**, pp. 205–212 (1952).

R.L. Leon and J.R. Scheibel, Current Status of the EPRI Acoustic Doppler Blade Monitor, *Proc. Third Conf. Incipient Failure Detection in Power Plants*, EPRI (1987).

R. Lerch, Electroacoustic Transducers Using Piezoelectric Polyvinylidenefluoride Films, *J. Acoust. Soc. Am.* **66** (4) pp. 952–954 (1979).

R. Lerch and G.M. Sessler, Microphones with Rigidly Supported Piezo Polymer Membranes, *J. Acoust. Soc. Am.* **67** (4), pp. 1379–1381 (April 1980).

W. Leung, H.J. Shaw, G.S. Kino, D.K. Winslow and L.T. Zitelli, Acoustic Phase Contrast Imaging with Electronic Scanning, pp. 151–156, *1976 Ultrasonics Symp. Proc.*, IEEE (1976).

A.P. Levitt and A.G. Martin, Ultrasonic Determination of Elastic Constants of Metals at Elevated Temperatures, *Nondestructive Testing* **18** (10), pp. 333–336 (1960).

P.A. Lewin and M.E. Schafer, Design of Piezoelectric Polymer Transducers for Time Delay Spectrometry Applications, pp. 721–724, *1987 Ultrasonics Symp. Proc.*, IEEE (1987).

P.A. Lewin and M.E. Schafer, Wideband Piezoelectric Polymer Acoustic Sources, *IEEE Trans. UFFC* **35** (2), pp. 175–184 (March 1988).

P.A. Lewin and J.M. Gilmore, Sensors for the Characterization of Extracorporeal Acoustic Shock Wave Devices, pp. 955–958, *1988 Ultrasonics Symp. Proc.*, IEEE (1988).

B. Lewis (ed.), *Bioacoustics—A Comparative Approach*, 520 pp., Academic Press (1983).

G.K. Lewis, Chirped PVDF Transducers for Medical Ultrasound Imaging, pp. 879–884, *1987 Ultrasonics Symp. Proc.*, IEEE (1987).

W.H. Lin and A.C. Raptis, Sound Scattering from a Thin Rod in a Viscous Medium, *J. Acoust. Soc. Am.* **79** (6), pp. 1693–1701 (1986).

Z.-C. Lin, H. Lee, G. Wade, M.G. Oravecz and L.W. Kessler, Holographic Image Reconstruction in Scanning Laser Acoustic Microscopy, *IEEE Trans, UFFC* **34** (3), pp. 293–300 (May 1987).

F.A. Lindemann and G.M.B. Dobson, Theory of Meteors and Density and Temperature of the Outer Atmosphere to Which it Leads, *Proc. Roy. Soc. London, Ser. A.*, Vol. **102**, pp. 411–439 (1923).

R.B. Lindsay, *Mechanical Radiation*, McGraw-Hill, New York (1960).

R.B. Lindsay, The Story of Acoustics, *J. Acoust. Soc. Am.* **39** (4) pp. 629–644 (1966).

R.B. Lindsay (ed.), *Acoustics: Historical and Philosophical Development*, Dowden, Hutchinson and Ross (1972).

R.B. Liptai, D.O. Harris, and C.A. Tatro, An Introduction to Acoustic Emission, pp. 3–10 in ASTM STP 505, *Acoustic Emission* (1972).

B.G. Liptak and R.K. Kaminski, Ultrasonic Instruments for Level and Flow, *Instrum. Techn.* **21** (9), pp. 49–50 (Sept. 1974).

T.A. Litovitz and E.H. Carnevale, Effect of Pressure on Sound Propagation in Water, *J. Appl. Phys.* **26** (7), pp. 816–820 (1955).

J.C. Livengood, T.P. Rona and J.J. Baruch, Ultrasonic Temperature Measurement in Internal Combusion Engine Chamber, *J. Acoust. Soc. Am.* **26** (5), pp. 824–830 (1954).

P.V. Lobachev and V.I. Myasnikov, Effect of the Supply-Pipe Surface Roughness on the Ultrasonic Flowmeter Readings, *Meas. Tech.* **23** (12), pp. 1135–1137 (1980).

C. Longstreet et al., Bendix Final Report, Contract F33615-71-C-1184 (1972).

J. Löschberger and V. Mágori, Ultrasonic Robotic Sensors with Lateral Resolution, pp. 547–552, *1987 Ultrasonics Symp. Proc.*, IEEE (1987).

W.R. Loosemore and A.H. Mustin, Apparatus for Providing Time Reference Signals, U.S. Patent 4,080,574 (March 21, 1978).

J.S. Love and D.B. Juanarena, Design and Application of Vibrating Cylinder Based Pressure Standards, pp. 337–345, *Proc. 34th Int'l. Instrum. Symp.*, ISA (1988).

G.K. Lucey, Jr., Resonant and Antiresonant Frequencies of Thick Discs and Thick Rods, *J. Acoust. Soc. Am.* **43** (6), pp. 1324–1328 (May 1968).

F. Luppe, G. Quentin and H. Überall, Observation of Surface Wave Modes of Franz and Stoneley Type on a Solid Cylinder in a Fluid, pp. 1098–1100, *1985 Ultrasonics Symp. Proc.*, IEEE (1985).

J.C. Luppens, P. Tucker and D.E. Gaucher, Water Quality Monitor Studied, *Offshore* **46** (6), pp. 46-48 (June 1986).

M. Lustenberger, Process for Detecting the Likelihood of Ice Formation, Ice Warning System for Carrying Out the Process and Utilization Thereof, U.S. Patent No. 4,570,881 (Feb. 18, 1986).

M. Luukkala and P. Meriläinen, Metal Plate Testing Using Airborne Ultrasound, *Ultrasonics* **11** (5), pp. 218–221 (Sept. 1973).

A. Lygre, P. Lunde, V. Berge and M. Vestrheim, Numerical Simulation of Ultrasonic Flowmeters, abstr., *Ultrasonics* **25** (6), p. 382 (November 1987).

L.C. Lynnworth, Applications of Ultrasonics to Ablation Studies, in: C.T. Morrow, L.P. Ely and M.R. Smith (Ed.), *Proc. 6th Symposium on Ballistic Missile and Aerospace Tech.*, **4**, 145–187, Academic Press, New York (1961).

L.C. Lynnworth, Ultrasonic Impedance Matching from Solids to Gases, *IEEE Trans. Sonics and Ultrasonics* **SU-12** (2), pp. 37–48 (June 1965).

L.C. Lynnworth and E.H. Carnevale, *Techniques for Mounting an Ultrasonic Device*, NASA CR-54979, esp. Fig. 37, Final Rpt., Contract NAS3-6211 (Feb. 1966).

L.C. Lynnworth and E.H. Carnevale, Ultrasonic Probe, U.S. Patent No. 3,302,044 (January 31, 1967).

L.C. Lynnworth, Ultrasonic Probes Using Shear Wave Crystals, Part I, Principles, *Materials Eval.*, **25**, (12), pp. 265–277 (1967a); Ultrasonic Testing Method, U.S. Patent No. 3,512,400 (May 19, 1970).

L.C. Lynnworth, Ultrasonic Probes Using Shear Wave Crystals, Part II, Applications, *Proc. Sixth Symposium on Nondestructive Evaluation of Aerospace and Weapons Systems Components and Materials*, San Antonio, Texas, pp. 323–361 (1967b). Western Periodicals Co., 13000 Raymer St., N. Hollywood, CA.

L.C. Lynnworth, Measuring Moduli to the Melting Point, *Materials Engineering* **68** (6), p. 8 (Dec. 1968).

L.C. Lynnworth, Rolling Ultrasonic Transducer, U.S. Patent No. 3,423,993 (January 28, 1969a).

L.C. Lynnworth, Ultrasonic Measurement, U.S. Patent No. 3,477,278 (Nov. 11, 1969b).

L.C. Lynnworth, E.H. Carnevale, M.S. McDonough and S.S. Fam, Ultrasonic Thermometry for Nuclear Reactors, *IEEE Trans. Nuclear Science* **NS-16** (1), pp. 184–187 (February 1969).

L.C. Lynnworth, Acoustical Techniques: Sound Ways to Measure Temperature, *Instrum. Techn.* **17** (4), cover, pp. 47–52 (April 1969c).

L.C. Lynnworth, Acoustical Nomograms for the Elastic Properties of Engineering Materials, *Ultrasonics* **7** (4), pp. 254–256 (October 1969d). See also: A. Zanker, Nomograph for the

Velocity of Sound in Solid Media, *Ultrasonics* **10** (4), pp. 180–181 (July 1972); J.F.W. Bell, A.E. Noble and T.N. Seth, Graphical Displays of Acoustic Properties of Solids, *Ultrasonics* **11** (4), pp. 174–181 (July 1973).

L.C. Lynnworth and J.J. Benes, Design Guide—Understanding the Many Methods of Measuring Temperature, *Machine Design*, **4** (26), cover, pp. 189–204 (Nov. 13, 1969).

L.C. Lynnworth and E.H. Carnevale, Ultrasonic Testing of Solids at Elevated Temperatures, *Proc. Fifth International Conference on Nondestructive Testing*, Montreal, Canada, pp. 300–307 (May 21–26, 1967); Queens Printer, Ottawa, Canada (1969).

L.C. Lynnworth, Ultrasonic Testing Method, U.S. Patent No. 3,512,400 (May 19, 1970a).

L.C. Lynnworth and B.J. Spencer, Ultrasonic Sensing System, U.S. Patent No. 3,514,747 (May 26, 1970).

L.C. Lynnworth, High Temperature Acoustics: Sixties' Research Spawns Seventies Applications, *Naval Research Reviews* **23** (7), pp. 1–22 (July 1970b); Noninvasive Ultrasonic Measurements for Process Control, in: *Proc. 7th International Congress on Acoustics* **4**, pp. 525–528 (August 18–26, 1971), Akademiai Kiado, Budapest.

L.C. Lynnworth, Dual Ultrasonic Sensors Employing Differing Modes of Ultrasonic Transmission, U.S. Patent No. 3,540,265 (Nov. 17, 1970c)

L.C. Lynnworth and D.R. Patch, New Sensors for Ultrasound: Measuring Temperature Profiles, *MTRSA* **10** (8), Cover, pp. 6–11, 40 (August 1970); with E.H. Carnevale, Ultrasonic Profile Measuring Apparatus, U.S. Patent No. 3,636,754 (January 25, 1972).

L.C. Lynnworth, Fluid Flowmeter, U.S. Patent No. 3,575,050 (April 13, 1971a).

L.C. Lynnworth, Dual Ultrasonic Sensors Employing a Single Mode of Ultrasonic Transmission, U.S. Patent No. 3,580,058 (May 25, 1971b).

L.C. Lynnworth, E.P. Papadakis, D.R. Patch, K.A. Fowler, and R.L. Shepard, Nuclear Reactor Applications of New Ultrasonic Transducers, *IEEE Trans. Nucl. Sci.* **NS-18** (1), 351–362 (1971).

L.C. Lynnworth and J.E. Bradshaw, Magnetostriction Transducers for Acoustic Emission, Impulse, Vibration and Noise Analysis, *Materials Research and Standards MTRSA* **11** (3), pp. 33–35 (1971).

L.C. Lynnworth and E.H. Carnevale, Ultrasonic Thermometry Using Pulse Techniques, pp. 715–732 in H.H. Plumb (ed.-in-chief), *Temperature—Its Measurement and Control in Science and Industry* **4**, ISA (1972).

L.C. Lynnworth and B.J. Spencer, Magnetostrictive Ultrasonic Transducer, U.S. Patent No. 3,633,424 (January 11, 1972).

L.C. Lynnworth, Torsional Wave Transduction in Magnetostrictive Wire, *Ultrasonics* **10** (5), pp. 195–197 (1972).

L.C. Lynnworth and N.E. Pedersen, Ultrasonic Mass Flowmeter, pp. 87–90, *1972 Ultrasonics Symp. Proc.*, IEEE (1972).

L.C. Lynnworth, N.E. Pedersen and E.H. Carnevale, USAAMRDL Technical Report 72–66, *Ultrasonic Mass Flowmeter for Army Aircraft Engine Diagnostics* (1973c).

L.C. Lynnworth, Attenuation Measurements Using the Pulse-Echo AB Method, *Materials Eval.* **31** (1), pp. 6–16 (1973a).

L.C. Lynnworth, Ultrasonic Measurement of Elastic Moduli in Slender Specimens Using Extensional and Torsional Wave Pulses, *J. of Testing and Eval., JTEVA* **1** (2), pp. 119–125 (March 1973b).

L.C. Lynnworth and N.E. Pedersen, Nonintrusive Dynamic Flowmeter, pp. 178–181, *1973 Ultrasonics Symp. Proc.*, IEEE (1973a).

L.C. Lynnworth, E.P. Papadakis and W.W. Rea, Ultrasonic Measurement of Phase and Group Velocity Using Continuous Wave Transmission Techniques, Final Report AMMRC CTR 74–20 (March 1974); pp. 533–536, *1973 Ultrasonics Symp. Proc.*, IEEE (1973b).

L.C. Lynnworth, Attenuation and Reflection Coefficient Nomogram, *Ultrasonics* **12** (2), pp. 72–73 (March 1974a).

L.C. Lynnworth, N.E. Pedersen and C.A. Carey, Nonintrusive (Noninterfering) Ultrasonic Techniques to Measure Gas Mass Flow Rates, Final Report, AEDC-TR-74-77, AD787675 (March 1974a).

L.C. Lynnworth and E.P. Papadakis (ed.), *Measurement of Elastic Moduli of Materials at Elevated Temperature*, AD-780 231/7GA, Final Rpt., ONR Contract N00014-73-C-0023, I.D. No. NR 384-320/04-27-72/468 (May 31, 1974).

L.C. Lynnworth, Ultrasonic Flowmeter Cell Designs for Liquids, pp. 678–681, *1974 Ultrasonics Symp. Proc.*, IEEE (1974b).

L.C. Lynnworth, N.E. Pedersen, E.P. Papadakis and J.H. Bradshaw, Nonintrusive Ultrasonic Measurement of Flow Velocity and Mass Flow Rate, pp. 917–924 in *Flow—Its Measurement and Control in Science and Industry* **I**, ISA (1974b).

L.C. Lynnworth, Industrial Applications of Ultrasound—A Review II. Measurements, Tests and Process Control Using Low-Density Ultrasound, *IEEE Trans. on Sonics and Ultrasonics* **SU-22** (2), pp. 71–101 (1975a).

L.C. Lynnworth, Clamp-On Ultrasonic Flowmeters-Limitations and Remedies, *Instrum. Techn.* **22** (9), pp. 37–44 (1975b).

L.C. Lynnworth, Area Averaging Ultrasonic Flowmeters, U.S. Patent No. 3,906,791 (Sept. 23, 1975c.)

L.C. Lynnworth and J.N.C. Chen, Energy Transmission Coefficients at Liquid/Solid Interfaces, pp. 575–578, *1975 Ultrasonics Symp. Proc.*, IEEE (1975).

L.C. Lynnworth, E.P. Papadakis and K.A. Fowler, Ultrasound Propagation Measurements and Applications, pp. 71–115, in W.J. McGonnagle (ed.), *Intl. Adv. in NDT*, Vol. **5**, Gordon and Breach (1977).

L.C. Lynnworth, Ultrasonic Reflections from Interfaces Experiencing Static vs Dynamic Friction, *Wear*, Vol. **41**, 195–199 (1977a).

L.C. Lynnworth, Slow Torsional Wave Sensors, pp. 29–34, *1977 Ultrasonics Symp. Proc.*, IEEE (1977b).

L.C. Lynnworth, Ultrasonic Measuring System with Isolation Means, U.S. Patent No. 4,004,461 (Jan. 25, 1977c).

L.C. Lynnworth, New Designs for Magnetostrictive Probes Using Extensional, Torsional and Flexural Waves, pp. 300–304, *1978 Ultrasonics Symp. Proc.*, IEEE (1978).

L.C. Lynnworth, Ultrasonic Flowmeters, Chap. 5 in W.P. Mason and R.N. Thurston (ed.) *Physical Acoustics* **14**, 407–525, Academic Press, NY (1979a).

L.C. Lynnworth, Liquid Level Measurements Using Longitudinal, Shear, Extensional and Torsional Waves, pp. 376–379, *1979 Ultrasonics Symp. Proc.*, IEEE (1979b).

L.C. Lynnworth, Slow Torsional Wave Densitometer, U.S. Patent No. 4,193,291 (March 18, 1980).

L.C. Lynnworth, N.E. Pedersen, J.E. Bradshaw, J.E. Matson, E.S. Johansson and T.H. Nguyen, *Advanced Fuel Flowmeter for Future Naval Aircraft*, NADC-80254-60, Final Rpt. (28 June 1981a).

L.C. Lynnworth, K.A. Fowler and D.R. Patch, Sealed, Matched Piezoelectric Transducer, U.S. Patent No. 4,297,607 (Oct. 27, 1981b).

L.C. Lynnworth, Ultrasonic Flowmeters, *Trans. Inst. MC* **3** (4), pp. 217–223 (Oct.-Dec. 1981); **4** (1), pp. 2–24 (Jan./-Mar. 1982a).

L.C. Lynnworth, Cryogenic Magnetostrictive Transducer, *IEEE Trans. Sonics and Ultrasonics*, **SU-29,** (6), p. 235 (July 1982b).

L.C. Lynnworth, Temperature Profiling Using Multizone Ultrasonic Waveguides, in: *Proceedings of the Sixth Symposium, Temperature—Its Measurement and Control in Science and Industry*, Vol. **5**, pp. 1181–1190, ISA (1982c).

L.C. Lynnworth, Ultrasonic Flowmeter Using Waveguide Antennas, U.S. Patent 4,336,719 (June 29, 1982d).

L.C. Lynnworth, J.L. Seger and J.E. Bradshaw, Ultrasonic System for Measuring Fluid Impedance or Liquid Level, U.S. Patent No. 4,320,659 (March 23, 1982a).

L.C. Lynnworth, Special Applications of Ultrasonic Contrapropagation Flowmeters, pp. 231–238, *Proc. of the 30th International Instrumentation Symposium*, ISA (1984).

L.C. Lynnworth and J.L. Adsmond, Clamp-On Flowmeter Surveys Powerplant Water Balance, *Power* **128** (1), pp. 118–119 (1984).

L.C. Lynnworth and T.H. Nguyen, Screw Pressure Coupler and Mode Converter, *NDT Comm.* **1**, 164–174 (1984).

L.C. Lynnworth, D.R. Patch and W.C. Mellish, Impedance-Matched Metallurgically Sealed Transducers, *IEEE Trans. Sonics and Ultrason.* SU-31 (2), pp. 101–104 (1984).

A.M. and L.C. Lynnworth, Calculated Turbulent-Flow Meter Factors for Nondiametral Paths Used in Ultrasonic Flowmeters, *J. Fluids Engineering* **107** pp. 44–48 (March 1985).

L.C. Lynnworth, J.M. Korba and D.R. Wallace, Fast Response Ultrasonic Flowmeter Measures Breathing Dynamics, *IEEE Trans. Biomedical Engineering* **BME-32** (7), pp. 530–535 (July 1985).

L.C. Lynnworth and T.H. Nguyen, Leaky Rayleigh Wave Clamp-On Flowmeter, pp. 519–524, *1985 Ultrasonics Symp. Proc.*, IEEE (1985).

L.C. Lynnworth and J.M. Ulte, Ultraschall-Durchflußmesser für Fackelgase, *Chemie-Technik* **15** (10) pp. 134–139 (Oct. 1986).

L.C. Lynnworth, T.H. Nguyen and J.A. True, Flowmeter Designs: Small ID; Small OD; Limited Length, Limited Access, pp. 582–587, *1986 Ultrasonics Symp. Proc.*, IEEE (1986).

L.C. Lynnworth, Engineering Aspects of Ultrasonic Process Control - Flow, Temperature, and Liquid Level Applications, Trans. *ASME JVASRD* **108** (1), pp. 69–81 (Jan. 1986).

L.C. Lynnworth, *Ultrasonic Waveguide Experiments Booklet*, Panametrics (May 1987).

L.C. Lynnworth, Method and Apparatus for Measuring Fluid Characteristics Using Surface Generated Volumetric Interrogation Signals, U.S. Patent No. 4,735,097 (April 5, 1988a).

L.C. Lynnworth, Buffer Rod Designs for Ultrasonic Flow Measurements at Cryogenic and High Temperatures, ±200°C, pp. 697–702 in: *Proc. 34th Int'l Instrum. Symp.*, ISA (1988b).

L.C. Lynnworth, Ultrasonic Transducer for High Temperature Applications, U.S. Patent No. 4,783,997 (Nov. 15, 1988c).

L.C. Lynnworth, Proposed Downhole Ultrasonic Quadrant Flow Tool: Technology Background and Design Considerations, pp. 78–83 in: M. Lippmann (ed.), *Proc. Technical Review "Advances in Geothermal Reservoir Technology—Research in Progress,"* Lawrence Berkeley Laboratory Report LBL-25635 (Sept. 1988d).

L.C. Lynnworth, S.A. Jacobson, J.M. Korba, T.H. Nguyen and J.A. True, Downhole Quadrant by Quadrant Flow Tool (QFT) for Geothermal Applications to 350 Deg Celsius, Phase 1 Final Rpt., Contract DE-AC02-88ER80587 (1989a).

L.C. Lynnworth et al., Ultrasonic Transducers, Deployment and Signal Processing Means for Cryofluids, Phase I Final Rpt., SBIR Contract NAS3-25371 (1989b).

R.A. McBane, R.L. Campbell, Jr, and E.G. DiBello, Acoustic Flowmeter Field Test Results, pp. 171–182, *Society of Petroleum Engineers Gas Technology Symposium Proc.* (1988).

R.D. McCarty, Speed of Sound in Natural Gas Mixtures, pp. 507–515, *Proc. AGA Int'l Symp. on Fluid Flow* (November 1986).

J.C. McDade, D.R. Pardue, A.L. Hedrich and F. Vrataric, Sound Velocity in Water Above 212°F, *J. Acoust. Soc. Am.* **31** (10), pp. 1380–1383 (October 1959). Anon., Diamond Ordnance Fuze Laboratories Interim Report No. 2 on the Development of an Experimental

Sonic Flowmeter System, R-51.2-56-4 (25 October 1956). Compare with Chavez et al. (1985).

R.D. McGunigle and M. Jennings, Ultrasonic Ablation Recession Measurement System, ISA paper ASI 75210, pp. 19–24 (1975).

R.J. McKee, Pulsation Effects on Various Gas Meters, pp. 107–110, *Proc. 1987 ASME Pipeline Engineering Symposium-ETCE,* Dallas, TX (1987).

R.B. McLellan and M. Yoshihara, The Thermodynamics of Dilute Solutions of Hydrogen in Pd and Substitutional Alloys, *Acta Met.* **35**, pp. 197–225 (1987).

R.C. McMaster (ed.), *Nondestructive Testing Handbook,* Vol. II, Sections 43–51, Ronald Press Co., New York (1959).

J.G. McMillan and R.H. Pamperin, Application of Fluidic Sensors for Measurement of Turbine Inlet Temperature, SAE Paper 720158 (Jan. 1972).

D.H. McQueen, Noise Generation in Bench-Top Ultrasonic Cleaners, *Ultrasonics* **26** (5), pp. 286–290 (Sept. 1988).

H.J. McSkimin, Measurement of Elastic Constants at Low Temperatures by Means of Ultrasonic Waves—Data for Silicon and Germanium Single Crystals, and for Fused Silica, *J. Appl. Phys.* **24** (8), pp. 988–997 (1953).

H.J. McSkimin, Ultrasonic Pulse Technique for Measuring Acoustic Losses and Velocities of Propagation in Liquids as a Function of Temperature and Hydrostatic Pressure, *J. Acoust. Soc. Am.* **29** (11), pp. 1185–1192 (November 1957).

H.J. McSkimin, Measurement of Ultrasonic Wave Velocities and Elastic Moduli for Small Solid Specimens at High Temperatures, *J. Acoust. Soc. Am.* **31** (3), pp. 287–295 (March, 1959).

H.J. McSkimin, Measurement of Dynamic Properties of Materials, U.S. Patent No. 2,996,058 (Dec. 27, 1960).

H.J. McSkimin, Ultrasonic Methods for Measuring the Mechanical Properties of Liquids and Solids, pp. 272–334, in: W.P. Mason (ed.), *Physical Acoustics* Vol. I, Part A, Academic Press (1964).

H.J. McSkimin and P. Andreatch, Jr., Measurement of Dynamic Shear Impedance of Low Viscosity Liquids at Ultrasonic Frequencies, *J. Acoust. Soc. Am.* **42** (1), pp. 248–252 (July 1967).

P. Macedo and T.A. Litovitz, Ultrasonic Viscous Relaxation in Molten Boron Trioxide, *Phys. Chem. Glasses* **6** (3), pp. 69–80 (1965).

P.A. Magnin, Doppler Effect: History and Theory, *Hewlett-Packard J.* **37** (6), pp. 26–31 (June 1986).

P.A. Magnin, A Review of Doppler Flow Mapping Techniques, pp. 969–978, *1987 Ultrasonics Symp. Proc.,* IEEE (1987).

P.A. Magnin, Doppler Ultrasound Maps Blood Flow, *SOMA* pp. 14–18 (July 1986).

V. Mágori, A Novel Ultrasonic Flow Sensor Based on Lambda Locked Loop Principle with Interdigital Transducers, pp. 525–530, *1985 Ultrasonics Symp. Proc.,* IEEE (1985).

R.J. Mahany and A.G. Johnson, Vortex Generating Device, U.S. Patent No. 4,312,236 (Janaury 26, 1982).

R.A. Makofski, J.T. Massey, F. F. Mark, F.B. Weiskopf, Jr., W.H. Guier, P.C. Walsh and F.F. Marshall, Means and Method for the Noninvasive Fragmentation of Body Concretions, U.S. Patent No. 4,610,249 (September 9, 1986).

J.T. Malone and D.K. Whirlow, Fluid Flow Measurement System, U.S. Patent No. 3,564,912 (Feb. 23, 1971).

A. Mandelis, Time-Delay-Domain and Pseudorandom-Noise Photoacoustic and Photothermal Wave Processes: A Review of the State of the Art, *IEEE Trans. UFFC* **33** (5), pp. 590–614 (September 1986).

U. Mann and E.J. Crosby, Flow Measurement of Coarse Particles in Pneumatic Conveyers, *Ind. Eng. Chem., Process Des. Dev.* **16** (1), pp. 9–13 (1977).

T.L. Mansfield, Probe for the Ultrasonic Inspection of Molten Aluminum, U.S. Patent No. 4,261,197 (April 14, 1981).

J.J. Markham, R.T. Beyer and R.B. Lindsay, Absorption of Sound in Fluids, *Rev. Mod. Phys.* **23** (4), pp. 353–411 (October 1951).

B.J. Marlow, D. Fairhurst and H.P. Pendse, Colloid Vibration Potential and the Electrokinetic Characterization of Concentrated Colloids, *Langmuir* **4**, pp. 611–626 (1988).

R.B. Martin and R.R. Haynes, *J. Amer. Cer. Soc.* **54** (8), pp. 410–411 (Aug. 1971).

R.W. Martin and D.W. Watkins, An Ultrasonic Catheter for Intravascular Measurement of Blood Flow: Technical Details, *IEEE Trans. Sonics and Ultrasonics* **SU-27** (6), pp. 277–286 (November 1980).

S.J. Martin, G.C. Frye, A.J. Ricco and T.E. Zipperrian, Measuring Thin Film Properties Using SAW Devices: Diffusivity and Surface Area, pp. 563–568, *1987 Ultrasonics Symp. Proc.*, IEEE (1987).

T. Marttinen and M. Luukkala, An Acoustic, Noncontacting Instrument to Measure Tension in a Moving Paper Web, pp. 553–556, *1985 Ultrasonics Symp. Proc.*, IEEE (1985).

W.P. Mason, Measurement of the Viscosity and Shear Elasticity of Liquids by Means of a Torsionally Vibrating Crystal, *Trans. ASME* **69**, pp. 359–370 (May 1947).

W.P. Mason, W.O. Baker, H.J. McSkimin and J.H. Hess, Measurement of Shear Elasticity and Viscosity of Liquids at Ultrasonic Frequencies, *Phys. Rev.* **75**, pp. 936–946 (March 15, 1949).

W.P. Mason, *Piezoelectric Crystals and Their Application to Ultrasonics*, Van Nostrand, New York (1950).

W.P. Mason, *Physical Acoustics and the Properties of Solids*, pp. 40–50, Van Nostrand, New York (1958).

W.P. Mason (ed.), *Physical Acoustics—Principles and Methods*, Vols. **1–5**, Academic Press, New York (1968—1972).

W.P. Mason, Multiple Reflection Ultrasonic Delay Lines, pp. 485–500 in: W.P. Mason (ed.), *Physical Acoustics* Vol. 1, Part A, Academic Press (1964).

W.P. Mason, Use of High Amplitude Strains in Studying Wear and Ultrasonic Fatigue in Metals, pp. 647–657 in: J.M. Wells, O. Buck, L.D. Roth, and J.K. Tien (ed.), *Proc. First International Conf. on Fatigue and Corrosion Fatigue Up to Ultrasonic Frequencies*, Metall. Soc. of AIME (1981).

D.P. Massa, An Automatic Ultrasonic Bowling Scoring System, *Sensors* **4** (10), pp. 20–26 (October 1987).

R.J. Massimino, J.D. Gough, G.T. Sterns and J. Martin, Jr., Gaseous Emboli Removal Efficiency in Arterial Screen Filters: A Comparative Study, JAMSECT **15** (2), pp. 25–34 (1983).

J.E. Matson, T.H. Nguyen and L.C. Lynnworth, Ultrasonic Measurement of Liquid Flow Using Clamp-On Rayleigh Wave Transducers, *Proc. IEEE IMTC/87 Instrumentation/Measurement Technology Conference*, pp. 197–206 (April 1987).

A.L. Matthews, K.A. Murphy, R.E. Rogers and R.O. Claus, Acoustic Fiber Waveguide Coupler, pp. 629–632, *1987 Ultrasonics Symp. Proc.*, IEEE (1987).

J.E. May, Jr., Guided Wave Ultrasonic Delay Lines, pp. 418–483, in: W.P. Mason (ed.), *Physical Acoustics* Vol. 1, Part A, Academic Press (1964).

A.M. Mayer, On an Acoustic Pyrometer, *Phil. Mag.* Ser. 4, **45**, pp. 18–22 (1873).

W.G. Mayer, Reflection and Refraction of Mechanical Waves at Solid-Liquid Boundaries, *J. Appl. Phys.* **34** (4), Part I, pp. 909–911 (April 1963).

W.G. Mayer, Energy Partition of Ultrasonic Waves at Flat Boundaries, *Ultrasonics* **3** (2), pp. 62–68 (April-June 1965).

D.M. Mechlenburg, Ultrasonic Sensor Prototype Detects Gas Concentrations, *Sensors* **2**, (9), pp. 17–19 (September 1985).

H. Medwin, Scattering from the Sea Surface, Chapter 3, pp. 57–89 in R.W.B. Stephens (ed.), *Underwater Acoustics*, Wiley-Interscience (1970).

G. Meltz, W.W. Morey, W.H. Glenn and J.D. Farina, In-Fiber Bragg-Gating Temperature and Strain Sensors, pp. 239–242, *Proc. 34th Int'l. Instrum. Symp.*, ISA (1988).

J. Miklowitz, *The Theory of Elastic Waves and Waveguides*, North Holland Pub. Co. (1978).

G.N. Miller, R.L. Anderson, S.C. Rogers, L.C. Lynnworth, W.B. Studley and W.R. Wade, High Temperature, High Pressure Water Level Sensor, pp. 877–881, *1980 Ultrasonics Symp. Proc.*, IEEE (1980).

J.R. Miller III, Operating Equation, Pressure, Temperature Coefficients and Other Characteristics of a Quartz Resonator Pressure Transducer, pp. 317–324, *Proc. 34th Int'l. Instrum. Symp.*, ISA (1988).

R.W. Miller, *Flow Measurement Engineering Handbook*, McGraw-Hill, N.Y. (1983). See also: 2nd Ed. (1989).

Y. Mitsuta, pp. 341–347, in *Flow—Its Measurement and Control in Science and Industry*, Vol. 1, R.B. Dowdell (ed.), Pittsburgh, PA: Instrument Society of America (1974). See also, J.C. Kaimal, J.T. Newman, A. Bisberg and K. Cole, pp. 349–359, ibid.

R.J. Moffat, Identifying the True Value—The First Step in Uncertainty Analysis, pp. 255–258, *Proc. 34th Int'l. Instrum. Symp.*, ISA (1988).

O.G. Molina, Ultrasonic Couplant Gel Compositions and Method for Employing Same, U.S. Patent No. 4,365,516 (December 28, 1982).

Molytek, Liquid Flowmeter Thermalpulse, *Meas. & Control News* **21** (6), Issue 126, p. 46 (December 1987).

J.-P. Monchalin and R. Héon, Laser Ultrasonic Generation and Optical Detection with a Confocal Fabry-Pérot Interferometer, *Materials Eval.* **44**, pp. 1231–1237 (September 1986).

C.E. Mongan, Method and Apparatus for Measuring Liquid Level, U.S. Patent No. 3,010,318 (November 28, 1961).

F.R. Montero de Espinosa, V. Pavia, J.A. Gallego-Juarez and M. Pappalardo, Fractured Piezoelectric Ceramics for Broadband Ultrasonic Composite Transducers, pp. 691–696, *1986 Ultrasonics Symp. Proc.*, IEEE (1986).

R.S. Moore and H.J. McSkimin, Dynamic Shear Properties of Solvents and Polystyrene Solutions from 20 to 300 MHz, pp. 167–242 in W.P. Mason and R.N. Thurston (ed.), *Physical Acoustics* **VI**, Academic Press (1970).

M.E. Motamedi and R.M. White, Preface, *IEEE Trans. UFFC* **34** (2), pp. 122–123 (March 1987).

C.D. Moriarty, Ultrasonic Flaw Detection in Pipes by Means of Shear Waves, *Trans. ASME* **73**, pp. 225–235 (1951).

R.J. Morin and D.E. Gaucher, In-Line Continuous Monitoring System Improves Water Plant EOR Performance, *Oil & Gas J.* **84** (28), pp. 104–105 (July 14, 1986).

T. Moriizumi, Y. Unno and S. Shiokawa, New Sensor in Liquid Using Leaky SAW, pp. 579–582, *1987 Ultrasonics Symp. Proc., IEEE* (1987).

J.A. Morris, C.P. Yakymyshyn and C.R. Pollock, Fiber Optic Fuel Level Sensors, *ISA Trans.* **26** (3), pp. 25–32 (1987).

P.M. Morse, *Vibration and Sound*, 2nd Ed., McGraw-Hill, New York (1948).

P.M. Morse and K.U. Ingard, *Theoretical Acoustics,* Mc-Graw-Hill (1968).

L.E. Morse, W.R. Shapton, D.L. Brown and E. Kuljanic, Applications of Pulse Testing for Determining Dynamic Characteristics of Machine Tools, presented at the 13th International Machine Tool Design and Research Conference, Univ. of Birmingham, England (Sept. 22, 1972); reprinted in *Hewlett-Packard Application Note 140–3* (Santa Clara, CA Div. HP).

M. Moshfeghi, NDT of Rods and Pipes Using Shear Wave Caustics, *Ultrasonics* **24** (1), pp. 19–24 (January 1986).

R. Motegi, S. Takeuchi, T. Sato, N. Chubachi, M. Takeuchi, Remote Measurement Method of Acoustic Velocity in Liquids by a Pair of Leaky Acoustic Waveguides, pp. 625–628, *1987 Ultrasonics Symp. Proc.*, IEEE (1987).

J. Muggli, Ultrasonic Ranging System for a Camera, U.S. Patent No. 4,199,246 (April 22, 1980).

A.C. Munce, Jr., L. Crawforth, W. Imaino and S. Wang, Determination of Case Depth in Small Sintered Parts Using Ultrasonic Waves, pp. 581 and 588–592, *1986 Ultrasonics Symp. Proc.*, IEEE (1986).

W.D. Munk, Ultrasonic Flowmeter Offers New Approach to Large-Volume Gas Measurement, *Oil & Gas J.* **80** (36), pp. 111–117 (Sept. 6, 1982).

J.F. Muratore and H.R. Carleton, Phase Spectroscopy in Lossy Media, pp. 1047–1051, *1985 Ultrasonics Symp. Proc.*, IEEE (1985).

M.J.P. Musgrave, On the Propagation of Elastic Waves in Aeolotropic Media, *Proc. Royal Society* (London) A*226*, 339 (1954); 356 (1954); A*236* 352 (1956).

K.S. Mylvaganam, R. Bφ and T. Folkestad, Ultrasonic Flare Gas Flowmeter with Combined CHIRP and CW Detection, CMI-no. 831401–8, Bergen (1985).

K.S. Mylvaganam, Low Frequency Ultrasonic Contrapropagating Transit Time System for Gas Flowmetering, pp. 938–946, *Ultrasonics International 87 Conf. Proc.*, Butterworths (1987).

K.S. Mylvaganam, T. Folkestad and R. Bφ, Fluenta FGM 100 Ultrasonic Gas Flowmeter, CMI-no. 871413, Bergen, 1987.

K.S. Mylvaganam, Ultrasonic Flowmeters Measure Flare Gas in the North Sea, *Oil & Gas J.* **86** (42) pp. 54–56 (October 17, 1988.)

K.S. Mylvaganam, High Rangeability Ultrasonic Gas Flowmeter for Monitoring Flare Gas, *IEEE Trans. UFFC* **36** (2), pp. 144-149 (1989).

K.S. Mylvaganam, R. Bφ and T. Folkestad, Low Frequency Ultrasonic Propagating Transit Time System for Gas Flowmetering, abstr., *Ultrasonics* **25** (6), p. 382 (November 1987).

W.P. Nagel, Microcomputers Come of Age for Open Channel Flow Measurement, *Measurements & Control* **26** (6), Issue 126, p. 236 (December 1987).

H. Nakano and S. Nagai, Microcomputer-Assisted System for Measurement of Ultrasonic Velocity, *Ultrasonics* **26** (5), pp. 256–259 (Sept. 1988).

M. Nakazawa, A. Ballato, and T. Lukaszek, An Ultralinear Stress-Compensated Temperature Sensor, *IEEE Trans. UFFC* **34** (2), pp. 270–277 (March 1987).

A.H. Nayfeh and D.E. Chimenti, Reflection of Finite Acoustic Beams from Loaded and Stiffened Half Spaces, *J. Acoust. Soc. Am.* **75**, pp. 1360–1368 (1984).

K.F. Neusen and I.C. Romer, Jr., Sound-Wave Attenuation Through Air Contained in Small Tubes, *J. Acoust. Soc. Am.* **46** (5) Part 2, pp. 1148–1152 (1969).

V.L. Newhouse, J.A. Cisneros, D. Censor and B. Goldberg, Doppler Spectrum Probing of Flows Transverse with Respect to Beam Axis, pp. 971–974, *1985 Ultrasonics Symp. Proc.*, IEEE (1985).

V.L. Newhouse, D. Censor, T. Vontz, J.A. Cisneros and B.B. Goldberg, Ultrasound Doppler Probing of Flows Transverse with Respect to Beam Axis, *Proc. IEEE Trans.* **BME-34**, No. 10, pp. 779–789 (October 1987).

W.H. Newman and P.P. Lele, A Transient Heating Technique for the Measurement of Thermal Properties of Perfused Biological Tissue, *J. Biomedical Eng.* **107**, pp. 219–223 (August, 1985).

T. H. Nguyen, L.C. Lynnworth, H. A. Runde and M. Merilo, High Temperature Clamp-On Ultrasonic Flowmeter, pp. 871–876, *1982 Ultrasonics Symp. Proc.*, IEEE (1982).

T.H. Nguyen and L.C. Lynnworth, Zigzag Flowcells, pp. 1041–1046, *1983 Ultrasonics Symp. Proc.*, IEEE (1983).

N.C. Nicholson and W.N. McDicken, Waveguides in Medical Ultrasonics, *Ultrasonics* **26** (1), pp. 27–30 (1988); with T. Anderson, ibid **27** (2), pp. 101–106 (1989).

J. Nikuradse, *VDI-Forschungsh.* **281**, pp. 1–44 (1926a).

J. Nikuradse, Ph.D. Thesis, Göttingen (1926b).

J. Nikuradse, *VDI-Forschungh.* **356**, pp. 1–36, esp. p. 20 (1932).

J. Nikuradse, *Laws of Flow in Rough Pipes*, PB 102594, NACA Techn. Memo. 1292, trans. from *VDI-Forschungsh.* (1933).

F.W. Noble, K. Abel and P.W. Cook, Performance and Characteristics of an Ultrasonic Gas Chromatograph Effluent Detector, *Anal. Chem.* **36** (8), pp. 1421–1427 (July 1964).

F.W. Noble, Dual Frequency Ultrasonic Fluid Flowmeter, *Rev. Sci. Instrum.* **39** (9), pp. 1327–1331 (1968).

M.E. Nolan, J.G. O'Hair and R. Teyssandier, The Measurement of High Pressure Natural Gas Flows Using the Four-Path Ultrasonic Flowmeter Developed by British Gas, pp. 809–822, *International Symposium on Fluid Flow Measurement Proc.*, AGA, 1986.

B.E. Noltingk, *Instrumentation Reference Book*, Butterworths (1988).

T. Nomura, S. Shiokawa, T. Moriizumi and T. Yasuda, Two Dimensional Mapping of SAW Propagation Constants by Using Fresnel-Phase-Plate Interdigital Transducer, pp. 621–626, *1983 Ultrasonics Symp. Proc.*, IEEE (1983).

T. Nomura and T. Yasuda, A Coupling Method for Nondestructive Evaluation Using an Interdigital Transducer, pp. 1060–1063, *1985 Ultrasonics Symp. Proc.*, IEEE (1985).

F.D. Norvelle, A Survey of Automatic Particle Counting Methods for Fluid Power Applications, *BFPR J.* **17** (2), pp. 295–302 (1984).

M.H. November, in ITT-Barton Bulletin MF-1 (1974); Electronic Density-Measuring System is Accurate, Versatile, *Oil & Gas J.* **70** (8), pp. 62–64 (Feb. 21, 1972).

M.H. November and L.D. Lyon, Densitometer, U.S. Patent 4,037,460 (July 26, 1977).

V.F. Nozdreva (ed.), *Soviet Progress in Applied Ultrasonics*, Volume 1, Ultrasound in Industrial Processing and Control, Consultants Bureau, New York (1964).

S.P. Nuspl, E.P. Szmania, J.A. Kleppe and P.R. Norton, Acoustic Pyrometry Applied to Utility Boilers, Paper 86-JPGC-PTC-2, presented at the Jt. ASME/IEEE Power Generation Conf., Portland, OR (October 19–23, 1986).

S.P. Nuspl, E.P. Szmania, J.A. Kleppe and P.R. Norton, Acoustically Measuring Boiler Furnace Exit Gas Temperature, *Proc. EPRI Conf. on Incipient Boiler Failure Detection in Power Plants*, Philadelphia, PA (March, 1987).

J.A. Nystuen, Rainfall Measurements Using Underwater Ambient Noise, *J. Acoust. Soc. Am.* **79** (4), pp. 972–982 (April 1986).

R.W. O'Brien, Electro-acoustic Effects in a Dilute Suspension of Spherical Particles, *J. Fluid Mech.* **190**, pp. 71–86 (1988).

J.G. O'Hair and M.E. Nolan, Ultrasonic Flowmeter, U.K. Patent Application GB 2 139 755A (14 November 1984).

C.B. Officer, *Introduction to the Theory of Sound Transmission, McGraw-Hill Series in the Geological Sciences*, McGraw-Hill, N.Y. (1958).

T. Oja, G.L. Petersen and D.W. Cannon, A Method for Measuring the Electrokinetic Properties of a Solution, U.S. Patent No. 4,497,207 (1985).

M. Okujima and S. Ohtsuki, pp. 457–460 in *Proc. 7th Int. Cong. Acoust.* **4** (1971).

A.A. Oliner, H.L. Bertoni and R.C.M. Li, A Microwave Network Formalism for Acoustic Waves in Isotropic Media, *Proc. IEEE* **60** (12), pp. 1503–1512 (December 1972).

A.A. Oliner, R.C.M. Li and H.L. Bertoni, Catalog of Acoustic Equivalent Networks for Planar Interfaces, *Proc. IEEE* **60** (12), pp. 1513–1518 (December 1972).

T. Ono (1948–1956). Japanese Patent 182549 (applied for: Jan. 26, 1948; public notice: Nov. 11, 1949; granted: March 27, 1950; published: Sept. 29, 1956).

T. Ono, M. Kohata and T. Miyamoto, Ultrasonic Phase-Sensitive Rangefinder with Double Modulation Doppler-Free Method for Shallow Seafloor Survey, pp. 480–483, *1984 Ultrasonics Symp. Proc.*, IEEE (1984).

E. Opik, *Atomic Collisions and Radiation of Meteors*, Harvard Reprints No. 100 (1933).

A.G. Ovchinnikov, Compensation for the Effect of Velocity Fluctuations in Ultrasonic Frequency Flowmeters, *Izmeritel'naya Tekhnika*, No. 11, pp. 40–41 (1986); English translation in: *Russian Ultrasonics* **17** (2), pp. 34–39 (1987).

D. Owen, A. Pawlowski and J. Powers, Capital cost savings of $50,000 achieved with sonic flowmeters, *Chemical Processing*, **47** (2), pp. 66–67 (February 1984).

Y. Ozaki, H. Sumitani, T. Tomoda, and M. Tanaka, A New System for Real-Time Synthetic Aperture Ultrasonic Imaging, *IEEE Trans. UFFC* **35** (6) pp. 828–838 (Nov. 1988).

PAR Scientific, Fully Automated Pulse-Echo Overlap System, *Ultrasonics* **25**, p. 252 (July 1987).

PCB Piezotronics, Dynamic Tester Hammers Out Structural Defects, *Machine Design* **47** (1), p. 38 (January 9, 1975).

R.F. Pacanowski, W.A. Bajek and G.S. Browne, A Unique Device for Gas Analysis, *Proc. Natural Gas Conf.*, pp. 155–170 (1986).

Y.H. Pao and W. Sachse, On the Determination of Phase and Group Velocities of Dispersive Waves in Solids, *J. Appl. Phys.* **48**, pp. 4320–4327 (1978).

E.P. Papadakis, Revised Grain-Scattering Formulas and Tables, *J. Acoust. Soc. Am.* **37** (4), pp. 703–710 (April 1965).

E.P. Papadakis, Ultrasonic Attenuation Caused by Scattering in Polycrystalline Metals, *J. Acoust. Soc. Am.* **37** (4), pp. 711–717 (April 1965).

E.P. Papadakis, Chap. 15, pp. 269–328, in: W.P. Mason, ed., *Physical Acoustics, Principles and Methods*, Vol. IV, Part B, Academic Press (1968).

E.P. Papadakis, Ultrasonic Attenuation in Thin Specimens Driven Through Buffer Rods, *J. Acoust. Soc. Am.* **44** (3), 724–734 (1968).

E.P. Papadakis, Traveling Wave Reflection Methods for Measuring Ultrasonic Attenuation and Velocity in Thin Rods and Wires, *J. Appl. Phys.* **42**, 2990–2995 (1971).

E.P. Papadakis, Buffer-Rod System for Ultrasonic Attenuation Measurements, *J. Acoust. Soc. Am.* **44**, pp. 1437–1441 (November 1968).

E.P. Papadakis, Diffraction Grating Dispersive Delay Lines Utilizing Anisotropic Propagation Media, *Ultrasonics*, pp. 102–104 (April 1970).

E.P. Papadakis and K.A. Fowler, Broadband Transducers: Radiation Field and Selected Applications, *J. Acoust. Soc. Am.* **50** (3), Part I, pp. 729–745 (1971).

E.P. Papadakis, Lens Equation for Focused Transducers, *Int'l J. NDT* **4**, pp. 195–198, Gordon and Breach (1972).

E.P. Papadakis, L.C. Lynnworth, D.R. Patch and E.H. Carnevale, Ultrasonic Thermometry for LMFBR Systems, Final Report, NYO 3906–13 (1972a).

E.P. Papadakis, L.C. Lynnworth, K.A. Fowler and E.H. Carnevale, Ultrasonic Attenuation and Velocity in Hot Specimens by the Momentary Contact Method with Pressure Coupling, and Some Results on Steel to 1200°C, *J. Acoust. Soc. Am.* **52** (3, Part 2), pp. 850–857 (1972b).

E.P. Papadakis, K.A. Fowler and L.C. Lynnworth, News Uses of Ultrasonic Spectrum Analysis, pp. 81–86, *1972 Ultrasonics Symp. Proc.*, IEEE (1972b).

E.P. Papadakis, *Ultrasonic Methods for Modulus Measurement in Paper, Tappi J.* **56** (2), pp. 74–77 (Feb. 1973).

E.P. Papadakis, The Measurement of Small Changes in Ultrasonic Velocity and Attenuation, *Crit. Rev. in Solid State Sciences CRC*, pp. 373–418 (August 1973).

E.P. Papadakis, Monitoring the Moduli of Polymers With Ultrasound, *J. Applied Physics* **45** (3), pp. 1218–1222 (March 1974).

E.P. Papadakis, K.A. Fowler, L.C. Lynnworth, A. Robinson and E.D. Zysk, Ultrasonic Measurements of Young's Modulus and Extensional Wave Attenuation in Refractory Metal Wires at Elevated Temperatures with Application to Ultrasonic Thermometry, *J. Appl. Phys.* **45** (6), pp. 2409–2420 (June 1974).

E.P. Papadakis, Ultrasonic Velocity and Attenuation: Measurement Methods with Scientific and Industrial Applications, pp. 277–374 in: W.P. Mason and R.N. Thurston (Eds.), *Physical Acoustics: Principles and Methods*, Vol. **XII**, Academic Press, NY (1976).

E.P. Papadakis, New, Compact Instrument for Pulse-Echo Overlap Measurements of Ultrasonic Wave Transit Times, *Rev. Sci. Instrum.* **47** (7), pp. 806–813 (1976).

E.P. Papadakis, On-Line Statistical Process Control with NDE and Computers, pp. 523–528, *1988 Ultrasonics Symp. Proc.*, IEEE (1988). See also, Pfeifer (1988) and Yehling et al. (1988).

N.A. Park and T.F. Irvine, Jr., The Falling Needle Viscometer: A New Technique for Viscosity Measurements, *Wärme- und Stoffübertragung* **18**, pp. 201–206 (1984).

R.L. Parker, J.R. Manning and N.C. Peterson, Application of Pulse-Echo Ultrasonics to Locate the Solid/Liquid Interface During Solidification and Melting of Steel and Other Metals, *J. Appl. Phys.* **58** (11), pp. 4150–4164 (December 1985).

J.M. Paros, Mounting System for Applying Forces to Load-Sensitive Resonators, U.S. Patent No. 4,384,495 (May 24, 1983a).

J.M. Paros, Isolating and Temperature Compensating System for Resonators, U.S. Patent No. 4,406,966 (Sept. 27, 1983b).

J.M. Paros, R.B. Wearn and J.F. Tonn, Mounting and Isolating System for Tuning Fork Temperature Sensor, U.S. Patent No. 4,706,259 (Nov. 10, 1987).

B. Paul, Predictions of Elastic Constants of Multiphase Materials, *Trans. Metall. Soc. AIME* **218**, pp. 36–41 (February 1960).

M.T. Paulsen and B. Birker, Apparatus for Measuring the Flow Quantity or Associated Parameters of a Liquid with Two Ultrasonic Transducers, Canadian Patent No. 1,131,755 (Sept. 14, 1982).

L.H. Pearson, W.J. Murri and D.S. Gardiner, Ultrasonic Detection of In-Plane Properties of Composite Laminates, pp. 983–989, *1985 Ultrasonics Symp. Proc.*, IEEE (1985).

H.P. Pendse and B.J. Marlow, A Novel Sensor for Electrokinetic Characterization of Concentrated Colloidal Slurries, pp. 73–79 in: *1987 Sensors Expo Proc.*, Helmers Publ. (1987).

N.E. Pedersen, L.C. Lynnworth and J.E. Bradshaw, Improved Ultrasonic Fuel Mass Flowmeter for Army Aircraft Engine Diagnostics, Eustis Final Report No. USAAMRDL-TR-75-8 (June 1975).

N.E. Pedersen, J.E. Bradshaw, L.C. Lynnworth and P.R. Morel, A New Ultrasonic Flowmeter for the Natural Gas Industry, pp. 293–318, in L.K. Irwin (ed.), *Flow Measurement in Open Channels and Closed Conduits*. NBS Spec. Publ. 484 (1977).

N.E. Pedersen, Method and Apparatus for Determining Fluid Flow, U.S. Patent 4,300,401 (Nov. 17, 1981).

N.E. Pedersen, J.E. Bradshaw, J.E. Matson and L.C. Lynnworth, Ultrasonic Flowmeter, U.S. Patent No. 4,308,754 (January 5, 1982).

Pennwalt, Piezo Film is Key to Unique Flowmeter in New Perkin-Elmer Monitoring Systems, News Release (May 1982).

Pennwalt, Kynar Piezo Film Technical Manual, 88 pages (1983).

J.M. Perdigão, N. Gazalet, J. Frohly and C. Bruneel, Coherent to Incoherent Ultrasonics Wave Conversion in Heterogeneous Media, *Ultrasonics* **25**, pp. 209–214 (July 1987).

P.J. Perkins, Aircraft Icing Sensors, *Sensors* **4**, (3), pp. 33–43 (March 1987).

C.R. Peterson, D.H. McFadden and J.A. LaPlante, Cutter Vibration Monitoring for Vertical Guidance of Mining Machines, *SME Preprint 85-571, 1985 Society of Mining Engineers (SME) Fall Meeting* (Oct. 1985)

C.R. Peterson, U.S. Patent No. 4,655,082, Mining Machine Having Vibration Sensor (April 7, 1987).

B. Pfau, Erweiterung des Meßbereiches eines Ultraschalldurchflußmessers unter Ausnutzung des Doppler-Effektes (Increasing the Measurement Range of an Ultrasonic Flowmeter Using the Doppler Effect), *Chemie-Ingenieur-Technik* **17**, pp. 1103–1109 (1970).

B. Pfau, Optimierung der Lage der Meßstrecke integrierender Durchflußmeßverfahren, *Arch. Tech. Mess. Bl. V* **1246-1**, pp. 21–24 (1973).

C.G. Pfeifer, SPC in the Process Industries, *Quality* **27**, pp. 38–40 (Dec. 1988).

L. Piché, F. Massines, G. Lessard and A. Hamel, Ultrasonic Characterization of Polymers as Function of Temperature, Pressure and Frequency, pp. 1125–1130 in *1987 Ultrasonics Symp. Proc.*, IEEE (1987).

J. Picht, Beitrag zur Theorie der Totalreflexion, *Ann. Physik* **3**, pp. 433–496 (1929). [5. Folge, Band 3, Heft 4]

A.D. Pierce, *Acoustics—An Introduction to Its Physical Principles and Applications*, 642 pp., McGraw-Hill, N.Y. (1981).

M. Platte, A Polyvinylidene Fluoride Needle Hydrophone for Ultrasonic Applications, *Ultrasonics* **23** (3), pp. 113–118 (May 1985).

D.I. Plaut and J.G. Webster, Ultrasonic Measurement of Respiratory Flow, *IEEE Trans. BME* **BME-27** (10), pp. 549–558 (October, 1980).

D.I. Plaut and J.G. Webster, Design and Construction of an Ultrasonic Pneumotachometer, *IEEE Trans. BME* **BME-27** (10) (October 1980).

T.J. Plona, L.E. Pitts and W.G. Mayer, Theoretical Similarities of Rayleigh and Lamb Modes of Vibration, *J. Acoust. Soc. Am.* **60** (2) pp. 374–377 (1976).

H.H. Plumb (ed.), *Temperature—Its Measurement and Control in Science and Industry*, ISA (1972).

H.H. Plumb and G. Cataland, Acoustical Thermometer, *Science* **150** (3693), pp. 155–161 (1965).

Polysonics, *Engineer's/User's Guidebook to Doppler Flow Measurement in Liquids* (1986).

S.C. Pomeroy, H.J. Dixon, M.D. Wybrow and J.A.G. Knight, Ultrasonic Distance Measuring and Imaging Systems for Industrial Robots, pp. 261–270 in: A. Pugh (ed.), *International Trends in Manufacturing and Technology, Robot Sensors*, Vol. 2—Tactile and Non-Vision, IFS (Publications) Ltd., UK (1986).

G.F. Potter and W.P. Stadig, Flowrate and Molecular Weight Determinations Enable Ultrasonic Probe to Pinpoint Leaks, *Chem. Proc.* **50** (3), pp. 110–112 (March 1987).

Proceedings of the 24th Annual Meeting of the American Institute of Ultrasound in Medicine and the 8th Annual Meeting of the American Society of Ultrasound Technical Specialists, Montreal, Canada, Volume 1 (August 27–31, 1979).

E.S. Prohaska and A.M. Lauletta, Jr., Re-Entry Vehicle Instrumentation Technology, Avco RAD-TM-61-46, AD363780 (December 20, 1961).

R. Prough, Sound Emission Level Detection, U.S. Patent No. 4,182,177 (Jan. 1980).

P.E. Purves and G. Pilleri, *Echolocation in Whales and Dolphins,* 280 pp., Academic Press (1983).

C.R. Randall, *Nat. Bur. Standards J. of Research,* **8** (January 1932).

A.C. Raptis and S.-H. Sheen, Ultrasonic Properties of Coal Slurries and Flow Measurements by Cross Correlation, *IEEE Trans. Sonics and Ultras.* **SU-28** (4), pp. 248–256 (1981).

J.W.S. Rayleigh, On Waves Propagated along the Plane Surface of an Elastic Solid, *Proc. London Math. Soc.* **17**, pp. 4–11 (1885).

R.J. Redding, Gas Detection, U.S. Patent No. 4,119,950 (October 10, 1978).

R.J. Redding, Sound Beams for Fire Alarm and Gas Detection Systems, *FIRE,* pp. 532–533 (March 1980).

R.J. Redding, Wide Range Flow Quantity Measurement Using Coherently Modulated Ultrasound, pp. 582–585, *1985 Ultrasonics Symp. Proc.,* IEEE (1985).

R.J. Redding, Gas Flow and Energy Measurement Using Coherently Modulated Ultrasound, pp. 621–624, *1987 Ultrasonics Symp. Proc., IEEE* (1987).

M.B. Reynolds, The Determination of the Elastic Constants of Metals by the Ultrasonic Pulse Technique, *Trans. ASM* **45**, pp. 839–861 (1953).

U. Riebel, Method of And an Apparatus for Ultrasonic Measuring of the Solids Concentration and Particle Size Distribution in a Suspension, U.S. Patent No. 4,706,509 (November 17, 1987).

V.M. Ristic, *Principles of Acoustic Devices,* Wiley, N.Y. (1983).

R.A. Roberts, Ultrasonic Beam Transmission at the Interface Between an Isotropic and a Transversely Isotropic Solid Half-Space, *Ultrasonics* **26** (3), pp. 139–147 (May 1988).

S.M. Rocha and R.D. Finch, Leak Detection Using Expansion Waves, *Sensors* **3** (2), pp. 6–10 (February 1986).

H.M. Roder, ASRDI Oxygen Technology Survey, Volume V: Density and Liquid Level Measurement Instrumentation for the Cryogenic Fluids Oxygen, Hydrogen and Nitrogen, NASA SP-3083 (1974).

S.C. Rogers and G.N. Miller, Ultrasonic Level, Temperature and Density Sensor, *IEEE Trans. Nucl. Sci.* **NS-29** (1) 665–668 (Feb. 1982).

R.S. Rogowski, J.S. Heyman and R.O. Claus, The Evolution of "Smart" Composite Materials, pp. 20–22, *NASA Tech Briefs* (Oct. 1988).

S.I. Rokhlin, T.K. Bolland and L. Adler, Reflection and Refraction of Elastic Waves on a Plane Interface Between Two Generally Anisotropic Media, *J. Acoust. Soc. Am.* **79** (4), pp. 906–918 (April 1986).

J.J. Ronchetto, Ultrasonic Thermometry in Oil Shale Retorts, UCRL-85845, *Proc. 1981 Symposium on Instrumentation and Control for Fossil Energy Processes,* ANL-81-62, Conf 810607 (1981).

M.S. Roos, Microparticle and Cell Characterization Using Acoustic Scattering, PhD Dissertation, Yale University (May 1983).

R.L. Rosenberg and L.A. Coldren, Flexible Capillary Ultrasonic Delay Lines, *IEEE Trans. Sonics and Ultrason.* **SU-24** (1), pp. 1–6 (1977).

W. Roth and I.R. Rich, *J. Appl. Phys.* **24**, pp. 940–950 (1953).

W. Roth, Flowmeters, U.S. Patent No. 3,188,862 (June 15, 1965).

R.A. Roy, Quantitative Particle Characterization by Scattered Ultrasound, Technical Memorandum No.5, Dept. of Mechanical Engineering, Yale Univ. (May 1987).

S.N. Rschevkin, *A Course of Lectures on The Theory of Sound,* Pergamon Press, New York (1963).

W.G.A. Russell-Cargill (ed. & publ.), *Recent Developments in Side Scan Sonar Techniques,* 141 pp., (1982).

A. Rutgers, Bemerkung zu den von Ultraschallwellen hervorgerüfenen Potential differenzen in Lösungen, *Physica* **5**, p. 46 (1938).

O. Rütten, Verfahren und Vorrichtung zum Messen von strömenden Flüssigkeits-, Gas- oder Dampfmengen, German Patent 520,484 (19 Sept. 1928).

A. Ryan and R. Postle, Application of Sonic Wave Theory to the Measurement of the Dynamic Elastic Moduli of Woven and Knitted Fabrics, *Textile Research Journal* **51** (11), pp. 732–740 (November 1981).

W. Sachse, *Experiments in Mechanical Vibrations and Acoustics*, esp. Ch. 3. Cornell University (1975).

W. Sachse and N.N. Hsu, Ultrasonic Transducers, Ch. 4, pp. 277–406 in W.P. Mason and R.N. Thurston, *Physical Acoustics* **14**, Academic Press (1979).

W. Sachse and K.Y. Kim, Ultrasonic Point-Source/Point-Receiver Materials Testing, pp. 311–319, in D.O. Thompson and D.E. Chimenti (ed.), *Review of Progress in Quantitative Nondestructive Evaluation* **6A** Plenum (1987).

R. Samuels, *Structured Polymer Properties*, pp. 41–51, 57–63, 117, 119, 154, 237–240 and 246–247, Wiley (1974).

F.H. Sanders, Transmission of Sound Through Thin Plates, *Canadian J. Research* **17**, Sec. A (9), pp. 179–193 (September 1939).

J. Saniie and N.M. Bilgutay, Quantitative Grain Size Evaluation Using Ultrasonic Back-scattered Echoes, *J. Acoust. Soc. Am.* **80** (6), pp. 1816–1824 (1986).

J. Saniie, T. Wang and N. Bilgutay, Statistical Evaluation of Backscattered Ultrasonic Grain Signals, *J. Acoust. Soc. Am.* **81** (1), pp. 400–408 (July 1988). See also, N.M. Bilgutay, X. Li and J. Saniie, Spectral Analysis of Randomly Distributed Scatterers for Ultrasonic Grain Size Estimation. *Ultrasonics* **27** (1), pp. 19–25 (Jan. 1989).

A.P. Sarvazyan and T.V. Chalikian, Development of a Multichannel Ultrasonic Interfero-meter for the Measurements Under High Pressures and Its Application to Biomolecular Studies, pp. 937–940, *1988 Ultrasonics Symp. Proc.*, IEEE (1988).

A. Sather, Ultrasonic Buffer-Rod Technique for the High Temperature Measurement of the Elastic Moduli of Short Specimens, *J. Acoust. Soc. Am.* **43** (6), pp. 1291–1294 (1968).

T. Sato, T. Sunada and S. Wadaka, Ultrasonic Imaging System Which Uses a Rotating M-Sequence Phase Disk and Correlation Analysis, *J. Acoust. Soc. Am.* **64** (4), pp. 1101–1104 (October 1978).

S. Satomura, Ultrasonic Doppler Method for the Inspection of Cardiac Functions, *J. Acoust. Soc. Am.* **29** (11), pp. 1181–1185 (November 1957).

T. Sawada and T. Kitamori, Detection of Ultrafine Particles in Ultrapure Water by Laser-Induced Breakdown Effect, pp. 711–716, *1988 Ultrasonics Symp. Proc.*, IEEE (1988).

M.J. Scelzo and W.D. Munk, Field Test of an Ultrasonic Flowmeter for Natural Gas Pipe-lines, pp. 111–114 in: E.J. Seiders (ed.), *Pipeline Engineering Symposium, 1987*, PD-Vol. 6, ASME (1987).

H. Schlichting, *Boundary Layer Theory*, Pergamon Press, London (1955).

D.W. Schmidt, Acoustical Method for Fast Detection and Measurement of Vortices in Wind Tunnels, pp. 216–228 *ICIASF '75 Record* (1975). See also D.W. Schmidt and P.M. Tillman, *J. Acoust. Soc. Am.* **47**, Part 2, pp. 1310–1324 (1970); *Acustica* **27**, pp. 14–22 (1972).

D.W. Schmidt, Akustische Messung der Zirkulation von Wirbeln u. von Zirkulationsver-teilungen bei Modelluntersuchungen in Windkanälen, Mitt. Max Plank Inst. Strömungs-forsch. u. AVA, Göttingen, Nr. 61 (1975).

J.S. Schoenwald and J.F. Martin, PVF$_2$ Transducer for Acoustic Ranging and Imaging in Air, pp. 577–580, *1983 Ultrasonics Symp. Proc.*, IEEE (1983).

J.S. Schoenwald and C.V. Smith, Jr., Two-Tone CW Acoustic Ranging Technique for Ro-botic Control, pp. 469–474, *1984 Ultrasonics Symp. Proc.*, IEEE (1984).

J.S. Schoenwald, Strategies for Robotic Sensing Using Acoustics, pp. 472–482, *1985 Ultrasonics Symp. Proc.*, IEEE (1985).

J.S. Schoenwald, M.S. Black, J.F. Martin, G.A. Arnold and T.A. Allison, Acoustic Range Sensing Servo Control: Improved Robot Positioning and Trajectory, *IEEE Trans. UFFC* **34** (2), pp. 225–231 (March 1987).

J. Seferis and R. Samuels, Coupling of Optical and Mechanical Properties in Crystalline Polymers, *Polymer Engineering and Science* **19** (14), pp. 975–994 (November 1979).

H. Seki, A. Granato and R. Truell, Diffraction Effects in the Ultrasonic Field of a Piston Source and Their Importance in the Accurate Measurement of Attenuation, *J. Acoust. Soc. Am.* **28** (2) pp. 230–238 (March 1956).

R.H.J. Sellin, *Flow in Channels*, pp. 19–21, Gordon and Breach, New York (1970).

C.M. Semrow, Ultrasound Acoustical Coupling Pad, U.S. Patent No. 4,556,066 (December 3, 1985).

P.K. Seshan, Reynolds Number Effects in Combustion Noise, *Combust. Sci. and Techn.* **49**, pp. 263–275 (1986).

G.M. Sessler and J.E. West, Electret Transducers: A Review, *J. Acoust. Soc. Am.* **53** (6), pp. 1589–1600 (June 1973).

N. Shaikh, D.E. Chimenti and A.H. Nayfeh, Leaky Rayleigh Waves on Surfaces with Laminar Microstructure, pp. 831–836, *1987 Ultrasonics Symp. Proc.*, IEEE (1987).

S.M. Shaker, Sensor Development for Military Unmanned Undersea Vehicles, *Sensors* **5** (3), pp. 5–13 (March 1988).

P.M. Shankar, J.Y. Chapelon and V.L. Newhouse, Fluid Pressure Measurement Using Bubbles Insonified by Two Frequencies, *Ultrasonics* **24** (6), pp. 333–336 (November 1986).

R.S. Sharpe, ed., *Research Techniques in Nondestructive Testing*, Vol. 1, Academic Press (1970); Vol. 2 (1974).

D.P. Shattuck and J. Nouhi, Focusing with an Optoacoustic Transducer, *IEEE Trans. UFFC* **35** (4), pp. 445–449 (July 1988).

S.H. Sheen and A.C. Raptis, Development of Acoustic Flow Instruments for Solid-Gas Pipe Flows, ANL Report ANL/FE-85-07, Argonne National Laboratory (May 1986); Active Ultrasonic Cross-Correlation Flowmeters for Mixed-Phase Pipe Flows, *ISA Trans.* **24** (2), pp. 53–58 (1985). A.C. Raptis and S.H. Sheen, Acoustic Cross-Correlation Flowmeter for Solid-Gas Flow, U.S. Patent 4,598,593 (July 8, 1986).

S.H. Sheen, J.P. Bobis, A.C. Raptis and M. Turgeon, Development, Evaluation, and Testing of an Active Ultrasonic Cross-Correlation Coal-Slurry Flowmeter, pp. 359–362, *Proc. 33rd Int'l Instrum. Symp.*, ISA (May 1987).

S.H. Sheen, K.J. Reimann and A.C. Raptis, Ultrasonic Techniques for Measurement of Coal Slurry Viscosity, *1988 Ultrasonics Symp. Proc.*, pp. 537–542, IEEE (1988).

K.D. Sheffler and G.S. Doble, Thermal Fatigue Behavior of T-111 and ASTAR 811C in Ultrahigh Vacuum, pp. 491–499 in ASTM STP 520, *Fatigue at Elevated Temperatures* (1972).

F.H. Shepard, Jr., Distance Measuring System, U.S. Patent No. 2,333,688 (Nov. 9, 1943).

G.G. Sherratt and E. Griffiths, The Determination of the Specific Heat of Gases at High Temperatures by the Sound Velocity Method, I-Carbon Monoxide, *Proc. Roy. Soc. (London), Ser. A* **147**, pp. 292–308 (1934).

F.D. Shields and J. Faughn, Sound Velocity and Absorption in Low-Pressure Gases Confined to Tubes of Circular Cross Section, *J. Acoust. Soc. Am.* **46** (1) Part 2, pp. 158–163 (1969).

A. Shiozaki, S. Senda, A. Kitabatake, M. Inoue and H. Matsuo, A New Modulation Method with Range Resolution for Ultrasonic Doppler Flow Sensing, *Ultrasonics* **17** (6), pp. 269–275 (1979).

J.K. Sidney, N.W. King and J. Coulthard, Cross Correlation Flow Measurements in Oil-Air Mixtures, *ACTA IMEKO 88*, ISA (1988).

M.G. Silk, *Ultrasonic Transducers for Nondestructive Testing*, Adam Hilger Ltd., Bristol UK (1984).

D. Silvermetz and G. Adams, Integrity Check for Ultrasonic Fluid Interface Sensing System, International Patent Publication No. WO 81/03701 (December 24, 1981).

J.H. Simmons and P.B. Macedo, High-Temperature Shear Ultrasonic Interferometer Using Sensitive Phase-Lock Detection System, *J. Acoust. Soc. Am.* **43** (6), pp. 1295–1301 (June 1968).

K.J. Skogerboe and E.S. Yeung, Quantitative Gas Chromatography without Analyte Identification by Ultrasonic Detection, *Analytical Chemistry* **56** (14), pp. 2684–2686 (1984).

D. Slepian (ed.), *Key Papers in the Development of Information Theory*, IEEE (1974).

J.W. Smalling, L.D. Braswell, L.C. Lynnworth and D.R. Wallace, Flare Gas Ultrasonic Flow Meter, pp. 27–38, *Proc. 39th Annual Symp. on Instrumentation for the Process Industries*, ISA (1984).

J.W. Smalling, L.D. Braswell and L.C. Lynnworth, Apparatus and Methods for Measuring Fluid Flow Parameters, U.S. Patent No. 4,596,133 (June 24, 1986).

H.I. Smith, Optical Contact Bonding, *J. Acoust. Soc. Am.* **37** (5), pp. 928–929 (May 1965).

H.I. Smith and M.S. Gussenhoven, Adhesion of Polished Quartz Crystals Under Ultrahigh Vacuum, *J. Appl. Phys.* **36** (7), pp. 2326–2327 (July 1965).

H.I. Smith and A.B. Smith, Transmission of Gigahertz Ultrasonic Waves through Optical-Contact Bonds at Room Temperature, *J. Acoust. Soc. Am.* **44** (6), pp. 1737–1738 (December 1968).

C.V. Smith, Jr. and J.S. Schoenwald, A Two-Tone Narrow Bandwidth Range-Rate Finding System, pp. 465–468, *1984 Ultrasonics Symp. Proc.*, IEEE (1984).

M.R. Smith and A.K. Dunhill, The Design and Performance of PVDF Transducers, pp. 675–679, *1987 Ultrasonics Symp. Proc.*, IEEE (1987).

P.W. Smith, Jr., and M.C. Junger, Sonic Wave Conductor, U.S. Patent No. 2,988,723 (September 5, 1961).

R.W. Smith, C.K. Day, and W.L. Kelly, High Temperature Ultrasonic Transducers for In-Sodium Service, pp. 699–702, *1974 Ultrasonics Symp. Proc.*, IEEE (1974).

W. Smith, Doppler Flow Measurement, *Meas. & Control* **22** (2), p. 122 (April 1987).

R.E. Snyder, Underwater Control Advances Unveiled at OTC '88, *Ocean Industry* **23** (5), pp. 34–39 (May 1988).

Z.W. Sochaczewski, C.A.E. Clay, and J.A. Morris, Development of a Turbine-Generator Thermal Performance Monitoring System, *Proc. Inst. Mech. Engineers* (UK) **195** (31), pp. 295–304 (1981).

J.H. Sondericker, Helium Flow Measurement Using Ultrasonic Technique, *Advances in Cryogenic Engineering* **29**, pp. 887–894, Plenum Press, NY (1984).

Southern Research Institute, The Thermal Properties of 26 Solid Materials to 5000°F or Their Destruction Temperature, ASD TR 62–765 (January 1963).

C.R. Sparks and R.G. Durke, Improving the Accuracy of Field Orifice Installations for Gas Flow Measurement, *Proc. 1987 ASME Pipeline Engineering Symposium-ETCE*, Dallas, TX, pp. 101–105 (1987).

J.H. Speake, R.G.C. Arridge and G. Curtis, Measurement of the Cure of Resins by Ultrasonic Techniques, *J. Physics D.* **7** (3), pp. 412–424 (February 11, 1974).

J.M. Speake and G. Curtis, Ultrasonic and Stress-Wave Emission NDT of Polymeric Adhesives, Plastics and Rubber, *Materials and Applications* (September 1976).

G.R. Speich, A.J. Schwoeble and W.C. Leslie, Elastic Constants of Binary Iron-Base Alloys, *Met. Trans.* **3** (8), pp. 2031–2037 (August, 1972).

S. Spinner, Elastic Moduli of Glasses at Elevated Temperatures by a Dynamic Method, *J. Am. Cer. Soc.* **39** (3), pp. 113–118 (1956).

S. Spinner and R.C. Valore, Jr., Comparison of Theoretical and Empirical Relations Between the Shear Modulus and Torsional Resonance Frequencies for Bars of Rectangular Cross Section, *J. Natl. Bur. Std.* **60** (5), pp. 459–464, Research Paper 2861 (May 1958).

R.M. Spriggs, Effect of Open and Closed Pores on Elastic Moduli of Polycrystalline Alumina, *J. Amer. Cer. Soc.* **45** (9), p. 454 (Sept. 1962).

D.O. Sproule, Ultrasonic Flaw Detecting Apparatus, U.S. Patent No. 2,972,069 (Feb. 14, 1961).

P.L. Squire, Piezoelectric PVdF Ultrasonic Transducers, *Sensors* **3** (7), pp. 12–16 (July 1986).

J.P. Stagg, Reagent Concentration Measurements in Metal Organic Vapour Phase Epitaxy (MOVPE) Using an Ultrasonic Cell, *Chemtronics* **3** (1), pp. 44–49 (1988).

J.W. Stansfeld, Application of On-Line Density Meters for Fiscal Measurement of Oil and Gas, *Petroleum Review* **36** (430), p. 18, 20, 22 (November 1982).

F.E. Stanke and C.J. Randall, Practical Forms of the Reciprocity Relations for Ultrasonic Transducers, pp. 725–730, *1987 Ultrasonics Symp. Proc.*, IEEE (1987).

R.G. Stearns and G.S. Kino, High Frequency Photoacoustics in Air, *IEEE Trans. UFFC* **34** (2), pp. 179–190 (March 1987).

B.J. Steblay, Measuring Mine Roof Bolt Strains, U.S. Patent No. 4,601,207 (July 22, 1986).

J.P. Steiner, E.S. Furgason and W.L. Weeks, Robust Deconvolution of Correlation Functions, pp. 1031–1035, *1987 Ultrasonics Symp. Proc.*, IEEE (1987).

R.W.B. Stephens (ed.), *Underwater Acoustics*, Wiley-Interscience, NY (1970).

R. Stoneley, Elastic Waves at the Surface of Separation of Two Solids, *Proc. Roy. Soc. London Ser. A* **106**, pp. 416–428 (1924).

J.S. Strachan, Wind Measurement, British Patent Application (February 3, 1987).

J.A. Stratton, *Electromagnetic Theory*, McGraw-Hill, pp. 330–340 (1941).

G.C. Straty and B.A. Younglove, Velocity of Sound in Saturated and Compressed Fluid Oxygen, *J. Chem. Thermodynamics*, **1973** (5), pp. 305–312 (1973).

F. Strouhal, Über eine besondere Art der Tonerregung, *Ann. Phys. Chem.*, **5**, p. 216 (1878).

W. Struszynski, *The Marconi Review*, XXII **134** 119–143 (1959).

J.W. Strutt (Lord Rayleigh), On The Theory of Resonance, *Philos. Trans.* **161**, pp. 77–118 (1870).

J. Stuehr and E. Yeager, The Propagation of Ultrasonic Waves in Electrolytic Solutions, pp. 351–462, in: W.P. Mason (ed.), *Physical Acoustics*, Vol. **II**, Part A, Academic Press (1965).

G.E. Stungis and S.L. Merker, A Structural Model of Reconstituted Tobacco Substantiated by Ultrasonic Interrogation, *Beiträge z. Tabakforschung* **8** (5) pp. 293–301 (March 1976).

C.G. Suits, The Determination of Arc Temperature from Sound Velocity Measurements (Parts I and II), *Physics* **6**, pp. 190–195 and (with H. Poritsky) pp. 196–202 (1935).

K.J. Sun and W.P. Winfree, Propagation of Acoustic Waves in a Copper Wire Embedded in a Curing Epoxy, pp. 439–442, *1987 Ultrasonics Symp. Proc.*, IEEE (1987).

Y. Sunthankar, A Novel Ultrasonic Radiator, *IEEE Trans. Sonics and Ultrasonics* **SU-20** (3), pp. 274–278 (July 1973).

Y. Sunthankar, A Novel Ultrasonic Radiator for Underwater Communication, pp. 428–431, *1977 Ultrasonics Symp. Proc.*, IEEE (1977).

V.A. Sutilov, *Physik des Ultraschalls*, Springer (1984).

H. Suzuki, H. Nakabori, and M. Yamamoto, pp. 115–138 in: C.G. Clayton (ed.), *Modern Developments in Flow Measurement*, Peregrinus, London (1972).

E.G.F. Sweet, Ultrasonic Density Meter with Damping, Canadian Patent No. 1,203,615 (April 22, 1986).

R.C. Swengel, Fluid Velocity Measuring System, U.S. Patent No. 2,746,291 (May 22, 1956).

C.A. Swoboda, D.R. Fredrickson, S.D. Gabelnick, P.H. Cannon, F. Hornstra, N.P. Yao, K.A. Phan and M.K. Singleterry, Development of an Ultrasonic Technique to Measure Specific Gravity in Lead-Acid Battery Electrolyte, *IEEE Trans. Sonics and Ultras.* **30** (2), pp. 69–77 (1983).

P.H. Sydenham, *Measuring Instruments; Tools of Knowledge and Control*, Part 1 of the IEE History of Technology Series, Peregrinus (1979).

J. Szilard, Ultrasound Penetration Through Very Thin Gas Layers Embedded in Solid Bodies, pp. 159–161, *Proc. Fourth International Conf. Non-Destructive Testing*, Butterworths, London (1964). See also, J. Szilard, Comments on A.V. Clark and H.H. Chaskelis, Measurement of Ultrasound Reflected from Ultra-Thin Defects, *Ultrasonics* **20** (1), page 35 (January 1982).

J. Szilard (ed.), *Ultrasonic Testing*, Wiley (1982).

J. Tabin, Theoretical Analysis of the Ultrasonic Doppler Flowmeter for Measurements of High Flow Velocities, *IEEE Trans. UFFC* **34** (4), pp. 467–471 (July 1987).

J. Tabin, Ultrasonic Velocity Metering Based on Transit Time Differentials, *J. Phys. E. Sci. Instrum.*, **20**, pp. 559–561 (1987).

Y. Takeda, Velocity Profile Measurement by Ultrasound Doppler Shift Method, *Int. J. Heat & Fluid Flow* **7** (4), pp. 313–318 (December 1986).

Y. Takeda, Measurement of Velocity Profile of Mercury Flow by Ultrasound Doppler Shift Method, *Nuclear Technology* **79**, pp. 120–124 (October 1987).

C.R. Tallman, Acoustic Gas Analyzer, *ISA Trans.* **17** (1), pp. 97–104 (1978).

X.M. Tang, M.N. Toksoz, P. Tarif and R.H. Wilkens, A Method for Measuring Acoustic Wave Attenuation in the Laboratory, *J. Acoust. Soc. Am.* **83** (2), pp. 453–462 (February 1988).

S. Tanisawa and H. Hirose, Dependence of Actively Generated Sound Noise Spectrum on Flow-Velocity, pp. 905–908, *1988 Ultrasonics Symp. Proc.*, IEEE (1988).

H.A. Tasman, H.E. Schmidt, J. Richter, M. Campana and G. Fayl, The TRESON Experiments: Measurement of Temperature Profiles in Nuclear Fuels by Means of Ultrasonic Thermometers, *High Temperatures—High Pressures* **9**, pp. 387–406 (1977).

H.A. Tasman, Nuclear Applications of Ultrasonic Thermometry, pp. 380–383, *1979 Ultrasonics Symp. Proc.*, IEEE (1979).

H.A. Tasman, M. Campana, D. Pel and J. Richter, Ultrasonic Thin- Wire Thermometry for Nuclear Applications, *Temperature—Its Measurement and Control in Science and Industry*, AIP, pp. 1191– 1196 (1982).

A. Taub and D. Schilling, *Principles of Communication Systems,* McGraw-Hill (1971).

W.E. Tefft and S. Spinner, Torsional Resonance Vibrations of Uniform Bars of Square Cross Section, *J. Natl. Bur. Std.* **65A** (3), pp. 167–171 (May-June 1961).

S.W. Tehon and C.R. Roberts, An Ultrasonic Machine Tool Datum, pp. 468–471, *1985 Ultrasonics Symp. Proc.*, IEEE (1985).

S.W. Tehon, R.F. Clark and C. Mayer, A High-Q Mechanical Resonator for Remote Mechanical Excitation, pp. 577–580, *1986 Ultrasonics Symp. Proc.*, IEEE (1986).

S.W. Tehon, Arrangement for Remote Ultrasonic Temperature Measurement, U.S. Patent No. 4,676,663 (June 30, 1987).

L.R. Testardi, J.T. Krause, and H.S. Chen, Large Anharmonicity of Amorphous and Crystalline Phases of a Pd-Si Alloy, *Phys. Rev. B* **8**, 4464–4469 (1973).

R.B. Thompson and G. Alers, Noncontact Ultrasonic Surface Wave Transducers and Their Application to Nondestructive Testing, pp. 6–19 in *Proc. 9th Symp. on NDE* (1973).

R.B. Thompson, G.A. Alers, D.O. Thompson and M.A. Tennison, Dispersion of Flexural Elastic Waves in Honeycomb Sandwich Panels, *J. Acoust. Soc. Am.* **57** (5), pp. 1119–1127 (May 1975).

Thomson CSF, Improvements in Or Relating to Ultrasonic Flowmeters, International Patent Specification 1338436 (publ. 21 Nov. 1973).

E.A. Thorne, The Measurement of High Temperatures by the Determination of the Velocity of Sound Waves in Materials, Paper P23 in *Proc. of the Fourth International Congress on Acoustics*, Copenhagen (1963).

G.A. Thorne and A.G. Johnson, Vortex Generating Device, U.S. Patent No. 4,312,237 (January 26, 1982).

P.D. Thorne, A Broadband Acoustic Source for Underwater Laboratory Applications, *IEEE Trans. UFFC* **34** (5), pp. 515–523 (September 1987).

E.J. Thrush, C.G. Cureton, J.M. Trigg, J.P. Stagg and B.R. Butler, Reactor Design and Operating Procedures for ImP Based MOCVD, *Chemtronics* **2** pp. 62–68 (June 1987).

R.N. Thurston, Elastic Waves in Rods and Clad Rods, *J. Acoust. Soc. Am.* **64** (1), pp. 1–37 (July 1978).

R.N. Thurston, In Memoriam—Warren P. Mason, 1900–1986, pp. xi–xvii in: W.P. Mason and R.N. Thurston (ed.), *Physical Acoustics* **18** (1988).

T. Tiearney, Creep Behavior of Binary Solid Solutions of Nickel with Molybdenum and Tungsten, PhD thesis, MIT (1978).

B.R. Tittmann, L.A. Ahlberg, J.M. Richardson and R.B. Thompson, Determination of Physical Property Gradients from Measured Surface Wave Dispersion, *IEEE Trans. UFFC* **UFFC-34** (5), pp. 500–507 (September 1987).

D.L. Tolliver and H.G. Schroeder, Particle Control in Semiconductor Process Streams, *Microcontamination* **1** (1), pp. 34–43 (June/July 1983).

J. Tornberg, M. Karras, E. Härkönen and O. Hirsimäki, Analysis of Ultrasonic Correlation Flowmeters for Pulp Suspension, *IEEE Trans. Sonics and Ultrasonics* **SU-30** (4), pp. 264–269 (July 1983).

J. Tornberg, M. Karras and E. Härkönen, Two Ultrasonic Flowmeters for Pulp Suspensions Using Advanced Signal Analysis, *Tappi J.* **67** (7), pp. 72–74 (July 1984)

J. Tornberg, Correlation Methods for Ultrasonic Flow Measurements in Paper Pulp Suspensions, Acta Universitatis Ouluensis, Series C, Technica No. 35, Electronica No. 8 (1986).

J. Tornberg, Correlation Methods for Ultrasonic Flow Measurements in Paper Pulp Suspensions, PhD thesis, U. of Oulu, Finland (August 1986).

P. Tortoli, G. Manes and C. Atzeni, Velocity Profile Reconstruction Using Ultrafast Spectral Analysis of Doppler Ultrasound, *IEEE Trans. Sonics and Ultrasonics* **SU-32** (4), pp. 555–561 (July 1985).

P. Tortoli, F. Andreuccetti, G. Manes and C. Atzeni, Blood Flow Images by a SAW-Based Multigate Doppler System, *IEEE Trans. UFFC* **35** (5), pp. 545–551 (Sept. 1988).

J. Toulouse, A Modified Version of the Phase Sensitive Technique for Measurements of Absolute Sound Velocity in Solids, pp. 407–410, *1987 Ultrasonics Symp. Proc.*, IEEE (1987).

G.E. Trahey, S.W. Smith and O.T. von Ramm, Speckle Pattern Correlation with Lateral Aperture Translation: Experimental Results and Implications for Spatial Compounding, *IEEE Trans. UFFC* **33**, pp. 257–264 (1986).

G.E. Trahey, J.W. Allison, S.M. Hubbard and O.T. von Ramm, Measurement of Local

Speckle Pattern Displacement to Track Blood Flow in Two Dimensions, pp. 957–961, *1987 Ultrasonics Symp. Proc.*, IEEE (1987a).

G.E. Trahey, J.W. Allison and O.T. Von Ramm, Angle Independent Ultrasonic Detection of Blood Flow, *IEEE Trans. Biomed. Eng'g*. **BME-34** (12), pp. 965–967 (Dec. 1987b).

G.E. Trahey, S.M. Hubbard and O.T. von Ramm, Angle Independent Ultrasonic Blood Flow Detection by Frame-to-Frame Correlation of B-Mode Images, *Ultrasonics* **26** (5), pp. 271–276 (Sept. 1988).

J.A. Traina, Reliability of a Continuous Stack Flow Monitor and Application in Improved Process/Control Equipment Operation, in: J.A. Jahnke (ed.), *Trans.: Continuous Emission Monitoring - Advances and Issues*, pp. 261–270, Air Pollution Control Association (APCA), Pittsburgh (October 1985).

J.A. Traina, Method and Apparatus for Ultrasonic Measurements of a Medium, U.S. Patent No. 4,630,482 (23 December 1986).

R.J. Trett, Acoustic Telemetry, A Comparison of Theory and Practice, Paper OTC 5691, Offshore Technology Conf. [OTC '88] (1988).

W. Trimmer, BPA Flowmeter Report, Oregon State University, Agricultural Engineering Dept. (1987).

W.L. Trimmer, A. Taylor and H. West, Ultrasonic Flow Meter Testing, "Planning Now for Irrigation and Drainage in the 21st Century," pp. 643–650, *Am. Soc. Civil Engineers Specialty Conf. Proc.*, Lincoln NE (July 1988).

E.H. Trinh and C.J. Hsu, Acoustic Levitation Methods for Density Measurements, *J. Acoust. Soc. Am.* **80** (6), pp. 1757–1761 (1986).

R. Truell, C. Elfbaum and B.B. Chick, *Ultrasonic Methods in Solid State Physics,* Academic Press, New York (1969); *IEEE Trans. UFEC.* **36** (3), pp. 326–331 (May 1989).

Y. Tsukahara, N. Nakaso, J. Kushibiki and N. Chubachi, An Instrument for Layer Thickness Measurement Using Pseudo-Sezawa Waves, pp. 1031–1035, *1986 Ultrasonics Symp. Proc.*, IEEE (1986).

L.Y. Tu, J.N. Brennan and J.A. Sauer, Dispersion of Ultrasonic Pulse Velocity in Cylindrical Rods, *J. Acoust. Soc. Am.* **27** (3), pp. 550–555 (1955).

D.B. Turner, Comparison of Three Methods for Calculating the Standard Deviation of the Wind Direction, *J. Climate and Appl. Meteor.* **25** (5), pp. 703–707 (May 1986).

N.S. Tzannes, Joule and Wiedemann Effects—The Simultaneous Generation of Longitudinal and Torsional Stress Pulses in Magnetostrictive Materials, *IEEE Trans. Sonics and Ultrasonics*, **SU-13** (2), pp. 33–41 (July 1966). See also A. Rothbart, A Torsional Magnetostrictive Delay Line **47** (6), pp. 1153–1154 (June 1959).

H. Überall, p. 1 in: W.P. Mason and R.N. Thurston, (eds.), *Physical Acoustics* **10**, Academic Press, New York (1973).

S. Uozumi, Method for Continuous Supersonic Inspection of Hot Steel Plates, U.S. Patent No. 3,625,051 (December 7, 1971).

B.L. Upchurch, E.S. Furgason, G.E. Miles and F.D. Hess, Ultrasonic Measurements for Detecting Damage on Agricultural Products, pp. 990–993, *1985 Ultrasonics Symp. Proc.*, IEEE (1985).

H.G. Urban (ed.), *Adaptive Methods in Underwater Acoustics*, Reidel Publ. Co. (1984).

W.H. van den Berg, M.H. Homs and A.B.M. Hoff, Development of an Electromagnetic Acoustic Transducer for Inspecting the Wall Thickness of Offshore Risers from the Inside, *Ultrasonics* **26** (1), pp. 14–22 (January 1988).

H.E. Van Valkenburg, Ultrasonic Inspection Device, U.S. Patent No. 2,667,780 (February 2, 1954).

H.E. Van Valkenburg, Ultrasonic Liquid Depth Indicator, U.S. Patent No. 2,787,160 (April 2, 1957).

H.E. Van Valkenburg and R.E. Sansom, Liquid Level Indicator, U.S. Patent No. 2,883,861 (April 28, 1959).

H.E. Van Valkenburg, Air-Coupled Ultrasound as a Production Inspection Technique for Aircraft Tires, pp. 55–60, *Proc. 1973 Symp. on Nondestructive Testing of Tires*, NTIAC 74–1, P.E.J. Vogel (Ed.) AMMRC, Watertown, MA (1973).

G.M. Varga, Jr., Vapor Detector, Esso Tech. Report RK-CR-74-8, GRU.1DJAE.74, AD780171 (February 15, 1974).

A. Venema, E. Nieuwkoop, M.J. Vellekoop, W.J. Ghijsen, A.W. Barendsz, and M.S. Nieuwenhuizen, NO_2 Gas-Concentration Measurement with a SAW-Chemosensor, *IEEE Trans. UFFC* **34** (2), pp. 148–155 (1987).

I.A. Viktorov, *Rayleigh and Lamb Waves*, Plenum Press, New York (1967).

D. Vilkomerson, pp. 283–316, in: P.S. Green, ed., *Acoustical Holography*, Vol. 5, Plenum Press (1974).

E. Villari, Change of Magnetization of Tension by Electric Current, *Am. Phys. Chem. Lpz.* **126**, pp. 87–122 (1865).

M.S. Vogler and J.G. Groetsch, Jr., Monitoring of Steel Plant Effluent, *Waterworld News* **5** (3), in press (May/June 1989).

T. von Kármán and H. Rubach, Über den Mechanismus des Flüßigkeits- und Luftwiderstandes (On the Mechanism of Fluid Resistance), *Phys. Z.* Vol. **13**, p. 49 (1912).

H.C.L. Vos and J.A. Vogel, An Ultrasonic Circular Array Transducer for Pipeline and Borehole Inspection, pp. 659–662, *1988 Ultrasonics Symp. Proc.*, IEEE (1988).

R.C. Waag, W.L. Rhoades and R. Gramiak, Instrumentation for Non-Invasive Cardiac Chamber Flow Rate Measurement, pp. 74–77, *1972 Ultrasonics Symp. Proc.*, IEEE (1972).

H. Wada, Reflection Characteristics of Longitudinal Waves in a Semi-Infinite Cylindrical Rod Connected to an Infinite Elastic Stratum, *J. Acoust. Soc. Am.* **75** (6) pp. 1777–1782 (June 1984).

H.N.G. Wadley, Sensors for Intelligent Processing of Materials, *J. Metals*, **38** (10) pp. 49–53 (Oct. 1986).

H.N.G. Wadley, S.J. Norton, F. Mauer and B. Droney, Ultrasonic Measurement of Internal Temperature Distribution, *Phil. Trans.* **A320**, pp. 341–361, Royal Society, London (1986).

D.R. Wallace, J.M. Korba, J.E. Matson and L.C. Lynnworth, Intervalometer Time Apparatus and Method, U.S. Patent No. 4,515,021 (May 7, 1985).

J. Waller, Doppler Flowmeters—Proper Practices, pp. 209–217 in: *Proc. of the 30th International Instrumentation Symp.*, ISA (1984).

M.W. Wambsganss, Understanding Flow-Induced Vibrations, Part II: Fluid/Structure Coupling; Design Considerations, *Sound & Vibr.* **11** (4), pp. 18–21 (April 1977).

H.S.C. Wang, Digital Processing of Doppler Sonar Signals, *J. Acoust. Soc. Am.* **61** (3) pp. 782–791 (March 1977).

K.Y. Wang (guest ed.), *Proceedings of the IEEE* **67** (4), pp. 452–675 (April 1979).

A.W. Warburton and L.C. Lynnworth, Hot Tapped Ultrasonic Flowmeter Ports in Hot Steel Pipe, pp. 55–64, *Proc. 29th International Instrumentation Symp.* ISA (1983).

A.W. Warner, M. Onoe and G.A. Coquin, Determination of Elastic and Piezoelectric Constants for Crystals in Class (*3m*), *J. Acoust. Soc. Am.* **42** (6) pp. 1223–1231 (1967).

K. Watanabe, Y. Takebayashi and D.M. Himmelblau, Volume Measurement of Deformed Materials in Tanks, *ISA Trans.* **27** (4), pp. 9–19 (1988).

H.A. Waterman, Determination of the Complex Moduli of Viscoelastic Materials with the Ultrasonic Pulse Method, Part I. *Kolloid-Z. u. Z. f. Polymere* **192**, pp. 1–8 (1963); Part II, ibid, pp. 9–16 (1963).

K.H. Waters, *Reflection Seismology: A Tool for Energy Resource Exploration*, John Wiley and Sons (1978).

J.W. Watson and F.A. White, Acoustic Measurement for Gas BTU Content, *Oil and Gas J.* **80** (14), pp. 217–225 (April 5, 1982).

K.A. Wear and R.L. Popp, Theoretical Analysis of a Technique for the Characterization of Myocardium Contraction Based Upon Temporal Correlation of Ultrasonic Echoes, *IEEE Trans. UFFC* **34** (3), pp. 368–275 (May 1987).

R.B. Wearn, Jr., and J.M. Paros, Measurements of Dead Weight Tester Performance Using High Resolution Quartz Crystal Pressure Transducers, pp. 259–266, *Proc. 34th Int'l. Instrum. Symp.*, ISA (1988).

G.M.B. Webber and R.W.B. Stephens, Transmission of Sound in Molten Metals, pp. 53–97 in: W.P. Mason and R.N. Thurston (ed.) *Physical Acoustics* Vol. 4, Part B, Academic Press (1968).

W. Welkowitz and S. Deutsch, *Biomedical Instruments: Theory and Design*, 297 pages, esp. pages 106–108, Academic Press, New York (1976).

S.W. Wenzel and R.M. White, Ultrasonic-Oscillator Position Sensor, pp. 611–614, *1987 Ultrasonics Symp. Proc.*, IEEE (1987).

S.W. Wenzel and R.M. White, Generalized Lamb-Wave Multisensor, pp. 563–567, *1988 Ultrasonics Symp. Proc.*, IEEE (1988).

D.F. White, A.E. Rodely and C.L. McMurtie, The Vortex Shedding Flowmeter, *Flow—Its Measurement and Control in Science and Industry*, Vol. **1**, pt. 2, pp. 967–974, ISA (1974).

D.N. White, Johann Christian Doppler and His Effect—A Brief History, *Ultra. in Med. and Biol.* **8** (6), pp. 583–591 (1982).

F.E. White and D.C. Teas (ed.), References to Contemporary Papers on Acoustics, *J. Acoust. Soc. Am.*, Supplement 2, Vol. **77** (Summer 1985); Supplement 2, Vol. **79** (Summer 1986).

R.M. White and F.W. Voltmer, Direct Piezoelectric Coupling of Surface Elastic Waves, *Appl. Phys. Lett.* **7** (12), pp. 314–416 (December 15, 1965).

H.K. Whitesel, D.A. Nordling and C.P. Nemarich, Online Wear-Particle Monitoring Based on Ultrasonic Detection, *Intech* **33** (6), pp. 53–57 (June 1986).

P. Whitten and D. Kitchen, Acoustic Pyrometry for Improved Boiler Operations and Maintenance, *Proc. Power Generation Conf.*, 6 pp. (Dec. 1988). See also, Kleppe (1989).

G. Wiedemann, Magnetic Investigations, *Pogg. Ann.* **117**, pp. 193–217 (1862).

D.T. Wilson, R.H. Tancrell and J. Callerame, PVF$_2$ Polymer Microprobe, pp. 506–510, *1979 Ultrasonics Symp. Proc.*, IEEE (1979); U.S. Patent 4,316,115 (February 16, 1982).

E.J. Winston, Advances in Doppler Sonar for Improved Accuracy, Range and Performance, pp. 199–218, *Proc. ION National Marine Navigation Meeting on Navigation Associated with Manned Deep Submergence Vehicles*, Institute of Navigation—Second Symposium (Nov. 1969).

I. Wolff, Air Speed Indicator, U.S. Patent 2,274,262 (February 24, 1942).

M.-C. Wu and W.P. Winfree, Absolute Displacement Measurements with Piezoelectric Transducers, pp. 1147–1150, *1987 Ultrasonics Symp. Proc.*, IEEE (1987).

R.J. Wyber, The Design of a Spark Discharge Acoustic Impulse Generator, *IEEE Transactions on Acoustics, Speech, and Signal Processing*, **ASSP-23** (2), pp. 157–162 (1975)

L.A. Xu, R.G. Green, A. Plaskowski and M.S. Beck, The Pulsed Ultrasonic Cross-Correlation Flowmeter for Two-Phase Flow Measurement, *J. Phys. E. Sci. Instrum.* **21** (4), pp. 406–414 (April 1988).

Q.C. Xu, A.R. Ramachandran, R.E. Newnham and R.H. Tancrell, Measurement of Complex Coefficients for Thick PVDF Polymer, pp. 663–666, *1987 Ultrasonics Symp. Proc.*, IEEE (1987).

J. Yamaguchi, Toyota's Big Six Employs Karman-Vortex Optical Airflow Meter, *Automotive Engineering* **94** (5), pp. 84–86 (May 1986).

M. Yamamoto and K. Ito, Ultrasonic Flowmeter System, U.S. Patent No. 3,237,453 (March 1, 1956).

T. Yano, M. Tone and A. Fukumoto, Range Finding and Surface Characterization Using High-Frequency Air Transducers, *IEEE Trans. UFFC* **34** (2), pp. 232–236 (March 1987).

C.E. Yeack-Scranton, Microscopic Fluctuations in Air Bearings Measured by Novel Piezoelectric Sensors, pp. 535–538, *1987 Ultrasonics Symp. Proc.*, IEEE (1987).

D.A. Yehling, D.A. Westerkamp and J. Colfer, SPC Software Directory, *Quality* **27**, pp. 62–68 (Dec. 1988).

M. Yoshihara and R.B. McLellan, *Acta Metall.* **29** p. 1277 (1981).

M. Yoshihara, R.B. McLellan and F.R. Brotzen, The High-Temperature Elastic Properties of Palladium Single Crystals, *Acta Metall.* **35**, 775–780 (1987).

M. Yoshihara, The Thermodynamics of Hydrogen in Palladium-Yttrium Solid Solutions, *Acta. Metall.* **36** (2), pp. 385–391 (1988).

W.T. Yost and J.H. Cantrell, Liquid Membrane Coupling Response of Submersible Electrostatic Transducer, pp. 693–696, *1987 Ultrasonics Symp. Proc.*, IEEE (1987).

C.A. Youngdahl and W.A. Ellingson, Development of Ultrasonic Techniques for Remote Monitoring of Erosive Wear in Coal-Conversion Systems, pp. 305–310, *1978 Ultrasonics Symp. Proc.*, IEEE (1978).

C.A. Youngdahl and W.A. Ellingson, Acoustic System for Monitoring Pressure Boundary Wear, presented at 1982 Symp. Instrum. Control for Fossil Energy Processes, Houston (June 1982).

C.A. Youngdahl and W. A. Ellingson, Nondestructive Monitoring of Erosive Wear in Transfer Lines and Cyclones at Synfuels Pilot Plants, in: *Corrosion 84*, paper No. 86, pp. 86/1 to 86/12, NACE (1986).

B.A. Younglove, Speed of Sound in Fluid Parahydrogen, *J. Acoust. Soc. Am.* **38** (3), pp. 433–438 (September 1965).

B.A. Younglove, Velocity of Sound in Liquid Propane, *Journal of Research of the National Bureau of Standards* **86** (2), pp. 165–170 (1981).

E.M. Zacharias, Jr., Process Measurements by Sound Velocimetry, *Instr. and Control Systems* **43** (9), pp. 112–113 (Sept. 1970).

E.M. Zacharias, Jr., The Sonic Interface Detector Meets Field Tests in Pipelining, *Oil & Gas J.* **68** (27), pp. 96–103 (July 6, 1970).

E.M. Zacharias, Jr., Sonic Detectors See Gasoline Interfaces, *Oil & Gas J.* **70** (34), pp. 79–81 (Aug. 21, 1972).

E.M. Zacharias, Jr. and R.A. Parnell, Jr., Measuring the Solids Content of Foods by Sound Velocimetry, *Food Techn.* **26** (4), pp. 160–166 (1972).

E.M. Zacharias, Jr., and D.W. Franz, Sound Velocimeters Monitor Process Streams, *Chem. Eng.* **80** (2), pp. 101–108 (Jan. 22, 1973).

E.M. Zacharias, Jr., M. W. Zacharias and R.A. Parnell, Jr., Measuring Sound Velocity and Bulk Modulus of Plastics, *Modern Plastics* **51** (5), pp. 88–90 (May 1974).

E.M. Zacharias, Ultrasonic Transit Time Flowmeters for Full Pipes, Using Wetted Transducers, pp. 179–191, *Proc. of the 30th International Instrumentation Symposium*, ISA (1984).

R. Zana and E. Yeager, Ultrasonic Vibration Potentials, *Mod. Aspects Electrochem.* **14**, pp. 1–60 (1982).

A.J. Zuckerwar and D.S. Mazel, Depth Gauge for Liquids Under High Pressure, LAR-13300, NASA Tech Briefs, p. 55 (January 1987).

Index

A

Ablation rate, 466–470, 539
Absorption
  classical, 89
  in water, 230, 279
  see also Attenuation
Acceleration
  as acceptance criterion, in flow velocity
    measurement, 361
  sensing of, 96, 100, 101, 283
Accuracy
  Gaussian quadrature flowmeter, 268–274
  helium thermometer, 372
  pulse echo overlap method, 200, 201
  quartz pressure gage, 570
  quartz thermometer, 413
  sing-around velocimeter, 197, 623
Acoustic emission, 82–84, 371
Acoustic impedance, *see* Impedance
Acoustic intensity, 121, 605–609
Acoustic noise, *see* Noise
Acoustic properties, tabulated, 227, 229,
    231, 233, 234
  see also Materials; Medium itself
Acoustic pyrometer, 369, 370, 383–392,
    506
  see also Ultrasonic, thermometer

Active methods, 17, 18, 61, 186
Adiabatic compression, volume gage
    utilizing, 450–452
Airborne ultrasound, 125, 126, 492, 504,
    525, 526, 591–605, 609–611
Alignment bar, segmented, 173
Amplitude
  as acceptance criterion, in flow velocity
    measurement, 361
  sensing of, 96, 100, 101
  see also Gap
Anemometer, sonic, 305–307, 373, 631
Angle beam transducer, NDT, early use on
    pipes, 260
  see also Transducers, clamp-on
Anisotropy
  index, *see* Birefringence
  influence on c and α, 74, 75
Annular transducer, 35, 612, 613
Aperture
  large, in high resolution system, 86
  synthetic, 149–151
Area averaging
  midradius chord and, 249, 285, 287, 292,
    293
  multiple chords in, 22, 251, 256, 261,
    265, 266, 268–274, 280, 293
  range-gated methods for, 30, 31, 255